Communications in Computer and Information Science   729

Commenced Publication in 2007
Founding and Former Series Editors:
Alfredo Cuzzocrea, Xiaoyong Du, Orhun Kara, Ting Liu, Dominik Ślęzak,
and Xiaokang Yang

More information about this series at http://www.springer.com/series/7899

Guoliang Chen · Hong Shen
Mingrui Chen (Eds.)

Parallel Architecture, Algorithm and Programming

8th International Symposium, PAAP 2017
Haikou, China, June 17–18, 2017
Proceedings

 Springer

Editors
Guoliang Chen
Nanjing University of Posts
  and Telecommunications
Nanjing, Jiangsu
China

Mingrui Chen
Hainan University
Haikou, Hainan
China

Hong Shen
Sun Yat-sen University
Guangzhou, Guangdong
China

ISSN 1865-0929   ISSN 1865-0937 (electronic)
Communications in Computer and Information Science
ISBN 978-981-10-6441-8   ISBN 978-981-10-6442-5 (eBook)
DOI 10.1007/978-981-10-6442-5

Library of Congress Control Number: 2017955727

Printed on acid-free paper

This Springer imprint is published by Springer Nature
The registered company is Springer Nature Singapore Pte Ltd.
The registered company address is: 152 Beach Road, #21-01/04 Gateway East, Singapore 189721, Singapore

Preface

The Eighth International Symposium on Parallel Architectures, Algorithms, and Programming (PAAP 2017) was held during June 17–18, 2017 in Haikou, China. The conference received 192 submissions and after rigorous reviews, 50 high-quality full papers and 7 short papers were included in this volume. The acceptance rate of full papers was around 28%. These contributions from diverse areas of parallel computing have been categorized into two sections, namely: Algorithms and Programming, and Parallel Architectures.

The aim of PAAP 2017 was to provide a forum for scientists and engineers in academia and industry to present their research results and development activities in all aspects of parallel architectures, algorithms, and programming techniques. The technical program of the conference also included three keynote speeches delivered by internationally well-known scholars, Prof. Guo Liang Chen, Prof. Manu Malek, and Prof. Hong Shen, in addition to the invited sessions of accepted papers.

On behalf of the Organizing Committee, we thank the Hainan Association for Science and Technology and the Hainan Province Computer Federation for its sponsorship and logistics support. We also thank the Hainan Instructive Committee of University Computer Education, the Hainan Instructive Committee on Information, the Hainan Instructive Committee of Computer and Electronic Information in Vocational Education and the Guangzhou AVA Electronic Technology Co., Ltd. for co-sponsorship. We thank the members of the Organizing Committee and the Program Committee for their hard work. We are very grateful to the keynote speakers, invited session organizers, session chairs, reviewers, and student helpers. Last but not least, we thank all the authors and participants for their great contributions that made this conference possible.

June 2017

Guoliang Chen
Hong Shen
Mingrui Chen

Organization

General Chairs

Guoliang Chen   Nanjing University of Posts and Telecommunications, China
Rajkumar Buyya   Univ. of Melbourne, Australia

Program Chair

Hong Shen   Sun Yat-sen Univ., China and Univ. of Adelaide, Australia

Program Co-chairs

Mingrui Chen   Hainan Univ., China
Jiannong Cao   Hong Kong Polytechnic Univ., China
Manu Malek   Stevens Institute of Technology, USA
James J. Park   Seoul National Univ. of Science and Technology, South Korea

Program Committee

Hamid Arabnia   University of Georgia, USA
Han-Chie Chao   National Ilan University, Taiwan
Ling Chen   Yangzhou University, China
Yawen Chen   Otago University, New Zealand
Zhenhua Duan   Xidian University, China
Karl Fuerlinger   Ludwig-Maximilians Univ., Germany
Satoshi Fujita   Univ. of Hiroshima, Japan
Longkun Guo   Fuzhou University, China
Yijie Han   University of Missouri-Kansas City, USA
Shi-Jinn Horng   National Taiwan Univ. of Science and Technology, Taiwan
Ching-Hsien Hsu   Chung Hua University, Taiwan
Haping Huang   Nanjing Univ. of Posts and Telecommunications, China
Jiwu Huang   Shenzhen University, China
Zhiyi Huang   Otago University, New Zealand
Mirjana Ivanovic   University of Novi Sad, Serbia
Cruz Izu   University of Adelaide, Australia
Graham Kirby   University of St. Andrews, UK
Jianhuang Lai   Sun Yat-Sen University, China
Francis Lau   Univ. of Hong Kong, Hong Kong, China
Laurent Lefevre   Inria, France
Kenli Li   Hunan University, China
Xiaoming Li   Peking University, China
Yamin Li   Hosei University, Japan

Yidong Li	Beijing Jiaotong University, China
Sangsong Liang	Univ. College London, UK
Haixiang Lin	Delft Univ. of Technology, Netherlands
Xiaola Lin	Sun Yat-Sen University, China
Alex X. Liu	Michigan State University, USA
Kezhong Lu	Shenzhen University, China
Manu Malek	Stevens Institute of Technology, USA
Rui Mao	Shenzhen University, China
Teruo Matsuzawa	JAIST, Japan
Koji Nakano	Hiroshima University, Japan
Lionel M. Ni	Hong Kong Univ. of Science and Technology, Hong Kong, China
Ge Nong	Sun Yat-Sen Univ., China
Yi Pan	Georgia State University, USA
Marcin Paprzycki	Warsaw School of Social Psychology, Poland
Depei Qian	Beihang University, China
Zhao Qiu	Hainan University, China
Marcel C. Rosu	IBM TJ Watson, USA
Frode Eika Sandnes	Oslo Univ. College, Norway
Yingpeng Sang	Sun Yat-sen University, China
Neetesh Saxena	Georgia Tech., USA
Xiaojun Shen	Univ. of Missouri-Kansas City, USA
Oliver Sinnen	Univ. of Auckland, New Zealand
Guangzhong Sun	Univ. of Science and Technology of China, China
Shaohua Tang	South China Univ. of Technology, China
Haitao Tian	Sun Yat-Sen Univ., China
Hui Tian	Beijing Jiaotong University, China
Ye Tian	Univ. of Science and Technology of China, China
Teofilio Gonzalez	UC Santa Barbara, USA
Benjamin W. Wah	Chinese Univ. of Hong Kong, Hong Kong, China
Yan Wang	Suzhou University, China
Andrew Wendelbourne	Univ. of Adelaide, Australia
Jigang Wu	Guangdong Univ. of Technology, China
Chao-Tung Yang	Tunghai University, Taiwan
Fangguo Zhang	Sun Yat-Sen Univ., China
Haibo Zhang	Otago University, New Zealand
Xianchao Zhang	Dalian Univ. of Technology, China
Yunquan Zhang	Institute of Software, China
Wei Zhao	University of Macau, China
Siqing Zheng	Univ. of Texas at Dallas, USA
Cheng Zhong	Guangxi University, China

Local Arrangements Chair

Zhao Qiu   Hainan University, China

Publication Chairs

Yingpeng Sang   Sun Yat-sen University, China
Hui Tian   Beijing Jiaotong University, China

Publicity Chair

Chunyang Ye   Hainan University, China

Finance and Registration Chair

Ping Huang   Hainan University, China

Contents

Algorithms and Programming

Ford Motor Side-View Recognition System Based on Wavelet Entropy
and Back Propagation Neural Network and Levenberg-Marquardt
Algorithm 3
    Wen-Juan Jia, Shuihua Wang, Huimin Lu, Ying Shao, Elizabeth Lee,
    and Yu-Dong Zhang

Intrusion Detection Based on Self-adaptive Differential Evolution Extreme
Learning Machine with Gaussian Kernel 13
    Junhua Ku and Bing Zheng

Prediction for Passenger Flow at the Airport Based on Different Models 25
    Xia Liu, Xia Huang, Lei Chen, Zhao Qiu, and Ming-rui Chen

Election Based Pose Estimation of Moving Objects. 41
    Liming Gao and Chongwen Wang

A Novel Topology Reconfiguration Backtracking Algorithm
for 2D REmesh Networks-on-Chip 51
    Na Niu, Fang-Fa Fu, Hang Li, Feng-Chang Lai, and Jin-Xiang Wang

User Behaviour Authentication Model Based on Stochastic Petri Net
in Cloud Environment 59
    Peng Li, Cheng Yang, He Xu, Ting Fung LAU, and Ruchuan Wang

Performance Prediction of Spark Based on the Multiple Linear
Regression Analysis 70
    Lu Dong, Peng Li, He Xu, Baozhou Luo, and Yu Mi

CTS-SOS: Cloud Task Scheduling Based on the Symbiotic
Organisms Search 82
    Zhenpeng Liu, Xiaodan Liu, Yawei Dong, Xuan Zhao, and Bin Zhang

Exploration of Heuristic-Based Feature Selection on Classification
Problems 95
    Qi Qi, Ni Li, and Weimin Li

AGSA: Anti-similarity Group Shilling Attacks 108
    Peng Wang, Lingtao Qi, Haiping Huang, Feng Li, and Congxiang Yu

H_∞ Filtering Design for a Class of Distributed Parameter Systems
with Randomly Occurring Sensor Faults and Markovian
Channel Switching   117
  Huihui Ji and Baotong Cui

The Study of the Seabed Side-Scan Acoustic Images Recognition
Using BP Neural Network   130
  Hongyan Xi, Lei Wan, Mingwei Sheng, Yueming Li, and Tao Liu

Node Localization of Wireless Sensor Network Based on Secondary
Correction Error   142
  Xiaoxu Ma, Wenju Liu, and Ze Wang

Optimizations of the Whole Function Vectorization Based on SIMD
Characteristics   152
  Yingying Li, Yuchen Gao, Dong Wang, Yanbing Li, and Jinlong Xu

A Stacked Denoising Autoencoders Based Collaborative Approach
for Recommender System   172
  Baojun Niu, Dongsheng Zou, and Yafeng Niu

Research on Adaptive Canny Algorithm Based on Dual-Domain Filtering . . .   182
  Xiajiong Shen, Xiaoyu Duan, Daojun Han, and Wanli Yuan

A Dynamic Individual Recommendation Method Based
on Reinforcement Learning   192
  Daojun Han, Xiajiong Shen, Tian Gan, and Ruiqing Cai

Research on the Pre-distribution Model Based on Seesaw Model   201
  Mingshan Xie, Yanfang Deng, Yong Bai, Mengxing Huang,
  and Zhuhua Hu

An Efficient Filtration Method Based on Variable-Length Seeds
for Sequence Alignment.   214
  Ruidong Guo, Haoyu Cheng, and Yun Xu

An Optimized Fusion Method for Double-Wearable-Wireless-Band
Platform on Cloud-Health Application   224
  Wenchao Xu, Yanbo Liu, Yanqin Yang, Xiaoshuang Ning, Tianxing Chu,
  and Hongzhi Song

Research on Concept Drift Detection for Decision Tree Algorithm
in the Stream of Big Data   237
  Shangdong Liu, Lili Lu, Yongpan Zhang, Tong Xin, Yimu Ji,
  and Ruchuan Wang

Review of Various Strategies for Gateway Discovery Mechanisms
for Integrating Internet-MANET 247
  Lin Yang, Zhijie Han, Rui Wang, and YongHang Yan

Research on Extraction Algorithm of Palm ROI Based on Maximum
Intrinsic Circle 258
  Gang Liu and Jing Zhang

Power Adaptive Routing Scheme with Energy Hole Avoidance
for Underwater Acoustic Networks 268
  Xian-yi Chen and Guo-lan Lin

A Lightweight Algorithm for Computing BWT from Suffix Array in Disk. . . 279
  Jing Yi Xie, Bin Lao, and Ge Nong

H_∞ Filtering in Mobile Sensor Networks with Missing Measurements
and Quantization Effects 290
  Xueming Qian and Baotong Cui

A Cost-Effective Wide-Sense Nonblocking k-Fold Multicast Network 301
  Gang Liu, Qiuming Luo, Cunhuang Ye, and Rui Mao

The Design of General Course-Choosing System in Colleges
and Universities 311
  Chunmin Qiu, Shaojie Du, and Bailu Zhao

Research on Vectorization Technology for Irregular Data Access 321
  Wang Qi, Han Lin, Yao Jinyang, and Liu Hui

A New Simple Algorithm for Scrambling. 335
  Xing Zeng, Xiulai Li, Yali Luo, and Mingrui Chen

The Framework of Relative Density-Based Clustering 343
  Zelin Cui and Hong Shen

Efficient Algorithms for VM Placement in Cloud Data Center 353
  Jiahuai Wu and Hong Shen

Weighted One-Dependence Forests Classifier 366
  Guojing Zhong and Limin Wang

Research and Realization of Commodity Image Retrieval System Based
on Deep Learning 376
  Cen Chen, Rui Yang, and Chongwen Wang

An Improved Algorithm Based on LSB 386
  Yali Luo, Xiulai Li, Chaofan Chen, and Mingrui Chen

A Report on the Improvement of Information Technology Capability
of Teachers in Primary and Middle Schools in Hainan Province   393
  JingYu Luo, Zhao Qiu, JianZheng Hu, and XiaWen Zhang

The Research of the Airport Retail Layout Based on the Location Model. . . .   400
  Han-tao Yang

Research on Model and Method of Relevance Feedback Mechanism
in Image Retrieval.   410
  Xinying Li, Taijun Li, Feng Li, and Hongli Wu

Processing Redundancy in UML Diagrams Based on Knowledge Graph   418
  *Yirui Jiang, Yucong Duan, Mengxing Huang, Mingrui Chen, Jingbin Li,
  and Hui Zhou*

An Automatic Fall Detection System Based on Derivative Dynamic
Time Warping   427
  Hong Yang, Yanqin Yang, Wenchao Xu, and Yuxin Pang

On Signal Timing Optimization in Isolated Intersection Based
on the Improved Ant Colony Algorithm.   439
  Huang Min

Experiments on Neighborhood Combination Strategies for Bi-objective
Unconstrained Binary Quadratic Programming Problem   444
  *Li-Yuan Xue, Rong-Qiang Zeng, Wei An, Qing-Xian Wang,
  and Ming-Sheng Shang*

Porting Referential Genome Compression Tool on Loongson Platform.   454
  Zheng Du, Chao Guo, Yijun Zhang, and Qiuming Luo

Statistics of the Number of People Based on the Surveillance Video   464
  Zhao Qiu, ShiYao Lei, JianZheng Hu, and JingYu Luo

Differential Privacy in Power Big Data Publishing   471
  Ping Kong, Xiaochun Wang, Boyi Zhang, and Yidong Li

Parallel Architectures

Parallel Aligning Multiple Metabolic Pathways on Hybrid CPU
and GPU Architectures   483
  Yiran Huang, Cheng Zhong, Jinxiong Zhang, Ye Li, and Jun Liu

Speeding Up Convolution on Multi-cluster DSP in Deep Learning
Scenarios   493
  *Deng Wenqi, Yang Zhenhao, Lu Maohui, Wang Gai, Yang JiangPing,
  and Zheng Qilong*

Optimization Scheme Based on Parallel Computing Technology   504
    Xiulai Li, Chaofan Chen, Yali Luo, and Mingrui Chen

Research on Client Adaptive Technology Based on Cloud Technology   514
    Xiaojing Zhu

Resource Allocation and Energy Management Based on Particle
Swarm Optimization   522
    Gang Mei, Mingrui Chen, and Xing Zhen

A Parallel Clustering Algorithm for Power Big Data Analysis   533
    Xiangjun Meng, Liang Chen, and Yidong Li

Customized Filesystem with Dynamic Stripe Strategies
on Lustre-Based Hadoop   541
    Hongbo Li, Yuxuan Xing, Nong Xiao, Zhiguang Chen, and Yutong Lu

Research on Optimized Pre-copy Algorithm of Live Container Migration
in Cloud Environment   554
    Huqing Nie, Peng Li, He Xu, Lu Dong, Jinquan Song,
    and Ruchuan Wang

A Cost Model for Heterogeneous Many-Core Processor   566
    Yanbing Li, Qi Wang, Yingying Li, Lin Han, Yuchen Gao, and Qing Mu

Scalable K-Order LCP Array Construction for Massive Data   579
    Yi Wu, Ling Bo Han, Wai Hong Chan, and Ge Nong

A Load Balancing Strategy for Monte Carlo Method in PageRank Problem . . .   594
    Bo Shao, Siyan Lai, Bo Yang, Ying Xu, and Xiaola Lin

Experiences of Performance Optimization for Large Eddy Simulation
on Intel MIC Platforms   610
    Zhengxiong Hou, Chengwen Zhong, Christian Perez, Qing Zhang,
    and Yunlan Wang

Author Index   627

Algorithms and Programming

Ford Motor Side-View Recognition System Based on Wavelet Entropy and Back Propagation Neural Network and Levenberg-Marquardt Algorithm

Wen-Juan Jia[1], Shuihua Wang[1(✉)], Huimin Lu[2], Ying Shao[3],
Elizabeth Lee[4], and Yu-Dong Zhang[1,5(✉)]

[1] School of Computer Science and Technology, Nanjing Normal University,
Nanjing 210023, Jiangsu, China
{shuihuawang, yudongzhang}@ieee.org
[2] Department of Mechanical and Control Engineering,
Kyushu Institute of Technology,
Kitakyushu, Fukuoka Prefecture 804-8550, Japan
[3] John A. Paulson School of Engineering and Applied Sciences,
Harvard University, Cambridge, MA 02138, USA
[4] Department of Engineering Technology,
Chattanooga State Community College, Chattanooga, TN 37406, USA
[5] Department of Informatics, University of Leicester, Leicester LE1 7RH, UK

Abstract. (Aim) Automatic identification of the car manufacturer in the side-view position can be used for the intelligent traffic monitoring system. Currently, the side-view car recognition did not attract too much attention. (Method) We proposed a novel Ford Motor recognition system. We first captured the car image from the side view. Second, we used wavelet entropy to extract texture features. Third, we employed a back propagation neural network (BPNN) as the classifier. Finally, we employed the Levenberg-Marquardt algorithm to train the classifier. In the experiment, we utilized the 3×3-fold cross validation. (Result) This method achieved an overall accuracy of 80% in detecting Ford motors. (Conclusion) This method can detect Ford Motors from the side view effectively. In the future, it may also be used to detect cars of other brands.

Keywords: Wavelet entropy · Recognition · Ford motor · Back propagation neural network · Levenberg-Marquardt algorithm · Pattern recognition · Cross validation

1 Background

In China, the popularity of motors makes traffic accidents more frequent than ever before [1, 2]. In most accidents, how to extract images of the front of cars from surveillance video is difficult for us, but we can get the side view easily instead. In existing method, there is no such technique to detect the car brand from the side view of a car. Moreover, traditional manual recognition costs too much. Hence, we proposed a method hoping to solve this problem in real life.

© Springer Nature Singapore Pte Ltd. 2017
G. Chen et al. (Eds.): PAAP 2017, CCIS 729, pp. 3–12, 2017.
DOI: 10.1007/978-981-10-6442-5_1

In the advance of our research, we searched EI, SCI, Elsevier, Springer, IEEE, and other popular databases, but we did not found any relative articles. Therefore, our research is the first time to use the artificial intelligence to detect the car in the side view. In our research, we shot the cars image of side view to detect the Ford motors [3, 4] among different kinds of cars. The factors of sunlight, occlusion of different objects and the change of the weather all will affect the image processing. Hence, without considering above factors, we only shot the cars image in the condition of nature.

For the image feature extraction, some researchers used the Scale Invariant Feature Transform (SIFT) [5], but it will cause feature matching amplification errors, and thus the matching will be unsuccessful [6]. de Souza et al. used the method of Singular Value Decomposition (SVD) [7] to extract image feature sometimes. Nevertheless, this method is based on matrix operations, when the amount of image pixels is large, the image matrix will become large, which will decrease the efficiency of operating [8].

Therefore, we proposed the method of Wavelet Entropy (WE) to extract the feature from car images. Wavelet Entropy used in this paper plays an important role in image classification. It can extract the image features and make the image classification more accurate [9–11]. WE can greatly reduce or remove the correlation between the extracted features, which will cut down the difficulty during the procedure of image identification.

Then the back propagation neural network (BPNN) is used to identify the image of Ford motors from 20 car pictures. BP neural network can effectively control the scale of neural network based on number of extracted features of the images, and a small amount of the extracted features will improve the running speed of neural network.

2   Dataset and Methods

2.1   Dataset

We shot 20 pictures of car, including 4 pictures of Buick motors, 9 pictures of Ford motors, 2 pictures of Beijing Hyundai motors, 2 pictures of Shanghai Volkswagen motors and 3 pictures of Toyota motors.

The method we proposed in this paper includes the following steps as shown in Fig. 1: image preprocessing, feature extraction, NN-based classification.

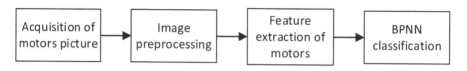

Fig. 1. Diagram of our method

2.2   Preprocessing

For the image shooting, we invited two senior cameramen, who used camera more than five years, to take the photos of five different kinds of motors. In the procedure of image preprocessing, we segmented each motor picture manually with the software of Photoshop. However, we did not take the factors, such as light duration, the weather and smog, into account for simplification in the preprocessing. When cameramen took photos, we measured the angle of the shot and set it to 90° to ensure that the shooting angle for the experiment is same. In addition, we shot 5 m away from the location of motors. Considering the background, which would interfere the feature extraction in the period of image preprocessing, we removed it.

For the segmented picture, we set the size of background to 300 × 300, and set the value of background color to zero. The twenty images have same size and background, so it won't affect the outcome. (See Fig. 2) Then we converted the initial images to gray-level images. (See Fig. 3) All we did above is necessary for us to detect the ford motors brand.

Fig. 2. Segment the motors from background and set the size of background to 300 × 300

Fig. 3. Converting the initial image into gray-level image

2.3 Principle of Wavelet Entropy

We use wavelet entropy to extract features of motors. Wavelet entropy is based on wavelet function and Shannon entropy [12, 13]. Continuous wavelet transform can be described as following mathematical formula

$$W_{\Psi}(a,c) = \int_{-\infty}^{+\infty} x(t)\Psi_{a,c}(t)\mathrm{d}t \qquad (1)$$

where

$$\Psi_{a,c}(t) = \frac{1}{\sqrt{|a|}}\Psi(\frac{t-c}{a}), a \neq 0, c \in R \qquad (2)$$

Here, a denotes the scale factor, and c denotes the translation factor (both real positive numbers). ψ is called wavelet basis function, which depend on the factors a and c. If the scale factor a and the translation parameter c are dispersed, the discrete wavelet transform (DWT) would be acquired [14]. Moreover, the DWT would be used to extract the feature of motors image in our method.

In our method, the high or low frequency information can be obtained by increasing or decreasing the scale factor a. We can analyze the profile or details of signals, and then get the analysis of local feature in different time scale and spatial scale [15, 16]. Hence, we can extract significant features that are helpful for us to identify the brand of motors from images. As shown in Fig. 4, the two-dimensional discrete wavelet transform is used twice, then we acquire seven sub-band [17]. LL2 sub-band represents discrete approximation component, and others represent discrete detail component. As we all know, the low frequency region LL contains most information of image, which characterize invariant features. Remaining sub-bands belong to high frequency region, the LH sub-band reflects vertical feature, HL sub-band keep the horizontal feature, HH sub-band keeps the diagonal detail of the image.

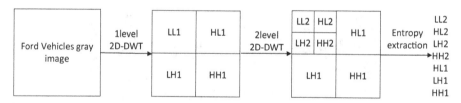

Fig. 4. Pipeline of calculating wavelet entropy

2.4   Image Classification Based on BPNN

The back propagation neural network (BPNN) is one of the multilayer feed-forward networks, which are trained typically via error back propagation algorithm [18]. As one of the neural network models, BPNN is widely used [19]. Through back-propagation, BPNN continuously tunes the biases and weights of the neural network, so the square sum error of the network will be minimized [20–22]. Input layer, hidden layer and output layer constitute the structure of BPNN model (see Fig. 5).

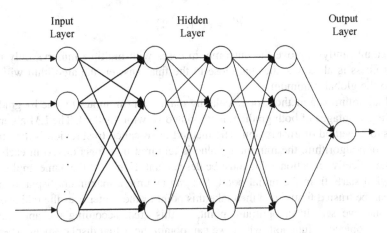

Fig. 5. Structure of a back propagation neural network

In BPNN, the learning procedure is composed of two phases: (i) forward propagation of information; and (ii) back propagation of error. The responsibility of the input layer neurons is to receive the messages from the outside, and transfer it to each neuron of middle layer, which is used to deal with interior information [23–25]. The middle layer is in charge of the transformation of information, and it can be designed to be a structure of a multi-hidden layer or single-hidden layer according to the need of the ability of information transform.

In our method, multi-hidden layer is used. After a further processing, the information from the last hidden layer to each neuron of output layer can finish a learning procedure of forward propagation process, and the output layer can transfer the outcome of information processing to the external [26]. When the actual output does not adapt to the desired layer, it enters to the stage of the back propagation of error. The error is corrected by the output layer, according the way of the gradient descent of the error, then the error is back propagated to the hidden layer and input layer, layer by layer. The process of positive and negative propagation of information is the process of constant adjustment of weights at each layer. It is a process of neural network learning and training as well. This process is carried out until the error of network output is reduced to an acceptable range or predefined learning times.

2.5 Levenberg-Marquardt Algorithm

Well-known as the damped least-squares (DLS), Levenberg-Marquardt (LM) algorithm is used to solve the problems of non-linear least squares. As a kind of numeric minimization algorithms, the procedure of LM algorithm is iterative. The user has to make an initial guess for the parameter vector β before starting a minimization. In order to avoid the situation of only one minimum value, an unacquainted standard guess like

$$\beta^T = (1, 1, \cdots 1) \tag{3}$$

It will significantly improve the outcome; in cases with multiple minima, only if the original guess is already somewhat close to the final answer, the algorithm will converge to the global minimum.

LM algorithm is a method for calculating maximum (minimum) value by gradient. It has the advantages of both gradient method and Newton method. The LM algorithm belongs to a method of trust region. The method of so-called trust region is that, in the optimization algorithm, the minimum value is required in a function, so in each iteration, the objective function value must be descendant. Hence, as the name implies, the trust region starts from the initial point. We can assume a maximum displacement s, which can be trusted firstly, and then take this point as the center, s as the radius of the region, and we search an optimum point, in this area, according the approximate function of objective function, where we can obtain the actual displacement. Then we can calculate the real value of objective function. If this value can satisfy the conditions for the descent of objective function values, the displacement is reliable, and we can calculate iteratively according this rule; if it can't satisfy the conditions, we should reduce the area of trust region, then solve it.

3   Experiments and Results

In the experiment, our method are conducted based on the MATLAB R2016 (a) (The Mathworks, Natick, MA, USA) with neural network toolbox. The computations of our method were performed on a personal computer with 3.2 GHz Core i5-3470 CPU, and 4 GB memory, under the operating system of Windows 7. However, the experiment can perform on any computer with MATLAB installed.

3.1   Experiment Design

The main programs are showed on Table 1. We divide our programs into three stages: (i) 2D-DWT, (ii) entropy extraction, and (iii) BPNN classification.

3.2   Detection Performance

Our method is compared with DWT + Cubic SVM, DWT + Disabled PCA, DWT + Complex Tree, DWT + Linear Discriminant, DWT + Logistic Regression, DWT + Linear SVM, DWT + Fine Gaussian SVM based on 3 × 3-fold cross validation.

Table 2 shows that the accuracy of our approach achieves 80%, the accuracy of our proposed method is better than DWT + Cubic SVM, DWT + Disabled PCA, DWT + Complex Tree, DWT + Linear Discriminant, DWT + Logistic Regression, DWT + Linear SVM, DWT + Fine Gaussian SVM. Our proposed method is outstanding compared with above methods.

Table 1. Pseudocode of the proposed strategy

Parameter: M, total image number
Input: M Motor images.
Put the target data into a matrix $T[M \times 1]$.

Stage 1. 2D-DWT
    Read all image files.
    Apply the DWT over j-th image, and extract seven subbands.
    $X[256 \times 256 \times M]$ saves all the coefficients.

Stage 2. Extract entropy from wavelet subband coefficients.
    Extract entropy of the coefficient subband.
    Save the entropy into $E[7 \times M]$.

Stage 3. BPNN Classification through 3×3 cross validation.
Divide randomly the input data $E[7 \times M]$ and target data $T[M \times 1]$ into 3 different folds.

---

for $j = 1 : 3$
    Use the j-th fold for test, and the rest two folds are combined as the training set.
    Record the classification results over j-th fold.
end

Output:
    Summarize the classification results over each fold.
    Report the overall accuracy.

Table 2. Classification Comparison

Method	Accuracy	Image	Fold
WE + BPNN + LM (Our)	80%	20	3 × 3 fold
DWT + Cubic SVM	60%	20	3 × 3 fold
DWT + Disabled PCA	50%	20	3 × 3 fold
DWT + Complex Tree	50%	20	3 × 3 fold
DWT + Linear Discriminant	50%	20	3 × 3 fold
DWT + Logistic Regression	45%	20	3 × 3 fold
DWT + Linear SVM	45%	20	3 × 3 fold
DWT + Fine Gaussian SVM	50%	20	3 × 3 fold
DWT + Medium Gaussian SVM	65%	20	3 × 3 fold
DWT + Quadratic Discriminant	75%	20	3 × 3 fold

3.3 Convergence Curve

In our experiment, the best overall accuracy of our method achieves 80%. Figure 6 shows its converge curves within seven epochs. In the final epoch, the gradient equals to 4.9983e-06, the learning rate (mu) equals to 1e-08, validation fail times equal to 6 at epoch 7.

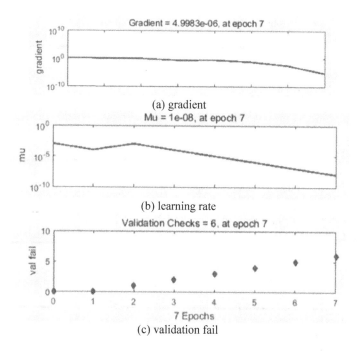

Fig. 6. The convergence curve of result at epoch 7

3.4 Confusion Matrix Result

The confusion matrix result of 3×3 fold cross validation of our method is illustrated in Table 3. Five Ford motorcars were correctly classified as Ford, and 4 Ford motorcars were wrongly misclassified as Non-Ford motorcars. In contrast, all 11 Non-Ford motors were identified correctly.

Table 3. One-Time different fold cross validation (here Fo represents Ford brand, No represents Non-Ford brand)

Prediction	Actual	
	Fo	No
Fo	5	0
No	4	11

3.5 Computation Time

In our experiment, we chose the method Wavelet Entropy and BPNN with Levenberg-Marquardt Algorithm. The computation time of our method is 2.917716 s. We found the computation time of different folds is closed to each other.

4 Conclusion

From our research, we can find that our method can be used to detect Ford motors, but cannot classify different motors into different brands. For example, we cannot classify the Ford motors, Buick motors, Beijing Hyundai motors, Shanghai Volkswagen motors and Toyota motors at the same time. Therefore, our next task is to find a better combination of methods to solve this problem. In addition, the computation time should be reduced furthermore by some advanced wavelet transforms. Another problem is the small-size of our dataset. We shall either collect more images or use data augmentation techniques.

Acknowledgement. This paper is supported by Program of Natural Science Research of Jiangsu Higher Education Institutions (16KJB520025, 15KJB470010), Natural Science Foundation of Jiangsu Province (BK20150983), Natural Science Foundation of China (61602250), Open fund of Key Laboratory of Guangxi High Schools Complex System and Computational Intelligence (2016CSCI01).

References

1. Andrieux, A., Vandanjon, P.O., Lengelle, R., Chabanon, C.: New results on the relation between tyre-road longitudinal stiffness and maximum available grip for motor car. Veh. Syst. Dyn. **48**, 1511–1533 (2010). Article ID: Pii 927059222
2. Babanoski, K., Ilijevski, I., Dimovski, Z.: Analysis of road traffic safety through direct relative indicators for traffic accidents fatality: case of republic of macedonia. Promet Traffic Transp. **28**, 661–669 (2016)
3. Dyrud, M.A.: The case of ford motor company. J. Eng. Technol. **33**, 10–21 (2016)
4. Bernaciak, M.: Paradoxes of internationalization. British and German trade unions at ford and general motors 1967–2000. Br. J. Ind. Relat. **51**, 2 (2013)
5. Neelima, A., Singh, K.M.: Perceptual hash function based on scale-invariant feature transform and singular value decomposition. Comput. J. **59**, 1275–1281 (2016)
6. Ghoualmi, L., Draa, A., Chikhi, S.: An ear biometric system based on artificial bees and the scale invariant feature transform. Expert Syst. Appl. **57**, 49–61 (2016)
7. de Souza, J.C.S., Assis, T.M.L., Pal, B.C.: Data compression in smart distribution systems via singular value decomposition. IEEE Trans. Smart Grid **8**, 275–284 (2017)
8. Nayak, M.R., Bag, J., Sarkar, S., Sarkar, S.K.: Hardware implementation of a novel water marking algorithm based on phase congruency and singular value decomposition technique. AEU Int. J. Electron. Commun. **71**, 1–8 (2017)
9. Phillips, P., Dong, Z., Yang, J.: Pathological brain detection in magnetic resonance imaging scanning by wavelet entropy and hybridization of biogeography-based optimization and particle swarm optimization. Prog. Electromagn. Res. **152**, 41–58 (2015)
10. Sun, P.: Pathological brain detection based on wavelet entropy and Hu moment invariants. Bio-Med. Mater. Eng. **26**, 1283–1290 (2015)
11. Zhou, X.-X.: Comparison of machine learning methods for stationary wavelet entropy-based multiple sclerosis detection: decision tree, k-nearest neighbors, and support vector machine. Simulation **92**, 861–871 (2016)

12. Mooij, A.H., Frauscher, B., Amiri, M., Otte, W.M., Gotman, J.: Differentiating epileptic from non-epileptic high frequency intracerebral EEG signals with measures of wavelet entropy. Clin. Neurophysiol. **127**, 3529–3536 (2016)

13. Wachowiak, M.P., Hay, D.C., Wachowiak-Smolikova, R., DuVal, D.J., Johnson, M.J.: Analyzing multiresolution wavelet entropy of ECG with visual analytics techniques. In: Canadian Conference on Electrical and Computer Engineering, Vancouver, Canada (2016)

14. Wang, S.-H.: Single slice based detection for Alzheimer's disease via wavelet entropy and multilayer perceptron trained by biogeography-based optimization. Multimed. Tools Appl. (2016). doi:10.1007/s11042-016-4222-4

15. Gorriz, J.M., Ramírez, J.: Wavelet entropy and directed acyclic graph support vector machine for detection of patients with unilateral hearing loss in MRI scanning. Front. Comput. Neurosci. **10** (2016). Article ID: 160

16. Lu, H.M.: Facial emotion recognition based on biorthogonal wavelet entropy, fuzzy support vector machine, and stratified cross validation. IEEE Access **4**, 8375–8385 (2016)

17. Wu, L.: A hybrid method for MRI brain image classification. Expert Syst. Appl. **38**, 10049–10053 (2011)

18. Ilangkumaran, M., Sakthivel, G., Nagarajan, G.: Artificial neural network approach to predict the engine performance of fish oil biodiesel with diethyl ether using back propagation algorithm. Int. J. Ambient Energy **37**, 446–455 (2016)

19. Karimi, R., Yousefi, F., Ghaedi, M., Dashtian, K.: Back propagation artificial neural network and central composite design modeling of operational parameter impact for sunset yellow and azur (II) adsorption onto MWCNT and MWCNT-Pd-NPs: isotherm and kinetic study. Chemom. Intell. Lab. Syst. **159**, 127–137 (2016)

20. Ji, G.: Fruit classification using computer vision and feedforward neural network. J. Food Eng. **143**, 167–177 (2014)

21. Feng, C.: Feed-forward neural network optimized by hybridization of PSO and ABC for abnormal brain detection. Int. J. Imaging Syst. Technol. **25**, 153–164 (2015)

22. Wu, J.: Fruit classification by biogeography-based optimization and feedforward neural network. Expert Syst. **33**, 239–253 (2016)

23. Wu, L.: Weights optimization of neural network via improved BCO approach. Prog. Electromagn. Res. **83**, 185–198 (2008)

24. Zhang, Y.: Stock market prediction of S&P 500 via combination of improved BCO approach and BP neural network. Expert Syst. Appl. **36**, 8849–8854 (2009)

25. Naggaz, N., Wei, G.: Remote-sensing image classification based on an improved probabilistic neural network. Sensors **9**, 7516–7539 (2009)

26. Lu, Z.: A pathological brain detection system based on radial basis function neural network. J. Med. Imaging Health Inform. **6**, 1218–1222 (2016)

Intrusion Detection Based on Self-adaptive Differential Evolution Extreme Learning Machine with Gaussian Kernel

Junhua Ku[✉] and Bing Zheng

Department of Information Engineering,
Hainan Institute of Science and Technology, Haikou 571100, China
kujunhua@163.com, 512049181@qq.com

Abstract. In our everyday life, intrusion detection system(IDS) becomes a promising area of research in the domain of security. With the rapid development of network-based services, IDS can detect the intruders who are not authorized to the present computer system, so IDS has emerged as an essential component and an important technique for network security.

In order to conquer the disadvantage of the traditional algorithm for single-hidden layer feedforward neural network (SLFN), an improved algorithm, called extreme learning machine (ELM), is proposed by Huang et al. However, ELM is sensitive to the neuron number in hidden layer and its selection is a difficult-to-solve problem. ELM is an interested area of research for detecting possible intrusions and attacks. In this paper, we propose an improved learning algorithm named self-adaptive differential evolution extreme learning machine with Gaussian Kernel (SaDE-KELM) for classifying and detecting the intrusions. We compare our methods with commonly used ELM, DE-ELM techniques in classifications. Simulation results show that the proposed SaDE-KELM approach achieves higher detection accuracy in classification case.

Keywords: Extreme learning machines · Differential evolution extreme learning machines · Self-adaptive differential evolution extreme learning machines · Intrusion detection · Network security

1 Introduction

Intrusion into computer networks and systems is a major threat in today's network centric world. Few most prevalent intrusion attacks include Denial-of-Service (DoS) attacks, Distributed-Denial-of-Service(DDoS) attacks, probing based attacks and account takeover attacks. Intrusion detection identifies computer attacks by observing various records processed on the network. Intrusion detection models are classified into two variants, misuse detection and anomaly detection systems. Misuse detection can discover intrusions based on a known pattern also known as signatures [1]. Anomaly detection can identify the malicious activities by observing the deviation from normal network traffic pattern [2]. Hence anomaly detection can identify new anomalies. The difficulty with the current developmental techniques is not only the high false positive rate, but also the low false negative rate.

© Springer Nature Singapore Pte Ltd. 2017
G. Chen et al. (Eds.): PAAP 2017, CCIS 729, pp. 13–24, 2017.
DOI: 10.1007/978-981-10-6442-5_2

Recently, extreme learning machine (ELM) [3, 4], was developed to improve the efficiency of SLFNs. ELM is different from the conventional learning algorithms for neural networks (such as BP algorithms [5]), which can easily deal with difficulties in manually tuning control parameters (learning rate, learning epochs, etc.) and/or local minima. ELM is fully automatically implemented without iterative tuning, and tends to provide better generalization performance at extremely fast learning speed. Furthermore, It was popular for its fast training speed by means of utilizing random hidden node parameters and calculating the output weights with least square algorithm [6–10]. However, in ELM, the number of hidden nodes is assigned in advance, the hidden node parameters are randomly chosen and they stay the same during the training phase. Many non-optimal nodes may exist and contribute less in minimizing the cost function. Moreover, in [11] Huang et al. indicated that ELM was inclined to require more hidden nodes than conventional tuning-based algorithms [12, 13] in many cases.

Differential evolution (DE) [14] which is a simple but powerful population-based stochastic direct searching technique is a frequently used method for selecting the network parameters [15–17]. In [15], DE is directly adopted as a training algorithm for feed forward networks where all the network parameters are encoded into one population vector and the error function between the network approximate output and the expected output is used as the fitness function to evaluate all the populations. However, Subudhi and Jena [16] have indicated that using the DE approach alone for the network training may lead to a decrease in convergence. Therefore, in [17], DE-ELM based on DE and ELM has been developed for SLFNs. Using the DE method to optimize the network input parameters and the ELM algorithm to calculate the network output weights, DE-ELM has shown several promising features. It not only ensures a more compact network size than ELM, but also has better generalization performance.

However, in the above DE based neural network training algorithms, the trial vector generation strategies and the control parameters in DE have to be manually chosen. Therefore, we propose the SaDE-KELM for SLFNs. In SaDE-KELM, the hidden node learning parameters are optimized by the self-adaptive differential evolution algorithm. We verify our SaDE-KELM using the data originated from the 1998 DARPA Intrusion Detection Evaluation Program 1999 [18], and deliberate to regard as a common benchmark for evaluating intrusion detection techniques [19–21].

The rest of the paper is organized as follows. In Sect. 2, a brief introduction to ELM and SaDE are given. In Sect. 3, we introduce model of proposed SaDE-ELM algorithm in detail. In Sect. 4, we present the dataset we use in our SaDE-KELM-based intrusion detection technique. In Sect. 5, we ran experiments for detecting intrusion in network traffic data and obtained the performance comparisons among ELM-based techniques, DE-ELM-based techniques and SaDE-KELM techniques. In Sect. 6, we conclude and summarize our results.

2 Background

As we know that ELM is very efficient and effective for SLFNs' training algorithms. In this section, we gave a review of ELM in brief. At the same time we briefly reviewed SaDE approach and proposed our SaDE-ELM techniques in the end.

2.1 Extreme Learning Machine (ELM)

For N arbitrary distinct samples (x_j, t_j), where $x_j = [x_{j1}, x_{j2}, \cdots, x_{jn}]^T \in \mathbb{R}^n$, $t_j = [t_{j1}, t_{j2}, \cdots, t_{jm}]^T \in \mathbb{R}^m$, SLFNs with L hidden nodes and activation function $g(x)$ are

$$\sum_{i=1}^{L} \beta_i g_i(x_j) = \sum_{i=1}^{L} \beta_i g_i(w_i \cdot x_j + b_j) = o_j \ (j = 1, 2, \ldots, N) \tag{1}$$

where $w_i = [w_{i1}, w_{i2}, \ldots, w_{in}]T$ is the weight vector connecting the ith hidden node and the input nodes, $\beta_i = [\beta_{i1}, \beta_{i2}, \ldots, \beta_{im}]^T$ is the weight vector connecting the ith hidden node and the output nodes, b_i is the threshold of the ith hidden node, $w_i \cdot x_j$ denotes the inner product of w_i and x_j, $g(x)$ is activation function and Sigmoid, Sine, Hardlim and other functions are commonly used. The output nodes are chosen linear in this paper, and $o_j = [o_{j1}, o_{j2}, \ldots, o_{jm}]^T$ is the jth output vector of the SLFNs [22].

The SLFNs with L hidden nodes and activation function $g(x)$ can approximate these N samples with zero error. It means $\sum_{j=1}^{L} \|o_j - t_j\| = 0$ and there exist β_i, w_i and b_i such that

$$\sum_{i=1}^{L} \beta_i g_i(x_j) = \sum_{i=1}^{L} \beta_i g_i(w_i \cdot x_j + b_j) = t_j \ (j = 1, 2, \ldots, N) \tag{2}$$

The equation above can be expressed compactly as follows:

$$\mathbf{H}\beta = \mathbf{T} \tag{3}$$

where $H(w_1, w_2, \cdots, w_L, b_1, b_2, \cdots, b_L, x_1, x_2, \cdots, x_L)$

$$= [h_{ij}] \begin{bmatrix} g(w_1 \cdot x_1 + b_1) & g(w_1 \cdot x_1 + b_2) & \cdots & g(w_1 \cdot x_1 + b_L) \\ g(w_1 \cdot x_2 + b_1) & g(w_2 \cdot x_2 + b_2) & \cdots & g(w_L \cdot x_2 + b_L) \\ \vdots & \vdots & & \vdots \\ g(w_1 \cdot x_N + b_1) & g(w_2 \cdot x_N + b_1) & \cdots & g(w_L \cdot x_N + b_L) \end{bmatrix}_{N \times L}$$

$$\beta = \begin{pmatrix} \beta_{11} & \beta_{12} & \cdots & \beta_{1m} \\ \beta_{21} & \beta_{22} & \cdots & \beta_{2m} \\ \vdots & \vdots & \vdots & \vdots \\ \beta_{L1} & \beta_{L2} & \cdots & \beta_{Lm} \end{pmatrix} \text{ and } T = \begin{pmatrix} t_{11} & t_{12} & \cdots & t_{1m} \\ t_{21} & t_{22} & \cdots & t_{2m} \\ \vdots & \vdots & \vdots & \vdots \\ t_{N1} & t_{N2} & \cdots & t_{Nm} \end{pmatrix}$$

We call the matric $\mathbf{H}$ the hidden layer output matrix of the neural network. Meanwhile, with respect to inputs $x_1, x_2, \ldots, x_N$, it is easy to see that the ith hidden node output is the ith column of H.

Because L is far smaller than N in actual problem, here in particular it is rare condition that L equals to N [23]. That is to say, in ELM algorithm, the important thing is to find a least-square solution of this linear system as follow

$$\hat{\beta} = \mathbf{H}^{\dagger}\mathbf{T} \tag{4}$$

where $\mathbf{H}^{\dagger}$ is the Moore-Penrose Generalized inverse of matrix $\mathbf{H}$, depending on the singularity of $\mathbf{H}^{\mathbf{T}}\mathbf{H}$ or $\mathbf{H}\mathbf{H}^{\mathbf{T}}$, we can obtain $\mathbf{H}^{\dagger} = (H^{T}H)^{-1}H^{T}$ or $H^{T}(HH^{T})^{-1}$ Then the output function of ELM can be modeled as follows.

$$f(\mathbf{x}) = h(\mathbf{x})\beta = h(\mathbf{x})\mathbf{H}^{\dagger}\mathbf{T} \tag{5}$$

It's getting more stable to introduce a positive coefficient into ELM. For example, $\mathbf{H}^{\mathbf{T}}\mathbf{H}$ is nonsingular, the coefficient $1/C$ is added to the diagonal of $\mathbf{H}^{\mathbf{T}}\mathbf{H}$ in the calculation of the output weights β. So, the corresponding function of the regularized ELM is:

$$f(x) = h(x)\beta = h(x)\mathrm{H}^{\mathrm{T}}\left(\frac{E}{C} + \mathrm{HH}^{\mathrm{T}}\right)^{-1}\mathrm{T} \tag{6}$$

Moreover, we know that many nonlinear activation and kernel functions can be used in ELM. Let $\Omega_{\text{ELM}} = \mathrm{HH}^{\mathrm{T}} : \Omega_{\text{ELM}i,j} = h(x_i)h(x_j) = K(x_i, x_j)$, the output function can be written as:

$$f(x) = h(x)\mathrm{H}^{\mathrm{T}}\left(\frac{E}{C} + \mathrm{HH}^{\mathrm{T}}\right)^{-1}\mathrm{T} = \begin{bmatrix} K(x,x_1) \\ \vdots \\ K(x,x_N) \end{bmatrix}^{T} \left(\frac{E}{C} + \Omega_{\text{ELM}}\right)^{-1}\mathrm{T} \tag{7}$$

The hidden layer feature mapping $h(x)$ need not to be known, and instead its corresponding kernel $K(u,v)$ can be computed. In this way, the Gaussian kernel is used, $K(u,v) = \exp(-\gamma\|u-v\|^2)$ [14].

2.2 Self-adaptive Differential Evolution

Differential evolution (DE), proposed by Storn and Price in 1995, is a powerful evolutionary algorithm [24]. There are three parameters in DE, which are the population size NP, mutation scaling factor F and crossover rate CR. To overcome the limitations of choosing the parameters in DE, Brest et al. [25] proposed a parameter adaptation technique to choose the mutation scaling factor F and crossover rate CR namely SaDE which performs better than the basic DE. In general, SaDE is composed of three main steps: mutation, crossover, and selection [26].

We consider the following optimization problem:

$$\text{Minimize} f(x), x_i \epsilon R_D$$

where $x_i = [x_{i1}, x_{i2}, \cdots, x_{iD}]^{T}, i = 1, 2, \cdots, NP$ is a target vector of D decision variables. During the mutation operation, mutant vector v_i is generated by mutation strategy in the current population:

$$v_i = x_{r1} + F \cdot (x_{r2} - x_{r3}) \tag{8}$$

where r_1, r_2, r_3 are randomly mutually exclusive integers in the range $[1, NP]$, and $r_1 \neq r_2 \neq r_3 \neq i$.

Following mutation, trial vector u_i is generated between x_i and v_i during crossover operation where the most widely used operator is the binomial crossover performed as follows:

$$u_{ij} = \begin{cases} v_{ij}, & if\,(rndreal(0,1) < CR\ or\ j = j_{rand}), \\ x_{ij}, & otherwise \end{cases} \tag{9}$$

where j_{rand} is a integer randomly chosen in the range $[1, D]$, and rndreal(0, 1) is a real number randomly generated in (0, 1). Finally, to keep the population size constant during the evolution, the selection operation is used to determine whether the trial or the target vector survives to the next generation according to one-to-one selection:

$$x_i = \begin{cases} u_i, & if\,(f(u_i)f(x_i)) \\ x_i, & otherwise \end{cases} \tag{10}$$

where $f(x)$ is the optimized objective function. During the evolution, F and CR are adaptively tuned to improve the performance of DE for each individual

$$F_{i,G+1} = \begin{cases} F_l + rand_1 \cdot F_u & if\,(rand_2 < \tau_1) \\ F_{i,G} & otherwise \end{cases} \tag{11}$$

$$CR_{i,G+1} = \begin{cases} rand_3 & if\,(rand_4 < \tau_2) \\ CR_{i,G} & otherwise \end{cases} \tag{12}$$

where $F_{i;G+1}$ and $CR_{i;G+1}$ are the mutation scaling factor and crossover rate for i individual in G generation respectively, $rand_j = 1;2;3;4$ are randomly chosen from (0, 1), τ_1 and τ_2 both valued 0.1 which is used to control the generation of F and CR, F_l valued 0.1 and F_u is valued 0.9. In the first generation, F and CR are initialized to 0.5.

3   Model of Proposed SADE-KELM Algorithm

Since the ELM may not reach the optimal result in classification or regression, a hybrid approach integrated SaDE and ELM with gaussian kernel namely SADE-KELM algorithm to optimize the input weights and hidden biases is able to obtain better generalization performance than ELM. In SaDE-KELM, we proposed SaDE-KELM for SLFNs by incorporating the SaDE to optimize the network input weights and hidden node biases and ELM with gaussian kernel to derive the network output weights.

Given a set of training data and L hidden nodes with an activation function g(·), we summarize the SaDE-KELM algorithm in the following steps.

Step 1. Initialization

A set of NP vectors where each one includes all the network hidden node parameters are initialized as the populations of the first generation

$$\theta_{k,G} = \left[w_{1,k,G}^T, \cdots, w_{L,k,G}^T, b_{1,k,G}^T, \cdots, b_{L,k,G}^T \right] \tag{13}$$

where w_j and b_j ($j = 1, \ldots, L$) are randomly generated, G represents the generation and $k = 1, 2, \ldots, NP$.

Step 2. Calculations of output weights and RMSE

Calculate the network output weight matrix and root mean square error (RMSE) with respect to each population vector with the following equations, respectively.

$$\beta_{k,G} = H_{k,G}^\dagger T \tag{14}$$

$$\text{RMSE}_{k,G} = \sqrt{\frac{\sum_{i=1}^{N} \left\| \sum_{j=1}^{L} \beta_j g(w_{j,k,G}, b_{j,k,G}, x_i) - t_i \right\|}{m \times N}} \tag{15}$$

Then use the value of RMSE to calculate the new best population vector $\theta_{k,G+1}$ with the following equation.

$$\theta_{k,G+1} = \begin{cases} u_{k,G+1} & \text{if } (\text{RMSE}_{\theta_{k,G}} - \text{RMSE}_{u_{k,G+1}}) > \varepsilon \cdot \text{RMSE}_{\theta_{k,G}} \\ u_{k,G+1} & \text{if } \left| \text{RMSE}_{\theta_{k,G}} - \text{RMSE}_{u_{k,G+1}} \right| < \varepsilon \cdot \text{RMSE}_{\theta_{k,G}} \\ & \text{and} \left\| \beta_{u_{k,G+1}} \right\| < \left\| \beta_{\theta,G} \right\| \\ \theta_{i,G} & \text{otherwise} \end{cases} \tag{16}$$

where ε is the preset small positive tolerance rate. In the first generation, the population vector with the best RMSE is stored as $\theta_{\text{best},1}$ and $\text{RMSE}\theta_{\text{best},1}$.

All the trial vectors $u_{k,G+1}$ generated at the $(G + 1)$th generation are evaluated using Eq. (11). The norm of the output weight $\|\beta\|$ is added as one more criteria for the trial vector selection as pointed out in [23] that the neural networks tend to have better generalization performance with smaller weights.

The three operations mutation, crossover and selection are repeated until the preset goal is met or the maximum learning iterations are completed. At last we calculate the output weigh $\beta = [\beta_{i1} \beta_{i2} \cdots \beta_{iL}]^T$ with equation $\beta = H^\dagger T$.

4 Intrusion Detection Using SADE-KELM

In this section, we describe the dataset that we use for our numerical studies, and our SaDE-KELM approach to classification of intrusions in the data.

4.1   Dataset Description

The dataset we use is from the 1998 DAPRA intrusion detection program [27]. Four main categories of attacks were simulated:

(1) DoS: denial of service attack
(2) R2L: unauthorized access from a remote machine
(3) U2R: unauthorized access to local root previledges
(4) Probing: surveillance and other probing

In the intrusion detection simulation, the dataset was labeled with 22 attack types falling into the four categories shown in Table 1. The feature list and its descriptions are in Tables 2, 3 and 4.

Table 1. Attack type

Denial of Service	User to Root	Remote to User	Probing
Back	Perl	FTP Write	IP Sweep
Neptune	Buffer Overflow	Guess Password	Nmap
Land	Load Module	Imap	Port Sweep
Teardrop	Rootkit	Multihop	Satan
Ping of Death		Phf	
Smurf		Spy	
		Warezclient	
		Warezmaster	

Table 2. Basic features of individual tcp connections

Feature name	Description	Type
Duration	Length (number of seconds) of the connection	Continuous
protocol_type	Type of the protocol	Discrete
service	Network service on the destination	Discrete
src_bytes	Number of data bytes from source to destination	Continuous
dst_bytes	Number of data bytes from destination to source	Continuous
flag	Normal or error status of the connection	Discrete
land	1 if connection is from/to the same host/port, 0 otherwise	Discrete
wrong_fragment	Number of "wrong" fragments	Continuous
urgent	Number of urgent packets	Continuous

4.2   Intrusion Detection System Using SaDE-ELM

We use a ELM method to classify the data to provide a comparison benchmark, our SaDE-KELM IDS has the following steps.

Table 3. Content features within a connection suggested by domain knowledge

Feature name	Description	Type
Hot	Number of "hot" indicators	Continuous
Num failed logins	Number of failed login attempts	Continuous
Logged in	1 if successfully logged in, 0 otherwise	Discrete
Num compromised	Number of "compromised" conditions	Continuous
Root shell	1 if root shell is obtained, 0 otherwise	Discrete
Su attempted	1 if "su root" command attempted, 0 otherwise	Discrete
Num root	Number of "root" accesses	Continuous
Num file creations	Number of file creation operations	Continuous
Num shells	Number of shell prompts	Continuous
Num access files	Number of operations on access control files	Continuous
Num outbound cmds	Number of outbound commands in an ftp session	Continuous
Is hot login	1 if the login belongs to the "hot" list, 0 otherwise	Discrete
Is guest login	1 if the login is a "guest"login, 0 otherwise	Discrete

Table 4. Traffic features computed using a two-second time window

Feature name	Description	Type
Count	Number of connections to the same host as the current connection in the past two seconds	Continuous
Serror rate	% of connections that have "SYN" errors	Continuous
Rerror rate	% of connections that have "REJ" errors	Continuous
Same srv rate	% of connections to the same service	Continuous
Diff srv rate	% of connections to different services	Continuous
Srv count	number of connections to the same service as the current connection in the past two seconds	Continuous
Srv serror rate	% of connections that have "SYN" errors	Continuous
Srv rerror rate	% of connections that have "REJ" errors	Continuous
Srv diff host rate	% of connections to different hosts	Continuous

Step 1. Converting the raw TCP/IP dump data into machine readable form.

Step 2. SaDE-KELM and ELM are trained on normal data and different types of attacks. For the binary classification case, the data has 41 features and falls into 2 classes: normal and attack; for the multi-class classification case, the data has 41 features and falls into 23 classes: normal and 22 types of attack. The model is trained in a large program which can test immediately after the training completed. According to SaDE-KELM theory that has been introduced above, we can summarize the following steps.

For N arbitrary distinct samples (x_i, t_i), $i = 1, \ldots N$, and hidden nodes and activation function $g(x)$:

(2.1) A set of *NP* individual parameter vectors $\theta_{k,G}$ ($k = 1, 2 \dots NP$), where each one includes all the network hidden node parameters are initialized as the populations of the first generation;

(2.2) In the case of $g(x)$ and L are invariable run the three operations including mutation, crossover and selection to produce the new population, and the process is repeated until the stop condition is completed.

(2.3) Changing the type of $g(x)$ and increase the number of hidden nodes L gradually from one to find the most suitable $g(x)$ and L to construct an optimal forecasting model with the best testing accuracy;

(2.4) Calculating the output matrix H;

(2.5) Calculating the output weights $\beta = H^{\dagger}H$.

Step 3. Testing phase: ELM, DE-ELM and SaDE-KELM are used to predict the type of each data point in the testing dataset, and their performances are compared.

Experiments have shown that if the number of values in an attribute is not too large, this coding is more stable than using a single number. The simulation of the three algorithms on all datasets are carried out using MATLAB 2013a on a machine with an Intel Core 2 Duo, 2.26 GHz CPU and 4 GB RAM.

5 Simulation Results

The datasets being tested are 2000, 4000, 8000 connection data chosen randomly from the dataset downloaded from the website [18]. We split them equally into training data and testing data. Simulation results including average testing accuracy and corresponding 95% confidence interval are given in Table 4.

In order to test the relationship between SaDE-KELM and the number of hidden layer, according to the different number of hidden layer nodes, we made classification tests using ELM, DE-ELM and SaDE-KELM respectively. Simulation results are given in Table 5.

Table 5. Performance comparison results

Dataset Size	ELM		DE-ELM		SaDE-KELM	
Training/Testing	Accuracy (%)	95% Confidence Interval (%)	Accuracy (%)	95% Confidence Interval (%)	Accuracy (%)	95% Confidence Interval
1000/1000	99.32	99.08–99.47	99.33	99.15–99.51	99.55	99.05–99.65
2000/2000	99.10	98.82–99.23	99.24	98.90–99.44	99.47	99.25–98.58
4000/4000	99.07	98.79–9.28	99.18	99.01–99.26	99.35	99.11–99.65

Figure 2 show the time spent by ELM, DE-ELM and SaDE-KELM when testing the same size of dataset. It can be seen that the training time and testing time spent by SaDE-KELM increase sharply when the size of data increases. In comparison, ELM and DE-ELM increase slowly when the number of data increases. Eventually, DE-ELM starts consuming more time for both training and testing than ELM.

Prediction of test samples (SaELM) with Nmin = 5, Nmax = 120 and NInterval = 5 for the classification problem can be seen from Fig. 1. A clear time consumption comparison can be seen from Fig. 2. From the results, we can conclude that ELM performs better than DE-ELM and SaDE-KELM in terms of speed. To increase accuracy, we can implement SaDE-KELM. This shows that our proposed SaDE-ELM methods have better scalability than ELM and DE-ELM when classifying network traffic for intrusion detection.

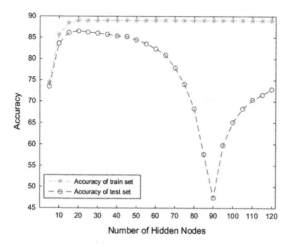

Fig. 1. Prediction of test samples (SaDE-KELM) with N_{min} = 5, N_{max} = 120 and $N_{Interval}$ = 5 for the classification problem.

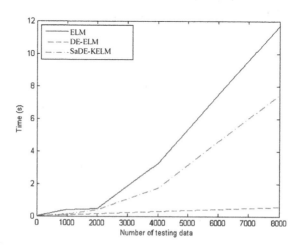

Fig. 2. Testing time comparison

6 Conclusion

ELM is sensitive to the neuron number in hidden layer and its selection is a difficult-to-solve problem. Recently, Extreme Learning Machine has been widely applied in IDS. It's an important and efficient way to improve the performance of IDS, However, the conventional feedback relevance schemes could not give considerations to the both accuracy and speed. To combine KELM(Extreme Learning Machine with Gaussian Kernel) with the SaDE method, we can overcome the limitation of the conventional problems. In this paper, we have made a comparison by the use of ELM, DE-ELM and SaDE-KELM for intrusion detection. For the SaDE-KELM, By incorporating the SaDE to optimize the network hidden node parameters and employing the KELM to derived the network output weights. Obviously, the proposed SaDE-KELM can obtain higher accuracy. Simulation results show that the proposed SaDE-KELM approach achieves higher detection accuracy in classification case.

Acknowledgement. This research was supported by key science research project of Education Department of Hainan province (Hnky2017ZD-20).

References

1. Ilgun, K., Kemmerer, R.A., Porras, P.A.: State transition analysis: a rule-based intrusion detection approach. IEEE Trans. Softw. Eng. **21**(3), 181–199 (1995)
2. Ikram, S.T., Cherukuri, A.K.: Improving accuracy of intrusion detection model using PCA and optimized SVM[J]. CIT. J. Comput. Inf. Technol. **24**(2), 133–148 (2016)
3. Huang, G.B., Zhu, Q.Y., Siew, C.K.: Extreme learning machine: a new learning scheme of feedforward neural networks. In: Proceedings of International Joint Conference on Neural Networks (IJCNN2004), vol. 2, no. 25–29, pp. 985–990
4. Huang, G.B., Zhu, Q.Y., Siew, C.K.: Extreme learning machine: theory and applications. Neurocomputing **70**(1–3), 489–501
5. Espana-Boquera, S., Zamora-Martínez, F., Castro-Bleda, M.J., et al.: Efficient BP algorithms for general feedforward neural networks. In: International Work-Conference on the Interplay Between Natural and Artificial Computation (pp. 327–336). Springer, Heidelberg (2007)
6. Thatte, G., Mitra, U., Heidemann, J.: Parametric methods for anomaly detection in aggregate traffic. IEEE/ACM Trans. Netw. **19**(2), 512–525 (2011)
7. Qin, M., Hwang, K.: Frequent episode rules for internet anomaly detection. In: Proceedings of the Network Computing and Applications, Third IEEE International Symposium. Washington, DC, USA: IEEE Computer Society, pp. 161–168 (2004)
8. He, X., Papadopoulos, C., Heidemann, J., Mitra, U., Riaz, U.: Remote detection of bottleneck links using spectral and statistical methods. Comput. Netw. **53**, 279–298 (2009)
9. Streilein, W.W., Cunningham, R.K., Webster, S.E.: Improved detection of low-profile probe and denial-of-service attacks. In: Proceedings of the 2001 Workshop on Statistical and Machine Learning Techniques in Computer Intrusion Detection (2001)
10. Cortes, C., Vapnik, V.: Support-vector networks. Mach. Learn. **20**, 273–297 (1995)
11. Huang, G.-B., Wang, D.H., Lan, Y.: Extreme learning machines: a survey. Int. J. Mach. Lean. Cybern. **2**(2), 107–122 (2011)
12. Tandon, G.: Weighting versus pruning in rule validation for detecting network and host anomalies. In: Proceedings of the 13th ACM SIGKDD international. ACM Press (2007)

13. Liao, Y., Vemuri, V.R.: Use of k-nearest neighbor classifier for intrusion detection. Comput. Secur. **25**, 439–448 (2002)

14. Storn, R., Price, K.: Differential evolution—a simple and efficient heuristic for global optimization over continuous spaces. J. Glob. Optim. **11**(4), 341–359 (2004)

15. Ilonen, J., Kamarainen, J.I., Lampinen, J.: Differential evolution training algorithm for feedforward neural networks. Neural Process. Lett. **17**, 93–105 (2003)

16. Subudhi, B., Jena, D.: Differential evolution and levenberg marquardt trained neural network scheme for nonlinear system identification. Neural Process. Lett. **27**, 285–296 (2008)

17. Zhu, Q.-Y., Qin, A.-K., Suganthan, P.-N., Huang, G.-B.: Evolutionary extreme learning machine. Pattern Recog. **38**(10), 1759–1763 (2005)

18. KDDCUPdataset: http://kdd.ics.uci.edu/databases/kddcup99/kddcup99.html (1999)

19. Mukkamala, S., Sung, A.: Detecting denial of service attacks using support vector machines. In: Proceedings of the 12th IEEE International Conference on Fuzzy Systems (2003)

20. Luo, M., Wang, L., Zhang, H., Chen, J.: A research on intrusion detection based on unsupervised clustering and support vector machine. In: Qing, S., Gollmann, D., Zhou, J. (eds.) ICICS 2003. LNCS, vol. 2836, pp. 325–336. Springer, Heidelberg (2003). doi:10.1007/978-3-540-39927-8_30

21. Kim, D.S., Park, J.S.: Network-Based Intrusion Detection with Support Vector Machines. In: Kahng, H.-K. (ed.) ICOIN 2003. LNCS, vol. 2662, pp. 747–756. Springer, Heidelberg (2003). doi:10.1007/978-3-540-45235-5_73

22. Lin, Y., Lv, F., Zhu S., Yang, M., Cour, T., Yu, K., Cao, L., Huang, T.S.: Large-scale image classification: fast feature extraction and SVM training. In: Proceedings of the 24th IEEE Conference on Computer Vision and Pattern Recognition (CVPR), pp. 1689–1696 (2011)

23. Huang, G.-B., Chen, L., Siew, C.K.: Universal approximation using incremental constructive feedforward networks with random hidden nodes. IEEE Trans. Neural Netw. **17**(4), 879–892 (2006)

24. Storn, R., Price, K.: Differential evolution-A simple and efficient heuristic for global optimization over continuous spaces. J. Global Optim. **11**, 341–359 (1997)

25. Brest, J., Greiner, S., Bŏskovíc, B., Mernik, M., Žumer, V.: Self-adapting control parameters in differential evolution: A comprehensive study on numerical benchmark problems. IEEE Trans. Evol. Comput. **10**, 646–657 (2006)

26. Wu, J., Cai, Z.H.: Attribute weighting via differential evolution algorithm for attribute weighted naive bayes (WNB). J. Comput. Inf. Syst. **7**, 1672–1679 (2011)

27. Stolfo, S., Fan, W., Lee, W., Prodromidis, A., Chan, P.K.: Costbased modeling for fraud and intrusion detection: results from the JAM project. In: Proceedings of DARPA Information Survivability Conference and Exposition, vol. 2, pp. 130–144 (2002)

Prediction for Passenger Flow at the Airport Based on Different Models

Xia Liu[1(✉)], Xia Huang[1], Lei Chen[1], Zhao Qiu[2(✉)],
and Ming-rui Chen[2]

[1] Sanya Aviation and Tourism College, Sanya 572000, Hainan, China
paolo_lx@qq.com
[2] Information and Technology School, Hainan University,
Haikou 570228, Hainan, China

Abstract. Correctly predicting the passenger flow of an air route is crucial for the construction and development of an airport. Based on the passenger flow data of Sanya Airport from 2008 to 2016, ARMA Model, Grey Prediction GM (1, 1) Model and ARMA-improved Regression Model were adopted for data fitting. Upon verification, the average absolute percentage error of such three models was 4.19%, 4.20% and 1.97% respectively with high prediction precision. As a result, the passenger flow at Sanya Airport is predicted to reach 20 million within two years.

Keywords: Passenger flow · ARMA model · Grey prediction model · RE-ARMA model · Prediction

1 Introduction

Correctly predicting the passenger flow of an airport is crucial for the capacity arrangement, future development and construction as well as the functional planning of an airport. Therefore, correctly predicting the passenger flow of an airport has become a significant subject for airport operation management and the fundamental basis for the effective allocation of airport resources. There are numerous passenger flow prediction methods. In terms of single model prediction, Paper [1–6] adopted Holt-winters Model, ARMA Model, One-Variable Regression Model, Grey Model and Multivariable Regression Model, etc. to predict the passenger flow. Although the prediction precision is high, the single model still features some uncertainty in long-term prediction; besides, the prediction model can be further improved and the prediction precision can be further increased. In terms of combined model prediction, Paper [7–11] combined ARIMA Model and BP Neural Network Model, the Multivariable Regression Model and the Time-Series Model, which has improved the prediction precision and the prediction result was superior to that achieved by the single model. In terms of the model comparison, Paper [12–14] compared the prediction precision of the Multivariable Regression Model, the One-Variable Regression Model, the ARIMA Model and GM (1,1). Upon comparison, the prediction precision of the Multivariable Regression Model was higher. In addition, in Paper [15], the means of WiFi was used to access to mobile phone MAC address to get passenger flow data and existing

G. Chen et al. (Eds.): PAAP 2017, CCIS 729, pp. 25–40, 2017.
DOI: 10.1007/978-981-10-6442-5_3

prediction methods were compared. ARMA model was established to predict passenger flow of short time. The results showed that the mean we used in this paper had high accuracy of prediction, MAPE was 2.8771, and the means can be used for the passenger flow prediction of exhibition well. In paper [16], this article studied the short-term prediction methods of sectional passenger flow, and selected BP neural network combined with the characteristics of sectional passenger flow itself. The empirical research showed that, the combining data characteristics of sectional passenger flow with the BP neural network had good prediction accuracy.

2   Data Descriptive Analysis

Rooted in the statistical data on the official website of Sanya Tourism Development Committee and the authentic data provided by Sanya Phoenix Airport, the monthly passenger numbers of the airport from 2008 to 2016 were selected for descriptive analysis and the sequence chart of passengers at the airport is shown as in Fig. 1.

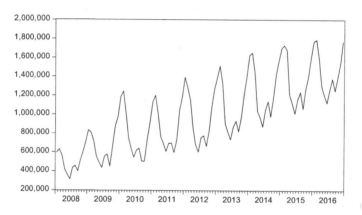

Fig. 1. Sequence chart of monthly passenger number

Indicated in Fig. 1, the monthly passenger numbers are evidently associated with the season, which is categorized by less passengers in July, August and September of each year and obviously more passengers to Sanya in January, February and March. It is pertinent to the local weather conditions. Statistical data in respect of the annual passenger flow at Sanya Airport from 2008 and 2016 are shown in Fig. 2.

As shown in Fig. 2, the annual passenger numbers at Sanya Airport from 2008 to 2016 remained increase from 6,008,308 in 2008 to 17,850,199 in 2016 with an annual increase of 12.86%. Statistical data of the monthly passenger numbers at Sanya Airport from 2008 to 2016 are indicated in Fig. 3.

As indicated in Fig. 3, more passengers came to Sanya in such five months as January, February, March, November and December of each year, all beyond 1 million people with less in May, June and September, which reflects obvious seasonal effect and conforms to the sequence chart in Fig. 1.

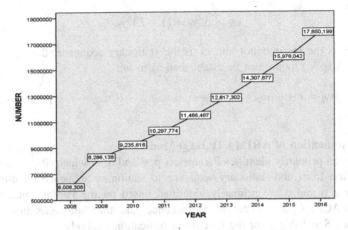

Fig. 2. Annual passenger numbers at sanya airport from 2008 to 2016

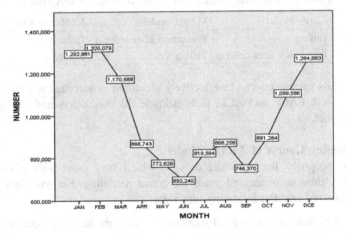

Fig. 3. Monthly passenger numbers at sanya airport from 2008 to 2016 on average

3   Discussion of Prediction Methods

3.1   ARMA Model

ARMA model, also known as the Autoregressive Integrated Moving Average Model (ARIMA Model), is one of the simple and practical models for time sequence analysis; besides, it features high prediction precision. ARMA model only considers the single variable instead of referring to the economic theory, trying to seek the rule of historical tendency of such single variable, and thus realize prediction by means of this rule.

3.1.1   Form of ARIMA Model

Consider the integrated sequence y_t can transform non-stationary sequence to stationary sequence by means of d-order difference, that is, $y_t \sim I(d)$; thus

$$\omega_t = \Delta^d y_t = (1 - L)^d y_t \qquad (1)$$

in which, L is the lag operator and ω_t is the stationary sequence, that is, $\omega_t \sim I(0)$. Thus ARMA (p, q) model can be established as to ω_t:

$$\omega_t = c + \phi_1 \omega_{t-1} + \cdots + \phi_p \omega_{t-p} + \varepsilon_t + \theta_1 \varepsilon_{t-1} + \cdots \theta_q \varepsilon_{t-q} \qquad (2)$$

3.1.2   Identification of ARIMA (P,D,Q) Model

ARIMA model primarily identifies Parameters p, d and q, in which, Parameter d is the transformation from non-stationary sequence to stationary sequence by d-order difference while p and q are primarily identified based on the Partial Autocorrelation Function (PACF) Chart of stationary sequence and the Autocorrelation Function (ACF) Chart. See Table 1 for the specific identification methods:

Table 1. Judgement of the identification chart of ARIMA model

	AR(p) model	MA(q) model	ARMA(p,q) model
ACF	Tailing	Truncation after order-q	Tailing
PACF	Truncation after order-p	Tailing	Tailing

The model's parameters can be effectively identified in accordance with Table 1 in combination with PACF and ACF; then estimate the parameters and finally get the prediction model.

3.1.3   Modeling Course of ARMA Model

Box-Jenkins proposed the modeling thought which may exert wide effect on the non-stationary time sequence and guide the actual modeling. His modeling thoughts can be categorized into the following four steps:

(1) Test the stationarity of original sequence. In case the sequence fails to fulfil the stationarity conditions, the difference transformation can be adopted. With the order of integration as d, the d-order difference or other transformations such as the logarithmic difference transformation can be adopted to enable the sequence to fulfil the stationarity conditions;

(2) Describe the statistics that can describe the sequence characteristics by computing, such as the autocorrelation function and the partial autocorrelation function to confirm order-p and q in ARMA Model, and choose as few parameters as possible in the initial estimation;

(3) Estimate the model's unknown parameters, inspect the parameters' significance and the rationality of the model itself;

(4) Carry out diagnosis and analysis to verify the conformity between the model and the data as actually observed.

3.2   Grey Prediction Model

For the grey system theory proposed by the scholar Deng Julong, with the uncertain system of "small sample and poor information" highlighting "partially known information and partially unknown information" as the research object, efforts have been made to generate and develop this partially known information to extract valuable information, achieve correct description and effective monitoring of the system's operation and evolution rules [11]. Considering the passenger's quantity on the flight conforms to the characteristics of grey system, the adoption of grey prediction model to predict the passenger's quantity of some air route is of high pertinence.

3.2.1   Establishment of the Model

Firstly, suppose the time sequence $X^{(0)}$ has n observations, $X^{(0)} = \{X^{(0)}(1), X^{(0)}(2), \ldots, X^{(0)}(n)\}$, which generates the new sequence $X^{(1)} = \{X^{(1)}(1), X^{(1)}(2), \ldots, X^{(1)}(n)\}$ through accumulation; and then the corresponding differential equation of GM (1,1) model is:

$$\frac{dX^{(1)}}{dt} + aX^{(1)} = \mu \tag{3}$$

Equation (11) is called the winterization equation or shadow equation, among which, ∂ is the development grey number and μ is the internal generation control grey number.

Secondly, we set $\hat{\alpha}$ as the parameter vector to be estimated $\hat{\alpha} = \begin{pmatrix} a \\ \mu \end{pmatrix}$, which can be solved by the lest square method:

$$\hat{\alpha} = \left(B^T B\right)^{-1} B^T Y_n \tag{4}$$

wherein

$$B = \begin{bmatrix} -\frac{1}{2}[X^{(1)}(1) + X^{(1)}(2)] & 1 \\ -\frac{1}{2}[X^{(1)}(2) + X^{(1)}(3)] & 1 \\ \ldots & \ldots \\ -\frac{1}{2}[X^{(1)}(n-1) + X^{(1)}(n)] & 1 \end{bmatrix}, Y_n = \begin{bmatrix} X^{(0)}(2) \\ X^{(0)}(3) \\ \ldots \\ X^{(0)}(n) \end{bmatrix} \tag{5}$$

the prediction model can be obtained by solving the differential equation:

$$\hat{X}^{(1)}(k+1) = \left[X^{(0)}(1) - \frac{\mu}{a}\right]e^{-ak} + \frac{\mu}{a}, k = 0, 1, 2\ldots, n \tag{6}$$

then the following can be obtained by regressive reduction as per Eq. (6):

$$\hat{X}^{(0)}(k+1) = \hat{X}^{(1)}(k+1) - \hat{X}^{(1)}(k) \tag{7}$$

The $X^{(0)}$ grey prediction model of the original sequence is as follows:

$$\hat{X}^{(0)}(K+1) = (1 - e^a)\left[X^{(0)}(1) - \frac{\mu}{\alpha}\right]e^{-ak}, k = 0, 1, 2 \ldots n \tag{8}$$

3.2.2 Model Test

The tests for grey prediction model mainly include residual test, correlation test and posterior variance test.

(1) Residual test: it calculates the absolute error $\Delta^{(0)}(i) = X^{(0)}(i) - \hat{X}^{(0)}(i)$ $(i = 1, 2, \ldots, n)$ and relative error $\Phi(i) = \frac{\Delta^{(0)}(i)}{X^{(0)}(i)} \times 100\% \, (k = 1, 2, \ldots, n)$ between the original sequence and the grey prediction sequence. The less relative tolerance is, the higher model precision will be.

(2) Correlation test: it calculates the correlation between $X^{(0)}$ and $\hat{X}^{(0)}$ according to the former correlation calculation method. It is qualified if the correlation is higher than 0.6.

(3) Posterior variance test: firstly, it calculates the standard difference of the original sequence $X^{(0)}$:

$$S_1 = \sqrt{\frac{\sum\limits_{i=1}^{n}[X^{(0)}(i) - \bar{X}^{(0)}]^2}{n - 1}} \tag{9}$$

then the standard difference of the absolute error sequence is:

$$S_2 = \sqrt{\frac{\sum\limits_{i=1}^{n}\left[\Delta^{(0)}(i) - \bar{\Delta}^{(0)}\right]^2}{n - 1}} \tag{10}$$

Then calculate the variance ratio $c = \frac{S_2}{S_1}$, finally we calculate the small error probability $p = \left\{\left|\Delta^{(0)} - \bar{\Delta}^{(0)}\right| < 0.6745 \cdot S_1\right\}$. The model accuracy can be confirmed according to the following prediction accuracy classification table.

Table 2. Table of prediction accuracy classification

SET value	VR C value	PP classification
>0.95	<0.35	A
>0.80	<0.5	B
>0.70	<0.65	C
≤0.70	≥0.65	F

If the residual test, the correlation test and the posterior variance test all say qualified, then the model as established can be used for prediction. This paper adopted the grouping prediction to predict the monthly passenger numbers of the airline passengers and finally calculate the prediction precession.

3.3   Regression and ARMA Model

3.3.1   Definition and Form of the Model

In case the time sequence model in respect of the regression model's error term can be further established, it is referred to as the combined model of regression and ARMA. The combination of such two analysis methods sometimes can achieve better prediction result than either method thereof.

If there's the following regression model:

$$y_t = \beta_0 + \beta_1 x_t + u_t \tag{11}$$

In which, x_t is the explaining variable, y_t is the explained variable and u_t is the random error term, the presumed conditions are usually fulfilled. When u_t has the autocorrelation, an effective application of the time sequence analysis is to establish ARMA model of the residual sequence u_t; then replace u_t with ARMA model. The combined model of regression and time sequence is:

$$y_t = \beta_0 + \beta_1 x_t + \Phi^{-1}(L)\Theta v_t \tag{12}$$

In which, $u_t = \Phi^{-1}(L)\Theta v_t$ or $\Phi(L)u_t = \Theta(L)\Theta v_t$ while v_t is the error term fulfilling all the conditions. The variance of v_t usually differs from u_t.

3.3.2   Importance of the Model

In case of understanding from the perspective regression model, the model has further described the error term not in conformity with the presumed conditions by the time sequence model; in case of understanding from the perspective time sequence model, the model has just established the time sequence model in respect of the stationary and random sequence without any uncertainty after having eliminated the certainty effect of explaining variable from the explained variable as far as the regression part is concerned.

4   Demonstration and Prediction of the Model

4.1   Prediction Based on ARMA

Carry out fitting and prediction of the said monthly passenger numbers by means of the traditional time sequence ARMA Model. Before the model establishment, the passenger numbers at Sanya Airport present obvious association with the seasonal factor, which can be seen in Fig. 1. Make seasonal adjustment of the monthly passenger numbers by means of the X12 seasonal adjustments. The index result of each season can be obtained, as shown in Table 3:

Carry out unit root test of the monthly passenger numbers after adjustment and the test result is shown in Table 4:

Table 3. Monthly seasonal indexes

Time	Seasonal index	Time	Seasonal index	Time	Seasonal index
Jan	1.4107	May	0.7899	Sep	0.7184
Feb	1.4337	Jun	0.6917	Oct	0.8685
Mar	1.2421	Jul	0.8170	Nov	1.0662
Apr	0.8964	Aug	0.8569	Dec	1.2213

As indicated in Table 4, the passenger number sequence after adjustment is stationary, therefore, ARMA(p,q) Model shall be established.

Table 4. Unit root test result of the passenger numbers after adjustment

Variable	ADF test value	P value	1% Critical value	5% Critical value	10% Critical value	Test result
NUM	−4.0852	0.0089	−4.0561	−3.4524	−3.1517	Stationary

4.1.1 Model Identification

Draw ACF and PACF charts of NUM to identify p and q. The result is shown in Fig. 4.

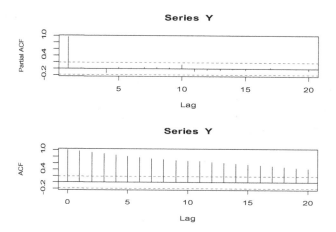

Fig. 4. ACF and PACF CHSRTS OF NUM

As indicated in Fig. 4: ACF and PACF, the PACF Chart corresponding to Y indicates "truncation" on Phase 2 while ACF Chart indicates "trailing", on this basis, establish the initial ARMA (2,1) in respect of the monthly passenger numbers at the airport. Upon repetitive debugging, the result of the final ARMA (2,1) Model is shown in Table 5:

As indicated in Table 5 under the significant level of 0.05, the model's general likelihood-ratio test is significant. All coefficients of ARIMA (2,1) are significant and the specific equation can be obtained:

$$Y_t = 999383.4 + 0.933343Y_{t-2} + 1.0689\varepsilon_{t-1} \qquad (13)$$

Table 5. ARMA result

Variable	Coefficient	Std. error	t-Statistic	Prob.
C	999383.4	121539.6	8.222699	0.0000
AR (2)	0.933343	0.014391	64.85657	0.0000
MA (1)	1.068920	0.052988	20.17291	0.0000
R-squared	0.978673	Mean dependent var		999373.1
Adjusted R-squared	0.978259	S.D. dependent var		307865.7
S.E. of regression	45394.04	Akaike info criterion		24.31204
Sum squared resid	2.12E+11	Schwarz criterion		24.38742
Log likelihood	−1285.538	Hannan-Quinn criter.		24.34259
F-statistic	2363.318	Durbin-Watson stat		1.990768
Prob (F-statistic)	0.000000			

4.1.2 Model Test

We tested the said model's residues and drew the residual ACF and PACF charts while observing whether the model has withdrawn all valid information about the passenger number sequence. The result is shown in Fig. 5:

Figure 5 shows both ACF and PACF of the residual sequence after the standard deviation, indicating the said model has efficiently withdrawn all the information of the

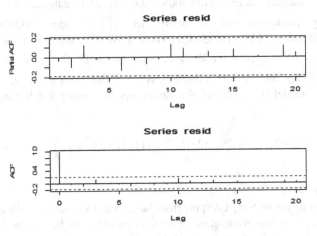

Fig. 5. Residual ACF and PACF charts

sequence. Then the established ARMA (2,1) Model was utilized to predict the monthly passenger numbers within the airport.

4.1.3 Model Prediction

Based on the establishment and test of a series of models, the monthly passenger numbers at the airport were predicted by means of the established ARMA (2,1) Model, the sequence Y_1' was obtained and finally the prediction sequence was achieved by multiplying the sequence with the seasonal index:

$$YF_1 = Y_1' * S_j \qquad (14)$$

In which, YF_1 indicates the prediction sequence of the monthly passenger numbers and S_j indicates the seasonal index. Finally, we calculated the predicted average absolute percentage error.

$$MAPE = \frac{1}{n}\sum_{i=1}^{n}\left|\frac{YF_{1,i} - Y_i}{Y_i} \times 100\right| \qquad (15)$$

Based on the said ARMA (2,1) as established, we predicted the monthly passenger numbers at Sanya Airport, that is, $MAPE = 4.20\%$.

4.2 Grey Prediction Model

Figure 1 shows significant seasonal trends in the passenger numbers at Sanya Airport from 2008 to 2016. Now, the corresponding passenger numbers of each month from 2008 to 2016 are considered as the original sequence to establish the grey prediction model and carry out grouping prediction. In this way, the influence of seasonal effect on prediction can be effectively avoided.

4.2.1 Model Establishment

Consider the passenger numbers from 2008 to January 2016 as the case to establish the specific grey prediction model. Firstly, let $X_1^{(0)} = \{595365, 833416, 1186933,$ $1132612, 1391203, 1384494, 1629585, 1700486, 1769993\}$, then solve the differential equation $\frac{dX^{(1)}}{dt} + aX^{(1)} = \mu$, solve the parameter vector by the least square method $\hat{a} = (B^T B)^{-1} B^T Y_n$. The estimation result is $\alpha_1 = -0.0893$ and $\mu_1 = 892413.5$. Then, the prediction model in respect of the passengers in January for Beijing-Sanya air route is:

$$\hat{X}_1^{(1)}(k+1) = 128297.12e^{0.0893k} - 111119.12, \quad k = 0, 1, 2\ldots, n \qquad (16)$$

4.2.2 Model Test

The prediction model of Sanya Airport from January to December can be achieved by the same way. The test parameters of various models can be obtained by R3.2.3 software. The test results are shown in Table 6:

Table 6. Test parameters of the model

Month	Prediction precision	Precision grade	Posterior ratio
Jan	94.30%	Good	0.1264
Feb	94.03%	Good	0.1686
Mar	94.43%	Good	0.1408
Apr	96.56%	Good	0.0849
May	95.76%	Good	0.0727
Jun	96.67%	Good	0.0709
Jul	98.09%	Good	0.0544
Aug	97.22%	Good	0.0663
Sep	98.13%	Good	0.0521
Oct	93.63%	Good	0.1525
Nov	94.45%	Good	0.1688
Dec	96.48%	Good	0.1015

In Table 6, the grouping prediction models are all qualified, the prediction precision of various months is good, the average prediction absolute percentage error $MAPE = 4.19\%$ and the general prediction precision is good.

4.3   Improved Regression Model Based on ARMA

Consider such three indexes as the aircraft movements (S), the luggage (LUG) and the passenger load factor (PLF) as the explaining variables and the regression as the explained variable to establish the regression model; consider the data from 2008 to 2015 as the training sample set to establish the regression model, and then consider the data in 2016 as the test data set to assess the predictive ability of the said regression model as established.

The regression results achieved by the regression model established on the basis of the training sample data set are shown in Table 7:

Table 7. Regression results

Variable	Coefficient	Std. error	t-Statistic	Prob.
C	−680558.4	66037.22	−10.30568	0.0000
S	108.3510	3.988959	27.16272	0.0000
LUG	38.28764	2.857268	13.40009	0.0000
PLF	7725.121	815.9222	9.467962	0.0000
R-squared	0.995658	Mean dependent var		926541.1
Adjusted R-squared	0.995517	S.D. dependent var		356865.9
S.E. of regression	23894.93	Akaike info criterion		23.04149
Sum squared resid	5.25E+10	Schwarz criterion		23.14834
Log likelihood	−1101.992	Hannan-Quinn criter.		23.08468
F-statistic	7032.522	Durbin-Watson stat		0.729701
Prob(F-statistic)	0.000000			

According to Table 7, the general regression equation is significant under the significance level of 0.05. The significance test P value of the regression coefficients including the aircraft movements (S), the luggage (LUG) and the passenger load factor (PLF) is all below 0.05, indicating significant effect on the passenger number. The regression equation as obtained is:

$$PN_t = -680558.4 + 108.351S_t + 38.28764LUG_t + 7725.121PLF_t \qquad (17)$$

The autocorrelation test results of the said model as established are shown in Table 8.

Table 8. Heteroscedasticity test result of the regression equation

F-statistic	31.91232	Prob. F(2,90)	0.0000
Obs*R-squared	39.83215	Prob. Chi-Square(2)	0.0000

According to Table 8, the residues fit by the regression model established under the significant level of 0.05 has autocorrelation. Establish ARMA (2,1) Model in respect of residues of the said regression model as established, the results of ARMA (2,1) Model are shown in Table 9:

Table 9. ARMA model result of residues of the regression model

Variable	Coefficient	Std. error	t-Statistic	Prob.
C	1445.440	5878.567	0.245883	0.8063
AR (2)	0.482377	0.111351	4.332048	0.0000
MA (1)	0.586721	0.100847	5.817915	0.0000
R-squared	0.402296	Mean dependent var		93.48460
Adjusted R-squared	0.389160	S.D. dependent var		23752.69
S.E. of regression	18564.21	Akaike info criterion		22.52725
Sum squared resid	3.14E+10	Schwarz criterion		22.60842
Log likelihood	−1055.781	Hannan-Quinn criter.		22.56004
F-statistic	30.62470	Durbin-Watson stat		1.965690
Prob(F-statistic)	0.000000			

According to Table 9 under the significant level of 0.05, the general equation and the coefficient of variables are significant. The ARMA (2, 1) Model can be expressed as:

$$\text{Re}sid_t = 1445.44 + 0.4823\text{Re}sid_{t-2} + 0.586721\varepsilon_{t-1} \qquad (18)$$

For ARMA (2,1) as established for predicting the training sample set by means of the said regression model, the average absolute percentage error of the predicted result by fitting the regression residuals $MAPE = 1.67\%$. See Table 10 for the prediction results of passenger numbers in 2016 by virtue of the improved regression model based on ARMA.

Table 10 shows close predicted and actual values of the passenger number at the airport in 2016 as achieved by the improved regression model based on ARMA, which signifies sound prediction ability of such improved regression model.

Table 10. Predicted passenger numbers in 2016

Date	Actual passenger number	Regression predicted value	Residual predicted value	Predicted passenger number
Jan-16	1769993	1711902	35269	1747171
Feb-16	1791933	1713822	27795	1741616
Mar-16	1597919	1565934	17863	1583798
Apr-16	1300443	1290536	14208	1304743
May-16	1200182	1179557	9432	1188989
Jun-16	1124740	1094448	7644	1102092
Jul-16	1258735	1209606	5347	1214953
Aug-16	1377864	1303784	4474	1308258
Sep-16	1251006	1200880	3369	1204248
Oct-16	1382230	1322471	2942	1325413
Nov-16	1534817	1494671	2410	1497082
Dec-16	1771423	1676498	2202	1678700

5   Comparison Analysis and Conclusion

5.1   Comparison Analysis

See Fig. 6 for the comparison result of the predicted value and the actual passenger number achieved by three models:

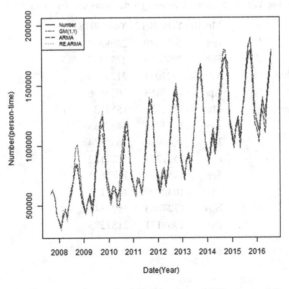

Fig. 6. Comparison of predicted value of different models

We calculated the average absolute percentage error of three models and the result is shown in Table 11:

Table 11. Mape of three models

Model	ARMA	GM (1,1)	RE-ARMA (2,1)
MAPE	4.20%	4.19%	1.67%

According to Table 11, the prediction precision of ARMA Model and the improved GM (1, 1) Model is not much differed; but the Regression RE-ARMA (2,1) Model improved on the basis of ARMA Model features more precise prediction in comparison with other two models, which signifies better prediction of passenger flow at the airport. However, as it's impossible to obtain the aircraft movements (S), the luggage (LUG) and the passenger load factor (PLF) in 2017, we cannot predict the future. On the one hand, as this paper adopted the grouping prediction and the grey model covered fewer small samples and insufficient data, the grey prediction model as established shall be of great pertinence to the passenger prediction at the airport; on the other hand, in order to predict the passenger flow at the airport in the future two years, the ARMA Model featured wide span of prediction time, in other words, it was equivalent to prediction without condition. Thus, more error would be available. In this sense, the grey prediction featured better pertinence. Based on these two reasons, this paper adopted the grey prediction model to predict the passenger numbers at Sanya Airport in 2017 and 2018. The results are shown in Table 12 and Fig. 7:

Table 12. Prediction of passenger numbers in 2017 and 2018

Month	Year 2017	Year 2018
Jan	2019930	2208522
Feb	2060756	2253897
Mar	1922051	2129173
Apr	1458458	1625856
May	1374338	1553137
Jun	1285796	1468938
Jul	1446974	1635224
Aug	1597231	1824162
Sep	1454895	1687671
Oct	1630190	1862173
Nov	1727863	1896843
Dec	1969131	2157273

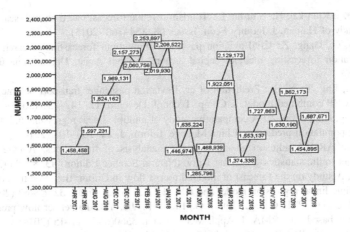

Fig. 7. Prediction of passenger numbers in 2017 and 2018

6 Conclusion

Based on a series of modeling and analysis mentioned above as well as comparison of models as adopted, the grey prediction model applicable to small samples prediction with high precision was adopted. Such model may precisely predict the passenger flow at Sanya Airport in the future two years, which will be of guiding and practice significance to the airport's development. Moreover, as it is simple and easy to understand its prediction method and results, such model can be promoted to the passenger flow prediction in association with traffic at scenic spots.

References

1. Xia, L., Lei, C., Yuan-hui, L., et al.: Sanya airport passenger flow forecast based on combination forecast method. J. Comput. Syst. Appl. **08**, 23–28 (2016)
2. Xia, L., et al.: Prediction for air route passenger flow based on a grey prediction model C, pp. 185–190. IEEE (2016)
3. Jin-jing, W., Xing-yong, H.: Forecasting and analysis of jiangsu province's logistics demand based on multiple linear regression model. J. Acta Agric. Shanghai **04**, 62–68 (2015)
4. Bang-ju, H.: The prediction for civil airport passenger throughput based on multiple linear regression analysis. J. Math Pract. Theor. **04**, 172–178 (2013)
5. Chen, P., Wu, L., Song, H.: A forecast of the inbound tourist's number of anhui province based on ARIMA model. J. Anhui Agric. Univ. **01**, 32–35+116 (2012). (Social Sciences Edition)
6. Ya-li, Q.: Analysis of the number of scenic spots based on grey model. J. Stat. Decis. **17**, 114–117 (2013)
7. Jing, Z., Jun, C.: Combination forecast model based on Arima and BP-Neural network. J. Stat. Decis. **04**, 29–32 (2016)
8. Ze-pei, D., Ling, Z.: Application of ARMA model and regression model to pesticide usage forecasting. J. Chin. Agric. Sci. Bullet. **31**, 304–307 (2014)

9. Bin-ru, Z., Xian-kai, H., Shu-lin, L.: Tourism revenue forecast based on web search data: a case study of Hainan. J. Inquiry Econ. Issues **08**, 154–160 (2015)
10. Yu-bao, C., Gang, Z.: Civil aviation passenger throughput forecasting research based on combination forecasting method: capital airport. J. Civil Aviat. Univ. China **02**, 59–64 (2014)
11. Yao, Y., Jing, T., Yi, L.: Prediction of civil aviation passenger transport volume based on ARIMA-BP combined model. J. Comp. Technol. Develop. **12**, 147–151 (2015)
12. Shao-ting, H., Yu-xin, Z.: A comparative study of multiple linear regression and ARIMA in chinese population prediction. China Manage. Inform. J. **22**, 100–103 (2014)
13. Yan, D.: ARIMA adjustment and regression analysis of time series: A case study of passenger traffic statistics. J Qiqihar Univ. (Natural Science Edition) **03**, 82–85 (2010)
14. Wei, J.: A study on the forecast of inbound tourist flow in Guilin: Based on the comparison of multiple linear regression model and GM (1,1). J. Times Finan. **32**, 65–67 (2014)
15. Chen, L.L., He, P.H., Cao, L., Liu, S.G., Liu, D.P., Jia, Y.J.: Passenger flow prediction of exhibition based on ARMA. J. Appl. Mech. Mater. **3560**(667), 11–15 (2014)
16. Li, Q., Qin, Y., Wang, Z., Zhao, Z., Zhan, M., Liu, Y., Li, Z.: The research of urban rail transit sectional passenger flow prediction method. J. Intel. Learn. Syst. Appl. **5**(4), 227 (2013)

Election Based Pose Estimation
of Moving Objects

Liming Gao and Chongwen Wang[✉]

School of Software, Beijing Institute of Technology, Beijing, China
wcwzzw@bit.edu.cn

Abstract. In this work, a key-points based method is presented to track and estimate the pose of rigid objects, which is achieved by using the tracked points of the object to calculate the attitude changes [1]. We propose to select a few points to represent the posture of the object and maintain efficiency. A standard feature point tracking algorithm is applied to detect and match feature points. The presented method is able to overcome key-points' errors as well as decrease the computational complexity. In order to reduce the error caused by feature points detection, we use the tacked key-points and their relation with the target center to get the most reliable tracking result. To avoid introducing errors, the model will maintain the features generated in initialization. Finally, the most reliable candidates will be picked out to calculate the pose information, and the small amount of key-points with highly accuracy can ensure real-time performance.

Keywords: Tracking · Positioning · Key-points · Voting · Online-learning

1 Introduction

Feature points is widely used in computing the position and orientation of the object. Because the key points can be used to describe both location and feature information of the targets. It is easy for us to rebuild the target with this information. Particle filter and median filter can get the target location by using the color information, and perform the tracking task in real time. But it's hard to get the pose or scale changes with only one tracked part. To estimate these changes, trackers should generate more than one clusters to represent the different parts of the tracked objects, which means we need a few key frames with high credibility to build this clusters [2]. In theory, we can get the scale and angle changes from only four points, however, in reality it's hard to get the ideal points to calculate this information by the reason of the feature correspondences between two frames may introduce errors and reduce the accuracy of estimation. Simple feature point matching algorithm cannot get ideal points we need. Different method was applied to improve the accuracy, such as Nonlinear Mean Shift [3], by using clustered hypotheses, the final result can get form the inliers. But it cannot be applied in tracking system where noises and occlusion may occur in every frame.

There has been a lot of progress in online mode learning algorithms. But the problems caused by occlusions or scale changes of the object are still a challenge. When the object is unknown before performing a tracking task, it will be hard to

G. Chen et al. (Eds.): PAAP 2017, CCIS 729, pp. 41–50, 2017.
DOI: 10.1007/978-981-10-6442-5_4

evaluate the weight of the features we get from the tracking results. To realize the online learning task, the object's model will be updated at each successful tracked frame and the model will be continuously improved with time going on [4, 5]. When partial occlusions occur and we don't have the evaluation of the new features' quality, the tracker may still get a successful detection with outer features. In some extreme cases the model can be updated with errors which may eventually lead to the failure of tracking tasks. So, the online evaluation of the features is very import in model updating. There has been some research about how to select the suitable new feature by introduce a classification process to measure the quality the features [5, 6]. These approaches focus on updating the model correctly and tracking in long term.

The keys to achieve a good performance in tracking are having a fast and stable model initialization and a high reliability feature generator. The typical key-points detecting and matching can do simple tracking and pose estimate jobs. We want to employ this in complex environments, and make it possible to get target's pose and location in real-time and the process of getting suitable feature should not add significant overhead to the tracking process. It's very hard to express targets' local changes from a global model, but key-points detectors can get local descriptions form image data using key-points and build a part-based model easily. We argue that the amount of key points of the object we got is not important while limited number of points with high reliability can both reduce the computing time and improve accuracy. We propose to build the object's model using a key-points detector from the first frame and get the appearance changes from reliable local descriptions.

Our contribution are as follows, we present an object model with the ability to get the target's location and evaluate the local descriptions generated by model. Although it relies on standard key-points detector, we improve the credibility by building the correlation between key-points. The second contribution is a quick method to vote for an evaluation threshold to get key-points with good feature to rebuild the object's appearance that is tracked. The third contribution is an outlier key-points identifier, as learning the target's feature often introduce error, we update the appearance model only if the feature is matched with forehead points to make the tracking process more stable.

2 Related Work

There are several approaches to getting the object's appearance representation based on local features. By using certain parts of the object to represent its features, the targets can be figured out from the background features [7, 8]. By combining the local feature information and geometrical relationship between them, the model can handle occlusions and scale changes of the targets, without adding extra calculation. The part base features have the ability to distinguish themselves form the background and other features, each feature should be unique to rebuild the geometry relations. Selecting the feature manually or based on background features and existing features can solve this problem. Feature points detection method such as SURF SIFT and ORB can do that in very efficient way and can give each point a unique descriptor [9, 10]. Thanks to the fast key-points generator we can get the target's local features without the scale and rotation's influence and save enough time for selecting the right points from points sets to reconstruct Geometrical Relationship.

There are a lot of related work on generating the target's model online [3, 4]. To achieve online learning, a method of measuring the weight of the features is required to decide which features should be added or whose weight need to change. By describe the feature as a binary vector instead of high dimensional feature vector, the model can quickly classify the key-points in a new image [4]. In particle filter the updating work can be done by the resampling process. A stationary-property-based model can handle sudden changes, and the model will be gradually improved when the target maintains static [2]. Through the model updating part, we can get a method to evaluate the quality of features in the new image. After the evaluation process the good features can be found to rebuild the target appearance. In reference [4], each key-points we can get a weight by means of combining the correspondence generation and transformation estimation into one learning framework.

In reference [1], the pose of the object can be found from an image when we can detect and match at least four or more key-points. And POSIT part can deal with the noise of the image through its iteration part. Less key-points with height quality can reduce iterations. Just like online learning algorithm, by using a geometric correlation point model, we can evaluate the points detected in the new image [4, 6]. By computing the new detected key-points relation with the model we can get each point's weight. Beside the geometric relation, voting can also help us on selecting the points. The matched points can vote for the center of the object according to the pre-learn model. The points with the majority voting will be collected as successful matched points.

3   Approach

In this section, we propose a key-points based object tacking and pose estimation method. We use the geometrical relation to evaluate the key-points. According to the model we can estimate the target's center. As the center is voted by key-points, weights of votes from each key-points will be available after finish voting. The scale and rotation changes will be estimated with new selected key-points.

3.1   Appearance Model

We assume that the object that we are going to track is a rigid object. So, we use a model based on key-points. We initial the model whit the image I_1 which we get from fist frame, by selecting the target with a bounding box A. Then a key-points detector is applied to get feature points P from the image. Location K and description d are contained in each feature.

$$P = \begin{bmatrix} (k_{11}, d_{11}) & \cdots & (k_{11}, d_{11}) \\ \cdots & \cdot & \cdots \\ (k_{11}, d_{11}) & \cdots & (k_{11}, d_{11}) \end{bmatrix} \tag{1}$$

The target's appearance model can be obtained from the key-points in area A, while the other points are the background. In this step, we can get the foreground and the background key-points set.

$$Foreground = Bouding_{(A \cap P)} \tag{2}$$

$$Background = \sim Bouding_{(A \cap P)} \tag{3}$$

Besides the key-points we can also get the center of the target from A. Besides the points information generated from the image, we also use the geometric relations between points. Combine all this information we got from A, the model of the target is built up.

In the following frames, the same key-point detector is used to generate new key-points information, the hamming distance between new and original key-points' description d is used to figure out the foreground and background points.

$$Foreground, Background = hamming_{match}\left(d_{org}, d_{new}\right) \tag{4}$$

After all the new key-points are matched, we can roughly estimate the target's location. As updating the target's feature while running may introduce errors. The original foreground information will not be updated, and the key-points which are matched in outer area will be added to foreground. Therefore, model's accuracy will gradually increase when the background stays unchanged or changes slowly, and can also get good tracking results when background information changes rapidly.

3.2 Target Estimation

We use geometry relation to estimation the target's scale and rotation changes. Each key-point is related by distance and angle to others, we use two matrices distances between two points $D_{i,j}$ and the angle between x axis and the line between two points a_{ij} to save this info. By using the Foreground and Background data set we can quickly separate newly detected key-point into two parts. And the area matched with foreground is used to estimate the target location and other related information.

$$D_{ij} = \left(\left(x_i - x_j\right), \left(y_i - y_j\right)\right) \tag{5}$$

$$a_{ij} = atctan\left(\left(y_i - y_j\right)/\left(x_i - x_j\right)\right) \tag{6}$$

Although the foreground key-points is already successful matched, some points may contain errors. A voting process is used to over com these errors. First the center of the target is estimated. Use the distance matrices D_0 in first frame and D_i to estimate the scale changes. Euclidean distance may loss some three-dimensional transformation information, the two dimensions' information is used to estimate scale changes. The changes in X axis dx_{0i} and changes in Y axis dy_{0i}.

$$dx_{0i}, dy_{0i} = D_0/D_i \tag{7}$$

And we use the angles generated from each pair of points to estimate angles changes. Then we use the mean value of these change to get the new center of the target. Each key-points in foreground will come up with a $center_{ij}$ by using these changes.

$$dRa_{0i} = a_0 - a_i \tag{8}$$

$$center_{i,j} = center_{org} + (dx_{ij}, dy_{ij}) * a_{ij} \tag{9}$$

Then we can get a set of new potential centers. When the background of the target changes, the original background's key-point cannot identify the foreground simply by matched points and the mean value of the centers may still not reliable, and the reliability of a point is also directly influencing the center predicting part. So, we need to come up with a reliable center as a standard to evaluate the key-points. We use the distance between the candidate centers to perform a hierarchical clustering.

$$center = cluster \begin{bmatrix} center_{11} & \cdots & center_{1n} \\ \cdots & \cdots & \cdots \\ center_{m1} & \cdots & center_{mn} \end{bmatrix} \tag{10}$$

As shown in Fig. 1, in first frame the target is pick out by a bounding box, in the following frames the matched key-points will be spited into foreground and background points. The green points are the new selected foreground and the red is the background.

Fig. 1. Distinguish the foreground and background key-points

According to the cluster result the original candidate points set can be divided into several small points sets. And the subset with the largest number of votes, will be the final center of the target, and the key-points voting for this subset is the candidate key-points we need to estimate the object attitude. The object's new post can be easily got from combining the final center value as well as the scale and rotation changes.

3.3 Model Update

As the target's appearance model is built on the first frame information and the key-points detection may make mistakes. The error made in the model initialization step may affect the detect result in following frames. It's necessary to come up with a method to update the features.

Instead of adding new key-points information to the appearance model, we update the feature vector of original key-points, because updating the model of the target may introduce new errors. Slow changes and mistakes made in the model update stage may reduce the tracking accuracy in long term. It's important to choose the right value to update features.

We maintain a list of successful tracked foreground image and its key-points. In target estimation stage, each point will come up with a center, using the center value we perform a cluster to get the most possible center and most reliable points. And we can also get the points sets that voting for other centers. Using the centers' distance to the estimate center to measure the key-points quality, we assume that points set with the longest distance and minimum members is the outer points. And the inner key-points will be used to calculate the object size and angle changes.

After we got the target's new pose and new features we should come up with a method to decide whether the tracking result is right or not, as the estimation part can only get the potential position, not the tracking accuracy. Before the model updating part, we should use a method to calculate the tracking result accuracy in real time. Thus, we use the image hash, by using the tracked object's pixels' information, we can get the hash code of the target. For example, the PHASH, first resize the image to 8 * 8 size, and compare each pixel's gray value we can get a sequence with 64 Boolean values as the image hash code. The normal image hash method can't distinguish the rotation of the target, which can be used to check our tacking results' accuracy, especially the estimate of the rotation. The tracked object's area can be reshaped into a normal rectangle by using the estimate rotate angle. Using the reshaped image, the tracking results area and angle information can be checked at the same time.

The total number of the key-points that we used is less than 200, so the model updating can be finished in real-time.

4 Experiments

The method we proposed can get the object's position, rotation and scale changes. And it can also deal with partial or full occlusions of the object. To check the accuracy of the method, we employ a set of synthetic images, where the tracking target area is picked out by a bounding box from the background image. And by changing the box's scale position and rotation in the background image, we can get the real changes of the object. The visual comparison of real and calculated values is as shown in Fig. 2.

Fig. 2. The experiments on different type of objects show that rigid objects with more corner features will have better tracking performance.

Because we can get the rotation, scale changes and translation information from the test dataset, we can use stricter measure method. Getting the target's position is not the only function of the tracking task as the changes of the object appearance is also very important. Updating the target object's appearance model needs more detailed information rather than the center of the object. Every change of the object is very important, and any wrong estimation may result in tracking failure or introducing errors to the object's appearance model. We need to check the rotation, transition and scale changes to evaluate the detail performance of the method. In order to get the comprehensive performance, we use the coincident area as the measurement of the error. In a tracking task, wrong estimation of the object will lead to a result that object area is not complete coincidence with the estimate area (We use the bounding box to pick out the target). As 90 and 180° rotation may mix up the result, we take the angular deviation greater than 90° as the missing of the tracking target.

In order to evaluate the tracking results A as accuracy in an easier way, we use the bounding box in each frame to figure the target out from the background as the original target area $Area_t$, and in the flowing frames, we use the tracking result to rebuild the bounding box $Area_c$.

$$A = |Area_t, Area_c|/Area_t \tag{11}$$

Figure 3 is the image we use to test the tracking performance. In Fig. 4 we compared the same key-points detect algorithm with different confidence intervals by accuracy.

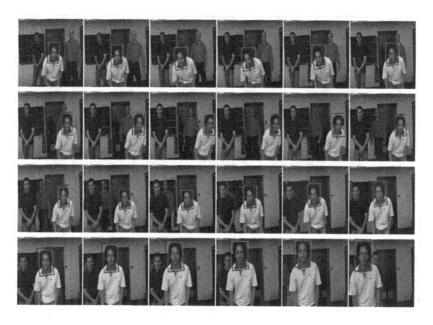

Fig. 3. The image sequence we used to test the different confidence thresholds' affection on the tracking results.

The experiment results show that our method is able to use the key-point information to estimate the target appearance changes with high accuracy. The method can still maintain relative high accuracy than simple key-points matching when the target appearance changes. With different confidence interval in updating part and different key-points detecting algorithm the method can be applied to various long term tracking situation.

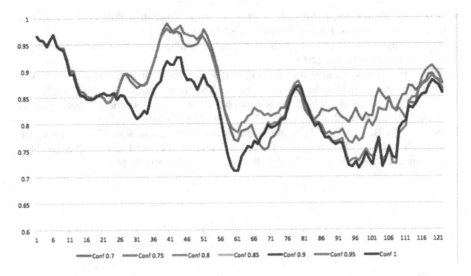

Fig. 4. Different confidence value is applied in the model updating part. The more information we accept (or the lower confidence we set) the less error will be made in the following frames.

5 Conclusion

In this paper, a long-term tracking method based on the election of the key-points with high accuracy is presented. By measuring the image similarity between tracked object and the original information, the candidate's feature is gradually updated, without losing the original appearance features. In the future, we will improve our method to deal with objects with frequently changing appearance whose similarity features are harder to picked out.

References

1. Dementhon, D.F., Davis, L.S.: Model-based object pose in 25 lines of code. Int. J. Comput. Vis. **15**(1), 123–141 (1995)
2. Mikami, D., Otsuka, K., Yamato, J.: Memory-based particle filter for face pose tracking robust under complex dynamics. In: IEEE Conference on Computer Vision and Pattern Recognition, pp. 999–1006. (2009)
3. Subbarao, R., Genc, Y., Meer, P.: Nonlinear mean shift for robust pose estimation. In: Eighth IEEE Workshop on Applications of Computer Vision. IEEE Computer Society, p. 6. (2007)
4. Hare, S.: Efficient online structured output learning for keypoint-based object tracking. In: IEEE Conference on Computer Vision and Pattern Recognition, pp. 1894–1901. (2012)
5. Grabner, M., Grabner, H., Bischof, H.: Learning Features for Tracking, pp. 1–8. (2007)
6. Grabner, H., Grabner, M., Bischof, H.: Real-time tracking via on-line boosting. In: Proceedings 17th British Machine Vision Conference, vol. 1, pp. 47–56. (2006)
7. Kumar, S., Hebert, M.: Multiclass discriminative fields for parts-based object detection. In: Snowbird Learning Workshop. (2004)

8. Bergtholdt, M., Kappes, J., Schmidt, S., et al.: A study of parts-based object class detection using complete graphs. Int. J. Comput. Vis. **87**(1), 93–117 (2010)

9. Bay, H., Tuytelaars, T., Gool, L.V.: SURF: speeded up robust features. Comput. Vis. Image Underst. **110**(3), 404–417 (2006)

10. Rublee, E., Rabaud, V., Konolige, K., et al.: ORB: an efficient alternative to SIFT or SURF. In: IEEE International Conference on Computer Vision, pp. 2564–2571. IEEE (2011)

11. Leibe, B., Leonardis, A., Schiele, B.: Robust object detection with interleaved categorization and segmentation. Int. J. Comput. Vis. **77**(1), 259–289 (2008)

12. Ma, W., Ma, B., Zhan, X.: Kalman particle PHD filter for multi-target visual tracking. In: Zhang, Y., Zhou, Z.-H., Zhang, C., Li, Y. (eds.) IScIDE 2011. LNCS, vol. 7202, pp. 341–348. Springer, Heidelberg (2012). doi:10.1007/978-3-642-31919-8_44

A Novel Topology Reconfiguration Backtracking Algorithm for 2D REmesh Networks-on-Chip

Na Niu, Fang-Fa Fu$^{(\boxtimes)}$, Hang Li, Feng-Chang Lai,
and Jin-Xiang Wang

Micro-Electronics Center, Harbin Institute of Technology, Harbin 150001, China
fff1984292@hit.edu.cn

Abstract. This paper presents a fault-tolerant topology reconfiguration back-tracking algorithm to tolerate faulty cores in 2D REmesh based (reconfigurable mesh based) Networks-on-Chip. This new algorithm can be dynamically reconfigured to support irregular topologies caused by faulty cores in a REmesh network without destroying the integrity of topologies. In addition, the proposed reconfigure method has a high-level fault-tolerance capability and therefore it is capable to tolerate more faulty components in more complicated faulty situations without additional hardware costs. The reliability performance and fault-tolerance capability of the reconfiguration backtracking algorithm in a 2D REmesh network are evaluated through appropriate simulations. The experimental results show that in different sizes of topologies (the max size is 7×8), when less than 10.7% faulty cores occur, more than 91.5% successful reconfiguration rate can be achieved. In addition, in the 7×8 REmesh, when the faulty core reaches 7, the successful reconfiguration rate has reached 61.49%, which enhanced 9.74% compared with the TRARE algorithm.

Keywords: Topology reconfiguration · Backtracking algorithm · NOCs · Fulty cores · REmesh

1 Introduction

Multiprocessor systems-on-chips (MPSoCs) have emerged in the past decades as an important class of very large scale integration (VLSI) systems. An MPSoC is a system-on-chip that uses multiple programmable processors as system components [1]. Traditional on chip buses can no longer sustain the increasing demand for communication between these processors. At the same time, the need to calculate large amounts of information in a short period of time is growing every day. In the implementation of high processing speed of the processor, the on-chip network (NOC) has become the computer architecture and the main areas of Internet research [3]. We notice that, to manage the exchange of data between the Processing Elements (PEs), Network-on-Chip (NOC), which promotes high bandwidth communication and scalability, is to be the optimal way [2]. However, as the integration technology advancing, silicon features approach the atomic scale, which results in a more sensitive and susceptible environment for complex systems such as system-on-chips (SoC) and chip

© Springer Nature Singapore Pte Ltd. 2017
G. Chen et al. (Eds.): PAAP 2017, CCIS 729, pp. 51–58, 2017.
DOI: 10.1007/978-981-10-6442-5_5

multiprocessor (CMP). Like any other systems, NOCs, are prone to failure, with the increasing number of network cores on a chip [3]. In orders to overcome the gap, the fault-tolerant algorithms in NOCs have become more important. Fault tolerance is the ability of a system to work properly in presence of faults [5]. Architecture designers proposed a variety of fault-tolerant methods to tolerate possible faults in the operation of the programs. The fault-tolerant methods can be divided into two categories: software-based fault tolerant method and hardware-based fault tolerance method. Faults are divided into two categories: transient and permanent. Tackling the persistent faults is of particular importance [3]. For this purpose, an algorithm that tolerates core failures (Core-level fault-tolerance) can be very beneficial for NOCs.

Core-level fault tolerance using "m-out-of-n" scheme [6–9]: there are M processor core on the chip, but to maintain the system to work normally only N cores are needed. The rest cores are kept as redundant cores, when a working core fails to work, one of the redundancy cores will replace the faulty core to maintain the system. Replacing faulty cores not only deforms the 2D mesh topology into irregularity but also increases the distance between various cores [9]. The fault-tolerance algorithm without performance degradation is necessary. More precisely, the algorithm should ensure the same topology scale as big as the original topology can be reconfigured with the rest cores when some faulty cores occur. The integrity of the physical network is crucial to the performance of systems; it has influence on both task communication time and link bandwidth. In this paper, we propose a novel topology reconfiguration backtracking algorithm which tolerates faulty cores with no performance degradation and ensures the integrity of the network. The reconfiguration backtracking algorithm performed in this paper is on the foundation of a 2D REmesh based architecture.

2   Reconfiguration 2D Mesh Architecture

In this paper, our target NOCs platform is a 2D REmesh [4] (reconfigurable mesh) based architecture, which is proposed by our laboratory and consists of various resources and network elements. Conventional 2D mesh without redundancy cores, which is lack of flexibility. When faulty cores accrue, the integrity of topology is destroyed. In 2D mesh with redundancy cores network,

Despite the virtual topology can present an unabridged topology of network to the software level, the large distinction between actual topology and virtual topology enhances overheads of upper software level.

Figure 1 presents a 3 × 3 topology architecture with 3 redundant cores. In this architecture, one core could be connected to anyone of its surrounding routers, besides one router could connect with one core at least. To the cores or the routers, REmesh network platform has no need to add extra ports. It only consumes one more column and row of routers [4]. When some faulty cores accrued, the running 2D mesh integral topology is going to be destroyed, which will make the communications crash down. But when the same case happened in the REmesh network, the new-build 2D mesh architecture could be reconstructed without destroying the integrity of the network.

Distinctly, the REmesh architecture not only keeps the integrity of a 2D mesh topology but also advances the possibility of high fault tolerant capability. With all

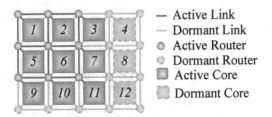

Fig. 1. REmesh architecture

these advantages mentioned above, in this paper, the fault-tolerant backtracking algorithm we proposed is based on the 2D REmesh architecture model.

3 Fault-Tolerant Backtracking Algorithm

3.1 Problem Formulation

In the $P \times Q$ processor cores REmesh network, only $R \times T$ are necessary to maintain the system operating normally. The required router size is $R \times T$; the complete router size is $(P + 1) \times (Q + 1)$; the other $(P \times Q - R \times T)$ cores are redundant. We should select $R \times T$ cores from the total $P \times Q$ processor cores to sustain the integrity of the $R \times T$ network. In this new topology, each core is connected to one router, and so does the router.

3.2 Fault-Tolerant Backtracking Algorithm

This section presents a fault-tolerant topology reconfiguration backtracking algorithm for 2D REmesh NOCs that tolerates core-level faults. The new method draws the ideas of backtracking algorithm and greedy algorithm. The latter algorithm chooses the local optimal solution to approximate the global optimal solution. Figure 2 shows the flowchart of the proposed algorithm. In this proposed algorithm, the routers are categorized into 3 groups according to the number surrounding cores. Group 1 incorporates the router that has no corresponding core via the network; group 2 is the router that has only one correspondence core; group 3 includes the router that has more than one correspondence core and we call it more-link router. In our research, the predefined search order for both routers and cores, is top left, top right, bottom left, and bottom right.

As is shown in Fig. 2, the fault-tolerant backtracking algorithm starts to choose a $R \times T$ virtue framework and scans the routers in predefined search order in the choosing $R \times T$ virtue framework.

In the process of connecting, the fault-tolerant backtracking algorithm starts to scan whether there is a router in group 1 firstly or not. If not, the algorithm begins to search routers in group 2 and 3. When coming across the router elements in group 2, the router will connect its sole corresponding core. If there is no router element in group 2 and only exists router elements in group 3, the fault-tolerant backtracking algorithm will

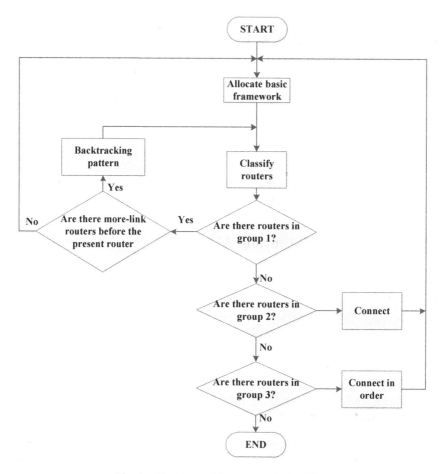

Fig. 2. Flowchart of the proposed algorithm

connect the router, in the predefined search order, to one of its corresponding cores. Each time, if a core is connected, all the routers are re-categorized.

When the group 1 router accrues in the scanning, and before this router there is no more-link router, the current virtue framework will be discarded right now. However, if there is at least one more-link router before the current router in the process of scanning, the current virtue framework will not be discarded right now. This case always happens in the event that a part of the routers are connected. The algorithm will enter the backtracking pattern, which should reconsider the previous connection routing modes. Then choose the other core to connect the previous router, which must be the more-link router. If the first step backtracking more-link router has been tested with all the surrounding cores, there is still no corresponding core element can be found to the present router via the present virtue framework, then the previous two step more-link router will be added in the algorithm. This connecting process will be iteratively performed until the previous eight step more-link router in the present basic router framework is analyzed. If, running out of the backtracking pattern, the router still could not find corresponding core element in the present virtue framework, a new R × T framework in the first step will be made. This sorting and connection process is repeated until all the connection conditions of the router in the basic router framework have been analyzed, so that the reconfigured solution may eventually be found.

4 Experimental Results

To evaluate fault-tolerance capabilities of fault-tolerant backtracking algorithm, appropriate simulation with core networks of 4 × 5, 5 × 6, 6 × 7and 7 × 8 were conducted. In all constructed NOCs, Assume that the redundant cores are on the most rightside columns. The reconfiguration success rate test has been performed from one to the total number of redundant cores. We randomly generate 1 million fault cores and inject them into the 2D REmesh network in each case.

Figure 3 shows the successful reconfiguration rate in different topology sizes with various numbers of faulty cores. As is shown in Fig. 3, successful reconfiguration rate could be reached to more than 91.5% when less than 10.7% faulty cores occur. It indicates using the fault-tolerant topology reconfiguration backtracking algorithm; the REmesh network not only tolerates more faulty cores but also preserves an integrity 2D REmesh topology. Otherwise, with faulty cores' number increasing, the rate of successful reconfiguration is decreasing. As implicated in the architectures of 7 × 8, the reconfiguration success rate decreases to 61.49% when the network has 7 faulty cores. More precisely, when the faulty cores' number is equal to the redundant cores' max number, which is in the extreme situation, the possibility to resume a 2D mesh topology is decreasing.

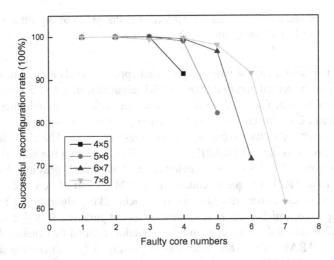

Fig. 3. Successful reconfiguration rate in different topology sizes with various numbers of faulty cores

Figure 4 compares the successful reconfiguration rate of the proposed approach with TRARE in different topology networks, where the faulty cores' number is equal to the redundant cores' max number,. As can be seen in this figure, with the same faulty cores' number in the same network architecture sizes, the proposed algorithm outperformed at least 4.8% in successful reconfiguration rate. when the faulty cores reaches 7 in the 7 × 8 REmesh network, the successful reconfiguration rate has reached 61.49%, which enhanced 9.74% compared with the TRARE algorithm.

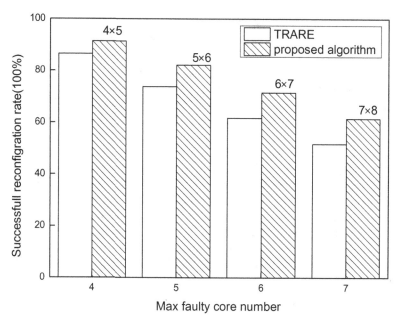

Fig. 4. Successful reconfiguration rate of TRARE and the proposed algorithm in different topology networks when the faulty core number reaches max

Figure 5 implies the reason why the proposed approach leads to a more significant improvement in the reconfiguration than TRARE algorithm in a 4 × 5 network when the faulty cores reach 3. The routers and cores are cited n different coordinates, which R means router and C means core. As can be seen in this Fig. 5(a), when core C33, C41, C42 go faulty, router R42 has no corresponding cores to connect. The virtue framework will be discarded according to TRARE algorithm. However, with the same faulty case, the new algorithm will go back to the previous more-link router's position. As is shown in Fig. 5(b), router R24 changes to connect core C24 instead of core C23. The connection method of each core with fault-tolerant backtracking algorithm in this case is shown in Fig. 5 (b),which leads to a successful reconfiguration. It is due to the proposed algorithm can find the crucial router connection method that makes the framework failure in TRARE algorithm. For instance, in Fig. 5 (a), connecting the crucial router R24 to core C23 leads router R42 has no corresponding cores to connect, and the present framework to be discarded in TRARE algorithm. If the crucial router connection method can be changed, a new connection method for the framework might be found. That is why we add the backtracking pattern. The proposed algorithm, as can be seen in Fig. 5, outperformed the TRARE algorithm in terms of topology reconfiguration. As shown in Table 1, the details of the connection methods with TRARE and our proposed backtracking algorithm.

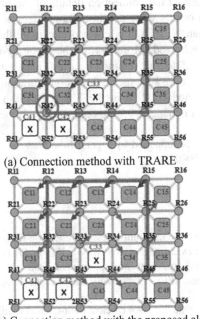

(a) Connection method with TRARE

(b) Connection method with the proposed algorithm

Fig. 5. Connetion methods with different algorithms

Table 1. The connection methods with TRARE and our proposed backtracking algorithm

Order	TRARE	Router status	The proposed	Router status
1	R12-C11	More-link	R12-C11	More-link
2	R13-C12	More-link	R13-C12	More-link
3	R14-C13	More-link	R14-C13	More-link
4	R15-C14	More-link	R15-C14	More-link
5	R22-C21	More-link	R22-C21	More-link
6	R23-C22	More-link	R23-C22	More-link
7	**R24-C23**	**More-link**	**R24-C24**	**More-link**
8	**R33-C32**	**One-link**	**R25-C15**	**More-link**
9	**R32-C31**	**One-link**	**R32-C31**	**More-link**
10	**R42**	**None-link**	**R42-C32**	**One-link**
11	**End**		**R33-C23**	**One-link**
			Continue to connect	

5 Conclusion

A fault-tolerant backtracking algorithm is first presented for 2D REmesh NOCs in this paper. The proposed backtracking algorithm can not only reconfigure the irregular topologies caused by faulty cores in a REmesh network without destroying the integrity of topologies, but also has a high-level fault-tolerance capability compared with TRARE algorithm. We presented the connection details of the proposed algorithm in the paper. The experimental results show that in different sizes of topologies (the max size is 7×8), when less than 10.7% faulty cores occur, more than 91.5% successful reconfiguration rate can be achieved. Moreover, in the topology networks, where the faulty cores' number is equal to the redundant cores' max number, the proposed algorithm outperformed at least 4.8% in successful reconfiguration rate. The successful reconfiguration rate has reached 61.49% in the 7×8 REmesh network, when the faulty cores reaches 7, which enhanced 9.74% compared with the TRARE algorithm.

Acknowledgement. This work is supported by a grant from National Natural Science Foundation of China (NSFC, No.61504032).

References

1. Wolf, W., Jerraya, A.A., Martin, G.: Multiprocessor System-on-Chip (MPSoC) technology. IEEE Trans. Comput. Aided Des. Integr. Circuits Syst. **27**(10), 1701–1713 (2008)
2. Feng, C., et al.: Addressing transient and permanent faults in NOCs with efficient fault-tolerant deflection router. IEEE Trans. Very Large Scale Integr. Syst. **21**(6), 1053–1066 (2013)
3. Akbar, R., Etedalpour, A.A., Safaei, F.: An efficient fault-tolerant routing algorithm in NOCs to tolerate permanent faults. J. Supercomput. **72**(12), 1–22 (2016)
4. Wu, Z.X., et al.: Exploration of a reconfigurable 2D mesh network-on-chip architecture and a topology reconfiguration algorithm. In: IEEE International Conference on Solid-State and Integrated Circuit Technology, pp. 1–3. IEEE (2012)
5. Dally, W.J., et al.: The reliable router: a reliable and high-performance communication substrate for parallel computers (1994)
6. Chou, C.L., Marculescu, R.: FARM: Fault-aware resource management in NOCs-based multiprocessor platforms. In: Design, Automation and Test in Europe Conference and Exhibition IEEE Xplore, 1–6 (2011)
7. Zhang, L., et al.: On topology reconfiguration for defect-tolerant NOCs-based homogeneous manycore systems. **17**(9):1173–1186 (2009)
8. Kohler, A., Schley, G., Radetzki, M.: Fault tolerant network on chip switching with graceful performance degradation. IEEE Trans. Comput. Aided Des. Integr. Circuits and Syst. **29**(6), 883–896 (2010)
9. Zhang, J.Y., et al.: A novel reconfiguration strategy for 2D mesh-based NoC faulty core tolerance. In: IEEE International Conference on Solid-State and Integrated Circuit Technology, pp. 1–3. IEEE (2012)

User Behaviour Authentication Model Based on Stochastic Petri Net in Cloud Environment

Peng Li[1,2(✉)], Cheng Yang[1], He Xu[1,2], Ting Fung LAU[3],
and Ruchuan Wang[1,2]

[1] School of Computer Science,
Nanjing University of Posts and Telecommunications, Nanjing 210003, China
lipeng@njupt.edu.cn
[2] Jiangsu High Technology Research Key Laboratory for Wireless Sensor
Networks, Nanjing 210003, Jiangsu Province, China
[3] School of Mathematics and Information Security, Royal Holloway,
University of London, Surrey TW200EX, UK

Abstract. Cloud Computing has been developing rapidly and has impacted various aspects of our daily life. However, the growth of cloud service raises concerns about its security. This paper will be focussing on user identity authentication issues in the cloud environment. Through analysing and classifying user behaviours, we propose a Stochastic Petri net-based User Behaviour Authentication model (SPUBA), which uses the behaviour of a user while logging in and browsing to analyse user behaviour credibility. Also, in order to quantify user behaviour credibility, we have modified a K-modes algorithm to solve user habitual behavioural standard, proposed an algorithm for calculating user behaviour credibility. The user operational behaviours simulations has been performed in the cloud environment to analyse the execution time of the proposed model. The results regarding the detection and false positive rates shown are better than current models.

Keywords: Cloud computing · Stochastic Petri Net · User's behaviour trustworthy degree · User behaviour authentication model

1 Introduction

Cloud computing industry grows rapidity. With the computing resources allocation flexibility, to increase the computing resources utilisation rate, and optimised overall operations cost, these characteristics driving many organisations to cloud adoption. In cloud environments, users access cloud services they need on demand basis. Despite cloud services bring users with great convenience and benefits, if cloud service providers cannot ensure their service are secure, user's benefits may have big impact. Since cloud computing services was launched, its security issues have become one of the most important factors in affecting its healthy development [1, 2].

According to the report of Gartner, a leading information technology research and advisory company, organizations of all sizes are moving their critical business and workloads into the cloud. The survey report shows that many IT organizations want to

© Springer Nature Singapore Pte Ltd. 2017
G. Chen et al. (Eds.): PAAP 2017, CCIS 729, pp. 59–69, 2017.
DOI: 10.1007/978-981-10-6442-5_6

increase their investment in cloud computing in the next few years, and this tendency indicates the urgent need for IT organisations to develop specific cloud security strategies for cloud computing, to adapt the rapid development of cloud computing [3]. From the global cloud computing development perspective, many cloud service providers are analysing and researching about cloud computing security issues actively. Cloud computing security has already become one of the state key focusses in information construction projects, and is a key factor in promoting cloud computing technology development in China [4, 5].

The second part of the paper will be a brief summary on existing researches about user behavioural analysis and related works; The third part is uses a random Stochastic Petri net (SPN) to establish a user's behaviour analyse model, and proposing user's behaviour trustworthy level algorithm; The fourth part is to prove this paper proposed model and method through simulation experiment, and analyse the experiment results; the last part of paper is wrap up and conclusion.

2 Related Work

In the research of the user behavioural analysis papers, there have been a variety of research methods. Examples of such include (i) keystroke biometrics, (ii) algorithms based on abnormal behaviour, (iii) algorithms based on authentication model, and (iv) algorithms based on user behaviour credibility.

(i) Keystroke biometrics

Tang et al. [6] have proposed a way to authenticate user's identity using mouse and keyboard keystroke logging. They set up a user training environment, and perform authentication by classifying and analysing data obtained from training sessions with the aid of Support Vector Machine (SVM) algorithm. The research [7] proposed a 3D classification model, which used as an extra protection layer after the user has logged in. It is compatible with different keyboard layouts and able to update data by constantly analysing user's keystroke in browsing sessions.

(ii) Algorithms based on abnormal behaviour

Zhao et al. [8] proposed an association rule mining algorithm based on Long Sequence Frequent Pattern (LSFP). It improves the conventional association rules, and user behaviour records can be obtained from the computer log.

Chiu et al. [9] have proposed another method based on frequent itemset. It uses two different kinds of frequency algorithms to look for abnormal factors and to identify abnormal behaviours through setting minimum thresholds. Anis Ahmed-Nacer et al. [10–12] proposed two solutions in preventing man-in-the-middle attacks. It integrates 3G authentication protocol and Transport Layer Security (TLS) to come up with an algorithm for detecting malicious behaviour, so to ensure security in message exchanges.

(iii)   Algorithms based on authentication model

Junfeng et al. [13] used multi-partite graphs to analyse the issue of user behaviour credibility. Observation points were set in the process of cloud services to construct Cloud Behaviour Multipartite Graphs (CG) and User Behaviour Multipartite Graphs (UG), and results revealed that the model has a higher detection rate for malicious users. Chen et al. [14–16] proposed a user behaviour authentication scheme, to further categorise user behaviours. It uses Stochastic Petri Net to model the behaviour authentication process, and simplify the Stochastic Petri Net model, to analyse the false positive rate of the model. The research gave a detailed modelling and rationalised analysis to user behaviour. Lu et al. [16] proposed a User Behaviour Credibility Authentication Model, which standardise and categorise the behavioural data collected and uses Stochastic Petri Net to construct an authentication model. With detailed descriptions of the model, the credibility worked out by the model could then be compared with conventional credibility authentication models.

(iv)   Algorithms based on user behaviour credibility

Jun-Jian et al. [17] proposed a user's behaviour integrated weight algorithm. By using entropy method and analytic hierarchy process (AHP) to get objective weight and subjective weight, respectively, an integrated weight could be calculated by using objective deviation and subjective deviation. Hosseini et al. [18] proposed a scoring method to evaluate the credibility of the user in the cloud environment.

This paper will first divide user behaviour analysis into two phases, then construct two behavioural authentication models using Stochastic Petri Net, for computing credibility of user behaviour. The models and algorithm are then tested by using simulations to prove their effectiveness. This work aims to propose a rationalised model, and to quantify user behaviour in cloud environment, and to give an algorithm on credibility calculations based on different types of user's behaviour, offering cloud service providers a more precise and intuitive result.

3   User Behaviour Authentication Model

3.1   Classification in User's Behaviours Under Cloud Environment

The credibility of user's identity determines the right of which to use cloud services. In cloud environments, cloud service providers will provide a user with an initial credibility value. This is dynamic and will change according to user's behaviours. Since a user can perform a wide variety of tasks within the cloud environment, service providers have to be able to collect behavioural data from different phases of the browsing session. For the ease of analysis on user's behaviour, the behaviour information should be collected through three perspectives, and to classify them into the following three categories:

(i) User habitual behaviour

IP address, operating systems used and the location used for login process, duration of the browsing session, number of incorrect login attempts.

(ii) User abnormal behaviour

The amount of resources downloaded by the users, the number of virtual machines used, RAM size, storage space size, the bandwidth of the network. These behaviours are constrained by the cloud service providers and can be obtained from the operation log file.

(iii) User malicious behaviour

Malicious acts to cloud platforms such as password cracking, TCP flooding, Trojan horse, virus attacks and IP spoofing. Information is obtained from the feedback of the intrusion detection system.

Information included in the types of user's behaviour are termed as "user behaviour evidence". A set of behaviour evidence will form a record of user behaviour. In conducting analysis on user's credibility, the differences in user behaviour evidence may lead to the case of behaviours marked as untrustworthy. We, therefore, construct Stochastic Petri net-based User Behaviour Authentication models (SPUBA), to analyse and authenticate the types of user's behaviours.

3.2 Stochastic Petri Net

Petri Net is a mathematical tool used in describing and modelling information processing systems. It can precisely describe the system properties on parallelisation, asynchronous and uncertainty, and have intuitive descriptions in graphic modelling. By correlating the delays in the firing of transitions and random variables, will give each transition a firing rate, and this forms Stochastic Petri Net (SPN). Apart from functionalities of Petri Nets, SPN can work with Continuous-time Markov Chains (MC) under certain conditions, so to analyse standards on the properties of system models [19].

3.3 Stochastic Petri Net-Based User Behaviour Authentication Model (SPUBA)

This paper proposed the use of SPN to analyse user's behaviour in the cloud environment. When a user logs into the cloud server, habitual behaviour will be analysed to determine whether he has sufficient credibility for accessing the cloud server. Once the user is on the server, a second user behaviour analysis will be conducted to determine the credibility of the user's identity.

Phase 1 Analysis
Process in Phase 1 analysis includes conventional identity authentication and user habitual behaviour authentication. Using SPN, an analysis model is constructed as shown in Fig. 1.

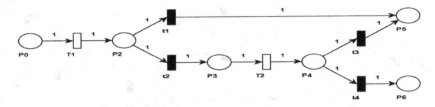

Fig. 1. Phase 1 user behavior authentication model

Table 1: Meaning of places in Phase 1 user behaviour authentication model

Table 1. Illustrates the meaning of each place in Fig. 2:

Place	Meaning
P_1	User request state, this is the initial status of the user
P_2	Result from conventional credentials authentication
P_3	Credentials successfully authenticated
P_4	Behavior analysis results
P_5	The user is untrustworthy, login failed
P_6	The user is trustworthy, login succeeded

Phase 2 Analysis

After Phase 1 analysis, the user will reach P_6 if they have granted permissions to retrieve resources in a cloud server. Here, user behaviour in the cloud server need to be analysed and authenticated again, to determine the credibility of its identity. The user behaviour authentication model can be built using SPN:

Authentication of user behaviour in cloud server mainly involves user abnormal behaviour and user malicious behaviour. Table 2 illustrates the meaning of each place in Fig. 2:

Table 2. Meaning of places in Phase 2 user behavior authentication model

Place	Meaning
P_7	Initial state, marking the start of authentication
P_8, P_9	Set of user abnormal behavior, set of user malicious behaviour
P_{10}, P_{11}	Behaviour determination result
P_{12}	Abnormal behaviour detected
P_{13}	Credibility level of user abnormal behaviour
P_{14}	User's identity is trustworthy
P_{15}	User's identity is untrustworthy

3.4 User's Behaviour Trustworthy Degree Calculations

After analysing user's behaviour by using modelling, the next stage is to illustrate the method for calculating credibility level of the user, which is time transitions T_2 and T_6 in user behaviour authentication model.

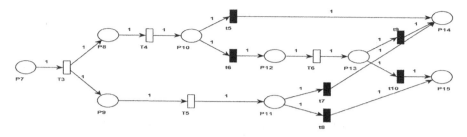

Fig. 2. Phase 2 user behavior authentication model

3.4.1 User Habitual Behaviour Standard

Definition: The set of behaviour evidence used by the user frequently in login stage, and those behaviours have a higher probability of happening.

During the analysis process at time transition T_2, it needs to analyse user's behaviour trustworthy degree by using user habitual behaviour standard. It is, therefore, an essential and important step to obtain user habitual behaviour standard, when performing behaviour analysis. This paper uses K-modes algorithms in obtaining user habitual behaviour standard.

A user has n behaviour records $X = \{X_1, X_2, X_3, \ldots, X_n\}$, with each behaviour record X_i described by m behaviour properties, which is the behaviour evidence properties of user habitual behaviour $E = \{E_1, E_2, E_3, \ldots, E_m\}$. Hence, $X_i = \{X_{i,1}, X_{i,2}, X_{i,3}, X_{i,m}\}$.

According to the ideas of K-modes algorithm, k cluster centres are chosen as initial cluster centres. Based on the definition of user habitual behaviour standard, each property needs to pick two cluster centres, to divide each property into two parts. One of the cluster centres, HC, represents the common behaviour evidence property cluster centre, and is termed as "habitual centre point". Another cluster centre AC is the cluster of uncommon behaviour evidence property, termed as "auxiliary centre point". The intensity of habitual centre is larger than that of the auxiliary centre. Therefore, user behaviour cluster centre $C_1 = \{HC_1, HC_2, \ldots, HC_m\}$ represents user habitual behaviour standard, while $C_2 = \{AC_1, AC_2, \ldots, AC_m\}$ represents deviated behaviour standard.

After picking the centres, each behaviour record will be compared with the centres by using dissimilarity H:

$$H(X_i, C_1) = \sum_{j=1}^{m} (W_j * h(X_{i,j}, HC_j)); \tag{8}$$

$$H(X_i, C_2) = \sum_{j=1}^{m} W_j * h(X_{i,j}, AC_j); \tag{9}$$

Wj in formula (9) represents the impact weight of the jth behaviour evidence property in the entire behaviour. Behaviour evidence impact is an important factor in the process of solving for credibility, and AHP is used as the method in computing behaviour evidence weight.

AHP is a method simulating human thinking and to break down complicated problems into layers.

(1) Building behaviour layer model

This model has 3 layers. The bottom layer is formed by behaviour evidence properties, the middle layer is the three types of user behaviour, and the top layer is the credibility of user's behaviour

(2) Construct a judgement matrix using 9 point ratio
(3) Obtain the eigenvectors to perform consistency test. The test is used to test the correctness of the judgement matrix constructed
(4) The judgement matrix has to be reconstructed if the consistency test fails.

Based on the four steps above, the weight of each behaviour evidence can be calculated as Wj.

When solving for dissimilarity h, after nominalising numeric data (such as time, frequency), its distance to the centres is calculated as:

$$h(X_{i,j}, HC_j) = \sqrt{(X_{i,j} - HC_j)^2};\tag{10}$$

For non-numeric values, h is defined as:

$$h(X_{i,j}, HC_j) = \begin{cases} \left|1 - \frac{count(X_{i,j})}{count(HC_j)}\right| & X_{i,j} \neq HC_j \\ 0 & X_{i,j} = HC_j \end{cases};\tag{11}$$

When the property is same as the cluster centre, their dissimilarity is 0; if they are different, the dissimilarity is calculated by the ratio $\frac{count(X_{i,j})}{count(HC_j)}$, based on the central tendency property of habitual behaviours.

3.4.2 User's Behaviour Trustworthy Degree

Trustworthy degree is to quantify user behaviour, and its value has a direct impact in determining whether the cloud service provider trusts the user's visit. After obtaining the user behaviour cluster centres, user habitual behaviour standard is then obtained and enables us to find out the dissimilarity value between the user and habitual behaviour $H(X, C_1)$. The larger the dissimilarity is, the least trustworthy the behaviour is. We define User's behaviour Trustworthy Degree (UTD) as:

$$UTD = \lambda \times (1 - H); \lambda \in (0, 1);\tag{12}$$

λ in the formula represents the user credibility impact factor and is determined by user's historical behaviour. If history records reveal that the UTD is low, the λ value will be lower than that of a normal user.

In Phase 2 model analysis, since the category and properties of each behaviour varies, analysis standard of trustworthy degrees such as the amount of resources allowed to use and size of storage space, are provided by the cloud service provider in

T_6 time transition. User behaviours in cloud server are then analysed by correlating the behaviour standards. The behaviour dissimilarity H is calculated as:

$$H = \sum_{j=1}^{m} W_j * h(X_{i,j}, S_j); \tag{13}$$

$$h(X_{i,j}, S_j) = \begin{cases} \sqrt{(X_{i,j} - S_j)^2} & X_{i,j} > S_j \\ 0 & X_{i,j} \leq S_j \end{cases}; \tag{14}$$

$X_{i,j}$ in the formulas above is the value of the jth property after normalising behaviour property data, S_j represents the jth property's standard value after nominalisation, which is provided by the cloud service provider. W_j is the impact weight of each property. We then can solve for user's behaviour trustworthy degree in the cloud environment by the UTD formula.

Based on the UTD formula $\in (0, 1)$, trustworthy degree can be divided into 5 types: {(0,0.2], (0.2,0.6], (0.6,0.8], (0.8,0.9], (0.9,1]}, which represents {very untrustworthy, untrustworthy, slightly trustworthy, trustworthy, very trustworthy} respectively. Confidence thresholds can then be set using these standards.

4 Simulation and Results

Simulations were performed on a cloud platform built on Hadoop and Spark [20]. User behaviours such as operation behaviour and attacking behaviour are simulated on the platform. An embedded behaviour collection software is used to collect behavioural information in the process, data is then passed through the model proposed in this paper to obtain results and are then analysed.

First, the experiment simulates the behaviours of six cloud users, and to calculate their respective credibility value. Users 1 and 2 are to simulate untrustworthy operation behaviours, with user 1 changing the devices used to login and intentionally provide incorrect passwords for multiple times, and user 2 downloading a huge amount of resources from the cloud server and continuously uploading data to the server. Users 3 and 4 are to simulate normal browsing behaviours to the cloud servers. User 5 has the source of the attack, which it should be detected by the intrusion detection system of the cloud server. User 6 has a history of untrustworthy behaviours, the λ value is set to be 0.85 and it is browsing normally for this time. The simulations knew the habitual behaviour standard and behaviour property impact weight for each user, the calculation results is summarised in Table 3:

Table 3. User behaviour credibility

User	1	2	3	4	5	6
Phase 1	0.1698	0.2294	0.2383	0.2608	0.2579	0.2026
Phase 2	/	0.2999	0.5324	0.5829	/	0.4525
Credibility value	0	0.5293	0.7707	0.8437	0	0.6551

To prove the effectiveness of the model proposed in this paper, they are analysed from the perspectives of detection rate and false positive rate.

The detection rate of the model is the probability of correctly identifying an untrustworthy behaviour in the authentication process by the model. False positive rate is the probability of a user with low credibility rate being classified as an untrustworthy user.

SPUBA model proposed here is compared with UBCA model [16] in the simulations. 10% of the user is set as untrustworthy in the first simulation, and the behaviour data is classified into different situations. Simulation data is generated randomly based on a fixed threshold, and history impact factor λ is included in some data. Normal users have $\lambda = 1$ and abnormal users have λ randomly generated between 0.8 and 0.9. To simplify the analysis process, it is assumed to have three types of user habitual behaviour standard, all users follow these three habitual behaviour standards and behavioural data is generated within a given range. The UTD formula gives the user behaviour trustworthy degree, and the threshold is set to be 0.6. By simulating the situations for ten times, the detection rate and false positive rates of the model are shown in the figures below.

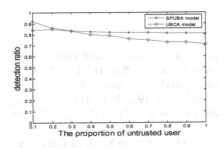
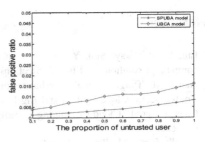

Fig. 3. Detection rate of the model

Fig. 4. False positive rate of the model

From Fig. 3, it can be seen that when the ratio of untrustworthy user increases, detection rate decreases. When comparing the SPUBA model against UBCA model, SPUBA model's overall detection rate of is higher than that of the UBCA model, and SPUBA model has a better authentication model and credibility algorithm than UBCA model. And SPUBA model algorithm considers the impact of user historical behaviour, which restricts fluctuation to user's credibility level. It is hard for an untrustworthy user to gain the trust from the cloud server in any normal operation.

In Fig. 4, with a false positive rate of around 1%, it can be claimed that the false positive rate of SPUBA model is low and steady. When compared with UBCA model, this research divided the authentication process into two phases, which is equivalent to conduct dual authentication on user's identity from two different aspects. This enhances the stability of the authentication process, and at the same time, lowering the false positive rate.

5 Conclusions

The model proposed here are only addressing the issues on user's identity authentication. There are further challenges in cloud environments such as security on privacy data, which require other tools for addressing them. The model proposed also have some deficiencies, including: (1) user's historical behaviour must be analysed before analysing user's current behaviour, (2) establishing initial credibility value for a new user could be hard.

Acknowledgments. The subject is sponsored by the National Natural Science Foundation of P. R. China (Nos. 61373017, 61572260, 61572261, 61672296, 61602261), the Natural Science Foundation of Jiangsu Province (Nos. BK20140886, BK20140888, BK20160089), Scientific & Technological Support Project of Jiangsu Province (Nos. BE2015702, BE2016777, BE2016185), China Postdoctoral Science Foundation (Nos. 2014M551636, 2014M561696), Jiangsu Planned Projects for Postdoctoral Research Funds (Nos. 1302090B, 1401005B), Jiangsu High Technology Research Key Laboratory for Wireless Sensor Networks Foundation (No. WSNLBZY201508).

References

1. Liu, Y., (Lindsay) Sun, Y., et al.: A survey of security and privacy challenges in cloud computing: solutions and future directions. J. Comput. Sci. Eng. **9**(3), 119–133 (2015)
2. Nanda, M., Tyagi, A., et al.: Hindrances in the security of cloud computing. In: International Conference of Cloud System and Big Data Engineering, pp. 193–198 (2016)
3. Chen, C., et al.: Cloud Computing Security Architecture. Science Press, Beijing (2014). (in Chinese)
4. Hu, T., Feng, M., Tang, H.: Security Technology and Application in Cloud Computing. Publishing House of Electronics Industry, Beijing (2012). (in Chinese)
5. Wei, D., Zilong, J.: A study on analysis engine for large-scale user behavior based on cloud computing. Int. J. Multimed. Ubiquitous Eng. **9**(12), 37–48 (2014)
6. Tang, H., Mantao, W.: User identity authentication based on the combination of mouse and keyboard behavior. Int. J. Secur. Appl. **10**(6), 29–36 (2016)
7. Senthil Kumar, T., Suresh, A., Karumathil, A.: Improvised classification model for cloud based authentication using keystroke dynamics. In: Frontier and Innovation in Future Computing and Communications (2014)
8. Zhao, C., Tu, S., Chen, H., et al.: Efficient association rule mining algorithm based on user behavior for cloud security auditing. In: IEEE International Conference of Online Analysis and Computing Science, pp. 145–149. IEEE (2016)
9. Chiu, C.Y., Yeh, C.T., Lee, Y.J.: Frequent pattern based user behavior anomaly detection for cloud system. In: Technologies and Applications of Artificial Intelligence, pp. 61–66. IEEE (2013)
10. Ahmed-Nacer, A., Ahmed-Nacer, M.: Strong authentication for mobile cloud computing. In: 13th International Conference on New Technologies for Distributed Systems. IEEE (2016)
11. Bakar, K.A.A., Haron, G.R.: Adaptive authentication based on analysis of user behavior. In: Science and Information Conference, London (2014)

12. Brosso, I., La Neve, A., et al.: A continuous authentication system based on user behavior analysis. In: 2010 International Conference on Availability, Reliability and Security. IEEE (2010)

13. Junfeng, T., Xun, C.: A cloud user behavior authentication model based on multi-partite graphs. J. Comput. Res. Develop., 51(10), 2308–2317 (2014). (in Chinese)

14. Chen, Y., Tian, L., Yang, Y.: Modeling and analysis of dynamic user behavior authentication scheme in cloud computing. J. Syst. Simul., 23(11): 2302–2307 (2011). (in Chinese)

15. Yarui, C., Liqin, T.: Modelling and performance analysis of user behavior authentication using stochastic Petri Nets. In: 2012 International Conference on Industrial Control and Electronics Engineering, pp. 1421–1425. IEEE (2012)

16. Lu, X., Xu, Y.: An user behavior credibility authentication model in cloud computing environment. In: 2nd International Conference on Information Technology and Electronic Commerce, pp. 271–275. IEEE (2014)

17. Jun-Jian, L., Li-Qin, T.: User's behavior trust evaluate algorithm based on cloud model. In: 2015 Fifth International Conference on Instrumentation and Measurement, Computer, Communication and Control, pp. 556–561. IEEE (2015)

18. Hosseini, S.B., Shojaee, A., Agheli, N.: A new method for evaluating cloud computing user behavior trust. In: 2015 7th International Conference on Information and Knowledge Technology, pp. 1–6. IEEE (2015)

19. Fan, L.J., Wang, Y.Z., et al.: Privacy Petri Net and privacy leak software. J. Comput. Sci. Technol., 30(6): 1318–1343 (2015). (in Chinese)

20. Jie, F.: Mining analysis on user search behavior based on hadoop. Int. J. Database Theory Appl. 9(3), 81–86 (2016)

Performance Prediction of Spark Based on the Multiple Linear Regression Analysis

Lu Dong[1], Peng Li[1,2(✉)], He Xu[1,2], Baozhou Luo[1], and Yu Mi[1]

[1] School of Computer Science and Technology,
Nanjing University of Posts and Telecommunications, Nanjing 210003, China
lipeng@njupt.edu.cn
[2] Jiangsu High Technology Research Key Laboratory for Wireless Sensor
Networks, Jiangsu Province, Nanjing 210003, China

Abstract. It is crucial to evaluate performance of a cloud platform and determine the main factors influencing the property. Moreover, the analysis results of related performance indicators can be applied to making theoretical predictions about the performance status of the cloud platform. This work mainly focuses on researching the interrelations between the performance indicators based on the Spark technology of the cloud platform and the load performance of the cluster, and furthermore makes effective predictions for the load performance. Firstly, we put forward the analytic frameworks of Spark performance analysis, the specific indicators analysis as well as the prediction models towards the cluster load. Secondly, with respect to the evaluation indicators, we explore the basis for their selections as well as their concrete implications, and then objectively, accurately calculate the correlation formula between the practically produced performance parameters and the load performance of the cluster when the Spark cluster performs the batch applications utilizing the MLR (Multiple Linear Regression) method, and, therefore, determine the main factors impacting the load performance. Finally, we predict the load value utilizing the Spark indicator analysis and the load prediction model. The results indicate that accuracy is up to 92.307%. Consequently, the solution presented in this paper predicts the cluster load value with effetioncy.

Keywords: Cloud computing · Spark · Indicator analysis · Performance prediction · Multiple linear regression

1 Introduction

Spark is a fast and general engine for large-scale data processing, and it extends the MapReduce calculation model which has been widely used. The advantage of the Spark is that it can calculate in the memory so that the efficiency of algorithm has been obviously improved. Spark brings forward a new data structure called RDD to cache the data. In this way, there is no need to load data from disk in each iterative operation so that it improves the speed of the iteration operation in Spark, and users can cache or reuse these intermediate results stored in RDD. Meanwhile, this new data structure may cause several performance issues such as the memory bottleneck and the increasing of cluster load [1].

© Springer Nature Singapore Pte Ltd. 2017
G. Chen et al. (Eds.): PAAP 2017, CCIS 729, pp. 70–81, 2017.
DOI: 10.1007/978-981-10-6442-5_7

To solve the problem on the Spark such as memory bottleneck and improve the performance of cluster load, this paper designs the performance analysis framework of Spark platform, and puts forward the indexes analysis and load prediction model, which obtains the correlation formulas between cluster load and performance indicators by using MLR and predicts the value of cluster load through indexes analysis and load prediction model.

2   Related Works

As to the performance analysis of the cloud platform, many scholars elaborate their evaluation indexes, evaluation methods and evaluation results from the perspective of their own understanding. From the perspective of the application, with regard to the four typical applications of SQL query, Streaming application, Machine Learning, Graph computation, Li and Tan *et al.* [2] designed the application program of the SparkBench, which can composite the data base suitable for the four Spark typical applications from the real data. Through the direct test for operating every Spark application, it can explore the obvious characteristics of every application program, resource consumption and the influences of communication pattern of dataflow [3] on work time. This paper focuses on revealing the resource usage characteristics of four application programs, and does not describe the basis for selecting evaluation indexes and index information in detail.

From the perspective of performance comparison, it is well known that in the respect of the operation of the large data processing, the performances of Spark and Hadoop has a strong comparability. Gu and Li [4] selected a typical iterative algorithm PageRank [5] to evaluate the system performance of Spark and Hadoop through the experiments. The results reveal that during the whole iterative computation process, when Spark is allocated enough memory, Spark is superior to Hadoop. Spark is a cloud computing platform which needs to consume large amounts of memory, although at the beginning the testing dataset does not occupy the whole cluster performance. However, the produced intermediate results during each iterative computation process would occupy the whole cluster memory, and further result in the bottleneck of Spark cluster memory [6].

From the perspective of theoretical exploration, an angle of mathematical analysis model is put forward in the article [7] to evaluate the performance of a cloud work-station system, which adopted CPU, memory, network bandwidth and other influence index. Utilizing the comprehensive fuzzy evaluation model [8] to evaluate the influence of different parameter values of the applications on the system comprehensively. This rigorous mathematical theory in this paper can be conveniently referred. But the work does not describe the interaction relationship between indicators and the performance of the system.

This work studied the former contributions as to the performance evaluation in different aspects of the cloud platform, and analyzed that there are less researching works about the correlation between performance index and cluster performance. Therefore, in this paper, we proposes the prediction model of the performance analysis and cluster load performance.

3 System Model

This paper designs the performance analysis framework of Spark to evaluate the performance of Spark platform, as shown in Fig. 1.

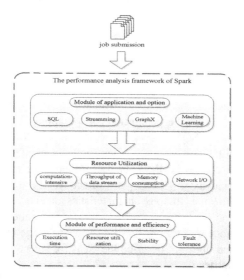

Fig. 1. The performance analysis framework of spark

The whole performance analysis framework is classified into three major categories:

1. Module of application and option [9–11]. For the Spark platform, the most important projecting attached part, big data processing and analysis application, includes four types: (1) SQL interactive query is the interface for the Spark platform to process the structured and unstructured data, just like the Hive of the Hadoop, which implements the SQL query with RDD; (2) Stream computing is the application interface for the Spark to process the data immediately, which describes the DStream as an abstract representation; (3) GraphX helps the Spark achieve the basic graph calculation function with graph algorithm and diagram programming framework; (4) Machine learning is the library with functions which are provided for machine to learn in the Spark, including classification, clustering, regression, crosscheck and so on. The above four application interfaces can help to process the data.

2. Module of resource occupation [12, 13]. The main characteristic of resource occupation can be divided into the following points: (1) computation-intensive: when executing the application of the interactive query, it needs to consume significant CPU for data search and sorting. (2) The throughput of data stream: when performing the stream computing, the transfer of the real-time data will bring the changes in data throughput capacity. (3) Memory consumption: the Spark was

designed as a Large distributed computing framework based on memory computing at the beginning, so the memory should be the key resource. (4) Network I/O: it is the objective factor that can affect the speed of accessing data and optimize the cluster deployment solution and the clustering performance of the Spark.

3. Module of performance and efficiency [14–16]. Of all the resource occupancy, it needs to grasp the major project to evaluate the performance of the Spark. (1) Execution time: including the big data working time, the system response time, the job response time and so on. (2) Resource utilization: including CPU utilization, disk utilization, memory utilization, I/O utilization and so on. (3) Stability: including the rate of packet loss, system stability and so on. (4) Fault tolerance: including the multi-copy files, the fault tolerance of applications, the probabilities of node downtime and so on.

Base on the performance analysis framework of the Spark, in order to obtain the correlation formula between cluster load and performance indicators and predict the value of cluster load, this paper designs the indexes analysis and load prediction model as shown in Fig. 2.

Fig. 2. Spark indicator analysis and the load prediction model

After the application type selection, this model drives the corresponding performance data, researches the selecting principles of the performance indicator, collects and processes the relevant indicators about cluster load, and obtains the correlation formula between cluster load and performance indicators by using MLR. Then, it predicts the value of cluster load, finally, verifies the accuracy of the examined result. The accuracies are limited within ±0.5, or else the application type selection needs to be executed again.

4   Selection of Performance Indicators

Table 1 is a simple description of the big data 5 V characteristics.

Table 1. The 5 V characteristics of big data

Characteristic	Meaning
Volume	Data reaches 100 TB, to tens or hundreds of PB, or even EB scale
Variety	It includes various formats and forms of data
Velocity	Many large data need to be dealt with in a certain time limit
Veracity	The results of the processing need to ensure a certain accuracy
Value	It contains a lot of huge business value.

The characteristics are shown in Table 2.

Table 2. The classification of index types

Type	Characteristic
Data transaction intensive	The feature of data intensive transaction is that it is necessary to consume large memory resources and to store, read and write operations are cumbersome
Data flow throughput	The characteristic of data stream throughput is that the network transmission is relatively large, and the network throughput increases with the rising of applied load
Compute intensive	Computing intensive features that there are a large number of CPU operations, in other areas of resources is not obvious

Based on the above three types of resource occupation, we select the physical indicators in Fig. 3 to quantify the load performance of Spark, and classify them into hardware resources and virtual resources.

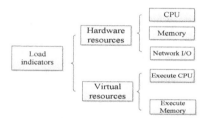

Fig. 3. Evaluation indexes of spark load

The performance indicators related to CPU are shown in Table 3.

Table 3. The CPU performance index

Collection	The meaning of the collected index
CPU	User: Time ratio of User process consumes CPU
	Nice: The time ratio of user state CPU after using the nice weighted process allocation.
	System: Time ratio of system state
	Wait: CPU complete disk write time
	Idle: The time ratio of idle CPU
	Execute CPU: CPU number of job execution

The meaning of performance indicators related to memory, as shown in Table 4:

Table 4. The memory performance index

Collection	The meaning of the collected index
Memory	Use: Total number allocated to cache
	Cache: Cache, located between the CPU and the main memory
	Buffer: Buffer occupancy. A region that is used to store data at different speeds or different priorities
	Swap: It takes disk space as part of the use of virtual memory, known as Swap consumption
	Execute memory: The allocation size of cluster execution memory

The meaning of performance indicators related to network are shown in Table 5:

Table 5. The network performance index

Collection	The meaning of the collected index
Network	In:Slave node network input
	Out:Slave node network output

5 MLR Index Analysis and Prediction Test

5.1 Definition of MLR Analysis

On the basis of the determined performance indexes, MRL index analysis is proposed between the load performance of the Spark cluster and CPU, memory, and network index. Their analytical is defined by formula 1.

$$Y_{load} = \beta_0 + \beta_1 X_1 + \beta_2 X_2 \ldots + \beta_p X_p + \varepsilon \tag{1}$$

Where Y_{load} is the load value of the cluster, β_i ($= 1, 2,\ldots,$ p) is the regression coefficients of the corresponding indexes, and ε is the constant term of the error.

It sets the nth independent observations of the $(X_1, X_2, \ldots, X_p, Y)$ as $(x_{i1}, x_{i2}, \ldots, x_{ip}, y_i)$, $i = 1, 2, \ldots, n$, and it can be expressed as the formula 2 matrix form.

$$y = \begin{bmatrix} y_1 \\ y_2 \\ \ldots \\ y_n \end{bmatrix}, \quad \beta = \begin{bmatrix} \beta_0 \\ \beta_1 \\ \ldots \\ \beta_p \end{bmatrix}, \quad X = \begin{bmatrix} 1 & x_{11} & \ldots & x_{1p} \\ 1 & x_{21} & \ldots & x_{2p} \\ \ldots & \ldots & \ldots & \ldots \\ 1 & x_{n1} & \ldots & x_{np} \end{bmatrix}, \quad \varepsilon = \begin{bmatrix} \varepsilon_1 \\ \varepsilon_2 \\ \ldots \\ \varepsilon_n \end{bmatrix} \tag{2}$$

As a result, the MLR analysis formula 1 can be expressed as formula 3.

$$Y = X\beta + \varepsilon \tag{3}$$

Where Y is n-dimensional vector consisted by the nth response variables, and X is the designed matrix, X and Y are mentioned in formula 2. β is -dimensional vector, ε is n-dimensional.

5.2   MLR Significance Test

The significance test aims at examining that whether it is reasonable for the data set to do regressive computation. Here, $H_0 : \beta_0 = \beta_1 = \ldots = \beta_p = 0$, while not all the data of $H_1 : \beta_0, \beta_1, \ldots, \beta_p$ is   equal   to   zero.   Under the   condition   of $H_0 : \beta_0 = \beta_1 = \ldots = \beta_p = 0$, the statistics

$$F = \frac{SS_R/p}{SS_E/(n - p - 1)} \sim F(p, n - p - 1) \tag{4}$$

SS_R is explained sum of squares and SS_E is the residual sum of squares, namely the square of the difference between the actual value and predictive value. The square of the correlation coefficient is defined as formula 5.

$$R^2 = \frac{SS_R}{SS_T} \tag{5}$$

R^2 is utilized to measure the degree of correlation between Y and $X_1, X_2, \ldots, X_p$, $SS_T = \sum_{i=1}^{n} (y_i - \bar{y})^2$ is the sum of total square of deviations.

5.3   Calculation of MLR Correlation

When the testing data meets the condition of the model proposed in the paper, ultimately, the MRL analysis aims at calculating the estimated value β_i of the corresponding weight β between the cluster load and CPU, memory as well as internet indexes, so as to obtain the correlation formula between the cluster and performance indexes, namely the least square function of formula 6.

$$Q(\beta) = (y - X\beta)^T (y - X\beta) \tag{6}$$

In case of the minimum value of β, the least squares estimation is as formula 7.

$$\hat{\beta} = (X^T X)^{-1} X^T y \tag{7}$$

The empirical regression equations is obtained as formula 8.

$$\hat{Y} = \hat{\beta}_0 + \hat{\beta}_1 X_1 + \ldots + \hat{\beta}_p X_p + \hat{\varepsilon} \tag{8}$$

$\hat{\varepsilon} = y - X\hat{\beta}$ is the residual vector. It sets the $\sigma^2 = \hat{\varepsilon}^T\hat{\varepsilon}/(n - p - 1)$ as the least squares estimate of σ^2.

The variance of β is estimated as formula 9.

$$Var(\beta) = \sigma^2 (X^T X)^{-1} \tag{9}$$

Correspondingly, the standard deviation of $\hat{\beta}$ is as formula 10.

$$sd(\hat{\beta}_i) = \hat{\sigma}\sqrt{c_{ii}}, \quad i = 0, \ldots, p \tag{10}$$

c_{ii} is the square of the ith element on the diagonal of $C = (X^T X)^{-1}$.

At last, the correlation formula is obtained as formula 11.

$$\hat{Y}_{load} = \hat{\beta}_0 + \hat{\beta}_1 X_1 + \ldots + \hat{\beta}_p X_p + \hat{\varepsilon} \tag{11}$$

5.4 Accuracy Analysis of the Prediction Model

It can be obtained the predictions of the cluster load from the combinations of Spark index analysis, load prediction model and the correlation formula 11, to test the accuracy of the prediction model, we compare the prediction with the real value as formula 12.

$$S_i = \sqrt{(y_i - \hat{y}_i)^2} \quad i = 1, 2, \ldots, n \tag{12}$$

n is the total prediction number, S_i is the difference between the prediction and the real value, y_i is the real value, $\hat{y}_i$ is the prediction value, when $S_i < 0.5$, it can be considered as the accurate prediction. It sets the number of S_i matching the criteria is m, $m \leq n$, and then the accuracy of the prediction model η can be expressed as formula 13.

$$\eta = \frac{m}{n} * 100\% \tag{13}$$

6 Simulation Experiment

6.1 Experiment Description

The experimental steps are as follows:

1. It configures and starts the Spark cluster, then, it discusses the generation of the data and index selection with BigDataBench, a benchmark program for big data developed by CAS.

2. Base on the performance indicators collected in the step one, it filters the indicators generated by Spark cluster within one hour, using the cluster monitoring software, Ganglia.
3. It obtains the correlation formula between cluster load and performance indicators by using MLR.
4. It fits the real data with the predictions of the cluster load which is generated by indexes analysis and load prediction model.

6.2   Analysis of the Experimental Results

It utilizes the step-wise method to carry out regression analysis on performance data, then it gets the table of model summary and regression analysis of variance as shown in Table 6.

Table 6. Model summary

R	R^2	Regulated R^2	Deviation of standard estimates
0.873	0.761	0.746	1.6832

It can come to this conclusion from Table 6: the fitting degree of the original data and the linear fitting equation established in this paper, or the degree of linear equation can react to real data is the modulation of R^2 that equals to 74.6% > 50%, it can be seen that the reaction degree of the original data that the linear fitting equation can reach is better (Table 7).

Table 7. Analysis of variance in regression

Analysis model	Sum of the squares	Degree of freedom	Mean square	F	sig.
Regression	261.398	4	65.349	47.876	0.000
Residual	81.898	60	1.365		
Total	343.296	64			

The regression analysis table has an original hypothesis. The assumption is that, all the independent variables have no significant effect on the dependent variables. That is to say, the CPU, memory and network index selected in the experiment will not have a significant impact on the load performance of Spark cluster. Based on this hypothesis, the F value is 47.876, it corresponds to the level of significance of sig. = 0.000, far less than 0.05. That is to say, all of the independent variables can not have a significant impact on the dependent variable, the probability of occurrence of this event is 0, so the original assumption is not established, consequence, the conclusion is an inverse proposition. That is, at least one independent variable in this paper has a significant impact on the dependent variable, it can be concluded that the regression equation is valid. The state of the data is described by a residual diagnostic table, as it's shown in the Table 8:

Table 8. Statistics of residual error

	Min	Max	Ave	SD	N
Prediction	3.0486	12.8004	5.1886	2.02097	65
Residual error	−2.94780	2.43162	0.000	1.13122	65
Standard prediction	−1.059	3.766	0.000	1.000	65
Standard residual error	−2.523	2.081	0.000	0.968	65

Plots of the regression standardized residual error are shown as Figs. 4, 5 and 6.

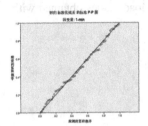 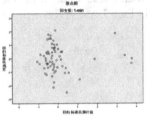 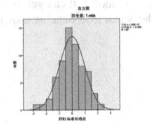

Fig. 4. Plot of regression standardized residual

Fig. 5. Residual scatterplot

Fig. 6. Histogram of residual

As shown in Fig. 4, the data residuals are linear. Also, Fig. 6 is consistent with the normal distribution. Figure 5 is the residual scatter plot, where the point of the basic hash at 0 point and distributed irregularly. The above conditions show that the fitting equation is more suitable for the original data. It satisfies the establishment of regression equation, and can be used to solve the corresponding parameters.

In the experiment, it gets the index weight parameters which can meet the model's significant effect, as shown in Table 9:

Table 9. Parameters of index weight

Index	Weight parameter β	T test	Level of significance
Constant	8.395	0.673	0.504
x_1: User	0.280	3.615	0.001
x_2: Nice	0.113	0.507	0.005
x_3: System	−0.617	−2.723	0.000
x_4: Wait	0.144	0.325	0.007
x_6: Use	0.057	1.439	0.006
x_7: Cache	0.255	-3.868	0.000
x_8: Buffer	0.868	6.148	0.000
Dependent variables: load			

It can be obtained the prediction formula 14 for the experiment platform:

$$
\begin{aligned}
y_{load} = {} & 0.280 * x_1 + 0.113 * x_2 - 0.617 * x_3 + 0.144 * x_4 - 0.030 * x_5 + 0.057 \\
& * x_6 + 0.255 * x_7 + 0868 * x_8
\end{aligned}
\tag{14}
$$

It can be seen from the correlation coefficient that the Buffer, Cache index and System index of CPU have a significant impact on the cluster load. Therefore, for the Spark cluster in this experiment, the reasonable use of memory Buffer and Cache can help to adjust the cluster load.

Finally, combined with the formula 14 and Spark index analysis and load forecasting model, it can get the predicted load value. It fits the load values obtained with the actual load data, as shown in Fig. 7.

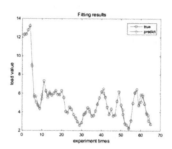

Fig. 7. Fitting results

In the fitting graph, it can be seen that the predicted values obtained by Spark index analysis and load forecasting model can accurately reflect the true value. And in a number of experiments, by using formula 13 the accuracy of the prediction model is calculated up to 92.307%. Therefore, the Spark index analysis and load forecasting model proposed in this paper can effectively predict the load value of the cluster.

Acknowledgments. The subject is sponsored by the National Natural Science Foundation of P. R. China (Nos. 61373017, 61572260, 61572261, 61672296, 61602261), the Natural Science Foundation of Jiangsu Province (Nos. BK20140886, BK20140888, BK20160089), Scientific & Technological Support Project of Jiangsu Province (Nos. BE2015702, BE2016777, BE2016185), China Postdoctoral Science Foundation (Nos. 2014M551636, 2014M561696), Jiangsu Planned Projects for Postdoctoral Research Funds (Nos. 1302090B, 1401005B), Jiangsu High Technology Research Key Laboratory for Wireless Sensor Networks Foundation (No. WSNLBZY201508), Research Innovation Program for College Graduates of Jiangsu Province (SJZZ16_0148).

References

1. Mesbahi, M.R., Hashemi, M., Rahmani, A.M.: Performance evaluation and analysis of load balancing algorithms in cloud computing environments. In: Second International Conference on Web Research, pp. 145–151. IEEE (2016)

2. Li, M., Tan, J., Wang, Y., et al.: SparkBench: a comprehensive benchmarking suite for in memory data analytic platform Spark. In: ACM International Conference on Computing Frontiers, pp. 1–8. ACM (2015)
3. Mershad, K., Artail, H., Saghir, M., et al.: A mathematical model to analyze the utilization of a cloud datacenter middleware. J. Netw. Comput. Appl. **59**(3), 399–415 (2014)
4. Gu, L., Li, H.: Memory or time: performance evaluation for iterative operation on Hadoop and Spark. In: IEEE International Conference on High Performance Computing and Communications and 2013 IEEE International Conference on Embedded and Ubiquitous Computing, pp. 721–727. (2013)
5. Villalpando, L.E.B., April, A., Abran, A.: Methodology to determine relationships between performance factors in hadoop cloud computing applications. In: International Conference on Cloud Computing and Services Sciences, pp. 375–386. (2014)
6. Sha, L., Ding, J., Chen, X., et al.: Performance modeling of openstack cloud computing platform using performance evaluation process algebra. In: International Conference on Cloud Computing and Big Data, pp. 49–56. IEEE (2015)
7. Expósito, R.R., Taboada, G.L., Ramos, S., et al.: Evaluation of messaging middleware for high-performance cloud computing. Pers. Ubiquit. Comput. **17**(8), 1709–1719 (2013)
8. Grandhi, S., Wibowo, S.: Performance evaluation of cloud computing providers using fuzzy multiattribute group decision making model. In: International Conference on Fuzzy Systems and Knowledge Discovery, pp. 130–135. IEEE (2015)
9. Villalpando, L.E.B., April, A., Abran, A.: Performance analysis model for big data applications in cloud computing. J. Cloud Comput. **3**(1), 1–20 (2014)
10. Prieto, M., Tanner, P., Andrade, C.: Multiple linear regression model for the assessment of bond strength in corroded and non-corroded steel bars in structural concrete. Mater. Struct. **49**(11), 4749–4763 (2016)
11. Pavón-Domínguez, P., Jiménez-Hornero, F.J., Ravé, E.G.D.: Evaluation of the temporal scaling variability in forecasting ground-level ozone concentrations obtained from multiple linear regressions. Env. Monit. Assess. **185**(5), 3853–3866 (2013)
12. Khedher, O., Jarraya, M.: Performance evaluation and improvement in cloud computing environment. In: International Conference on High Performance Computing and Simulation, pp. 650–652. IEEE (2015)
13. Ataş, G., Gungor, V.C.: Performance evaluation of cloud computing platforms using statistical methods. Comput. Electr. Eng. **40**(5), 1636–1649 (2014)
14. Gong, L., Xie, J., Li, X., et al.: Study on energy saving strategy and evaluation method of green cloud computing system. In: IEEE, Conference on Industrial Electronics and Applications, pp. 483–488. IEEE (2013)
15. Goga, K., Terzo, O., Ruiu, P., et al.: Simulation, modeling, and performance evaluation tools for cloud applications. In: Eighth International Conference on Complex, Intelligent and Software Intensive Systems, pp. 226–232. IEEE (2014)
16. Li, L., Rong, M., Zhang, G.: An internet of things QoE evaluation method based on multiple linear regression analysis. In: International Conference on Computer Science and Education, pp. 925–928. IEEE (2015)

CTS-SOS: Cloud Task Scheduling Based on the Symbiotic Organisms Search

Zhenpeng Liu[1,2,3], Xiaodan Liu[1], Yawei Dong[2], Xuan Zhao[1],
and Bin Zhang[3(✉)]

[1] College of Electronic Information Engineering,
Hebei University, Baoding, Hebei, China
[2] School of Computer Science and Technology,
Hebei University, Baoding, Hebei, China
[3] Center for Information Technology, Hebei University, Baoding, Hebei, China
zb@hbu.edu.cn

Abstract. Cloud task scheduling affects the overall operating efficiency of the cloud platform. Thus, how to effectively use resources in the cloud environment and make massive tasks to implement a reasonable and efficient scheduling becomes more crucial. Firstly, the mathematical model of cloud task computing was reconstructed by adding the expected completion time to the task. Secondly, on the basis of the completion time as the fitness function, the task priority was dynamically adjusted by user satisfaction, which was added to reduce the user's completion time and improve the user's satisfaction. Thirdly, aiming at the continuous search space, a cloud task scheduling algorithm based on the Symbiotic Organisms Search (CTS-SOS) was proposed. Not only does the CTS-SOS have fewer specific parameters, but also take a little time complexity. Through using the CloudSim toolkit package, the CTS-SOS algorithm was compared with Round Robin algorithm of the CloudSim and ACO algorithm. Experimental results show that CTS-SOS can provide a better optimization and scheduling of resources, reduce the makespan effectively, and improve the efficiency of processing tasks and user's satisfaction.

Keywords: Task scheduling · Cloud computing · Symbiotic Organisms Search · CloudSim

1 Introduction

Nowadays, cloud computing is an increasingly technological application in the IT industry [1]. Cloud computing is a new business computing model, which is through the establishment of computing virtualization and storage technology to provide reliable service [2, 3]. In recent decades, task scheduling has attracted increasing attention and become a very challenging research field. However, the cloud task scheduling problem has traditionally been an NP complete problem. Now there are many meta-heuristic algorithms used to solve the task scheduling problems, such as Ant Colony Optimization (ACO) [4, 5], Genetic Algorithm (GA) [6–8], Particle Swarm Optimization (PSO) [9, 10].

© Springer Nature Singapore Pte Ltd. 2017
G. Chen et al. (Eds.): PAAP 2017, CCIS 729, pp. 82–94, 2017.
DOI: 10.1007/978-981-10-6442-5_8

In order to address the problem of scheduling tasks to resources, in this paper, a Cloud Task Scheduling Algorithm based on Symbiotic Organisms Search (CTS-SOS) is proposed. Symbiotic Organisms Search [11] is a new meta-heuristic optimization algorithm proposed by Cheng and Prayogo in 2014, which solves complex numeric calculating problems. It is analogous to most meta-heuristic algorithms imitating the natural phenomena. For example, GA simulates natural evolutionary processes [12], PSO algorithm simulates behaviors of flock foraging [9, 13], and ACO algorithm imitates the foraging behavior of real ant colony [14, 15]. SOS simulates the symbiotic interactions between a pair of organisms to search for survival of the fittest organism. The algorithm was designed originally to solve the problem of numerical optimization in continuous search space, which is applicable to the optimization problem that usually can get a satisfactory solution.

First, this paper introduced Symbiotic Organisms Search algorithm briefly, then established the model and objective function of cloud task scheduling. Then, on the basis of the completion time as the fitness function, the task priority was dynamically adjusted by user satisfaction, which was added to reduce the user's completion time and improve the user's satisfaction. Following this, SOS algorithm was added to the cloud computing which needed reasonable encoding of the tasks and discretization to adapt to the discrete search space in the cloud, thus the author proposed a Cloud Task Scheduling algorithm based on the Symbiotic Organisms Search (CTS-SOS). The simulation experiment results show that CTS-SOS algorithm can effectively improve the efficiency of cloud task scheduling and reduce the total execution time (i.e. makespan) of tasks.

2 Symbiotic Organisms Search

Similar with other meta-heuristic algorithms, SOS also iteratively uses a candidate solution to search for the optimal solution in the search space [11]. This algorithm mainly includes three phases:

(1) Mutualism phase

Two organisms have contact with each other so that both sides gain benefits. At the i-th iteration, $X_j (j \neq i)$ which is selected randomly from the ecosystem interacts with X_i, their new candidate solutions are obtained by formula (1) and (2) separately.

$$X_{inew} = X_i + \text{rand}\,(0, 1) * (X_{best} - Mutual\_Vector * BF_1) \qquad (1)$$

$$X_{jnew} = X_j + \text{rand}\,(0, 1) * (X_{best} - Mutual\_Vector * BF_2) \qquad (2)$$

$$Mutual\_Vector = X_i + X_j/2 \qquad (3)$$

where BF_1 and BF_2 are factors of interest randomly identified as 1 or 2; X_{best} represents the best organism who has the highest fitness in the current ecosystem.

(2) Commensalism phase

Similar to the mutualism phase, it also select X_j $(j \neq i)$ randomly interacted with X_i in the ecosystem. In this case, one side X_i can benefit from the interaction.

Nevertheless, the organism X_j is neither beneficial nor affected. Hence, the new candidate solution of X_i is shown in formula (4).

$$X_{inew} = X_i + \text{rand}\,(-1, 1) * \left(X_{best} - X_j\right) \tag{4}$$

(3) Parasitism phase

In this phase, creating a parasite (Parasite_Vector) artificially from organism X_i increases the diversity of the ecosystem to prevent premature convergence. The formula (5) is Parasite_Vector.

$$\text{Parasite\_Vector} = \text{rand}\,(0, 1) * X_i \tag{5}$$

Hence, at each phase, each organism interacts with other organisms randomly. This process is repeated until the termination condition is satisfied.

3   Task Scheduling Model in Cloud Computing Environment

To make more efficient use of computing resources and meet the needs of users in cloud computing, a reasonable task scheduling is crucial. According to the real cloud environment, this paper gives a detail description of the problem to be solved.

3.1   The Establishment of Task Scheduling Model

To simplify the complexity of the problem and establish an effective task scheduling model, we make the following assumptions: Tasks submitted by the users are indivisible Meta-task; furthermore, each task owns independent operation and does not run a priority; The number of tasks submitted by users in cloud computing is far greater than virtual machines' in cloud datacenter; The execution time of tasks in a virtual machine can be calculated according to the information processing speed (MIPS) of the virtual machine.

To establish the mathematical model of task scheduling facilitatedly, we make the following definition of the related parameters of the task and the virtual machine:

Define 3.1 Task set as:

$T = \{Task_1, Task_2, Task_3, \ldots, Task_i, \ldots, Task_m\} = \{Task_i \,|\, i > 0, i \in [1, m]\,\}$, where m is the number of tasks submitted by the users. $Task_i$ Represents the ith task in the task sequence. The feature of $Task_i$ is defined as $\{ID_i, task\_length_i, Time\_exp_i, P_i\}$, in which ID_i is the serial number of tasks and $task\_length_i$ is the instruction length of the task (unit: million instruction MI). And $Time\_exp_i$ refers to the user's expected completion time for the $Task_i$; P_i refers to the task priority.

Define 3.2 VM set as:

$VM = \{vm_1, vm_2, vm_3, \ldots, vm_j, \ldots, vm_n\} = \{vm_j \,|\, j > 0, j \in [1, n]\}$, where n is the number of virtual machines and vm_j denotes the jth virtual machine resource in the cloud environment. The feature of vm_j is defined as $\{ID_j, MIPS_j\}$, in which ID_j is the serial number of virtual machines and $MIPS_j$ is the information processing speed of virtual machines (unit: millions-of-instructions-per-second, mips).

Definition 3.3 Set the number of tasks for m, the number of virtual machines for n. Matrix m * n (Expect Time to Complete) denotes the execution time required to run the task on each computing resource (virtual machine) that can be calculated by formula (6).

$$ETC = \begin{pmatrix} ETC_{11} & ETC_{12} & ETC_{13} & ETC_{14} & \cdots & ETC_{1n} \\ ETC_{21} & ETC_{22} & ETC_{23} & ETC_{24} & \cdots & ETC_{2n} \\ \cdots & & & & \cdots & \\ ETC_{m1} & ETC_{m2} & ETC_{m3} & ETC_{m4} & \cdots & ETC_{mn} \end{pmatrix} \qquad (6)$$

where ETC_{ij} refers to the required execution time of $Task_i$ running on vm_j completely.

$$ECT_{ij} = \text{task\_length}_i / MIPS_j \qquad (7)$$

Accordingly, cssloud task scheduling is to schedule tasks submitted by one or a plurality of cloud users to virtual machines in cloud datacenter by scheduling algorithm.

3.2   The Selection of Fitness Function

Through the establishment of the cloud task scheduling model, we can transform cloud computing task scheduling problem into how to schedule and distribute multiple tasks to multiple virtual machines reasonably and make all the tasks completed in a relatively short time. So the task scheduling problem can be attributed to work towards an optimal allocation of scheduling a set of tasks to a set of computing resources to minimize the execution time.

Formula (8) expresses the time to accomplish all the tasks contained in vm_j, where T_j is the task set assigned to vm_j. Since, the sum of time required to complete all tasks is the maximum value of makspan of all virtual machines achieved by formula (8).

Fitness function showed in formula (10) is the reciprocal of the makespan which required by vms to complete all the tasks. The bigger the value of Fitness is, the lower makespan required is, and the more perfect the task scheduling algorithm is.

$$Time(j) = \sum_{i \in T_j} ETC(i,j), j = 1, 2, \ldots, n \qquad (8)$$

$$Total\_time = Max(Time(1), Time(2), \ldots, Time(j), \ldots, Time(n))$$
$$= \underset{j=1}{\overset{n}{Max}}(Time(j)) \qquad (9)$$

$$Fitness = 1/Total\_time \qquad (10)$$

4 Cloud Task Scheduling Based on the Symbiotic Organisms Search

Symbiotic Organisms Search algorithm was used to solve numerical optimization problems in the continuous search space. In order to be adapted to the discrete search space of the cloud environment, and apply to the task scheduling problem, we propose a cloud task scheduling algorithm based on Symbiotic Organisms Search (CTS-SOS). Therefore, encoding tasks and virtual machines and discreting SOS algorithm can map it to the discrete search space.

4.1 Task Encoding

In this paper, we adopt real_ coded mode for the task code. That is, m tasks submitted by the cloud users and n virtual machines in datacenters are numbered with 0 to (m − 1) and 0 to (n − 1) respectively. In the initial phase, we initialize the population of the ecosystem (Ecosystem) randomly. The number of organisms in the initial population is E and each of the elements in the ecosystem randomly takes the integer 0 to (n − 1). The ecosystem is expressed as follows:

$$Ecosystem = [X_1, X_2 \ldots, X_i, \ldots, X_{E-1}, X_E]^T \tag{11}$$

In the formula (11), each organism X_i ($i \in [1, E]$) represents a candidate solution, which is a single-dimensional array. The elements of the vector are natural numbers, and it ranges from [0, n − 1], where n is the number of virtual machines. $X_i[j]$($j \in [0, m − 1]$) indicates that the jth task is performed on the $X_i[j]$-th computing resource VM; this is, real number string corresponding each organism represents a kind of task scheduling scheme.

4.2 The Legalization of Encoding

Encoding results for the final operation must conform to the encoding requirements of the original problem that uniquely corresponding to a solution in solution space of the discrete problem. However, SOS algorithm applies only to the optimization problem of continuous search space originally. To meet the requests of encoding, we need to carry out the legalization of encoding.

Hence, to map the value of each element in the ecosystem to [0, n − 1], the values derived from the formula (1), formula (2), formula (4) and formula (5) need have the operation of taking absolute, rounding and taking remainders. The updating formulas of the improved CTS-SOS algorithm are shown as follows:

$$X_{inew} = \mod(\lceil |\{X_i + rand\,(0, 1) \, * \, (X_{best} \text{ - Mutual\_Vector} \, * \, BF_1)\}| \rceil, n) \tag{12}$$

$$X_{jnew} = \mod(\lceil \, |\{X_j + rand\,(0, 1) \, * \, (X_{best} \text{ - Mutual\_Vector} \, * \, BF_2)\}| \rceil \, , n) \tag{13}$$

$$X_{inew} = \text{mod}(\lceil \,|\, \{X_i + \text{rand}\,(-1,1)\, * \,(X_{best}-X_j)\} \,|\, \rceil, \text{n}) \tag{14}$$

$$\text{Parasite\_Vector} = \text{mod}(\lceil \,|\, \{\text{rand}\,(0,1)\, * \, X_i\} \,|\, \rceil, \text{n}) \tag{15}$$

4.3 The User Satisfaction Mechanism

In the initial stage of task scheduling, the task is given an initial priority, which can be called static priority. If in the scheduling process, the priority of each task remains unchanged, although simple, but can not dynamically respond to the characteristics of scheduling, easily lead to poor scheduling performance. Cloud computing is the most essential goal is to allow users to obtain satisfactory cloud services to obtain revenue. Therefore, we introduce the user satisfaction mechanism, according to the user satisfaction to determine the advantages and disadvantages of cloud services, so as to dynamically update the task priority.

Assuming that $Task_i$ is assigned to vm_j, the actual completion time of being scheduled for the vm_j:

$$
\begin{aligned}
Time\_fact_{ij} &= Wait_{ij} + ETC_{ij} \\
&= \sum_{i \in T_k} ETC\,(i,j) + ETC_{ij}\,(where,\, j = 1,2,\ldots,k \text{ - } 1)
\end{aligned}
\tag{16}
$$

Among them, T_k represents the task set on the vm_j; $Wait_{ij}$ indicates the queue waiting time of $Task_i$, that is, it is the completion time of previous task before it is scheduled to the virtual machine.

Therefore, we can measure the degree of user satisfaction based on the difference between the expected completion time and the actual completion time. To make the following definition:

Define 4.1 The $Time\_exp_i$ indicates that the expected completion time of $Task_i$, and S_i is used to represent the user's satisfaction. Then S_i can be expressed as:

$$
S_i = \begin{cases} 1, & Time\_fact_i \leq Time\_exp_i; \\ Time\_exp_i/Time\_fact_i, & Time\_fact_i > Time\_exp_i. \end{cases}
\tag{17}
$$

When setting the initial priority, all other tasks have different priorities, which can be set according to the order of tasks. The highest task priority is set to 1, the smaller the value, the higher the priority; the contrary, the greater the value, the lower the priority. We hope that the task can be dynamically updated according to the user's satisfaction. Therefore, we define the dynamic priority as follows:

$$
P_i = \begin{cases} P_i, & S_i = 1; \\ P_i - 1/S_i, & 1 < \dfrac{1}{S_i} \leq P_i - 1; \\ 1, & \dfrac{1}{S_i} > P_i - 1. \end{cases}
\tag{18}
$$

When the user satisfaction is 1, at this time, do not have to adjust the value; when the user satisfaction is less than 1, which is higher than the expected real time task completion time, the priority according to the satisfaction of the dynamic adjustment.

Different users have different service quality requirements for cloud resources, some users expect to be able to get real-time guarantee, some users need high performance computing services, but also some users pay attention to the cost of services, etc.

Therefore, this paper not only uses the time as the fitness function, and the introduction of user satisfaction change task priority, according to user satisfaction corresponding to individual organisms in the task of re sort, let the whole system satisfaction reached a relatively high state. Through this process, the ecological system can not only get shorter the total task completion time but also provide a high degree of user satisfaction for Symbiotic Organisms Search algorithm to retain a good biological.

4.4   The Basic Steps and Procedure of CTS-SOS

Assume that users submit a batch of tasks to the cloud; and assign tasks to the corresponding virtual machine processed by the task scheduler. The basic steps of the proposed CTS-SOS are as follows:

Step 1 Initialization: The size of the population is E, that is the number of cloud task scheduling solution. Use random method to generate initial ecosystem, and calculate the fitness of each organism in the current ecosystem according to the formula (10) to find the optimal organism X_{best}.

Step 2 Priority ranking: the current best organism will recalculate the priority of the task in accordance with the formula (18), According to the priority of the task in ascending order.

Step 3 Mutualism phase: select organisms X_j ($j \neq i$) randomly, and create new organisms X_{inew} and X_{jnew} If the fitness value of the new organism X_{inew} (X_{jnew}) is larger than the initial one X_i (X_j), X_{inew} (or X_{jnew}) will replace X_i (X_j); otherwise X_i (X_j) wouldn't have changed.

Step 4 Commensalism phase: Select X_j ($j \neq i$) randomly interacted with X_i in the ecosystem, and create new organism X_{inew} according to formula (15). If the fitness value of the new organism X_{inew} is better than the original one X_i, X_{inew} replace X_i; otherwise X_i wouldn't have changed.

Step 5 Parasitism phase: Create a Parasite_Vector by formula (16), and continue to select X_j ($j \neq i$) randomly as a carrier of Parasite_Vector. The fitness of parasites and organisms is calculated separately. If the fitness of organism X_j is lower than that of the parasite, X_j will be replaced by the Parasite_Vector; otherwise Parasite_Vector would not survive.

Step 6 If the condition of the loop termination is satisfied, then the optimal organism is returned; otherwise it skips to step2 to continue optimizing.

Step 7 According to the best organisms, the optimal scheduling scheme is obtained.

Through the analysis of the pseudo code on Table 1, the time complexity of the CTS-SOS is $O(n^2)$, which is lower than the time complexity $O(n^3)$ of the task scheduling algorithm based ACO in [3] significantly.

Table 1. The pseudo code of CTS-SOS

CTS-SOS Algorithm
1.Initialization: $Ecosystem = \{X_1, X_2, X_3,..., X_E\}^T$
2. inter←100;
3. X_{best}←0;
4. **do**
5.　　　　Interation=interation+1;
6.　　　　i←0;
7.　　　　For i to E by 1 do
8.　　　　For j←1, 2, 3, ... , E
9.　　　　if fitness(X_j)>fitness(X_{best}) Then X_{best} ← X_j end if
10.　　　　End for
11　　　　Sort tasks according to formula (3-10)
12.　　　　//Mutualism phase
13.　　　　Select X_j (j≠i) randomly ,and calculate X_{inew} 和 X_{jnew} by formula(13)(14)
14.　　　　if fitness(X_{inew})>fitness(X_i) Then X_i ← X_{inew}
15.　　　　End if
16.　　　　if fitness(X_{jnew})>fitness(X_j) Then X_j ← X_{jnew} end if
17.　　　　//Commensalism phase
18.　　　　Select X_j (j≠i) randomly ,and calculate X_{inew} by formula(15)
19.　　　　if fitness(X_{jnew})>fitness(X_j) Then X_j ← X_{jnew} end if
20.　　　　//Parasitism phase
21.　　　　Select X_j (j≠i) randomly , and calculate Parasite_Vector by formula(16)
22.　　　　if fitness(Parasite_Vector)>fitness(X_j) Then X_j←Parasite_Vector End if
23.　　　　End for
24. **While**(interation<=inter)

5 Simulation Experiments

5.1 Simulation Environment Settings

The softwares for the simulation experiments was based on CloudSim3.0. In order to verify the authenticity of the CTS-SOS algorithm and make the algorithm more persuasive, we used Robin Round (RR) algorithm, Ant Colony Optimization (ACO) and CTS-SOS to do comparing experiments under the same experimental conditions.

In order to improve the experiment, we will design two scenarios:

(1) There were 50 computing nodes (50 VMs) in the cloud computing environment, and the number of tasks is from 100, 100 to 1000. The simulation parameter setting of the experiment environment were presented in Table 2(a).

(2) There were 500 computing tasks (50 VMs) should be executed in the cloud computing environment, and the number of VMs is from 10, 20 to 50. The simulation parameter setting of the experiment environment were presented in Table 2(b).

Table 2. Parameters settings in CloudSim

(a)			(b)		
Type	Parameters	Values	Type	Parameters	Values
Cloudlet	Number of tasks	100–1000	Cloudlet	Number of tasks	500
	Length of tasks (MI)	1000–20000		Length of tasks (MI)	1000–20000
VM	Number of VM	50	VM	Number of VM	10–50
	Speed of VM (MIPS)	100–1000		Speed of VM (MIPS)	100–1000
	VM memory (GB)	1–4		VM memory (GB)	1–4
	Bandwidth (Mbps)	100–1000		Bandwidth (Mbps)	100–1000

5.2 Experimental Results and Analysis

To avoid the interference of random experimental data on the test results and more accurately reflect the performance of the algorithm, each algorithm was repeated 30 times in the test and the average value of the 30 tests was taken as the holistic makspan.

Scene 1, the number of virtual machines is certain, but the number of tasks scheduling changes: the experimental datas were sorted out, and the experimental results were shown in Figs. 1, 2 and 3.

From Figure 1, we can see that the algorithm set out in the present paper is superior to the RR algorithm and the ACO algorithm in the scheduling quality. The time consumed by the CloudSim's own polling algorithm is the highest; that is because it

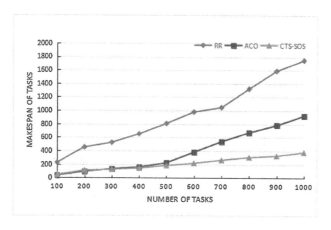

Fig. 1. Makespan of CTS-SOS, ACO and RR

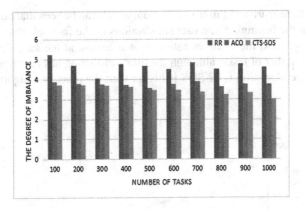

Fig. 2. The DI of CTS-SOS, ACO and RR

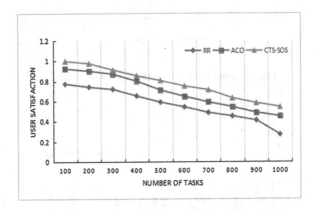

Fig. 3. User satisfaction of CTS-SOS, ACO and RR

does not take the processing speed of the virtual machines and the length of tasks into account. When the number of tasks ranges from 100 to 500, the makespan of CTS-SOS algorithm and ant colony algorithm is not much difference. However, when the number of tasks ranges from 100 to 500, the makespan of CTS-SOS was significantly shorter than that of the ant colony algorithm. That is to say, CTS-SOS algorithm is better than ACO algorithm in large scale tasks.

The Degree of Imbalance (DI) [16] is calculated by the formula (19).

$$DI = T_{\max} - T_{\min}/T_{avg} \tag{19}$$

where, $T_{\max}$, $T_{\min}$ 和 T_{avg} are the maximum, minimum and average execution time of all virtual machines. The smaller value of DI is, the better the balance between the virtual machines are. Figure 2 shows that CTS-SOS can make the cloud system to achieve a better balance of virtual machines.

Figure 3 is a comparison of user satisfaction, which can be seen that due to the small scale of tasks in the beginning, the user satisfactions of the four scheduling algorithms are relatively high, and the value of satisfaction is almost at the same level; with the increase of the number of tasks, although the user satisfaction has begun to decline, the satisfaction of the algorithm CTS-SOS is higher than the other two algorithms.

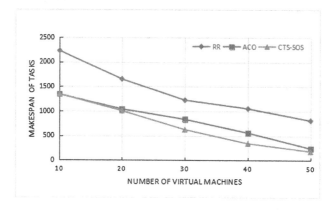

Fig. 4. Makespan of CTS-SOS, ACO and RR

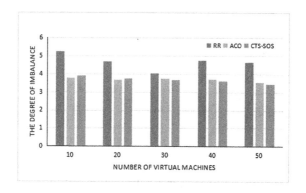

Fig. 5. The DI of CTS-SOS, ACO and RR

Scenario 2, the number of tasks is fixed, but the number of virtual machines is changing:the experimental datas were sorted out, and the experimental results were shown in Figs. 4, 5 and 6.

The experimental results showed that with the increase of the number of virtual machine, CTS-SOS, ACO and RR round robin task completion time were decreased, this was in the case of a certain number of tasks, the increase of virtual machines reduced the load of the original virtual machines. Moreover, the makespan of RR is much higher than that of CTS-SOS and ACO, mainly because it ignores the difference between virtual machines.

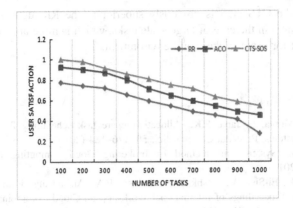

Fig. 6. User satisfaction of CTS-SOS, ACO and RR

In the comparison of user satisfaction, with the increase of the number of virtual machines, the three kinds of customer satisfaction were on the rise, but the trend was gradually slowing down, the gap between the satisfaction of the algorithm was gradually reduced and were close to 1. However, the user satisfaction of CTS-SOS algorithm was still higher than the other two algorithms. And in terms of load balancing, CTS-SOS and ACO load imbalance was also much lower than the RR, so the proposed algorithm had some advantages.

The advantage of CTS-SOS algorithm was so obvious, this was because the CTS-SOS used the user satisfaction mechanism to dynamically adjust the priority of the task. This gap was particularly evident in the case of fewer virtual machines, and the key was that when the number of was small, it will increase the number of tasks assigned to each VM, which leads to more tasks in the waiting state. With the increase of waiting time, the user's satisfaction mechanism was used to dynamically adjust the priority of the user's task, which would lead to the change of the task execution order, and made the scheduling change to a better satisfaction.

According to the analysis of the results of the experiment, we can see that the proposed CTS-SOS algorithm can get a good scheduling result in solving the independent task scheduling. It is not difficult to find, the algorithm CTS-SOS can not only reduce the task completion time, enable the user to obtain good user satisfaction, but also enable the system to achieve a relatively balanced load and high utilization rate of resources.

6 Conclusions

In this paper, a new meta-heuristic algorithm – Symbiotic Organisms Search is applied to the optimization problem of task scheduling in the cloud environment, and a cloud task scheduling algorithm based on the Symbiotic Organisms Search (CTS-SOS) has been proposed to optimize the scheduling objective. This paper has carried on the simulation test of the CTS-SOS, RR algorithm and ACO algorithm. Results show that performance of the CTS-SOS is obviously superior to the RR algorithm and ACO algorithm, especially in the case of large scale tasks, which is more able to adapt to the growing demand for users in the cloud environment.

References

1. Panda, S.K., Gupta, I., Jana, P.K.: Allocation-aware task scheduling for heterogeneous multi-cloud systems. Procedia Comput. Sci. **50**, 176–184 (2015)
2. Jackson, D.B.: System and method of brokering cloud computing resources, US, US9015324 (2015)
3. Tawfeek, M.A., El-Sisi, A., Keshk, A.E., Torkey, F.A.: An ant algorithm for cloud task scheduling. In: Proceedings of International Workshop on Cloud Computing and Information Security, pp. 169–172 (2013)
4. Gao, Y.Q., Guan, H.B., Qi, Z.W., Hou, Y., Liu, L.: A multi-objective ant colony system algorithm for virtual machine placement in cloud computing. J. Comput. Syst. Sci. **79**, 1230–1242 (2013)
5. Raju, R., Babukarthik, R.G., Chandramohan, D., Dhavachelvan, P.: Minimizing the makespan using hybrid algorithm for cloud computing. Adv. Comput. Conf. **7903**, 957–962 (2013)
6. Xu, Y.M., Li, K.L., Hu, J.T., Li, K.Q.: A genetic algorithm for task scheduling on heterogeneous computing systems using multiple priority queues. Inf. Sci. **270**, 255–287 (2014)
7. Jiang, Y.S., Chen, W.M.: Task scheduling for grid computing systems using a genetic algorithm. Kluwer Academic Publishers, Hingham (2015)
8. Dasgupta, K., Mandal, B., Dutta, P., Mandal, J.K., Dam, S.: A genetic algorithm (GA) based load balancing strategy for cloud computing. Procedia Technol. **10**, 340–347 (2013)
9. Awad, A.I., El-Hefnawy, N.A., Abdel_Kader, H.M.: Enhanced particle swarm optimization for task scheduling in cloud computing environments. Procedia Comput. Sci. **65**, 920–929 (2015)
10. Cai, Q., Shan, D.H., Zhao, W.T.: Resource scheduling in cloud computer based on improved particle swarm optimization algorithm. J. Liaoning Tech. Univ. (Natural Science) **5**, 93–96 (2016)
11. Cheng, M.Y., Prayogo, D.: Symbiotic organisms search: a new meta-heuristic optimization algorithm. Comput. Struct. **139**, 98–112 (2014)
12. Cuppini, M.: A genetic algorithm for channel assignment problems. Eur. Trans. Telecommun. **5**, 285–294 (1994)
13. Guan, T.T.: Application research of multi objective partice swarm optimization in logistics distribution. Nanchang University, Nanchang (2012)
14. Dorigo, M., Birattari, M., Stutzel, T.: Ant colony optimization. IEEE Comput. Intell. Mag. **1**, 28–39 (2006)
15. Li-Fen, L.I., Zhu, Y.L., Zhang, J.Y.: A cloud model based multiple ant colony algorithm for the routing optimization of WSN with a long-chain structure. Comput. Eng. Sci. **32**(11), 10–14 (2010)
16. Tawfeek, M., El-Sisi, A., Keshk, A., Torkey, F.: Cloud task scheduling based on ant colony optimization. Int. Arab J. Inf. Technol. **12** (2015)

Exploration of Heuristic-Based Feature Selection on Classification Problems

Qi Qi[1], Ni Li[2(✉)], and Weimin Li[1]

[1] College of Information Science and Technology,
Hainan University, Haikou 570228, China
qqi@hainu.edu.cn
[2] School of Mathematics and Statistics,
Hainan Normal University, Haikou 571158, China
nl_hainnu@163.com

Abstract. We present two heuristics for feature selection based on entropy and mutual information criteria, respectively. The mutual-information-based selection algorithm exploiting its submodularity retrieves near-optimal solutions guaranteed by a theoretical lower bound. We demonstrate that these heuristic-based methods can reduce the dimensionality of classification problems by filtering out half of its features in the meantime still improving classification accuracy. Experimental results also show that the mutual-information-based heuristic will most likely collaborate well with classifiers when selecting about a half size of features, while the entropy-based heuristic will help most in the early stage of selection when choosing a relatively small percentage of features. We also demonstrate a remarkable case of feature selection being used in classification on a medical dataset, where it can potentially save half of the cost on the diabetes diagnosis.

Keywords: Feature selection · Heuristic · Dimensional complexity · Classification · Machine learning

1 Introduction

Big data often comes with high dimensionality, which makes machine learning tasks difficult. Learning from higher dimensional datasets theoretically needs more samples than from lower ones, which will make learning tasks less efficient. The dimensionality of a problem is usually correlated with the feature size of its dataset. By selecting out a subset of features and using them in a learning task, one can reduce the dimensionality of the problem. The selecting process is often referred to feature selection [1]. In the context of classification problems in machine learning, it can improve the scalability of training and predicting processes, and increase resulted classifiers' accuracy by eliminating irrelevant or noisy attributes.

Methods of feature selection can be briefly divided into two categories, the filter-based and the wrapper-based [2, 3]. The filter-based approach depends on characteristics of training data to select features without any learning process. The wrapper-based approach applies a learning algorithm to evaluate selected features. It could return a better result than the filter-based approach, but it also incurs more

© Springer Nature Singapore Pte Ltd. 2017
G. Chen et al. (Eds.): PAAP 2017, CCIS 729, pp. 95–107, 2017.
DOI: 10.1007/978-981-10-6442-5_9

computational cost than the latter. Another categorization is based on the size of results, whether it will return all features but with different weights or only a subset of its. Respectively, they are called feature weighting and subset selection [4–6].

Selection problems also exist in other fields. For example, variable or model selection is a typical problem in statistics. The goal is to select a subset of variables from usually a linear regression model to maximize the predictive accuracy with the strongest effects of predictors [7, 8]. In mathematics, given a matrix A and an integer k, the column subset selection problem is to determine a permutation matrix P so that $AP = (A_1 A_2)$, in which A_1 has k columns which should be linearly independent. The matrix P can be seen as a ranking of the column attributes for the matrix A. Rank-Revealing QR (RRQR), a matrix factorization method [9, 10], is one of well-known methods to solve this problem.

In this paper, we introduce two heuristic-based methods for feature selection into the filter-based or subset selection category. The two heuristics are based on entropy and submodular mutual information, respectively. Mutual information were proved to be submodular functions [11]. The Submodularity reflects the intuitive property of diminishing returns, and it can be exploited to develop strongly polynomial time combinatorial algorithms with provable theoretical performance guarantees [12–14]. Authors in [15] demonstrated the advantage of the mutual information criterion over the entropy criterion in sensor placement problems in spatial monitoring applications.

Our contributions are as follows. First we designed two heuristic-based feature selection methods in the scenario of the Gaussian process model. Second we explored these methods under classification problems through carefully designed experiments, and demonstrated its performance and characteristics.

In the following sections, we first present an entropy-based greedy algorithm and a mutual-information-based approximate algorithm that retrieves near-optimal solutions by exploiting mutual information's submodularity. Then, we will explore their feature selection performance under classification problems in machine learning with a variety of datasets.

2 Heuristic-Based Selection Algorithms

In this section, we present two greedy algorithms that employ heuristics of entropy and submodular mutual information, respectively.

The task of feature selection is to select out a subset of features, also known as attributes, of a dataset. Considering each feature as a random variable, we assume that all features in a dataset form a joint multivariate Gaussian distribution. Then, any finite subset of these variables also have a joint Gaussian distribution. This model is also known as Gaussian Process (GP) [16].

A joint multivariate Gaussian distribution is namely:

$$P(\mathcal{X}_\mathcal{V} = x_\mathcal{V}) = \frac{1}{(2\pi)^{n/2} |\Sigma_{\mathcal{V}\mathcal{V}}|^{1/2}} e^{-\frac{1}{2}(x_\mathcal{V} - \mu_\mathcal{V})^T \Sigma_{\mathcal{V}\mathcal{V}}^{-1}(x_\mathcal{V} - \mu_\mathcal{V})}$$

where $\mathcal{V}$ denotes the whole set of feature variable indexes with $|\mathcal{V}| = n$, $\mu_\mathcal{V}$ is the mean vector, and $\Sigma_{\mathcal{V}\mathcal{V}}$ is the covariance matrix. If we take a subset $\mathcal{A}$ from $\mathcal{V}$, then it also satisfies that the random variable $\mathcal{X}_A \sim \mathcal{N}(\mu_A, \Sigma_{AA})$ where μ_A is a corresponding sub-vector of μ_V, and Σ_{AA} is a corresponding sub-matrix of Σ_{VV}. This consistency property is also called the marginalization property in GP. It also applies to the conditional probability $P(\mathcal{X}_U | \mathcal{X}_A = x_A)$ that is a joint probability distribution of random variables of feature subset U conditional on the values x_A at a selected subset A, assuming $U, A \subset V$. The conditional mean $\mu_{U|A}$ and variance $\sigma^2_{U|A}$ are given by:

$$\mu_{U|A} = \mu_U + \Sigma_{UA}\Sigma_{AA}^{-1}(x_A - \mu_A) \tag{1}$$

$$\sigma^2_{U|A} = \Sigma_{UU} - \Sigma_{UA}\Sigma_{AA}^{-1}\Sigma_{AU} \tag{2}$$

where μ_A is the mean vector of subset variable $\mathcal{X}_A$; Σ_{UU}, Σ_{AA}, Σ_{UA} and Σ_{AU} are corresponding sub-matrices of Σ_{VV}. For example, Σ_{UA} is formed by the U rows and the A columns in Σ_{VV}.

The idea here is to select a subset of feature variables that minimizes the uncertainty of probability distribution comprised of the rest of unselected feature variables. In the following sections, we will present two heuristic-based methods of selecting feature variables in the Gaussian Process scenario.

2.1 The Entropy-Based Heuristic

Given a selected subset A, the uncertainty of conditional probability $P(\mathcal{X}_i | \mathcal{X}_A)$ can be measured by the entropy:

$$\begin{aligned} H(\mathcal{X}_i | \mathcal{X}_A) &= - \iint P(x_i, x_A)\log P(x_i|x_A)dx_i dx_A \\ &= \frac{1}{2}\log\sigma^2_{i|A} + \frac{1}{2}(\log\pi + \log 2 + 1), \end{aligned} \tag{3}$$

Note that the entropy is a monotonic function of the variance $\sigma^2_{i|A}$, which can be evaluated ahead of time by Eq. (2).

Feature selection becomes a subset selection problem, where choosing a subset A out of the whole feature variable index set V, so that uncertainty of the joint probability distribution of the rest of unselected variables, denoted as $\mathcal{X}_{V\setminus A}$, will be minimized. Namely, the selection is made by minimizing the entropy $H(\mathcal{X}_{V\setminus A} | \mathcal{X}_A)$. It is also equivalent to find a subset A that maximizes $H(\mathcal{X}_A)$, as the chain rule for conditional entropy holds that $H(\mathcal{X}_{V\setminus A} | \mathcal{X}_A) = H(\mathcal{X}_V) - H(\mathcal{X}_A)$. The optimization problem turns out to be a NP-hard problem. The heuristic is to greedily select the next feature variable $y^*_{i+1} \in V\setminus A_i$ that has the highest conditional entropy given the current selected set A_i:

$$y^*_{i+1} = \mathrm{argmax}_{y_{i+1}} H(\mathcal{X}_{y_{i+1}} | \mathcal{X}_{A_i}), \tag{4}$$

The greedy algorithm is shown as in Algorithm 1.

Algorithm 1: Greedy algorithm for maximizing entropy $H(\mathcal{A})$

Input: covariance matrix $\Sigma_{\mathcal{VV}}$, selection size k
Output: selected subset $\mathcal{A}(\mathcal{A} \subseteq \mathcal{V}, \text{and } |\mathcal{A}| = k)$
1 **begin**
2 $\quad$ $\mathcal{A} \leftarrow \emptyset$
3 $\quad$ **for** $i = 1$ **to** k **do**
4 $\quad\quad$ **foreach** $y \in \mathcal{V} \backslash \mathcal{A}$ **do** $\delta_y \leftarrow \sigma^2_{y|\mathcal{A}}$
5
6 $\quad\quad$ $y^* \leftarrow \text{argmax}_{y \in \mathcal{V} \backslash \mathcal{A}} \delta_y$
7 $\quad\quad$ $\mathcal{A} \leftarrow \mathcal{A} \cup \{y^*\}$
8 $\quad$ **end**
9 **end**

Where k is the selection size, and $\sigma^2_{y|\mathcal{A}}$ is computed by Eq. (2). Because the log function is monotonic, $\sigma^2_{y|\mathcal{A}}$ is proportional to $H(\mathcal{X}_y| \mathcal{X}_\mathcal{A})$. That means choosing a variable at y that maximizes $H(\mathcal{X}_y| \mathcal{X}_\mathcal{A})$ is equivalent to finding such a y that maximizes $\sigma^2_{y|\mathcal{A}}$.

The calculation of $\sigma^2_{y|\mathcal{A}}$ is expensive. Let $|\mathcal{V}| = n$, there are n times of these computations when $i = 1$, and $(n - k + 1)$ times when $i = k$. Hence, Algorithm 1 has totally $\frac{(2n-k+1)k}{2}$ times of evaluations of $\sigma^2_{y|\mathcal{A}}$.

2.2   The Mutual-Information-Based Heuristic

Another heuristic for optimizing feature subset selection is mutual information, which was originally proposed by Caselton and Zidek in [17]. The mutual information of a subset at $\mathcal{A}$ denoted as $\text{MI}(\mathcal{A})$, which actually is an entropy reduction, is defined as following,

$$
\begin{aligned}
\text{MI}(\mathcal{A}) &= \text{I}(\mathcal{X}_{\mathcal{V}\backslash\mathcal{A}}; \mathcal{X}_\mathcal{A}) \\
&= H(\mathcal{X}_{\mathcal{V}\backslash\mathcal{A}}) - H(\mathcal{X}_{\mathcal{V}\backslash\mathcal{A}}| \mathcal{X}_\mathcal{A}) \\
&= H(\mathcal{X}_\mathcal{A}) - H(\mathcal{X}_\mathcal{A} | \mathcal{X}_{\mathcal{V}\backslash\mathcal{A}})
\end{aligned}
\tag{5}
$$

Compared with the entropy-based method, the mutual-information-based heuristic selects a subset $\mathcal{A}$ by maximizing the reduction of the entropy over the rest of the feature space $\mathcal{V} \backslash \mathcal{A}$ before and after selecting out $\mathcal{A}$. Selecting a feature subset such that,

$$
\mathcal{A}^* = \text{argmax}_{\mathcal{A} \subset \mathcal{V}} \text{MI}(\mathcal{A})
\tag{6}
$$

which is a NP-complete problem. A greedy algorithm developed in [15] selects a feature variable y maximizing the mutual information gain, namely:

$$\Delta_y = \text{MI}(\mathcal{A} \cup y) - \text{MI}(\mathcal{A}), \tag{7}$$

That is, it chooses the next feature variable that provides the maximal increase in the value of mutual information. In the scenario of Gaussian Process, the Δ_y can be further deduced as following:

$$
\begin{aligned}
\Delta_y &= \text{MI}(\mathcal{A} \cup y) - \text{MI}(\mathcal{A}) \\
&= H(y| \mathcal{A}) - H(y| \bar{\mathcal{A}}) \\
&= \frac{1}{2} \log_2 \left(\frac{\sigma^2_{y|\mathcal{A}}}{\sigma^2_{y|\bar{\mathcal{A}}}} \right)
\end{aligned}
$$

where $\bar{\mathcal{A}}$ denotes variable indexes in $\mathcal{V}$ excluding selected $\mathcal{A}$ and y.

A note about the mutual information gain Δ_y is that it is monotonically decreasing as the selected subset $\mathcal{A}$ gets bigger. It inspired an enhanced version of the greedy algorithm with lazy evaluation [15].

Algorithm 2 presents the mutual-information-based heuristic of greedily selecting feature variables in the scenario of GP modelling.

Algorithm 2: Greedy algorithm for maximizing mutual information gain $\text{MI}(\mathcal{A} \cup y) - \text{MI}(\mathcal{A})$ with lazy evaluation

---

Input: covariance matrix $\Sigma_{\mathcal{V}\mathcal{V}}$, selection size k
Output: selected subset $\mathcal{A}(\mathcal{A} \subseteq \mathcal{V})$, mutual information gains Δ

1 **begin**
2     $\mathcal{A} \leftarrow \emptyset$
3   **foreach** $y \in \mathcal{V}$ **do** $\Delta_y \leftarrow +\infty$; $\Phi_y \leftarrow 0$
4
5   **for** $i = 1$ **to** k **do**
6   **repeat**
7             $y^* \leftarrow \text{argmax}_{y \in \mathcal{V} \setminus \mathcal{A}} \Delta_y$
8   **if** $\Phi_{y^*} == i$ **then**
9   **break**
10   **else**
11                 $\bar{\mathcal{A}} \leftarrow \mathcal{V} - (\mathcal{A} \cup y^*)$
12                 $\Delta_{y^*} \leftarrow \frac{1}{2} \log_2 (\frac{\sigma^2_{y^*|\mathcal{A}}}{\sigma^2_{y^*|\bar{\mathcal{A}}}})$
13                 $\Phi_{y^*} \leftarrow i$
14   **end**
15   **until** 0
16         $\mathcal{A} \leftarrow \mathcal{A} \cup \{y^*\}$
17   **end**
18 **end**

Where Φ_{y^*} records in which iteration Δ_{y^*} is updated. The lazy evaluation reduces an amount of computation of Δ_y based on the insight that the sequence of the mutual information gains on a fixed y decreases as the subset $\mathcal{A}$ grows. It will select the y^* if the maximal Δ_{y^*} is updated in the current iteration, otherwise it will update Δ_{y^*} and Φ_{y^*} and will repeat the selection process.

When $|\mathcal{V}| = n$, Algorithm 2 has $2(n + k - 1)$ times of evaluations of either $\sigma^2_{y^*|\mathcal{A}}$ or $\sigma^2_{y^*|\bar{\mathcal{A}}}$ in the best case. This is more efficient and scalable than Algorithm 1 when n becomes very large.

Algorithm 2 is not only efficient, but also provides its solution with a theoretic bound in terms of the optimal solution. Although the mutual information function as in Eq. (5) is not monotonic increasing, it has still been proved to be a partially monotonic submodular function [15]. According to [18], a greedy algorithm, such as the Algorithm 2, optimizing a monotonic submodular function guarantees a solution with a theoretical performance lower bound of $(1 - 1/e)$OPT, where OPT represents the optimal solution value.

3　Experiments

We explore the heuristic-based selection methods above under classification problems in machine learning. The purpose is to evaluate these methods in feature selection problems, and to check whether they can help classification learning tasks achieve similar or even better predictive accuracy than using full feature sets.

For comparison, we also add two other popular selection methods. One is the rank-revealing QR (RRQR) used for matrix column subset selection, and the other is the ranker method in the data mining software Weka [19]. Our heuristic-based selection methods and the RRQR method belongs to the filter-based category, while the ranker method in Weka is a wrapper-based selection method. In experiments, feature subsets are first chosen by these selection methods, then resulted attribute-filtered datasets will be fed into classifiers in Weka to calculate classification rates.

3.1　Experimental Setting

To compare the feature selection methods as shown in Table 1, we recruited 11 different classifiers (as in Table 3), and 13 different data sets (as in Table 2). Our running experiments systematically sweep feature selection size from 1 to $N - 1$, in which N is the total number of features in a dataset, for each of the selection methods, classifiers and datasets.

Table 1. Feature selection methods used in experiments

ID	Method
1	Mutual-information-based heuristic(MI)
2	RRQR
3	Weka's ranker
4	Entropy-based heuristic

We programmed the mutual-information-based and entropy-based heuristics in MatLab, and used a public MatLab implementation of the RRQR provided in [20]. We chose the ranker method with a default attribute evaluation function "ReliefF" in Weka, which weights all features and returns a ranked list of its.

Datasets in Table 2 covers a variety of situations in terms of number of classes and features. For those not providing a test set, we divided the original datasets into two parts, the two-third of which for training and the rest for testing. Most of the datasets are from UCI's machine learning repository [21], the others are from the Libsvm data website [22].

Table 2. Datasets

ID	Dataset name	Class no.	Feature no.	Training no.	Testing no.
1	Australian credit	2	14	460	230
2	Diabetes	2	8	507	261
3	Glass	6	9	142	72
4	Liver disorders	2	6	230	115
5	Satimage	6	36	2217	1000
6	Vehicle	4	18	564	282
7	Breast cancer	2	9	455	227
8	German credit	2	24	667	333
9	Heart	2	13	180	90
10	Pen digits	10	16	2623	1225
11	Sonar	2	60	138	70
12	Wine	3	13	118	60
13	DNA	3	180	2000	1186

Classifiers used in experiments are shown in Table 3. Basically, we picked up one or two representatives in each of classifier categories in Weka, so that it covered a wide spectrum of classification methods. They were employed with their default settings provided in Weka during experiments. For more descriptions of the classifiers, please refer to [23, 24].

Table 3. Classifiers from weka

ID	Classifier
1	RBFNetwork
2	GaussianProcesses
3	SimpleLinearRegression
4	PaceRegression
5	SMOreg
6	KStar
7	AdditiveRegression
8	Bagging
9	RandomSubSpace
10	DecisionTable
11	M5P

Batch experiments were carried out to explore contributions of the listed feature selection methods to the performance of classification methods for each dataset systematically. Each classifier tried out all of the selection methods, and each selection method screened out all of possible selection sizes including a full feature set.

Experimental results are shown next. For convenience, feature selection methods, classifiers and datasets are represented by their ID numbers.

3.2   Experimental Results and Discussion

Table 4 summarizes the best classification rates for each dataset by corresponding combinations of feature selection methods, classifiers, and selected feature sizes in percentage. The mutual-information-based and entropy-based feature selection methods appear multiple times in the table, whereas using full features only appears one time. It shows that feature selection not only reduces the dimensionality of classification problems but also helps to improve classification accuracy.

Table 4. Summary of the combinations for the best classification scores of each dataset

DatasetID	Best rate	Selection method	Selection size (%)	ClassifierID
1	0.90	Entropy-based	35.7	4
2	0.74	Weka-ranker	37.5	9
3	0.72	Weka-ranker	55.6	6
4	0.71	RRQR	83.3	2
4	0.71	Entropy-based	83.3	2
5	0.85	Entropy-based	72.2	6
6	0.76	Weka-ranker	88.9	11
7	0.99	MI-based	55.6	2
7	0.99	RRQR	55.6	1
8	0.69	Entropy-based	54.2	8
9	0.89	MI-based	53.8	4
9	0.88	MI-based	53.8	5
10	0.94	Full-attribute	100.0	6
11	0.93	MI-based	56.7	2
12	0.99	RRQR	76.9	6
13	0.94	Weka-ranker	25.6	11

For a dataset and a classifier, the selection method leading to the best classification score with the smallest selection size was picked out as a winner. We collected winning counts for each of the feature selection methods. We also accommodated co-winning situations of near performances within a 0.5% range of classification accuracy, as well as a shared selection size.

The result of winning counts is summed up in Table 5. It shows that each of the selection methods takes up a variety of seats given a classifier. It can't tell which classifier is favoured by a particular selection method, or vice versa. But it shows clearly that classification with a full feature set wins much less than using a selected subset of features.

Table 5. Summary of winning counts of each selection method given a classifier

Classifier	MI	RRQR	Ranker	Entropy	FullAttrs
RBFNetwork	3	5	5	5	1
GaussianProcesses	4	4	3	5	2
SimpleLinearRegression	3	3	7	6	0
PaceRegression	3	3	3	4	2
SMOreg	4	3	6	7	0
KStar	2	2	7	4	0
AdditiveRegression	2	5	4	5	0
Bagging	4	5	2	4	0
RandomSubSpace	2	4	5	4	0
DecisionTable	2	3	5	7	0
M5P	4	3	3	5	0
Total count	33	40	50	56	5
Percentage %	17.9	21.7	27.2	30.4	2.7

Because the selection size is a key parameter of feature selection methods, Fig. 1 examines the distribution of winnings among different selection sizes given a selection method. There is a distinguishable difference between the mutual-information-based as

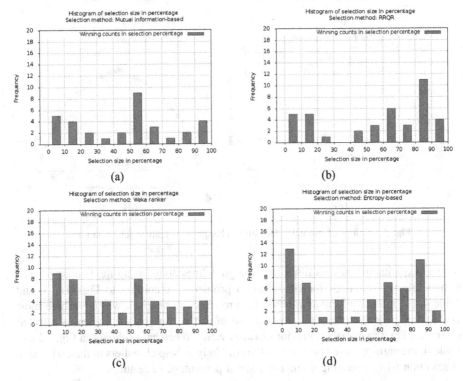

(a)

(b)

(c)

(d)

Fig. 1. Distribution of winnings among different selection sizes for each selection method

in Fig. 1a and the entropy-based as in Fig. 1d. The distribution in the first picture spikes at the middle range, while in the latter picture it scatters mostly at the two ends.

This finding reveals us a truth about the mutual information criterion. That is, the value of mutual information goes up first until about half of variables being selected, then it will go downward. Figure 2 draws a few pictures among the datasets to demonstrate the phenomenon. The number of points varies due to different feature numbers available in the datasets. Figure 3 shows for each dataset the selected feature sizes that achieve the maximal mutual information values. The numbers were converted into percentages for comparison convenience. It indicates that resulted selection sizes reside in the half size level.

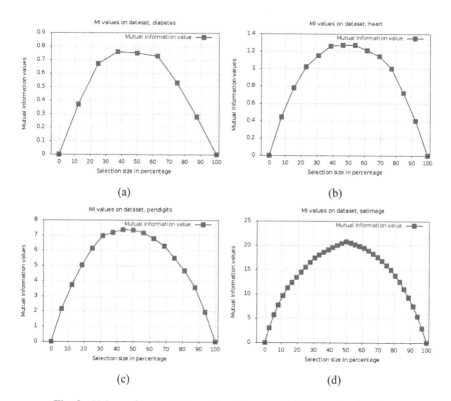

Fig. 2. Values of mutual information change with feature selection sizes

The experimental results disclose some guidelines about using the heuristic-based feature selection methods for classification problems. According to Figs. 1a, 3, and Table 4, the mutual-information-based feature selection heuristic will contribute most to classifiers when selecting out about a half of feature variable size when its selected subset scoring a maximal mutual information value. Whereas, as shown in Fig. 1d and Table 4, the entropy-based heuristic will most likely to help classifiers in the early stage of selection when choosing a relatively small percentage of features.

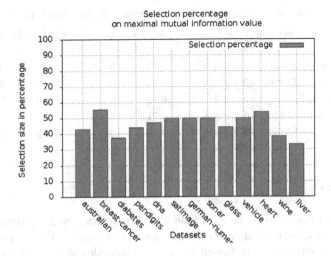

Fig. 3. Summary of feature selection sizes in percentage resulted in maximal mutual information gains for the datasets

Besides the observations above, we will also show a remarkable case for using the selection methods with classification on a medical dataset. It helps patients save their diagnostic costs. In the diabetes dataset for example, there are 8 features representing different diagnostic testings. Its individual costs are listed in Table 6. It looks like testing levels of the glucose and the insulin cost a lot more than the others.

Table 6. Costs of diagnostic testings in the diabetes dataset

Feature ID	Testing	Cost ($)
1	Times_pregnant	1.00
2	Glucose_tol	17.61
3	Diastolic_pb	1.00
4	Triceps	1.00
5	Insulin	22.78
6	Mass_index	1.00
7	Pedigree	1.00
8	Age	1.00

Table 7 sums up the result of how much it can save by selecting features against using full feature set for the classification of diabetes. It shows that the mutual-information-based selection method, which is a filter-based approach, achieves the same classification accuracy as using full features, in the meantime saving more than 50% of the diagnostic costs. It is remarkable because that it not only reduces patients' costs, but also can potentially help doctors improve the diagnostic accuracy.

Table 7. Cost savings by applying selection methods for classification of diabetes

Method	Selection size	Feature index	Accuracy	Classifier ID	Cost($)	Saving
Full-feature	100%	1 2 3 4 5 6 7 8	0.72	8	46.39	0%
Weka-ranker	37.5%	2 8 1	0.74	9	19.61	58%
MI-based	50%	4 8 2 7	0.72	8	20.61	56%
Entropy-based	62.5%	5 2 3 4 8	0.71	2	43.39	6%
RRQR	62.5%	5 2 3 4 8	0.71	2	43.39	6%

4 Summary

We introduced two heuristic-based feature selection methods, and explored their performance under classification problems for a number of datasets. Experimental results showed that feature selection helped reduce the dimensionality of the problems by improving classification accuracies with less number of features. It also showed that the mutual-information-based heuristic would contribute most to classifiers when selecting about a half size of features, while the entropy-based heuristic would most likely help in the early stage of the selection when choosing a relatively small percentage of features. We also demonstrated a remarkable case of feature selection for classification on a medical dataset.

Acknowledgement. This work was generously supported by the following funds: Hainan University's Scientific Research Start-Up Fund; Ministry of Education of China's Scientific Research Fund for the Returned Overseas Chinese Scholars; Hainan Province Natural Science Fund No. 20156243; China's Natural Science Fund Nos. 11401146, 11471135, 61462022, 61562017, 61562018, 61562019; Hainan Province's Major Science and Technology Project Grant No. ZDKJ2016015; Hainan Province's Key Research and Development Program Grant Nos. ZDYF2017010 and ZDYF2017128. This work was also supported by the State Key Laboratory of Marine Resource Utilization in the South China Sea, Hainan University.

References

1. Manning, C.D., Prabhakar Raghavan, H.S.: An Introduction to Information Retrieval. Cambridge University Press, Cambridge (2009)
2. Das, S.: Filters, wrappers and a boosting-based hybrid for feature selection. In: Proceedings of the eighteenth international conference on machine learning. pp. 74–81. Morgan Kaufmann Publishers Inc., San Francisco (2001)
3. Kohavi, R., John, G.H.: Wrappers for feature subset selection. Artif. Intell. **97**, 273–324 (1997)
4. Guyon, I., Elisseeff, A.: An introduction to variable and feature selection. J. Mach. Learn. Res. **3**, 1157–1182 (2003)
5. Yu, L., Liu, H.: Feature selection for high-dimensional data: a fast correlation-based filter solution. In: Fawcett, T., Mishra, N. (eds.) ICML, pp. 856–863. AAAI Press (2003)

6. Liu, H., Motoda, H.: Computational Methods of Feature Selection (Chapman & Hall/CRC data mining and knowledge discovery series). Chapman & Hall/CRC (2007)
7. George, E.I.: The variable selection problem. J. Amer. Statist. Assoc. **95**, 1304–1308 (1999)
8. Hastie, T., Tibshirani, R., Friedman, J.: The Elements of Statistical Learning: Data Mining, Inference, and Prediction. Springer, New York (2009)
9. Ipsen, I.C.F., Kelley, C.T.: Rank-deficient nonlinear least squares problems and subset selection. SIAM J. Numer. Anal. **49**, 1244–1266 (2011)
10. Gu, M., Eisenstat, S.C.: Efficient algorithms for computing a strong rank-revealing QR factorization. SIAM J. Sci. Comput. **17**, 848–869 (1996)
11. Krause, A., Singh, A., Guestrin, C.: Near-optimal sensor placements in gaussian processes: theory, efficient algorithms and empirical studies. J. Mach. Learn. Res. **9**, 235–284 (2008)
12. Iwata, S., Fleischer, L., Fujishige, S.: A combinatorial strongly polynomial algorithm for minimizing submodular functions. J. ACM **48**, 761–777 (2001)
13. Krause, A., Guestrin, C.: Near-optimal nonmyopic value of information in graphical models. In: Proceedings of the Twenty-First Conference Annual Conference on Uncertainty in Artificial Intelligence (UAI-05), pp. 324–331. AUAI Press, Arlington, Virginia (2005)
14. Krause, A., McMahan, B., Guestrin, C., Gupta, A.: Robust submodular observation selection. J. Mach. Learn. Res. **9**, 2761–2801 (2008)
15. Krause, A., Singh, A., Guestrin, C.: Near-optimal sensor placements in gaussian processes: theory, efficient algorithms and empirical studies. J. Mach. Learn. Res. **9**, 235–284 (2008)
16. Rasmussen, C.E., Williams, C.K.I.: Gaussian Processes for Machine Learning. The MIT Press, Cambridge, Massachusetts (2006)
17. Caselton, W., Zidek, J.: Optimal monitoring network designs. Stat. Prob. Lett. **2**(4), 223–227 (1984)
18. Nemhauser, G., Wolsey, L., Fisher, M.: An analysis of the approximations for maximizing submodular set functions. Math. Program. **14**, 265–294 (1978)
19. Hall, M., Frank, E., Holmes, G., Pfahringer, B., Reutemann, P., Witten, I.H.: The weka data mining software: an update. SIGKDD Explor. Newsl. **11**, 10–18 (2009)
20. Bischof, C.H., Quintana-Ortí, G.: Computing rank-revealing QR factorizations of dense matrices. ACM Trans. Math. Softw. **24**, 226–253 (1998)
21. Frank, A., Asuncion, A.: UCI machine learning repository. http://archive.ics.uci.edu/ml (2010)
22. Chang, C.C., Lin, C.J.: LIBSVM data: classification, regression, and multi-label. http://www.csie.ntu.edu.tw/~cjlin/libsvmtools/datasets/
23. University of Waikato, M.L.G. at: Weka 3: data mining software in java. http://www.cs.waikato.ac.nz/~ml/weka/index.html
24. Witten, I.H., Frank, E., Hall, M.A.: Data Mining: Practical Machine Learning Tools and Techniques, Third edn. Morgan Kaufmann, (2011)

AGSA: Anti-similarity Group Shilling Attacks

Peng Wang[1,2], Lingtao Qi[1,2], Haiping Huang[1,2(✉)] [iD], Feng Li[1,2],
and Congxiang Yu[1]

[1] Nanjing University of Posts and Telecommunications, Nanjing 210003, China
hhp@njupt.edu.cn
[2] Jiangsu High Technology Research Key Laboratory for Wireless Sensor
Networks, Nanjing 210003, China

Abstract. With the rapid development of e-commerce, the security issues of recommender systems have been widely investigated. Malicious users can benefit from injecting great quantities of fake profiles into recommender systems to reduce the frequency of undesired recommendation items. As one of the most important attack methods in recommender systems, the shilling attack has been paid considerable attention, especially to its model and the way to detect it. Although a multitude of studies have been devoted to shilling attack modeling and detection, few of them focus on group shilling attack. The attackers in a shilling group work together to manipulate the output of the recommender system. Based on the model of the loose version of Group Shilling Attack Generation Algorithm (GSAGenl), we design an anti-similarity group shilling attack model (AGSA). AGSA rationalizes the evaluation time interval of the group attack and strengthens the destructive powers of the group shilling attacks.

Keywords: Recommender systems · Anti-similarity group shilling attack model (AGSA) · Shilling attack · Security

1 Introduction

As the development of big data technology, recommender systems based on the collaborative filtering [1] are widely used in mobile Internet business (for example Taobao), social networking (for example Facebook), search engines (for example Google) and movie video recommendation sites (for example YouTube) [2]. However, due to openness and anonymity of recommendation systems, in order to reap huge profits, malicious users probably inject a large number of fake profiles into the recommender system to manipulate the recommendation results [3], and this kind of behavior is called "shilling attack". The user profiles record users' ratings data of various items, and these rating data indicate users' preferences, where the false user profile is possibly the attacked profile [4]. Shilling attack aims to facilitate target products recommended to real users, i.e. push attack [5], or prevent recommend systems from promoting target products to real users, i.e. nuke attack. Reliable evaluation on items should be constructed based on real users' preferences, which is also the foundation of collaborative filtering recommender systems. The real profile is likely to reveal true intention of the user, and meanwhile can offer the effective recommendations for their neighboring

G. Chen et al. (Eds.): PAAP 2017, CCIS 729, pp. 108–116, 2017.
DOI: 10.1007/978-981-10-6442-5_10

users [6]. Therefore, how to detect and resist the shilling attack has become a key issue of recommender systems, conversely how to strengthen the power of shilling attacks is always an invaluable asset of adversaries.

Aiming at these challenging problems, the main contribution of this paper can be summarized as follows: based on the model of GSAGenl [7], a model of anti-similarity group shilling attack (AGSA) is proposed, which rationalizes the evaluation time interval of the group attack and increases the intensity of attacks.

The rest of this paper is organized as follows. Section 2 introduces shilling attack methods. The anti-similarity group shilling attack model (AGSA) is described in Sect. 3. Experiment results and performance analysis are discussed in Sect. 4. And we conclude our work in Sect. 5.

2 Related Works

Shilling attack methods are mainly comprised of the following categories: the random attack, the average attack, the bandwagon attack and the segment attack, where the target item usually has the highest rating when encountering a push attack, or has the lowest rating when suffering a nuke attack. In the random attack, the injected profiles are filled with random rating values obeying a normal distribution. And meanwhile the normal distribution is determined by the mean and standard deviation of real rating values in the recommender database. All these items involved in the attacked profile are called as "filler items". In order to make attacked profiles to be more similar as real profiles, the average attack assigns the mean rating of some certain kinds of items to randomly chosen filler items and the highest (or lowest) rating to the target item corresponded to the push (or nuke) attacks. The bandwagon attack [8] is one of the evolution forms of random attack, and it adopts the Zipf's Law where only a small number of items arouse most of users' attention. In bandwagon attack, an attacker generates the profiles with high ratings to well-known popular items and the possible highest rating to the target item. The attacker usually makes the popular items as selected items, and the prevalence of selected items are commonly measured by the number of scoring. The segment attack selects these special users who are interested in a specific item segment (e.g., swimmers is in favor of racing goggles) and casters recommendations of the item segment to their pleasure. The segment attack offers the highest rating to the target item and item segment and the lowest rating of filler items.

From the perspective of attack model, the differences of four attack methods in structure are more intuitive. The purpose of attack model is to create the inveracious user profile by forging or tampering the user data and items' ratings in recommender systems [9], where the attacked profile is quantized into n groups of score vectors, as shown in Fig. 1. To clearly describe the attack mode, generally, the shilling attack vectors is divided into four parts, which are the target items (i_t), the selected items (I_S), the filler items (I_F) and the unrated items (I_ϕ).

In Table 1, the first column displays different attack methods; the second column means the push attack mode. For each attacked profile, the target item it is the attacked object, which is assigned with either the maximum or minimum rating value determined by the attack type. I_S is the set of selected items with special features during the

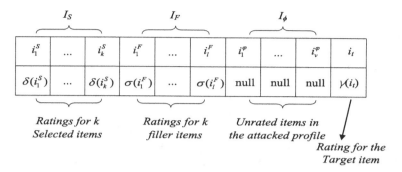

Fig. 1. The model of shilling attacks.

attack process. IF is the set of filler items usually chosen randomly. r_{min} denotes the minimum rating value, instead r_{max} represents the maximum rating value, r_{ran} is the random rating value, and r_{avg} stands for the average rating value. Table 1 lists the composition of the ratings of various attack modes aimed at different item set.

Table 1. The categories and modes of shilling attacks.

Attack method	Push
Random attack	$I_S = \emptyset; I_F = \{r_{ran}\}; i_t = r_{max}$
Average attack	$I_S = \emptyset; I_F = \{r_{avg}\}; i_t = r_{max}$
Bandwagon attack	$I_S = \{r_{max}\}; I_F = \{r_{max}\}; i_t = r_{max}$
Segment attack	$I_S = \{r_{max}\}; I_F = \{r_{min}\}; i_t = r_{max}$

3   Anti-similarity Group Shilling Attack Model

On the basis of GSAGenl, AGSA model can design the ingenious schedule of fake rating to achieve more concealed group attacks. The definition about AGSA model is introduced as follows:

Definition 1. Low time correlation. Suppose $B_i = \{b_1, \cdots, b_{n_i}\}$ is a group shilling attackers generated by *GSAGen$_l$*. $T_{i,j} = \{t_{1,j}, t_{2,j}, \cdots, t_{n_i,j}\}$ denotes the sequence of interval time of the target item j rated by group attackers in $B_i \cdot T_{i,j}$ can be considered low time correlation if it complies with two conditions: (1) if $\forall m \neq l, t_{m,j} \neq t_{l,j}$; (2) $\forall m, \forall j, t_{m,j} = Random(\min, \max)$.

In Definition 1, for instance, $t_{2,j}$ implies the time interval between the rating time point of the second group attacker b_2 and that of the first group attacker b_1 aiming at the target item j. The rating time interval sequence of attack group member is generated by a random function, which keeps the sequence from being detected by known time series detection algorithm due to the similarity of interval time. For the same item, the rating time interval of each group attacker is usually different from that of others. In condition (2), the maximum and minimum for the random function are respectively set according to the real user's rating time interval in recommendation systems.

The pseudo-code of AGSA is shown in Algorithm 1. $A_i = \{a_1, \cdots, a_{n_i}\}$ is the final attack group. $b_{m,j}$ implies that the score of item j which is rated by the mth attacker in group B_i. In steps 12–22, each profile is examined in terms of Definition 1.

Algorithm 1. Anti-similarity Group Shilling Attack Algorithm

1: **procedure** AGSA($B_i = \{b_1, \cdots, b_{n_i}\}$, $A_i = \{a_1, \cdots, a_{n_i}\}$)

2. $a_1 \leftarrow b_1, A_i \leftarrow \{a_1\}$
                /* Initialize an attack group A_i */

3.   **for** $j \leftarrow 1 : n$ **do**
                /* n is the number of all target items */

4.   **for** $m \leftarrow 1 : n_i$ **do**
                /* n_i denotes the number of attackers in B_i */

5.   **if** $b_{m,j} = \varnothing$ **then**
            /* If item j isn't rated by attacker b_m, then $t_{m,j} = \varnothing$. */

6. $t_{m,j} = \varnothing$

7.   **Else**

8. $t_{m,j} = Random(\min, \max)$

9.   **end if**

10.   **end for**

11.   **end for**

12.   **for** $j \leftarrow 1 : n$ **do**

13.   **for** $m \leftarrow 1 : n_i$ **do**

14.   **for** $l \leftarrow 1 : n_i$ **do**

15.   **if** $t_{m,j} = t_{l,j}$ **then**

16. $t_{m,j} = Random(\min, \max)$;

17.   Goto line 14

18.   **end if**

19.   **end for**

20. $A_i \leftarrow A_i \bigcup \{a_m\}$ /* Let the attacker who satisfies the definition1 join the attack group*/

21.   **end for**

22.   **end for**

---

4 Experimental Evaluation and Discussion

The MovieLens 100K data set is selected as the rating databases of recommender system, and it is widely used in simulation experiments of data mining and other fields due to the properties of real and aplenty. This data set is provided by the GroupLens research team at the University of Minnesota U.S, which includes 100000 ratings records of 943 users to 1682 movies. Each rating record includes a user ID, an item ID, a value of rating in the interval [1, 5] and a TimeStamp. The user ID and item ID start with 1 and increase by degree, the higher the rating indicates that the user more likes the movie, and the format of reference time is 1/1/1970 UTC.

4.1 Vulnerability Metric

The extent of damage of group attack is related with the attack model, the attack (file) size, the number of group attackers and the detection algorithm of recommender systems. In order to objectively evaluate the effectiveness of the group attack, we use the average prediction shift and the hit ratio to reflect the change of the recommender system after encountering the group attack.

The average prediction shift describes the prediction influence on normal user rating when recommender system are suffering group attacks, and the value aiming at a specific target item j can be figured out based on (1):

$$\bar{\Delta}_j = \frac{\sum\limits_{u \in U} O'_{u,j} - O_{u,j}}{|U|} \tag{1}$$

The average value of prediction shift of all items is calculated by (2):

$$\bar{\Delta} = \frac{\sum\limits_{j \in I} \bar{\Delta}_j}{|I|} \tag{2}$$

Wherein, $O'_{u,j}$ is the prediction rating to the target item j after shilling attack occurs, and $O_{u,j}$ is the normal prediction rating before shilling attack occurs. I is a collection of all items in the system, and U is the collection of all users. The more the average prediction shift is, the more fragile the recommender systems are when facing with shilling attacks.

Most of recommender systems provide users with a list of top-N recommended items, and the purpose of shilling attacks is to enable the target item to rank first in the recommended list. When the recommender systems encounter the shilling attacks, the prediction values of multiple items will be affected, which leads to the change of the ranking. Obviously, the higher the ranking of the target item, the better effectiveness of the shilling attack, and based on the situation, the hit ratio and expected top-N occupancy (ExpTop$N$) are proposed to evaluate the performance of shilling attacks.

Hit ratio measures the effect of attack profiles on top-N recommendations. The hit ratio of item j can be calculated by (3):

$$HR_j = \frac{\sum_{u \in U} H_{u,j}}{|U|} \tag{3}$$

The average hit ratio of all items can be figured out by (4):

$$\overline{HR} = \frac{\sum_{j \in I} HR_j}{|I|} \tag{4}$$

where $H_{u,j}$ represents the number of items j appearing in the recommended list of user u. $H_{u,j}$ is 0 means that item j did not appear in the recommended list of user u.

ExpTopN is defined as the expected number of occurrences of all target items in a top-N recommendation list. The expTopN of user u can be calculated by (5):

$$\Delta \exp TopN_u = \left(\exp TopN_u' - \exp TopN_u \right) / N \times 100\% \tag{5}$$

The average expTopN of all users can be figured out by (6):

$$\overline{\Delta \exp TopN} = \frac{\sum_{u \in U} \Delta \exp TopN_u}{|U|} \times 100\% \tag{6}$$

where $\exp TopN_u$ represents the target items appearing in the recommended list of user u before attack and $\exp TopN_u'$ after attack. N is the number of recommended items in recommendation list.

4.2 Analysis of Attack Effect

Firstly, it can be assumed that there does not exist any abnormal data in the MovieLens data set. Secondly, the attacked profiles are generated and injected into the MovieLens data set by some prevalent attack methods. Finally, a variety of detection algorithms are employed to distinguish the group attackers from the normal users.

Furthermore, the user-based collaborative filtering algorithm [10] will be adopted in the recommender system, which is the most widely used recommendation algorithm now. The number of neighbor users (knn) k is set to 50, and N is 40. As shown in following figures, 8 types of group attacks appear including the random attack, the average attack, the bandwagon attack, the segment attack, and their respective enhanced versions based on AGSA model.

Generally, the intensity of attacks relies on the attack size and the filler size, where the former is the ratio of the number of attacked profile injected into the system to the total user profiles, ranging from 1% to 15%; the latter represents the ratio of the number of rating items to the total items (i.e. the rating-density) in one attacked profile, ranging from 1% to 40%. 8 kinds of attacks with the attack sizes of 1%, 3%, 6%, 9%, 12% and 15% and the filler sizes of 1%, 5%, 15%,25% and 40% are compared.

In Fig. 2, it can be seen that the prediction shift of four enhanced attacks based on AGSA model rises faster than the original versions. This is because with the increase of attack size and filler size, AGSA model can attack two or more than two target items at the same time. And the more target item is, the greater prediction score after attack is, resulting in the fast increase of precision shift. In addition, bandwagon attack combined with AGSA model achieves the best attack effectiveness on the prediction shift in different attack size and filler size.

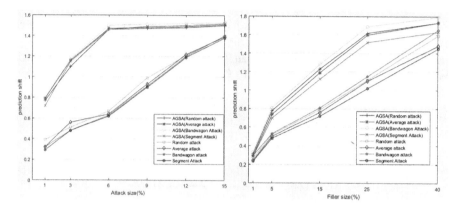

Fig. 2. The precision shift of general shilling attacks and AGSA.

In Fig. 3, it can be seen that the four enhanced attacks based on AGSA model also achieve better effectiveness than the original versions. Bandwagon attack combined with AGSA model achieves the best attack effectiveness on the prediction shift in different attack size and filler size. But when the attack size reaches 6% and filler size increases from 15% to 25%, the prediction shifts of four enhanced attacks step into the slow and steady growth period. This is because the hit ratio is above 0.9 and hard to grow.

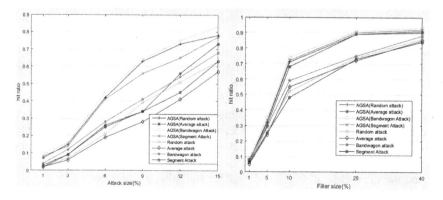

Fig. 3. The hit ratio of general shilling attacks and AGSA.

In Fig. 4, the expTopN of each attack rises with the increase of attack size and filler size. Four enhanced attacks based on AGSA model achieve better effectiveness than the original versions. It means that AGSA model can push target items to recommendation list more effectively.

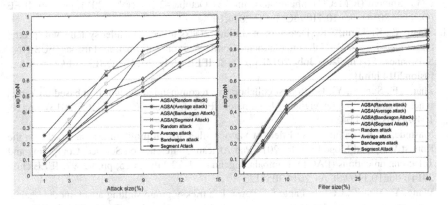

Fig. 4. The expTopN of general shilling attacks and AGSA.

5 Conclusions

Followed with the widespread applications of recommendation system in electronics business or social networks, the profile injection attacks attract considerable attentions. In order to strengthen the effectiveness of attacks, an anti-similarity group shilling attack model is designed, which facilitates the group attackers to keep out of detections based on Pearson correlation coefficient. In this paper, an AGSA attack algorithm is designed. The experiment shows that the AGSA attack algorithm is more effective than other algorithms. Even if the random, average, bandwagon, segment attacks based on AGSA achieve the more powerful attack efficiency.

Acknowledgment. This work was supported by the National Natural Science Foundation of P. R. China (Nos. 61672297, 61572260 and 61373138), the Key Research and Development Program of Jiangsu Province (Social Development Program, Nos. BE2016185 and BE2016177), Postdoctoral Foundation (Nos. 2015M570468 and 2016T90485), The Sixth Talent Peaks Project of Jiangsu Province (No. DZXX-017), Jiangsu Natural Science Foundation for Excellent Young Scholar (No. BK20160089), the Fund of Jiangsu High Technology Research Key Laboratory for Wireless Sensor Networks (WSNLBZY201516).

References

1. Bobadilla, J., Ortega, F., Hernando, A., Gutiérrez, A.: Recommender systems survey. Knowl.-Based Syst. **46**, 109–132 (2013). doi:10.1016/j.knosys.2013.03.012

2. Bhebe, W., Kogeda, O.P.: Shilling attack detection in collaborative recommender systems using a meta learning strategy. In: 2015 International Conference on Emerging Trends in Networks and Computer Communications (ETNCC), Namibia, Windhoek, 17–20 May 2015, pp. 56–61. IEEE (May 2015). doi:10.1109/ETNCC.2015.7184808

3. Fuguo, Z.: Analysis of profile injection attacks against recommendation algorithms on bipartite networks. In: 2014 International Conference on Management of e-Commerce and e-Government (ICMeCG), Shanghai, China, 31 October–2 November 2014, pp. 1–5. IEEE (October 2014). doi:10.1109/ICMeCG.2014.10

4. Zhang, Z., Kulkarni, S.R.: Detection of shilling attacks in recommender systems via spectral clustering. In: 2014 17th International Conference on Information Fusion (FUSION), Salamanca, Spain, 7–10 July 2014, pp. 1–8. IEEE (July 2014). [DBLP: db/conf/fusion/fusion2014.html]

5. Zhang, F., Sun, S., Yi, H.: Robust collaborative recommendation algorithm based on kernel function and Welsch reweighted M-estimator. IET Inf. Secur. 9(5), 257–265 (2015). doi:10.1049/iet-ifs.2014.0488

6. Masinde, N.W., Fatima, S.S.: Effect of varying filler-size in profile injection attacks on the Robust Weighted Slope One. In: 2014 Pan African Conference on Science, Computing and Telecommunications (PACT), Kampala, Uganda, 27–29 July, 2015, pp. 92–97. IEEE (July 2014). doi:10.1109/SCAT.2014.7055125

7. Wang, Y., Wu, Z., Cao, J., Fang, C.: Towards a tricksy group shilling attack model against recommender systems. In: Zhou, S., Zhang, S., Karypis, G. (eds.) ADMA 2012. LNCS, vol. 7713, pp. 675–688. Springer, Heidelberg (2012). doi:10.1007/978-3-642-35527-1_56

8. Noh, G., Kim, C.K.: RobuRec: robust Sybil attack defense in online recommender systems. In: 2013 IEEE International Conference on Communications (ICC), Budapest, Hungary, 9–13 June 2013, pp. 2001–2005. IEEE (June 2013). doi:10.1109/ICC.2013.6654818

9. Wang, Y., Zhang, L.: A comparative study of shilling attack detectors for recommender systems. In: 2015 12th International Conference on Service Systems and Service Management (ICSSSM), Guangzhou, China, June 22–24 2015, pp. 1–6. IEEE (2015). doi:10.1109/ICSSSM.2015.7170330

10. Hattori, S., Takama, Y.: Consideration about applicability of recommender system employing personal-value-based user model. In: 2013 Conference on Technologies and Applications of Artificial Intelligence, Taipei, Taiwan, 6–8 December 2013, pp. 282–287. IEEE (2013). doi:10.1109/TAAI.2013.63

H_∞ Filtering Design for a Class of Distributed Parameter Systems with Randomly Occurring Sensor Faults and Markovian Channel Switching

Huihui Ji[✉] and Baotong Cui

School of Internet of Things Engineering, Jiangnan University,
Wuxi 214122, People's Republic of China
jihuihui2009@163.com, btcui@jiangnan.edu.cn

Abstract. The paper is concerned with H_∞ filtering design for a class of stochastic distributed parameter systems with randomly occurring sensor faults over sensor networks with multiple communications. The channel switching is governed by a continuous-time Markovian process and the case of measurement failures is described by a stochastic variable which is satisfying the Bernoulli random distribution. Based on a Markovian switched Lyapunov-Krasovskii functional, delay-dependent conditions are achieved to guarantee the prescribed H_∞ performance. Finally, a practical simulation example is given to illustrate the validity of our results.

Keywords: H_∞ filtering · Distributed parameter systems · Randomly occurring failures · Markovian jump parameter · Time-delays

1 Introduction

For these years, filtering problem has attracted considerable attention in many fields, which is mainly due to its significant application in engineering [1–5]. The H_∞ filtering approach, which was developed by Elsayed and Grimble in [2], was to design an estimator to estimate the unknown states subject to the exogenous disturbance with bounded energy. Subsequently, massive achievements have been studied on the H_∞ filtering problem [3–5]. Nevertheless, it is worth pointing out that certain limitations inevitable occur in networked systems when the information is transmitted via the communication networks, which can lead to imperfections and constraints including time delays, external disturbances as well as packet dropouts. Consequently, plenty of effective methods have been provided to solve these challenging problems. For example, the random variable obeying the Bernoulli distribution was adopted to describe the stochastic nonlinearities in [6] and failures of the network sensors in [7], respectively. In addition, delay-partitioning method was employed in [8].

However, most of the results on filtering problems are modeled by ordinary differential equations (ODE), which may bring some drawbacks. Indeed, the behavior of most physical systems are depend on time as well as spatial position, such as thermal diffusion, fluid heat exchangers, chemical engineering. These spatiotemporal processes are

G. Chen et al. (Eds.): PAAP 2017, CCIS 729, pp. 117–129, 2017.
DOI: 10.1007/978-981-10-6442-5_11

distributed parameter systems (DPSs) which are described by partial differential equations (PDEs) or partial differential-integral equations (PDIEs). As far as we know, there are few results on the filtering problems of DPSs, for example, a method to design a specific filter was developed in [9] for single input single output (SISO) DPSs. Moreover, the problem of consensus controllers was investigated in [10] for distributed filters of DPSs. The consensus and adaptive consensus distributed filters problem was studied in [11] for DPSs. However, the filtering problem for Markovian switching DPSs with randomly occurring sensor faults and time delays has not been investigated so far. Therefore, the purpose of this paper is to make one of the first attempts to solve this problem.

In this paper, we mainly consider the H_∞ filtering problem for time-delayed DPSs over a sensor network with multiple communication channels, which is assumed to follow a continuous-time Markov process. In addition, the network-induced time-varying delays and randomly occurring sensor faults are considered in this paper. The case of measurement failures is described by a stochastic variable satisfying the Bernoulli random distribution. Based on a Markovian switched Lyapunov-Krasovskii functional, delay-dependent conditions are achieved to guarantee the prescribed H_∞ performance. Finally, the effectiveness of our results is demonstrated by a practical simulation example.

Notations: R, R_n and $R_{m \times n}$ stand for the set of all real numbers, n-dimensional Euclidean space, and the set of all real $m \times n$ matrices, respectively. I and 0 represent the appropriate dimension identity matrix and zero matrix, respectively. $*$ is used as an ellipsis for the terms with which are introduced by symmetry. $L_2([0, l]; R)$ is a Hilbert space of square integrable functions on $([0, l]; R)$. *diag*$\{\}$ denotes the diagonal matrix, and *col*$()$ denotes column vectors. For a symmetric matrix, $P > 0 (\geq 0)$ means that P is positive-(semi) definited.

2 Problem Formulation and Preliminary

We focus on spatially-distributed processes modeled by first-order hyperbolic PDE systems in one spatial dimension of the following form:

$$\begin{cases} \frac{\partial x(s,t)}{\partial t} = \phi(s) \frac{\partial x(s,t)}{\partial s} + Ax(s,t) + B_1 w(s,t) \\ y(s,t) = C\, x(s,t) + B_2 w(s,t) \\ z(s,t) = L\, x(s,t) \end{cases} \tag{1}$$

where $x(s, t) \in R^{n_x}$ stands for the state; $s \in [0, l], t \geq 0$; $y(s, t) \in R^{n_y}$ is the system measured output; $z(s, t) \in R^{n_z}$ is the controlled output, and $w(s, t) \in R^{n_z}$ represents the external disturbance signal, which belongs to $\mathcal{L}_2[0, \infty)$; A, B_1, B_2, C, L are known real constant matrices with appropriate dimensions.

The system (1) is subject to boundary and initial conditions, respectively, as following:

$$x(0, t) = x(l, t) = 0, \frac{\partial x(0, t)}{\partial t} = \frac{\partial x(l, t)}{\partial t} = 0, \quad t \in [0, +\infty), \tag{2}$$

and

$$x(s,0) = x_0(s). \tag{3}$$

Assumption 1. The term $\frac{\partial x(s,t)}{\partial s}$ satisfies the following conditions: $\| \frac{\partial x(s,t)}{\partial s} \| \le \alpha \| x(s,t) \|$, where α is a known positive constant.

We consider the switching communication channels. Let $\{c_t, t \ge 0\}$ denote the channel switching signal taking values in a finite set $\mathbb{S} = \{1, \ldots, M\}$. We assume the channel switching is governed by a continuous-time Markovian process with right continuous trajectories and transition probability matrix $\prod = \{\pi_{ij}\}$ given by

$$\mathbf{P}[c_{t+\Delta} = j | c_t = i] = \begin{cases} \pi_{ij}\Delta + o(\Delta), & j \ne i; \\ 1 + \pi_{ii}\Delta + o(\Delta), & j = i, \end{cases}$$

where $\Delta > 0$, $\lim\limits_{\Delta \to 0} = 0$, and $\pi_{ij} \ge 0$, for $j \ne i$, is the transition rate from mode i to mode j at time $t + \Delta$ and $\pi_{ij} = -\sum\limits_{j=1, j \ne i}^{M} \pi_{ii}$; For simplicity, denote $c_t = i, i \in \mathbb{S}$. We consider the filter with the following structure:

$$\begin{cases} \frac{\partial \hat{x}(s,t)}{\partial t} = \hat{A}_i\hat{x}(s,t) + \hat{B}_i\hat{y}(t_k), \\ \hat{z}(x,t) = \hat{C}_i\hat{x}(s,t) + \hat{D}_i\hat{y}(t_k) \end{cases} \tag{4}$$

with similar boundary and initial conditions as (2) and (3), respectively, where $\hat{x}(s,t)$ is the filter state, $\hat{z}(s,t)$ is the estimation of $z(s,t)$, and $\hat{A}_i, \hat{B}_i, \hat{C}_i, \hat{D}_i$ are the filter matrices which will be designed in this paper.

Let $t_k, k = 1, 2, \ldots$ be the updating instants of the network sensors, and $\tau_{c_{t_k}}(t), k = 1, 2, \ldots$ be the network-induced time-varying delay from the sensors to the filter at $t_k, k = 1, 2, \ldots$. We assume that $0 \le \tau_{c_{t_k}}(t) \le \tau_{M_{c_{t_k}}} \le \tau_M$ and the measured signal sent from the sensors with randomly occurring failures. Denote $\hat{y}(t_k) = \delta(t)y(t_k) + (1 + \delta(t))Ey(t_k)$ be the signal sent by sensors and $y(t_k) = y(t - \tau_i(t))$ be the signal received by filter, where $E = diag\{E_1, E_2, \ldots, E_{n_y}\}$ is the sensor failure function matrix. The stochastic variable $\delta(t) \in R$ is defined by a Bernoulli distributed sequence as following

$$\delta(t) = \begin{cases} 0, & \text{sensor failures happen}, \\ 1, & \text{sensor failures do not happen} \end{cases}$$

with $Pr\{\delta(t) = 1\} = \delta, Pr\{\delta(t) = 0\} = 1 - \delta$, where $\delta \in [0, 1]$ is a known constant. Then, the filter can be rewritten as

$$\begin{cases} \frac{\partial \hat{x}(s,t)}{\partial t} = \hat{A}_i\hat{x}(s,t) + \hat{B}_i[(\delta(t)y(t_k) + (1 - \delta(t))E)y(t - \tau_i(t))], \\ \hat{z}(x,t) = \hat{C}_i\hat{x}(s,t) + \hat{D}_i[(\delta(t)y(t_k) + (1 - \delta(t))E)y(t - \tau_i(t))]. \end{cases} \tag{5}$$

Define $\bar{x}(s,t) = [x^T(s,t), \hat{x}^T(s,t)]^T$, $\bar{w}(s,t) = [w^T(s,t), \hat{w}^T(s,t)]^T$, $\bar{z}(s,t) = z(s,t) - \hat{z}(s,t)$. Combining system (1) and filter (5), we obtain the filtering error system

$$
\begin{cases}
\frac{\partial \bar{x}(s,t)}{\partial t} = \varphi(s)\frac{\partial \bar{x}(s,t)}{\partial s} + \mathbb{A}\bar{x}(s,t) + \delta(t)\mathbb{C}_1 H\bar{x}(s, t - \tau_i(t)) + (1 - \delta(t))\mathbb{C}_2 \\
H\bar{x}(s, t - \tau_i(t)) + \mathbb{D}\bar{w}(s,t) + \delta(t)\mathbb{D}_1\bar{w}(s,t) + (1 - \delta(t))\mathbb{D}_2\bar{w}(s,t), \\
\bar{z}(x,t) = \mathbb{L}\bar{x}(s,t) + \delta(t)\mathbb{L}_1 H\bar{x}(s, t - \tau_i(t)) + (1 - \delta(t))\mathbb{L}_2 H\bar{x}(s, t - \tau_i(t)) \\
+ \delta(t)\mathbb{L}_3\bar{w}(s,t) + (1 - \delta(t))\mathbb{L}_4\bar{w}(s,t)
\end{cases} \tag{6}
$$

where $\varphi(s) = \begin{bmatrix} \phi(s) & 0 \\ 0 & 0 \end{bmatrix}$, $\mathbb{A} = \begin{bmatrix} A & 0 \\ 0 & \hat{A}_i \end{bmatrix}$, $\mathbb{C}_1 = \begin{bmatrix} 0 \\ \hat{B}_i C \end{bmatrix}$, $\mathbb{C}_2 = \begin{bmatrix} 0 \\ \hat{B}_i EC \end{bmatrix}$, $H = [I \ \ 0]$, $\mathbb{D} = \begin{bmatrix} B_1 & 0 \\ 0 & 0 \end{bmatrix}$, $\mathbb{D}_1 = \begin{bmatrix} 0 & 0 \\ 0 & \hat{B}_i B_2 \end{bmatrix}$, $\mathbb{D}_2 = \begin{bmatrix} 0 & 0 \\ 0 & \hat{B}_i EB_2 \end{bmatrix}$, $\mathbb{L} = [L \ \ -\hat{C}_i]$, $\mathbb{L}_1 = -\hat{D}_i C$, $\mathbb{L}_2 = -\hat{D}_i EC$, $\mathbb{L}_3 = [0 \ \ -\hat{D}_i B_2]$, $\mathbb{L}_4 = [0 \ \ -\hat{D}_i EB_2]$.

To obtain the main results of this paper, we need the following definition and lemmas.

Definition 1. The filtering error system (6) is said to satisfy the H_∞ performance under zero initial condition, if the following inequality holds

$$
\mathbb{E}\{\int_0^\infty \bar{x}^T(s,t)\bar{x}(s,t)dt\} < \gamma^2 \int_0^\infty \bar{w}^T(s,t)\bar{w}(s,t)dt
$$

for a given scalar $\gamma > 0$ and all non-zero $\bar{w}(s,t), t > 0$.

Lemma 1 [12]. For any matrix $\mathbb{M} > 0$, scalars $\tau > 0, \tau(t)$ satisfying $0 \le \tau(t) \le \tau$, vector function $\frac{\partial \bar{x}(s,t)}{\partial t} : [0, l] \times [-\tau, 0] \to R \times R^n$ such that the concerned integrations are well defined, then

$$
-\tau \int_{t-\tau}^t \frac{\partial x^T(s,\xi)}{\partial \xi} \mathbb{M} \frac{\partial x(s,\xi)}{\partial \xi} d\xi \le r^T(s,t)\hat{M}r(s,t),
$$

where $(x,t) = [x^T(s,t) \ \ x^T(s, t - \tau_i(t)) \ \ x^T(s, t - \tau_M)]^T, \hat{M} = \begin{bmatrix} -\mathbb{M} & \mathbb{M} & 0 \\ \mathbb{M}^T & -2\mathbb{M} & 0 \\ 0 & \mathbb{M}^T & -\mathbb{M} \end{bmatrix}$.

Lemma 2 [13]. For any vector $x \in R^n, y \in R^n$ and positive definite matrix S, the following inequality holds:

$$
2x^T y \le x^T S^{-1} x + y^T S y.
$$

Lemma 3 [14]. Given constant matrices A, B, C, where $A^T = A$ and $A^T = A > 0$, then $A + C^T B^{-1} C < 0$ if and only if $\begin{bmatrix} A & C^T \\ C & -B \end{bmatrix} < 0$, or $\begin{bmatrix} -B & C \\ C^T & A \end{bmatrix} < 0$.

3 Main Results

Theorem 1. For given $\gamma > 0$ and $\tau_M > 0$, the filtering error system (6) can satisfy the H_∞ performance γ with the given filter gains, if there exist symmetric positive-definite matrices $P_i, i \in \mathbb{S}, Q, R$, and positive definite matrix S such that the following LMI holds:

$$\Phi = \begin{bmatrix} \Phi_1 & \Phi_2 \\ * & \Phi_4 \end{bmatrix} < 0, \tag{7}$$

$$\Phi_1 = \begin{bmatrix} \Phi_{11} & \Phi_{12} \\ * & \Phi_{14} \end{bmatrix}, \Phi_2 = \begin{bmatrix} \Phi_{21} & \Phi_{22} \\ \Phi_{23} & \Phi_{24} \end{bmatrix},$$

$$\Phi_{11} = \begin{bmatrix} P_i S^{-1} P_i^T + 2P_i^T \mathbb{A} + \sum_{j=1}^{M} \pi_{ij} P_j + H^T Q H - H^T R H + \alpha^2 I & P_i(\delta \mathbb{C}_1 + (1-\delta)\mathbb{C}_2) + H^T R \\ * & -2R \end{bmatrix},$$

$$\Phi_{12} = \begin{bmatrix} 0 & 0 & P_i[\mathbb{D} + \delta \mathbb{D}_1 + (1-\delta)\mathbb{D}_2] \\ R & 0 & 0 \end{bmatrix}, \Phi_{14}$$

$$= \begin{bmatrix} -R - Q & 0 & 0 \\ 0 & \Phi^T(s)S\Phi(s) - I & 0 \\ 0 & 0 & -\gamma^2 I \end{bmatrix},$$

$$\Phi_{21} = \begin{bmatrix} \tau_M \mathbb{A} H^T R & 0 \\ \tau_M(\delta \mathbb{C}_1^T + (1-\delta)\mathbb{C}_2^T)H^T R & \tau_M[\delta(1-\delta)]^{\frac{1}{2}}(\mathbb{C}_1^T - \mathbb{C}_2^T)H^T R \\ 0 & 0 \end{bmatrix},$$

$$\Phi_{22} = \begin{bmatrix} \mathbb{L}^T & 0 \\ \delta \mathbb{L}_1^T + (1-\delta)\mathbb{L}_2^T) & \tau_M[\delta(1-\delta)]^{\frac{1}{2}}(\mathbb{L}_1^T - \mathbb{L}_2^T) \\ 0 & 0 \end{bmatrix},$$

$$\Phi_{23} = \begin{bmatrix} \tau_M \Phi^T(s)H^T R & 0 \\ \tau_M[\mathbb{D}^T + \delta \mathbb{D}_1^T + (1-\delta)\mathbb{D}_2^T]H^T R & \tau_M[\delta(1-\delta)]^{\frac{1}{2}}(\mathbb{D}_1^T - \mathbb{D}_2^T)H^T R \end{bmatrix},$$

$$\Phi_{24} = \begin{bmatrix} 0 & 0 \\ \delta \mathbb{L}_3^T + (1-\delta)\mathbb{L}_4^T & \tau_M[\delta(1-\delta)]^{\frac{1}{2}}(\mathbb{L}_3^T - \mathbb{L}_4^T) \end{bmatrix},$$

$$\Phi_4 = \begin{bmatrix} -R^{-1} & 0 & 0 & 0 \\ 0 & -R^{-1} & 0 & 0 \\ 0 & 0 & -I & 0 \\ 0 & 0 & 0 & -I \end{bmatrix}.$$

Proof. Consider the following Lyapunov-Krasovskii functional:

$$V(\bar{x}(s,t), i, t) = V_1(\bar{x}(s,t), i, t) + V_2(\bar{x}(s,t), i, t) + V_3(\bar{x}(s,t), i, t), \tag{8}$$

where

$$V_1(\bar{x}(s,t), i, t) = \bar{x}^T(s,t) P_i \bar{x}(s,t),$$

$$V_2(\bar{x}(s,t), i, t) = \int_{t-\tau_M}^{t} \bar{x}^T(s, \xi) H^T Q H \bar{x}(s, \xi) d\xi,$$

$$V_3(\bar{x}(s,t), i, t) = \tau_M \int_{-\tau_M}^{t} \int_{t+\theta}^{t} \frac{\partial \bar{x}^T(s,t)}{\partial \xi} H^T R H \frac{\partial \bar{x}(s, \xi)}{\partial \xi} d\xi d\theta,$$

Define the infinitesimal operator $\mathcal{L}$ of $V(\bar{x}(s,t), i, t)$ as follows:

$$\mathcal{L}V(\bar{x}(s,t), i, t) = \lim_{\mu \to 0^+} \frac{1}{\mu} \{\mathbb{E}\{V(\bar{x}(s, t+\mu), i, t+\mu)|t\} - V(\bar{x}(s,t), i, t)\}.$$

The derivative of (8) along the solution of system (6) can be obtained as

$$\mathbb{E}\{\mathcal{L}V_1(\bar{x}(s,t), i, t)\} = \mathbb{E}\{\frac{\partial \bar{x}^T(s,t)}{\partial t} P_i \bar{x}(s,t) + \bar{x}^T(s,t) P_i \frac{\partial \bar{x}(s,t)}{\partial t}\}$$

$$= \{2\bar{x}^T(s,t) P_i [\phi(s) \frac{\partial \bar{x}(s,t)}{\partial s} + A\bar{x}(s,t) + \delta(t)\mathbb{C}_1 H\bar{x}(s, t - \tau_i(t))$$

$$+ (1 - \delta(t))\mathbb{C}_2 H\bar{x}(s, t - \tau_i(t)) + \mathbb{D}\bar{w}(s,t) + \delta(t)\mathbb{D}_1\bar{w}(s,t) + (1 - \delta(t))\mathbb{D}_2\bar{w}(s,t)]\}, \tag{9}$$

$$\mathbb{E}\{\mathcal{L}V_2(\bar{x}(s,t), i, t)\} = \mathbb{E}\{\bar{x}^T(s,t) H^T Q H\bar{x}(s,t) - \bar{x}^T(s, t - \tau_M) H^T Q H\bar{x}(s, t - \tau_M)\}, \tag{10}$$

$$\mathbb{E}\{\mathcal{L}V_3(\bar{x}(s,t), i, t)\} = \mathbb{E}\{\tau_M^2 \frac{\partial \bar{x}^T(s,t)}{\partial t} H^T R H \frac{\partial \bar{x}(s,t)}{\partial t}$$

$$- \tau_M \int_{t-\tau_M}^{t} \frac{\partial \bar{x}^T(s, \xi)}{\partial \xi} H^T R H \frac{\partial \bar{x}(s, \xi)}{\partial \xi} d\xi\}, \tag{11}$$

By virtue of Lemma 1, it follows that

$$- \tau_M \int_{t-\tau_M}^{t} \frac{\partial \bar{x}^T(s, \xi)}{\partial \xi} H^T R H \frac{\partial \bar{x}(s, \xi)}{\partial \xi} d\xi \le \begin{bmatrix} \bar{x}(s,t) \\ H\bar{x}(s, t - \tau_i(t)) \\ H\bar{x}(s, t - \tau_M) \end{bmatrix}^T$$

$$\begin{bmatrix} -H^T R H & H^T R & 0 \\ R^T H & -2R & 0 \\ 0 & R^T & -R \end{bmatrix}^T \begin{bmatrix} \bar{x}(s,t) \\ H\bar{x}(s, t - \tau_i(t)) \\ H\bar{x}(s, t - \tau_M) \end{bmatrix}. \tag{12}$$

By Lemma 2, we have

$$2\bar{x}^T(s,t)P_i[\phi(s)\frac{\partial\bar{x}(s,t)}{\partial s} \leq \bar{x}^T(s,t)P_iS^{-1}P_i^T\bar{x}(s,t) + \frac{\partial\bar{x}^T(s,t)}{\partial s}\varphi^T(s)S\varphi(s)\frac{\partial\bar{x}(s,t)}{\partial s},$$

(13)

By Assumption 1, we have

$$0 \leq -\frac{\partial x^T(s,t)}{\partial s}\frac{\partial x(s,t)}{\partial s} + \alpha^2 x^T(s,t)x(s,t).$$

Note that

$$\mathbb{E}\{\tau_M^2\frac{\partial\bar{x}^T(s,t)}{\partial t}H^TRH\frac{\partial\bar{x}(s,t)}{\partial t}\} = \mathbb{E}\{\hat{v}^T(t)\hat{\Xi}\hat{v}(t)\},$$

where

$$\hat{\Xi} = \begin{bmatrix} \tau_M\mathbb{A}H^T \\ \tau_M(\delta\mathbb{C}_1^T + (1-\delta)\mathbb{C}_2^T)H^T \\ 0 \\ \tau_M\varphi^T(s)H^T \\ \tau_M[\delta(1-\delta)]^{\frac{1}{2}}(\mathbb{D}_1^T - \mathbb{D}_2^T)H^T \end{bmatrix} R \begin{bmatrix} \tau_M\mathbb{A}H^T \\ \tau_M(\delta\mathbb{C}_1^T + (1-\delta)\mathbb{C}_2^T)H^T \\ 0 \\ \tau_M\varphi^T(s)H^T \\ \tau_M[\delta(1-\delta)]^{\frac{1}{2}}(\mathbb{D}_1^T - \mathbb{D}_2^T)H^T \end{bmatrix}^T$$

$$+ \begin{bmatrix} 0 \\ \tau_M[\delta(1-\delta)]^{\frac{1}{2}}(\mathbb{C}_1^T - \mathbb{C}_2^T)H^T \\ 0 \\ 0 \\ \tau_M\varphi^T(s)H^T \end{bmatrix} R \begin{bmatrix} \tau_M\mathbb{A}H^T \\ \tau_M(\delta\mathbb{C}_1^T + (1-\delta)\mathbb{C}_2^T)H^T \\ 0 \\ 0 \\ \tau_M\varphi^T(s)H^T \end{bmatrix}^T,$$

and $\hat{v}(t) = col[\bar{x}^T(s,t),\bar{x}^T(s,t-\tau_j(t))H^T,\bar{x}^T(s,t-\tau_M)H^T,\frac{\partial\bar{x}^T(s,t)}{\partial t},\bar{w}(x,t)]^T$. Then, we have

$$\mathbb{E}\{\mathcal{L}V_3(\bar{x}(s,t),i,t)\} \leq v^T(t)(\Xi + \begin{bmatrix} \tau_M\mathbb{A}H^T \\ \tau_M(\delta\mathbb{C}_1^T + (1-\delta)\mathbb{C}_2^T)H^T \\ 0 \\ \tau_M\varphi^T(s)H^T \end{bmatrix} R \begin{bmatrix} \tau_M\mathbb{A}H^T \\ \tau_M(\delta\mathbb{C}_1^T + (1-\delta)\mathbb{C}_2^T)H^T \\ 0 \\ \tau_M\varphi^T(s)H^T \end{bmatrix}^T$$

$$+ \begin{bmatrix} 0 \\ \tau_M[\delta(1-\delta)]^{\frac{1}{2}}(\mathbb{C}_1^T - \mathbb{C}_2^T)H^T \\ 0 \\ 0 \end{bmatrix} R \begin{bmatrix} \tau_M\mathbb{A}H^T \\ \tau_M(\delta\mathbb{C}_1^T + (1-\delta)\mathbb{C}_2^T)H^T \\ 0 \\ 0 \end{bmatrix}^T)v(t),$$

(14)

where

$$v(t) = col[\bar{x}^T(s,t), \bar{x}^T(s,t-\tau_i(t))H^T, \bar{x}^T(s,t-\tau_M)H^T, \frac{\partial \bar{x}^T(s,t)}{\partial t}]^T, \varXi = \begin{bmatrix} \varXi_{11} & \varXi_{12} \\ * & \varXi_{22} \end{bmatrix},$$

$$\varXi_{11} = \begin{bmatrix} P_i S^{-1} P_i^T + 2P_i^T \mathbb{A} + \sum_{j=1}^M \pi_{ij} P_j + H^T Q H - H^T R H - \alpha^2 I & P_i(\delta \mathbb{C}_1 + (1-\delta)\mathbb{C}_2) + H^T R \\ * & -2R \end{bmatrix},$$

$$\varXi_{12} = \begin{bmatrix} 0 & 0 \\ R & 0 \end{bmatrix}, \varXi_{22} = \begin{bmatrix} -R-Q & 0 \\ 0 & \varphi^T(s)S\varphi(s) - I \end{bmatrix},$$

Obviously, (7) indicates $\mathbb{E}\{\mathcal{L}V(\bar{x}(s,t),i,t)\} < 0$. That is to say, the filtering error system with $\bar{w}(s,t) = 0$ is mean square stochastic stable.

Furthermore, by Lemma 3, we have

$$\mathbb{E}\{\mathcal{L}V(\bar{x}(s,t),i,t)\} + \mathbb{E}\{\tilde{z}^T(s,t)\tilde{z}(s,t)\} - \gamma^2 \bar{w}^T(x,t)\bar{w}(x,t) \leq$$

$$\mathbb{E}\{\hat{v}^T(t)(\tilde{\varXi} + \begin{bmatrix} \mathbb{L}^T \\ \delta \mathbb{L}_1^T + (1-\delta)\mathbb{L}_2^T) \\ 0 \\ 0 \\ \delta \mathbb{L}_3^T + (1-\delta)\mathbb{L}_4^T) \end{bmatrix} I \begin{bmatrix} \mathbb{L}^T \\ \delta \mathbb{L}_1^T + (1-\delta)\mathbb{L}_2^T) \\ 0 \\ 0 \\ \delta \mathbb{L}_3^T + (1-\delta)\mathbb{L}_4^T) \end{bmatrix}^T \tag{15}$$

$$+ \begin{bmatrix} 0 \\ \tau_M[\delta(1-\delta)]^{\frac{1}{2}}(\mathbb{C}_1^T - \mathbb{C}_2^T)H^T \\ 0 \\ 0 \\ \tau_M \varphi^T(s)H^T \end{bmatrix} I \begin{bmatrix} \tau_M \mathbb{A} H^T \\ \tau_M(\delta \mathbb{C}_1^T + (1-\delta)\mathbb{C}_2^T)H^T \\ 0 \\ 0 \\ \tau_M \varphi^T(s)H^T \end{bmatrix}^T)\hat{v}(t)\},$$

where

$$\tilde{\varXi}_1 = \begin{bmatrix} \tilde{\varXi}_{11} & \tilde{\varXi}_{12} \\ * & \tilde{\varXi}_{14} \end{bmatrix},$$

$$\tilde{\varXi}_{11} = \begin{bmatrix} P_i S^{-1} P_i^T + 2P_i^T \mathbb{A} + \sum_{j=1}^M \pi_{ij} P_j + H^T Q H - H^T R H - \alpha^2 I & P_i(\delta \mathbb{C}_1 + (1-\delta)\mathbb{C}_2) + H^T R \\ * & -2R \end{bmatrix},$$

$$\tilde{\varXi}_{12} = \begin{bmatrix} 0 & 0 & P_i[\mathbb{D} + \delta \mathbb{D}_1 + (1-\delta)\mathbb{D}_2] \\ R & 0 & 0 \end{bmatrix},$$

$$\tilde{\varXi}_{14} = \begin{bmatrix} -R-Q & 0 & 0 \\ 0 & \varphi^T(s)S\varphi(s) - I & 0 \\ 0 & 0 & -\gamma^2 I \end{bmatrix},$$

$$\tilde{\Xi}_2 = \begin{bmatrix} \tilde{\Xi}_{21} \\ \tilde{\Xi}_{22} \end{bmatrix}, \tilde{\Xi}_{21} = \begin{bmatrix} \tau_M \mathbb{A} H^T R & 0 \\ \tau_M(\delta \mathbb{C}_1^T + (1-\delta)\mathbb{C}_2^T)H^T R & \tau_M[\delta(1-\delta)]^{\frac{1}{2}}(\mathbb{C}_1^T - \mathbb{C}_2^T)H^T R \\ 0 & 0 \end{bmatrix},$$

$$\tilde{\Xi}_{23} = \begin{bmatrix} \tau_M \varphi^T(s)H^T R & 0 \\ \tau_M[\mathbb{D}^T + \delta \mathbb{D}_1^T + (1-\delta)\mathbb{D}_2^T]H^T R & \tau_M[\delta(1-\delta)]^{\frac{1}{2}}(\mathbb{D}_1^T - \mathbb{D}_2^T)H^T R \end{bmatrix},$$

By Lemma 3, $\Phi < 0$ yields

$$\mathbb{E}\{\mathcal{L}V(t)(\bar{x}(s,t),i,t)\} + \mathbb{E}\{\tilde{z}^T(x,t)\tilde{z}(x,t)dx\} - \gamma^2 \int_0^l \delta^T(x,t)\delta(x,t) < 0.$$

Then, it is not difficult to obtain that the error system (6) satisfies the H_∞ performance under zero initial condition. This completes the proof.

Based on the results of Theorem 1, the filter parameter design is given in the following theorem.

Theorem 2. For given $\gamma > 0$ and $\tau_M > 0$, the filtering error system (6) can satisfy the H_∞ performance γ with the given filter gains, if there exist symmetric positive-definite matrices $X_i, i \in \mathbb{S}, Y_i, i \in \mathbb{S}, P_i, i \in \mathbb{S}, Q, R$ and positive definite matrix S, matrices $\tilde{A}, \tilde{B}, \tilde{C}, \tilde{D}, W_i, i \in \mathbb{S}, U_{1i}, i \in \mathbb{S}$, nonsingular matrices $V_i, i \in \mathbb{S}$, such that the following LMI holds:

$$\hat{\Phi} = \begin{bmatrix} \hat{\Phi}_1 & \hat{\Phi}_2 \\ * & \hat{\Phi}_4 \end{bmatrix} < 0, \tag{16}$$

where

$$\hat{\Phi}_1 = \begin{bmatrix} \hat{\Phi}_{11} & \hat{\Phi}_{12} \\ * & \hat{\Phi}_{14} \end{bmatrix}, \hat{\Phi}_{11} = \begin{bmatrix} \hat{\Phi}_{111} & \begin{bmatrix} R + \delta \tilde{B}_i C + (1-\delta)\tilde{B}_i EC \\ \delta \tilde{B}_i C + (1-\delta)\tilde{B}_i EC \end{bmatrix} \\ * & -2R \end{bmatrix},$$

$$\hat{\Phi}_{111} = \begin{bmatrix} J_1 + 2X_i A & J_2 + \tilde{A}_i \\ J_3 Y_i A & J_4 + \tilde{A}_i \end{bmatrix}, \hat{\Phi}_{14} = \begin{bmatrix} -R-Q & 0 & 0 \\ 0 & \varphi^T(s)S\varphi(s) - I & 0 \\ 0 & 0 & -\gamma^2 I \end{bmatrix},$$

$$\hat{\Phi}_{12} = \begin{bmatrix} 0 & \begin{bmatrix} 0 \\ 0 \end{bmatrix} & \begin{bmatrix} X_i B_1 & \delta \tilde{B}_i B_2 + (1-\delta)\tilde{B}_i EB_2 \\ V_i B_1 & \delta W_i V_i^{-1} \tilde{B}_i B_2 + (1-\delta)W_i V_i^{-1} \tilde{B}_i EB_2 \end{bmatrix} \\ R & 0 & 0 \end{bmatrix},$$

$$\hat{\Phi}_2 = \begin{bmatrix} \hat{\Phi}_{21} & \hat{\Phi}_{22} \\ \hat{\Phi}_{23} & \hat{\Phi}_{24} \end{bmatrix},$$

$$\hat{\Phi}_{21} = \begin{bmatrix} \begin{bmatrix} \tau_M X_i A^T R \\ \tau_M V_i^T A^T R \\ 0 \end{bmatrix} & 0 \\ & 0 \end{bmatrix}, \hat{\Phi}_{23} = \begin{bmatrix} 0 & 0 \\ \tau_M \varphi^T(s) H^T R^T & 0 \\ \begin{bmatrix} \tau_M B_1^T R \\ 0 \end{bmatrix} & 0 \end{bmatrix},$$

$$\hat{\Phi}_4 = \begin{bmatrix} -R^{-1} & 0 & 0 & 0 \\ 0 & -R^{-1} & 0 & 0 \\ 0 & 0 & -I & 0 \\ 0 & 0 & 0 & -I \end{bmatrix},$$

$$\hat{\Phi}_{22} = \begin{bmatrix} \begin{bmatrix} X_i L^T - \tilde{C}_i \\ V_i^T L^T - W_i V_i^{-1} \tilde{C}_i \end{bmatrix} & 0 \\ -\delta C^T \tilde{D}_i^T - (1-\delta) C^T E^T \tilde{D}_i^T & \tau_M[\delta(1-\delta)]^{\frac{1}{2}}(-C^T \tilde{D}_i^T + C^T E^T \tilde{D}_i^T) \end{bmatrix},$$

$$\hat{\Phi}_{24} = \begin{bmatrix} 0 & 0 \\ 0 & 0 \\ -\delta B_2^T \tilde{D}_i^T - (1-\delta) B_2^T E^T \tilde{D}_i^T & \tau_M[\delta(1-\delta)]^{\frac{1}{2}}(-B_2^T \tilde{D}_i^T + B_2^T E^T \tilde{D}_i^T) & 0 \end{bmatrix},$$

$$\begin{bmatrix} J_1 & J_2 \\ J_3 & J_4 \end{bmatrix} = \mathbb{J}^T [P_i S^{-1} P_i^T + \sum_{j=1}^{M} \pi_{ij} P_j + H^T Q H - H^T R H + \alpha^2 I] \mathbb{J}.$$

Furthermore, the desired filter gains can be obtained as

$$\hat{A}_i = 2V_i^{-1} \tilde{A}_i V_i^{-1} W_i, \hat{B}_i = V_i^{-1} \tilde{B}_i, \hat{C}_i = V_i^{-1} \tilde{C}_i, \hat{D}_i = \tilde{D}_i. \tag{17}$$

Proof. Assume that $P_i = \begin{bmatrix} X_i & V_i \\ * & W_i \end{bmatrix}$, where $0 < W_i, i \in \mathbb{S}$ are a series of matrices, and $V_i, i \in \mathbb{S}$ are a series of nonsingular matrices. For any $Y_i, i \in \mathbb{S}$, let $Y_i = V_i W_i^{-1} V_i^T$.

In addition, we have $X_i - Y_i > 0$ and $P_i > 0$. To facilitate the calculation, we choose the following auxiliary variables:

$$\tilde{A}_i = V_i \hat{A}_i W_i^{-1} V_i^T, \tilde{B}_i = V_i \hat{B}_i^T, \tilde{C}_i = V_i \hat{C}_i^T, \tilde{D}_i = \hat{D}_i, \tag{18}$$

Construct the matrix

$$\mathbb{J}_i = \begin{bmatrix} I & 0 \\ 0 & W_i^{-1} V_i^T \end{bmatrix}. \tag{19}$$

Applying the congruent transformation $diag\{\mathbb{J}_i^T, I, I, I, I, I, I, I, I\}$ to the above inequality leads to the inequality (7) in Theorem 1, which means the error system (6) is mean square stochastic stable and the H_∞ performance under zero initial condition is satisfied. Now, this proof is completed.

Remark 1. The filtering problem is also discussed in [6, 7], which are all devoted into ODEs. Compared with [6, 7], we consider the filtering problem of DPSs which is filled with challenges. In addition, the results we obtained can be applied to the case with stochastic switching communication channels. One can choose different channel to fulfill the performance according to stochastic switching rule, which is more flexible and practical.

4 A Numerical Example

In this section, we give a practical numerical example to demonstrate the effectiveness of our proposed H_∞ filtering scheme. We consider the case that a large number of people with different hight chase each other in a line, which can be presented by the following continuous model:

$$\begin{cases} \frac{\partial x(s,t)}{\partial t} = 0.5\sin(s)\frac{\partial x(s,t)}{\partial s} + Ax(s,t) + B_1w(s,t) \\ y(s,t) = Cx(s,t) + B_2w(s,t) \\ z(s,t) = Lx(s,t) \end{cases} \tag{20}$$

where $x(s,t)$ represents the height of the people which is located in s at instant t. The model parameters are chosen by

$$A = -2; B_1 = 0.1; B_2 = 0.1; C = 0.8; L = 0.9.$$

Then, take the filter (5) with the following parameters $\hat{A}_1 = -4.4444; \hat{A}_2 = -4,2323;$ $\hat{B} = -0.3333; \hat{B}_2 = 0.4112;$ $\hat{C}_1 = -0.3333; \hat{C}_2 = -0.3644;$ $\hat{D}_1 = 0.1; \hat{D}_2 = 0.12;$ $\mathbb{E}\delta(t) = 0.4; \tau_1(t) = 0.1\sin(t); \tau_2(t) = 0.1\cos(t); E = 0.2; \Pi = \begin{bmatrix} 0.25 & 0.75 \\ 0.35 & 0.65 \end{bmatrix}.$

In order to check out the H_∞ performance with the given filter gains for error system (6), we pick up $P_1 = \begin{bmatrix} 1 & -0.6 \\ -0.6 & 0.8 \end{bmatrix}; P_2 = \begin{bmatrix} 1.2 & -0.3 \\ -0.3 & 1.1 \end{bmatrix}; Q = 0.1; R = 1;$ $S = 1; \gamma = 0.8.$

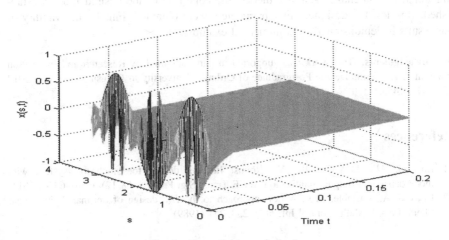

Fig. 1. The trajectory of x(s, t)

According to Theorem 2, and using Matlab LMI toolbox, it is not difficult to obtain $\hat{\Phi} < 0$, i.e., error system (6) can satisfy the H_∞ performance γ with the given filter gains.

Figures 1 and 2 show that the states $x(s, t)$, and $\hat{x}(s, t)$ of the error system satisfy the H_∞ performance.

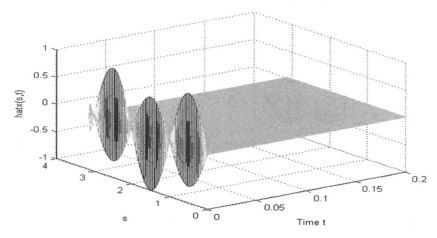

Fig. 2. The trajectory of $\hat{x}(s, t)$

5 Conclusion

In this paper, we concentrate on the H_∞ filtering problem for a class of time-delayed distributed parameter systems with randomly occurring sensor faults. A stochastic variable satisfying the Bernoulli random distribution is employed to describe the case of measurement failures. Based on a Markovian switched Lyapunov-Krasovskii functional and stochastic stability theory, the condition of the desired filter is established. Besides, the designed filter parameters were obtained. Finally, the validity of our results is demonstrated by a numerical example.

Acknowledgments. This work was supported in part supported by Research and Innovation Program of Academic Degree Postgraduate of Ordinary University in Jiangsu province in 2017 KYCX17_1455.

References

1. Liu, M., Ho, D.W.C., Niu, Y.: Robust filtering design for stochastic system with mode-dependent output quantization. IEEE Trans. Sig. Process. **58**(12), 6410–6416 (2011)
2. Elsayed, A., Grimble, M.J.: A new approach to the H_∞ design of optimal digital linear filters. IMA J. Math. Control Inf. **6**(2), 233–251 (1989)

3. Shen, B., Wang, Z., Shu, H., Wei, G.: Robust H_∞ finite-horizon filtering with randomly occurred nonlinearities and quantization effects. Automatica **46**(46), 1743–1751 (2010)
4. Karimi, H.R.: Robust H_∞ filter design for uncertain linear systems over network with network-induced delays and output quantization. Model. Ident. Control **30**(1), 27–37 (2009)
5. Wang, W.-W., Yang, J.-L., Yang, G.-H.: Quantised H_∞ filtering for networked systems with random sensor packet losses. IET Control Theory Appl. **4**(8), 1339–1352 (2010)
6. Wang, P.P., Che, W.W.: Quantized H_∞ filter design for networked control systems with random nonlinearity and sensor saturation. Neurocomputing **193**, 14–19 (2016)
7. Li, A., Yi, S., Wang, X.: New reliable H_∞ filter design for networked control systems with external disturbances and randomly occurring sensor faults. Neurocomputing **185**, 21–27 (2015)
8. Liu, Y., Wang, Z., Wang, W.: Reliable H_∞ filtering for discrete time-delay systems with randomly occurred nonlinearities via delay-partitioning method. Sig. Process. **91**(4), 713–727 (2011)
9. Alvarez, J.D., Normey-Rico, J.E., Berenguel, M.: Design of PID controller with filter for distributed parameter systems. IFAC Proc. Volumes **45**(3), 495–500 (2012)
10. Demetriou, M.A.: Spatial PID consensus controllers for distributed filters of distributed parameter systems. Syst. Control Lett. **63**(1), 57–62 (2014)
11. Demetriou, M.A.: Design of consensus and adaptive consensus filters for distributed parameter systems. Pergamon Press, Inc. (2010)
12. Park, P.G., Ko, J.W., Jeong, C.: Reciprocally convex approach to stability of systems with time-varying delays. Pergamon Press, Inc. (2011)
13. Wang, J.L., Yang, Z.C., Wu, H.N.: Passivity analysis of complex dynamical networks with multiple time-varying delays. J. Eng. Math. **74**(1), 175–188 (2012)
14. Gu, K., Chen, J., Kharitonov, V.: Stability of Time-Delay Systems. Birkhauser, Boston (2003)

The Study of the Seabed Side-Scan Acoustic Images Recognition Using BP Neural Network

Hongyan Xi, Lei Wan, Mingwei Sheng$^{(\boxtimes)}$, Yueming Li, and Tao Liu

Science and Technology on Underwater Vehicle Laboratory,
Harbin Engineering University, No.145 Nantong Avenue, Harbin, China
smwsky@163.com

Abstract. In recent years, mankind has made great achievements in the marine exploration. Ocean contains abundant resources, and the seabed has recorded amount of basic Earth information. Therefore, a complete study of the seabed can help to form a full appreciation of underwater environment. The study of the seabed recognition method, as the most basic work of the study of the seabed, is gradually gaining the attention of researchers. As a main marine exploratory tool, the side-scan sonar is fast, accurate and convenient for seabed information collection. In this paper, lots of seabed acoustic images were applied to extract the seabed substrate characteristics using the gray covariance matrix method. An improved BP neural network model was involved into classify and identify the seabed characteristics. In addition, several algorithms for BP neural network were proposed for testing the recognition accuracy of side-scan acoustic images and the convergence rate. The results show that although several algorithms were easy to fall into the minimum value during training, which can lead to slow convergence rate and unable to meet the recognition accuracy standard, the trainlm function had a faster convergence rate and higher recognition accuracy.

Keywords: Seabed side-scan acoustic images · Texture feature extraction · BP neural network · Recognition accuracy

1 Introduction

The seabed sediment type is an important marine environmental parameter. The distribution of the sediment type is of great significance to marine scientific research, marine engineering and national defense construction [1]. However, with the large-scale development of marine resources, the formation of the sea area to form a comprehensive grasps. Traditional seabed detection methods are labor intensive, high operating costs, and it is difficult to carry out large-scale seabed recognition.

Seabed side-scan acoustic image recognition is an important technical tool to understand the physical properties through the acoustic properties of seabed sediments, providing a quick and reliable method for the recognition of seabed types [2]. Sonar is often used for classification of seabed type, as there is a strong link between sonar backscatter and sediment characteristics of the seabed [3]. As an important feature of image, texture is widely used in image classification and retrieval areas. Texture can be seen as a repetitive pattern formed by a certain rule or primitive arrangement that

© Springer Nature Singapore Pte Ltd. 2017
G. Chen et al. (Eds.): PAAP 2017, CCIS 729, pp. 130–141, 2017.
DOI: 10.1007/978-981-10-6442-5_12

reflects the subsurface properties of the seabed and has been used in many studies for the classification of sonar images [4–7].

In recent years, the extensive application of neural networks has proposed a new method for the recognition of seabed subsets. Learning Vector Quantisation (LVQ) is a supervised learning algorithm of ANN that is found to be an effective tool and show good performance. The network was tried with a different size of hidden neurons and training data size to see the influence on classification [8]. A novel sampling strategy, namely polygon-based random sampling (PBRS), which maintains the complete independence of sampled data sets for training and testing, was proposed to generate more realistic landslide susceptibility maps. An ASTER image of the Candir catchment area which is located in western Antalya was selected for implementing the proposed approach using a support vector machine classification (SVM) algorithm [9, 10]. An adaptive genetic instance selection algorithm (AGISA) is proposed for underwater acoustic target classification. The AGISA is proposed to address the problem that the classification performance in classifying underwater acoustic targets declines and becomes unstable [11]. The original BP neural network has the shortcomings of slow convergence, low precision and easy to fall into the local minimum. An improved particle swarm optimization (PSO) algorithm is proposed to optimize the BP neural network. In this new algorithm, the PSO uses the improved adaptive acceleration factor and the improved adaptive inertia weight to improve the initial weighting and threshold of the BP neural network [12–16].

2 Principle and Algorithm

2.1 The Representing Method of Texture Features of Seabed Side-Scan Acoustic Images

As underwater detection equipment, the side-scan sonar is applied for submarine topography and underwater objects detection and recognition by scattered acoustic echo. The transducers of side-scan sonar are mounted on both sides of the ship with transmitting fan-shaped beams. In this paper, the texture features of the seabed side-scan acoustic images including four sorts of seabed sediment are extracted using gray covariance matrix. The contrast, the correlation, the energy and the homogeneity of each picture were extracted as characteristic sat 0°, 45°, 90°, and 135°, respectively (Fig. 1).

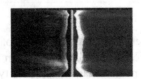

Fig. 1. Side-scan acoustic image

Texture feature extraction

Before the texture feature extracting, four sorts of seabed was chosen (Fig. 2), and the number of extracted features for each group of texture features is set as 80. The graycomatrix is used to generate the gray level co-occurrence matrices in the four directions of 0°, 45°, 90° and 135°, and the required texture feature is extracted by the graycoprops correlation function: Contrast, correlation, energy and homogeneity. The grayscale images of each selected substrate are 48 × 48 pixels, corresponding to a 48 × 48 matrix, set the gray level $L = 16$, keeping other parameters unchanged.

(a) First sort   (b) Second sort   (c) Third sort   (d) Fourth sort

Fig. 2. Four sorts of seabed side-scan acoustic images

The initial texture feature data extracted by the gray covariance matrix correlation function is more cluttered. For the four angles of 0°, 45°, 90° and 135°, the contrast, correlation, energy and homogeneity, the data were statistically compared to understand the difference in texture between the different seabed side-scan acoustic images.

- Contrast

Contrast refers to the measurement of different brightness levels between the brightest and the darkest in a dark area of the image. The contrast statistic of the gray matter images of the seabed is shown in Table 1, and the statistical histogram is shown in Fig. 3. It can be seen from the image that the contrast of the four sorts of gray scale images has similar distributions at 0°, 45°, 90° and 135°: the values from the first type of the substrate to the fourth type, the contrast feature followed by decreasing.

Table 1. Contrast data

Correlation	0°	45°	90°	135°
First sort	10.55	12.7	10.25	13.82
Second sort	7.89	11.97	8.16	11.85
Third sort	6.19	8.97	5.85	9.18
Forth sort	3.02	5.59	3.83	5.75

- Correlation

Correlation reflects the consistency of the gray levels of the pixels within the local range (Table 2), and the statistical histogram is shown in Fig. 4.

The distribution of the gray matter image of the four types of collars is opposite to that of the above four angles: from the first type of the substrate to the fourth type of sediment, the value of the correlation feature is gradually increased, contrast is more obvious.

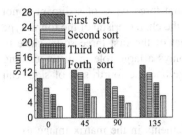

Fig. 3. Statistical histogram contrast

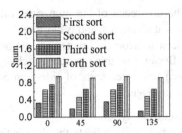

Fig. 4. Statistical histogram correlation

Table 2. Correlation data statistics

Correlation	0°	45°	90°	135°
First sort	0.33	0.19	0.35	0.13
Second sort	0.64	0.46	0.63	0.47
Third sort	0.76	0.65	0.77	0.64
Forth sort	0.95	0.91	0.94	0.91

- Energy

Energy reflects the uniformity of the small units distributed in the whole image and the texture thickness. The correlation statistic of the seabed side-scan acoustic images is shown in Table 3, and the statistical histogram is shown in Fig. 5.

Table 3. Energy data statistics

Correlation	0°	45°	90°	135°
First sort	0.01072	0.01056	0.01175	0.08829
Second sort	0.01036	0.00914	0.01014	0.07991
Third sort	0.01076	0.01044	0.01209	0.08546
Forth sort	0.01025	0.00918	0.01007	0.07972

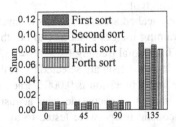

Fig. 5. Statistical histogram energy

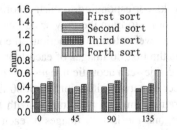

Fig. 6. Statistical histogram homogeneity

The difference of the numerical values of the first three types of sediment is not obvious at all angles, and is close to 0.01. However, the characteristic of the fourth type of side-scan acoustic images are almost 8 times that of the first three sorts, close to 0.08. Because of this near-abrupt change, energy plays a large role in distinguishing between first three sorts of side-scan acoustic image and the fourth sort of side-scan acoustic image.

- Homogeneity

The homogeneity reflects the distribution of the elements in the matrix image to the diagonal tightness. The homogeneity statistics of the seabed side-scan acoustic images are shown in Table 4, and the statistical histogram is shown in Fig. 6.

Table 4. Homogeneity data statistics

Correlation	0°	45°	90°	135°
First sort	0.389	0.373	0.392	0.362
Second sort	0.446	0.394	0.441	0.395
Third sort	0.478	0.435	0.486	0.432
Forth sort	0.707	0.651	0.689	0.649

It can be seen from the histogram that the statistical properties of homogeneity are similar to the statistical characteristics of energy as a whole, and the values are distinguished by the first three types of side-scan acoustic image and the fourth type of sediment.

In summary, the above-mentioned statistical characteristics of the texture: contrast, correlation, energy and homogeneity, can be convenient to sort the side-scan acoustic image recognition.

3 Recognition Testing and Results Analysis

Four texture feature parameters, such as contrast, correlation, energy and homogeneity, were extracted from the gray level co-occurrence matrix. BP neural network model is established with four texture feature parameters as input vectors.

There are twelve training functions of BP neural network algorithm used, and different convergence methods adopted. Therefore, the convergence rate of neural networks corresponding to each algorithm will be different.

In order to train the gray scale images of the four seabed side-scan acoustic image with the twelve training functions, the number of training images is set to 1 to 80, that means the first 80 images of each type network training, total training (4 × 80 =) 320 seabed side-scan acoustic images. As training, logistic and transig are selected to set the number of network iterations to 5000 times and the target precision is 0.0001. Each of the training images was trained with the first 80 images of each side-scan acoustic image, and the last 40 images of each side-scan acoustic image were tested to avoid duplication of the testing object and the training object. The testing results were subjected to multiple tests, it is found that BP neural network is not unique to the recognition of seabed sediment images, but is influenced by many factors. The following is a brief analysis of a single controlled trial to determine the conditions required for a formal statistical testing.

3.1   The Impact of the Number of Tests on the Recognition Accuracy

The basic parameters are invariant. Trainscg training function is used as an example. Each training function is trained with the first 80 images of each side-scan acoustic image, and the testing is performed with the last 40 images of each side-scan acoustic image.

The testing found that: each time the results of the operation of the neural network is not the same, and the testing results of the larger value span, does not meet the testing requirements. This is determined by the structure of the neural network. Therefore, in the testing, each training function should be tested several times to obtain the average of the side-scan acoustic image recognition accuracy to offset the randomness of a single testing. Based on the above conjecture, the trainscg function was used to testing each training function with the first 80 images of each side-scan acoustic image, and the testing was performed with the last 40 images of each side-scan acoustic image. The average recognition accuracy of the statistical algorithm, as a set of tests, the testing results is shown in Table 5.

Table 5. Testing Results

Group No.	Average recognition accuracy	Group No.	Average recognition accuracy
1	89.00%	2	88.69%
3	88.44%	4	86.63%
5	87.94%	6	88.88%
7	89.94%	8	89.50%
9	90.06%	10	88.56%

From the above table we can see that each group of side-scan acoustic image recognition accuracy is basically the same, fluctuations within the acceptable range, indicating that the recognition accuracy is basically stable. The results are shown in Fig. 7.

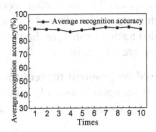

Fig. 7. Ten times testing average recognition accuracy line chart

Therefore, with 10 tests as a group, the average value of the statistical image recognition accuracy can basically offset the influence of the instability of the neural network on the recognition accuracy. The number variation of tests can be used as a standard testing.

3.2   The Effect of the Number Variation of Training Images on the Recognition Accuracy

According to the above analysis, the first 80 images of each side-scan acoustic image were trained and the last 40 images of each side-scan acoustic image were tested. The average recognition accuracy of the side-scan acoustic image was evaluated by 10 tests as a group, and the test results meet the requirements. But taking into account the meaning of the training by the neural network speculation, the difference of the number of test images will lead to the difference of the final recognition accuracy. So the following is the experiment of the number of test image: In the case of trainscg function, the initial parameters are unchanged, and the number of images for neural network training is set for the first 20, 40, 60 and 80 images of each side-scan acoustic image. The images of the testing neural network are still The final 40 images of the endoscopic images were not changed, and the same was tested for 10 times in each case, and the average recognition accuracy of the side-scan acoustic images was counted with 10 times. The testing data table is shown in Table 6.

Table 6. Testing data statistics

Number of testing images	Average recognition accuracy
20	72.75%
40	71.13%
60	71.38%
80	88.56%

The results show that as the number of testing images is 20, 40 and 60, the neural network has little effect on recognition accuracy of the side-scan acoustic image; as the number of testing images reaches 80, the recognition accuracy increases greatly.

3.3   Statistical Analysis of Network Testing Recognition Accuracy

According to the previous description, the recognition accuracy of the neural network on the gray scale image of each submarine is affected by the number of image training. This section will set up a detailed testing to analyze the twelve training functions of BP network. The parameter settings are shown in Table 7.

Training function testing based on gradient descent method
The gradient descent method is an optimization algorithm, often called the steepest descent method. It is one of the simplest and oldest methods of optimization of unconstrained problem solving.

Table 7. Preferences table

Parameter	Setting value
Number of training images	10–80
The total number of training images	40–320
Defined gray scale	16
Neural network activation function type	tansig, logsig
Maximum number of iterations of the BP network	5000
Target accuracy of BP network	0.0001
The number of each sort of testing images	81–120

- Gradient descent method–traingd training function

As the testing results show that the traingd function is applied to the network, it is still impossible to achieve the target accuracy by 5000 times, and the trend of the network is almost zero as the iteration is about 4500 times, and the convergence error of the network is very slow. This is because the training function traingd is a simple gradient descent training function, training speed is relatively slow and easy to fall into the local minimum situation.

- Other training function testing

Similar to traingd, traingdm, traingda, traingdx, trainrp the training function based on gradient descent cannot achieve the target precision in the preset 5000 iterations. Although the convergence rate is improved compared with the traingd function, the convergence rate is still over and slows, and the descending gradient has not reached the target. These data show that the gradient descent method is slow in close proximity to the minimum value, so it is not suitable for fast recognition of seabed side-scan acoustic images.

Training function testing based on conjugate gradient method

- Fletcher-Reeves conjugate gradient method–traincgf training function

As the number of samples subjected to the testing increased from 10 to 60, the recognition accuracy remained fluctuating. As the number of side-scan acoustic images starts from 60, the recognition accuracy of the neural network shows a steady upward trend. As the number of testing images reach to 80, the recognition accuracy reached the maximum, about 90%, the details are shown in Fig. 8.

- Ploak-Ribiere conjugate gradient method–triancgp training function

As the number of side-scan seabed acoustic images participating in the testing increased from 10 to 25, the recognition accuracy of the BP neural network shows a gentle upward trend. As the number of samples in the testing was between 25 and 55, the whole recognition accuracy is fluctuating in the upper and lower states, and the whole is not increased obviously. The number of the samples in the testing is 60, and the recognition accuracy of the neural network shows a steady upward trend. And the number of testing images reached 80, and the recognition accuracy reached the maximum, about 90%, the details is shown in Fig. 9.

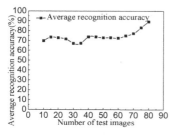

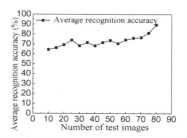

Fig. 8. Triancgf training function's trend of average recognition accuracy

Fig. 9. Triancgp training function's trend of average recognition accuracy

• Powell-Beale conjugate gradient method–traincgb training function
As the number of samples involved in the testing from 10 to 60, the recognition accuracy has been in a fluctuation, and the volatility gradually weakened. As the number of the samples in the test is 60, the recognition rate of the neural network shows a steady upward trend. Finally the number of test images reached 80, the recognition accuracy reached the maximum, about 87%. Details are shown in Fig. 10.

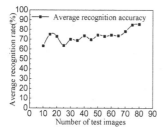

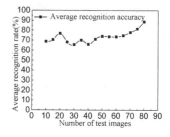

Fig. 10. Triancgb training function's trend of average recognition accuracy

Fig. 11. Trianscg training function's trend of average recognition accuracy

• Quantitative conjugate gradient method–trainscg training function
The testing results of trainscg training function and traincgb function are very similar. With the increasing of the number of samples, the recognition accuracy began to be steadily increased, ultimately reach to about 89%. Details are shown in Fig. 11.

Other training functions

• Quasi-Newton algorithm–trainbfg training function
Quasi-Newton algorithm is effective in solving nonlinear problems. It is mainly used in complex large-scale optimization. The trainbfg function recognition accuracy reaches to 88% and the details are shown in Fig. 12.

• Step-by-step algorithm–trainoss training function
The testing results of the trainoss function recognition accuracy reaches the maximum value of 88.19% as the number of testing images reaches 80. Details are shown in Fig. 13.

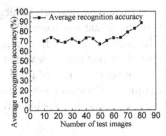 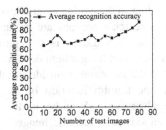

Fig. 12. Trianbfg training function's trend of average recognition accuracy

Fig. 13. Trianoss training function's trend of average recognition accuracy

- Levenberg-Marquardt algorithm–trainlm training function

The algorithm is mainly to use the gradient to find the extreme. As the neural network of the trainlm function is used, the recognition accuracy tends to rise on the whole (ignoring the local fluctuation). The more images are trained, the higher recognition accuracy is obtained. The recognition accuracy can reaches about 90%. Details are shown in Fig. 14.

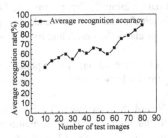

Fig. 14. Trianlm training function's trend of average recognition accuracy

4 Conclusion

In this paper, the following two major parts of the experimental data analysis: (1) the extraction of four sorts of seabed sediment texture characteristics, that is, contrast, correlation, energy and homogeneity were statistically analyzed; (2) the influence of the number of seabed side-scan acoustic images s and the number of training times on the neural network recognition accuracy is analyzed, and the comprehensive performance of twelve sorts of BP network training functions were tested, including the convergence speed and recognition accuracy. The results are shown as follows:

The training function (including traingd, traingdm, triangda, traingdx, trianrp) based on the gradient descent method is closer to the target value as training the network. The smaller the step size is, the slower the progress is, and the speed of convergence is extremely slow. This method is not suitable for the application of the pursuit of speed in the seabed recognition and classification.

According to the conjugate gradient training function (including traincgf, traincgp), each of its search directions are mutually conjugate. In the actual test found that the average time each testing is still more than 30 s. In addition to the five training functions based on the gradient descent method can reach the target accuracy within the specified 5000 iterations. Considering the factors of the convergence rate of the neural network, the trainlm function of the Levenberg-Marquardt algorithm has a faster velocity advantage and can be used as a BP neural network algorithm for identifying the seabed side-scan acoustic image.

Acknowledgment. This research work is supported by Major National Science and Technology Project (2015ZX01041101), and the National Natural Science Foundation of China (51609050, 51409059, 51509057).

References

1. Grabowski, R., Wharton, G.: Erodibility of cohesive sediment: The importance of sediment properties. Earth-Science **105**(3–4), 101–120 (2011)
2. Legendre, P., Ellingsen, K., Bjornbom, E., Casgrain, P.: Acoustic seabed classification. Can. J. Fish. Aqustic Sci. (2015)
3. Ahmed, K., Demsar, U.: Improved seabed classification from Multi-Beam Echo Sounder backscatter data with visual data mining. J. Coast. Conserv. **17**(3), 559–577 (2013)
4. Farrell, K., Harris, W., Mallinson, D., Culver, S.J.: Standardizing texture and facies codes for a process-based classification of clastic Sediment and rock. J. Sediment. Res. **82**(5–6), 364–378 (2012)
5. Lark, R., Dove, D., Green, S.L., Richardson, A.E.: Spatial prediction of seabed sediment texture classes by cokriging from a legacy database of point observations. Sed. Geol. **281**, 35–49 (2012)
6. Huang, Z., Siwabessy, J., Nichol, S., Anderson, T., Brooke, B.: Predictive mapping of seabed cover types using angular response curves of multibeam backscatter data: Testing different feature analysis approaches. Cont. Shelf Res. **61–62**(4), 12–22 (2013)
7. Chen, Q., Song, Z., Huang, Z., Hua, Y., Yan, S.: Contextualizing object detection and classification. IEEE Trans. Pattern Anal. Mach. Intell. **37**(1), 13–27 (2015)
8. Satyanarayana, Y., Naithani, S., Anu, R.: Seafloor sediment classification from single beam echo sounder data using LVQ network. Mar. Geophys. Res. **28**(2), 95–99 (2007)
9. San, B.: Anevalustion of SVM using polygon-based random sampling in landslide susceptibility mapping. Int. J. Appl. Earth Obs. Geoinf. **26**(1), 399–412 (2014)
10. Foody, G., Mathur, A.: Toward intelligent training of supervised image classifications: directing training data acquisition for SVM classification. Remote Sens. Environ. **93**(1–2), 107–117 (2004)
11. Dai, J., Yang, H.H., Wang, Y., Sun, J.C.: An adaptive genetic instance selection algorithm for underwater acoustic target classification. Tech. Acoust. **32**(4), 332–335 (2013)
12. Liu, T., Yin, S.: An improved particle swarm optimization algorithm used for BP neural and multimedia course-ware evaluation. Multimed. Tools Appl., 1–14 (2016)
13. Li, J.Y.: Bp neural network optimized by PSO and its application in function approximation. Adv. Mater. Res., 945–949, 2413–2416 (2014)
14. Wang, H.: Researching image demising model based PSO_TranlmBP. Math. Pract. Theory (2014)

15. Kuang,Y., Singh, R., Singh, S., Singh, S.P.: A novel macroeconomic forecasting model based on revised multimedia assisted BP neural network model and ant Colony algorithm. Multimed. Tools Appl., 1–22 (2017)
16. Wang, H.: Researching image demising model based PSO_TranlmBP. Math. Pract. Theory (2014)

Node Localization of Wireless Sensor Network Based on Secondary Correction Error

Xiaoxu Ma$^{(\boxtimes)}$, Wenju Liu, and Ze Wang

School of Computer Science and Software Engineering,
Tianjin Polytechnic University, Tianjin 300387, China
1656871571@qq.com

Abstract. Due to the large localization error of the range-free localization algorithm, a new node localization algorithm based on secondary correction error is proposed. Firstly, orthogonal polynomial fitting method, a mathematical model, is taken advantage of to correct the distance error. Moreover, subtraction first and then square, a strategy, is introduced to solve the equations. At the same time, the actual distance and distance error are taken as weighting factors to construct the weighted matrix to solve the unknown node coordinates. Finally, the redundant information obtained by solving the equations is employed to refine the coordinates of unknown nodes. Simulation experiment results in this paper are convincing evidence that our proposed algorithm can decrease positioning error and increase the positioning accuracy efficaciously.

Keywords: Orthogonal polynomial fitting · Distance error · Weighted matrix · Redundant information · Positioning error

1 Introduction

Wireless Sensor Network [1, 2] technology has been applied into many fields [3] for the last few years, so it is necessary to study deeply the node localization technology. The localization algorithm falls into two categories: one is range-based localization algorithm, the other is range-free localization algorithm [4–6]. The latter has attracted more and more attention because of its lower cost and lower energy consumption.

The actual distance between nodes is not required to measure precisely in range-free localization algorithm, so a large distance error will be generated. Moreover, the least square principle will accumulate error when solving the unknown node coordinates, leading to the secondary error of the positioning results. Reference [7] using compensation factor to correct distance error. In [8, 9], RSSI algorithm is introduced to adjust distance error. The drawback is that additional RSSI hardware devices need to be added to increase the node cost overhead. Reference [10] proposes a non-uniform non-linear model to reduce localization error. The algorithm in [11] specifies the communication radius of the node and estimate distance accurately, yet it increases the communication overhead. The various evolutionary algorithms are employed in some papers, the hybrid Genetic PSO algorithm, the bat algorithm and the hybrid bat-quasi-Newton algorithm are respectively exploited to determine the nodes coordinates in [12–14]. However, the cost of their improved accuracy is the increase in

© Springer Nature Singapore Pte Ltd. 2017
G. Chen et al. (Eds.): PAAP 2017, CCIS 729, pp. 142–151, 2017.
DOI: 10.1007/978-981-10-6442-5_13

computing time. The node coordinates can be solved by introducing coordinates correction strategy in [15] and subtraction firstly then square strategy in [16], however, both of them ignore the estimate phase error of the distance between nodes, as a consequence, the location effect is not distinct.

This paper proposes a novel localization algorithm which is based on secondary correction error. The paper discusses as following aspects: Sect. 2 presents the orthogonal polynomial fitting algorithm briefly. Section 3 describes the specific algorithm steps at length. Simulation results validate convincingly that proposed algorithm is capable to decrease positioning error effactually in Sect. 4, and eventually, Sect. 5 come to a conclusion of whole paper.

2 Orthogonal Polynomial Fitting Algorithm

A set of data points (x_i, y_i) $(i = 1, 2, ..., m)$ is given, and it generates a function by least square method:

$$\Phi(x) = a_0\varphi_0(x) + a_1\varphi_1(x) + \cdots + a_n\varphi_n(x) \tag{1}$$

Let the sum of deviation squares:

$$J(a_0, a_1, \cdots, a_n) = \sum_{i=1}^{m} \delta_i^2 = \sum_{i=1}^{m} (\Phi(x_i) - y_i)^2 \tag{2}$$

is minimum. The problem is transformed into solving the minimum points of $J(a_0, a_1, \cdots, a_n)$.

$$\begin{cases} a_0 \sum_{i=1}^{m} \varphi_0(x_i)\varphi_0(x_i) + \cdots + a_n \sum_{i=1}^{m} \varphi_n(x_i)\varphi_0(x_i) = \sum_{i=1}^{m} y_i\varphi_0(x_i) \\ a_0 \sum_{i=1}^{m} \varphi_0(x_i)\varphi_1(x_i) + \cdots + a_n \sum_{i=1}^{m} \varphi_n(x_i)\varphi_1(x_i) = \sum_{i=1}^{m} y_i\varphi_1(x_i) \\ \qquad\qquad\qquad\qquad \vdots \\ a_0 \sum_{i=1}^{m} \varphi_0(x_i)\varphi_n(x_i) + \cdots + a_n \sum_{i=1}^{m} \varphi_n(x_i)\varphi_n(x_i) = \sum_{i=1}^{m} y_i\varphi_n(x_i) \end{cases} \tag{3}$$

We convert (3) to (4)

$$\begin{pmatrix} (\varphi_0, \varphi_0) & \cdots & (\varphi_n, \varphi_0) \\ \vdots & \ddots & \vdots \\ (\varphi_0, \varphi_n) & \cdots & (\varphi_n, \varphi_n) \end{pmatrix} \begin{pmatrix} a_0 \\ \vdots \\ a_n \end{pmatrix} = \begin{pmatrix} (y, \varphi_0) \\ \vdots \\ (y, \varphi_n) \end{pmatrix} \tag{4}$$

The coefficient matrix is symmetric and positive definite, so unique solution will be obtained later. The algebraic polynomial is chosen as the base function in [17, 18] and a_k (k = 0, 1, ..., n) are obtained by solving n + 1equations. However, if data points are huge in size or the fitting order n is increased, the coefficient matrix is probably ill-conditioned and unstable. On this basis, orthogonal polynomial fitting algorithm [19, 20] is proposed, which is capable to avoid ill-conditioned coefficient matrix effectively.

Constructing an orthogonal polynomial family, the recurrence formula is as follows:

$$\begin{cases} \varphi_0(x) = 1 \\ \varphi_1(x) = (x - \alpha_0)\varphi_0(x) \\ \quad\vdots \\ \varphi_{k+1}(x) = (x - \alpha_k)\varphi_k(x) - \beta_k\varphi_{k-1}(x) \end{cases} \tag{5}$$

where

$$\alpha_k = \frac{(x\varphi_k, \varphi_k)}{(\varphi_k, \varphi_k)} \tag{6}$$

$$\beta_k = \frac{(\varphi_k, \varphi_k)}{(\varphi_{k-1}, \varphi_{k-1})} \tag{7}$$

Equation (4) can be simplified as:

$$\begin{pmatrix} (\varphi_0, \varphi_0) & \cdots & 0 \\ \vdots & \ddots & \vdots \\ 0 & \cdots & (\varphi_n, \varphi_n) \end{pmatrix} \begin{pmatrix} a_0 \\ \vdots \\ a_n \end{pmatrix} = \begin{pmatrix} (y, \varphi_0) \\ \vdots \\ (y, \varphi_n) \end{pmatrix} \tag{8}$$

From (8), we have:

$$a_k = \frac{(y, \varphi_k)}{(\varphi_k, \varphi_k)} = \frac{\sum_{i=1}^m y\varphi_k(x_i)}{\sum_{i=1}^m \varphi_k(x_i)\varphi_k(x_i)} \tag{9}$$

Finally, we plug (9) into (1) and get an orthogonal polynomial accordingly.

3 The Proposed Algorithm

3.1 Correcting the Distance Error by Orthogonal Polynomial Fitting Algorithm

The beacon nodes broadcast their information packets and all nodes obtain the minimum hop count with all beacons by the mechanism of distance vector routing. Each beacon determines the average distance per hop represented by $hopSize_i$ by Eq. (10) and broadcast it. Simultaneously, each beacon determines the estimated distance error between itself and other beacons represented by $dist_i$ by Eq. (11):

$$hopSize_i = \frac{\sum_{i \neq j} \sqrt{(x_i - x_j)^2 + (y_i - y_j)^2}}{\sum_{i \neq j} h_{ij}} \tag{10}$$

$$dist_i = hopSize_i \times h_{ij} - \sqrt{(x_i - x_j)^2 + (y_i - y_j)^2} \tag{11}$$

where (x_i, y_i) and (x_j, y_j) represent the coordinates of the beacon node i and j respectively, and h_{ij} is the minimum hop count between the beacon node i and j. There are several distance errors with different values correspond the same hop count value, accordingly, we assume that hop count is m, n is the number of distance error corresponding to the same hop count value is m, the average of n distance error is calculated by Eq. (12):

$$errdis_m = \frac{\sum_{i=1}^{n} dist_i}{n} \tag{12}$$

Taking $(hop_i, errdis_i)$ $(i = 1, 2, \ldots, m)$ as known data points (x_i, y_i), we get the orthogonal polynomial (1) which can fit the relationship between hop count and distance error.

Each unknown node selects the nearest beacon node and saves its hopSize value. Finally, each unknown node can reckon up the distance to all beacon nodes and correct distance error using orthogonal polynomial algorithm by Eq. (13):

$$d_{ik} = hopSize_i \times h_{ik} - \Phi(h_{ik}) \tag{13}$$

3.2   Calculating Unknown Node Coordinates

When the unknown node obtains the corrected distance, the least square method can be utilized to calculate its location. We set (x, y) and (x_i, y_i) are coordinates of unknown node and beacon node i respectively. The corrected distance represented by d_i from Eq. (13):

$$\begin{cases} \sqrt{(x - x_1)^2 + (y - y_1)^2} = d_1 \\ \sqrt{(x - x_2)^2 + (y - y_2)^2} = d_2 \\ \vdots \\ \sqrt{(x - x_n)^2 + (y - y_n)^2} = d_3 \end{cases} \tag{14}$$

In order to make the correction factor of the equations smaller and have a better positioning accuracy, subtraction first and then square, the strategy, is adopted [16] and simplify as follows:

$$-2(x_i + x_n)x - 2(y_i + y_n)y + 2S = d_i^2 + d_n^2 - (E_i + E_n) \tag{15}$$

S and E_i can be expressed as:

$$S = x^2 + y^2, \ E_i = x_i^2 + y_i^2 (i = 1, 2, \ldots, n) \tag{16}$$

Equation (15) can be denoted as AX = b, A, X and b are as:

$$A = -2 \begin{pmatrix} x_1 + x_n & y_1 + y_n & -1 \\ x_2 + x_n & y_2 + y_n & -1 \\ \vdots & \vdots & \vdots \\ x_{n-1} + x_n & y_{n-1} + y_n & -1 \end{pmatrix} \quad X = \begin{pmatrix} x \\ y \\ S \end{pmatrix}$$

$$b = \begin{pmatrix} d_1^2 + d_n^2 - (E_1 + E_n) \\ d_2^2 + d_n^2 - (E_2 + E_n) \\ \vdots \\ d_{n-1}^2 + d_n^2 - (E_{n-1} + E_n) \end{pmatrix} \tag{17}$$

The solution of AX = b is:

$$X = \left(A^T A \right)^{-1} A^T b \tag{18}$$

Meanwhile, the influence of different beacon nodes should be considered fully. The closer the distance and the smaller the distance error, in this way, the weights given is greater. Therefore, the weights of the unknown node k can be set:

$$w_{k,i} = \left(\frac{1}{\Phi(h_{ki})} \right)^2 + \left(\frac{1}{D_{ki}} \right)^2 \tag{19}$$

Then, we construct the weighted matrix:

$$W = \begin{pmatrix} w_{k,1} & 0 & \cdots & 0 \\ 0 & w_{k,2} & \cdots & 0 \\ \vdots & \vdots & \ddots & \vdots \\ 0 & 0 & \cdots & w_{k,n-1} \end{pmatrix} \tag{20}$$

Finally, the solution should be:

$$X = \left(A^T W^T W A \right)^{-1} A^T W^T W b \tag{21}$$

3.3   Refining Unknown Node Coordinates

We get (x, y) and S from (21), if

$$S = x^2 + y^2 \tag{22}$$

(x, y) is final unknown node location; else, x_1 and y_1 are given by:

$$\begin{cases} x_1 = ax \\ y_1 = ay \end{cases} \tag{23}$$

For x_1 and y_1 in (22) to get parameter a, x_1 and y_1 are obtained. Finally, we get the refining the coordinates:

$$\begin{cases} X = \frac{x_1 + x}{2} \\ Y = \frac{y_1 + y}{2} \end{cases} \qquad (24)$$

4 Simulation Analyses

We simulated DV-Hop [21] and two existed algorithms in [15, 16] by simulation platform of MATLAB. All sensor nodes are randomly and uniformly distributed within a square area of 100 m by 100 m. Simulation repeat 100 times and average the result. Each algorithm regards the average localization error LE_{avg} as the appraise standard, as follows:

$$LE_{avg} = \frac{\sum_{i=1}^{N-n} \sqrt{(X_i - x_i)^2 + (Y_i - y_i)^2}}{(N - n) \times R}$$

In the above equality, N denotes the total number of nodes. Parameter n denotes the number of beacon nodes and R represents the communication radius of the nodes. (X_i, Y_i) and (x_i, y_i) are actual coordinate and estimation coordinate of unknown node respectively.

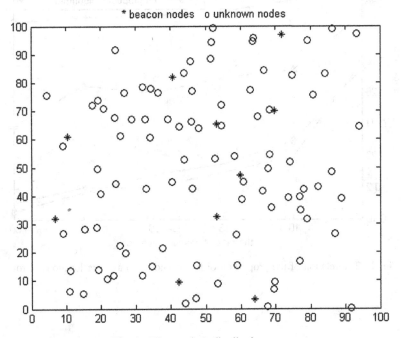

Fig. 1. The random distribution map

From the above, we can see that LE_{avg} depends on the value of N, n and R. The influence of these three aspects on LE_{avg} is discussed emphatically based on the simulation results respectively below.

When N is 100 and n is 10, the random distribution map is shown as Fig. 1:

4.1   The Influence Exerted by the Proportion of Beacon Nodes upon Locational Error

The total number N is set as 300, R is set as 30 m, and the proportion of beacon nodes is changed from 5% to 30%. Figure 2 shows the comparative curve of the average location error of four algorithms. As shown, with ratio of the beacon nodes is on the increase, the average location error of four algorithms decreased. Increasing the proportion of the beacon node will make hop count value more accurate. Simultaneously, the more beacon nodes, the more data points will be obtained and we will get more accurate result by orthogonal polynomial fitting method to correct the distance error, consequently, the comparative curve in Fig. 4 indeed can be confirmed that the performance of our proposed algorithm is superior to existing algorithms in [15, 16] evidently.

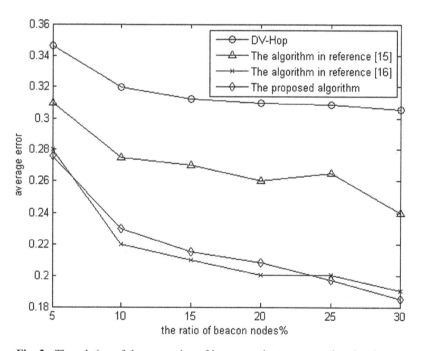

Fig. 2. The relation of the proportion of beacon nodes to average locational error

4.2 The Influence of Communication Radius upon Average Locational Error

N is set as 300, n is 30, R is changed from 15 m to 40 m. The comparative curve of the average location error of four algorithms can be showed in Fig. 3. As the chart demonstrates, the error of this algorithm in [16] declines firstly and then it rises. It is because increasing the communication radius will enhance the network connectivity and each unknown node neighbor beacon nodes also increases. However, when the communication radius increases to a certain value and it can't stop increasing, it will cause more unknown nodes are found the within the communication scope of the beacon node, and these unknown nodes is calculated the distance among other anchor nodes based on an average jump of the anchor nodes, thus, it will increase the error of distance estimation. With the increase of communication radius, the error will be more obvious. So if the positioning error of algorithm is improved, the error will be increased rather than declined. In this paper, the error which is generated by the revised algorithm is corrected. In this way, it stops the positioning error rising, but the increasing communication radius makes the hop count between beacon node is reduced. As a result, data samples between obtained hop counts and distance error are reduced. The fitting result is limited to the correction of the distance error and the optimization advantage is no longer obvious. All in all, we can see that the performance of our algorithm is superior to others evidently with communication radius is increased.

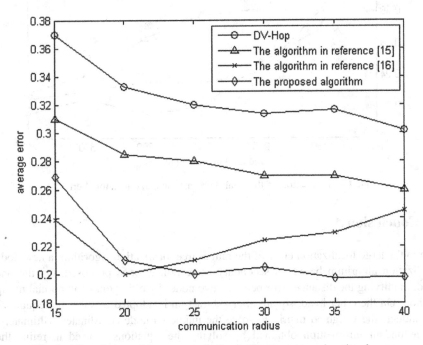

Fig. 3. The relation of the communication radius to average locational error

4.3   The Influence of the Total Number of Nodes upon Average Locational Error

The proportion of the beacon node is set as 10% of N, R is 30 m and N is changed from 100 to 400. With N value is on the increase, average locational error declines as demonstrated in Fig. 4. This is because when the area is fixed, increasing the total number of nodes will enhance node density and network connectivity which can make hop distance more accuracy. At the same time, the more beacon nodes, the more data points will be obtained and we will get more accurate results by orthogonal polynomial fitting method to correct the distance error, in the final analysis, the comparative curve in Fig. 4 indeed can be confirmed the performance of our presented algorithm is much superior to existing algorithms in [15, 16] evidently.

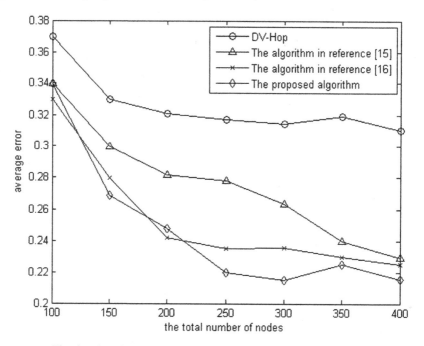

Fig. 4. The relation of the total number to average locational error

5   Conclusion

Due to the large localization error of the range- free localization algorithm, a new node localization algorithm based on secondary correction error is proposed. On the one hand, modifying the distance error between two nodes by orthogonal polynomial fitting method. On the other hand, square after subtraction is adopted to solve the equations and inducts the weighted matrix to solve the unknown node coordinates. Ultimately, the redundant information obtained by solving the equations is used to refine the coordinates of unknown nodes. Simulation experiment results are convincing evidence that our proposed algorithm is qualified to efficaciously decrease positioning error and increase the positioning accuracy.

References

1. Chaturvedi, P.: Wireless Sensor Networks-A Survey. International Conference on Recent Trends in Information (2014)
2. Rawat, P., Singh, K.D., Chaouchi, H., et al.: Wireless sensor networks: a survey on recent developments and potential synergies. J. Supercomput. 68(1), 1–48 (2014)
3. Kulkarni, R.V., Förster, A., Venayagamoorthy, G.K.: Computational intelligence in wireless sensor networks: a survey. IEEE Commun. Surv. Tutor. 13(1), 68–96 (2011)
4. Woo, H., Lee, S., Lee, C.: Range-free localization with isotropic distance scaling in wireless sensor networks. In: International Conference on Information Networking, pp. 632–636 (2013)
5. Wu, G., Wang, S., Wang, B.: A novel range-free localization based on regulated neighborhood distance for wireless ad hoc and sensor networks. Comput. Netw. 56(16), 3581–3593 (2012)
6. Meguerdichian, S., Slijepcevic, S., Karayan, V.: Localized algorithms in wireless ad-hoc networks: location discovery and sensor exposure. In: The ACM International Symposium, pp. 106–116 (2001)
7. Huang, C.H., Shen, J.: Improved DV-Hop positioning algorithm based on compensation coefficient. J. Softw. Eng. 9(3), 650–657 (2015)
8. Wu, N., Liu, F., Wang, S.: Algorithm for location nodes in WSN based on modifying hops and hopping distance. Microelectron. Comput. 32(1), 91–95 (2015)
9. Tian, S., Zhang, X., Liu, P.: A RSSI-Based DV-Hop algorithm for wireless sensor networks. In: International Conference on Wireless Communications, Networking and Mobile Computing. IEEE, pp. 2555–2558 (2007)
10. Jiang, M., Li, Y., Ge, Y.: An advanced DV-hop localization algorithm in wireless sensor network. Int. J. Control Autom. 32(8), 405–422 (2015)
11. Ma, S., Zhao, J.: Multi communication ranges DV-Hop localization algorithm for wireless sensor network. Chin. J. Sens. Actuators (2016)
12. Mehrabi, M., Taheri, H., Taghdiri, P.: An improved DV-Hop localization algorithm based on evolutionary algorithms. Telecommun. Syst. 64, 1–9 (2017)
13. Yang, X., Zhang, W.: An improved DV-Hop localization algorithm based on bat algorithm. Cybern. Inf. Technol. 16(1), 89–98 (2016)
14. Yu, Q., Sun, S., Xu, B.G.: Node localization of wireless sensor networks based on hybrid bat-quasi-Newton algorithm. J. Comput. Appl. 35(5), 1238–1241 (2015)
15. Qiu, F.M., Li, H.Z.: Improvement of DV-Hop localization algorithm for wireless sensor network. Comput. Eng. (2014)
16. Li, C.G., Liu, M.S., Sun, K.H.: Improved DV-Hop localization algorithm based on the weighted least square method. Microelectron. Comput. 33(1), 24–27 (2016)
17. Li, Y.F., Jiang, M., Ge, Y.: Improved DV-Hop localization algorithm based on curving fitting. Comput. Syst. Appl. 24(5), 118–123 (2015)
18. Huang, X.L., Mu, D.J., Li, Z.: Research on interpolation algorithm based on curve fitting for energy consumption of WSN. Modern Electron. Technique 39(1), 9–12 (2008)
19. Zhu, X.D., Tie-Ding, L.U., Chen, X.J.: Orthogonal polynomial curve fitting. J. East China Inst. Technol. (2010)
20. Xin, H.M., Xue, L., Liu, J.: Positioning optimization technique in non-line-of-sight environment. J. Netw. 6(2), 287–294 (2011)
21. Niculescu, D., Nath, B.: DV based positioning in Ad Hoc networks. Telecommun. Syst. 22(1), 267–280 (2003)

Optimizations of the Whole Function Vectorization Based on SIMD Characteristics

Yingying Li[(✉)], Yuchen Gao, Dong Wang, Yanbing Li, and Jinlong Xu

State Key Laboratory of Mathematical Engineering and Advanced Computing, Zhengzhou 450001, China
liyingying1005@163.com

Abstract. Vectorization for SIMD extensions is similar to programming for CUDA/OpenCL on GPU platforms. They are both Single Program Multiple Data (SPMD) programming models. However, SIMD extensions and GPU accelerators are different from each other in many aspects, such as memory access, divergence, etc. There are still optimization opportunities when using existing methods to implement vectorization for SIMD extensions. As a result, we propose a whole function vectorization optimization algorithm based on SIMD characteristics in this paper. First, we analyze some SIMD characteristics that may affect the whole function vectorization. These characteristics include instance versioning, instance regrouping and SIMD code optimization. We then implement a SIMD characteristics-based algorithm for whole function vectorization. In addition, we introduce a directive based method to help us fully exploit opportunities of this kind of vectorization. We choose nine benchmarks from multi-media and image processing applications to evaluate our technique. Compared with un-optimized codes, the speedup is 1.59 times faster in average on processor E5-2600 when the proposed technique is applied.

Keywords: SIMD extension · Whole function vectorization · SIMD characteristics · Code optimization

1 Introduction

The demand for higher performance and more efficient in modern processors has led to a wide adoption of SIMD (single-instruction multiple-data) vector units. Almost all the major vendors support vector instructions and the trend is pushing them to become more powerful [1]. SIMD extension instructions are quite common today in both high performance and embedded microprocessors [2]. SIMD widths have been following an upward trend: the 128-bit Streaming SIMD Extensions (SSE) of x86 architectures was augmented to 256-bit Advanced Vector Extensions (AVX); the new Intel Knights Landing processors architecture supports 512-bit SIMD (AVX-512). Thus, SIMD extensions are playing a more and more important accelerating role in the age of multicore.

Theoretically, manual vectorization is able to achieve the higher performance of SIMD vectorization [3]. However, it is usually supposed to be annoying and difficult to

© Springer Nature Singapore Pte Ltd. 2017
G. Chen et al. (Eds.): PAAP 2017, CCIS 729, pp. 152–171, 2017.
DOI: 10.1007/978-981-10-6442-5_14

generate platform-specific and efficient hand-written code [4]. Compiler-based automatic vectorization is one of the solutions to this problem. There are two main types of vectorization methods [5]. One is loop-based method, which can convert multiple iterations of a loop into single vector iteration [6]. The other is SLP (superword level parallelism) method, which targets straight-line code on repeated sequences of scalar instructions. They do not require sophisticated dependence analysis and can be applied in much more general cases. Furthermore, data level parallelism can also be performed at function level. Inline optimization can eliminate function calls, but may increase the compilation time. Therefore, function-level vectorization is necessary since a compiler should not rely on inline to eliminate function calls completely.

Function-level vectorization can achieve the performance of task-level parallelism while loop-based vectorization is often constrained by few iteration numbers, complex loop structures and other conditions. These restrictions are not suitable for function level vectorization because it exploits SIMD parallelism from another scope. Several instances combine when whole function vectorization implements, the dependences between instances have to be considered to ensure the correctness. There may be many function calls in a program; the order of function level vectorization can be determined by call graph. If SIMD parallelism in a program has been exploited from the scope of loop or basic block, it is difficult to exploit whole function vectorization further.

Whole function vectorization for SIMD extensions is similar to CUDA or OpenCL in GPU, both of them are Single Program Multiple Data (short for SPMD) programming models. However, huge differences exist between SIMD extensions and GPU in the architecture, such as memory access, divergence and so on. A whole-function vectorization transformation based on SSA form is presented for processors with SIMD instructions in [9]. SSA is a program representation in which each variable has exactly one static definition. SSA is particularly useful for vectorization because Φ- functions give the locations where blending code has to be placed. We maintain the SSA form during our transformation. The whole-function vectorization transformation consists of four phases, which are mask generation, select generation, partial CFG (control-flow graph) linearization and instruction vectorization respectively. For SIMD extensions, there is still huge optimization space when codes are generated by current method. Therefore, we reconsider code generation of whole function vectorization to achieve better performance on SIMD extensions. Specifically, this paper makes the following contributions.

- The SIMD characteristic of operations and basic blocks are analyzed.
- Code optimization of whole function vectorization implements based on the SIMD characteristic, which includes instance multi-versions, instance regroup and vectorization instruction optimization.
- The algorithm is implemented in the LLVM compiler infrastructure and evaluated using nine practical kernels. We show an average speed up of 1.59 times faster for these kernels compared to un-optimized one.

The context of the paper is organized as following. In Sect. 2 we briefly outline the whole function vectorization transformation. Section 3 presents the core contribution of this paper, SIMD property are analyzed and optimization based on SIMD property are proposed. Section 4 presents our experimental evaluation. Section 5 discusses related work and Sect. 6 concludes.

2   Whole-Function Vectorization Transformation

A whole-function vectorization transformation based on SSA form is presented for processors with SIMD instructions in [7]. SSA is a program representation in which each variable has exactly one static definition. SSA is particularly useful for vectorization because Φ- functions give the locations where blending code has to be placed. We maintain the SSA form during our transformation. The whole-function vectorization transformation consists of four phases, which are mask generation, select generation, partial CFG (control-flow graph) linearization and instruction vectorization.

2.1   Mask Generation

Loops are simplified to ensure that each loop has exactly one incoming edge and one back edge. This guarantees the existence of a unique loop header, a unique loop pre-header (the block from which the loop is entered) and a unique loop latch (the block from which an edge leads back to the header). The entry mask of a block is the disjunction of the masks of all incoming edges as well. In case of a loop header, the entry mask is a Φ-function with incoming values from the loop's pre-header and latch. The masks of the control-flow edges leaving a block are given by the block entry mask and a potential the mask of its single exit-edge is equal to the entry mask. If the exit branch is conditional, the exit mask of the true edge of the block is the conjunction of its entry mask and the branch condition. The exit mask of the false edge is the conjunction of the entry mask and the negated branch condition.

2.2   Select Generation

Linearization of control-flow is possible only if results of inactive instances are discarded. This is achieved by inserting blend operations at control-flow join points and loop latches. Each Φ-function in the original CFG that is not in a loop header is replaced by a select instruction. Φ-functions with n incoming values are transformed into series of $n - 1$ connected select instructions. Additionally, each loop requires result vectors in order to conserve the loop live values of instances that leave the loop early. Loop live values are those values that are live across loop boundaries. A value is defined as live across loop boundaries if it is used either in a subsequent iteration or outside the loop.

2.3   Partial CFG Linearization

After all mask and select operations are inserted, all control flow is effectively encoded by data flow and can thus be removed. To this end, the basic blocks have to be put into a sequence that preserves the execution order of the original CFG. If a block A executed before B in every possible execution of G, then A has to be in front of B in the flattened CFG G0. If the CFG splits up into two paths, one path is chosen to be executed entirely before the other. The ordering is determined by topologically sorting the blocks recursively over the loop tree of G.

2.4 Instruction Vectorization

After linearization, the actual transformation into vector code is applied. Vectorizing a single instruction is basically a one-to-one translation from the scalar instruction to its SIMD counterpart. This holds for all instructions except for function calls and memory operations. Operations that cannot be vectorized have to be duplicated into W scalar instructions.

We take the kernel of Mandelbrot as an example to illustrate whole function vectorization transformation. The scalar code is shown in Fig. 1(a).

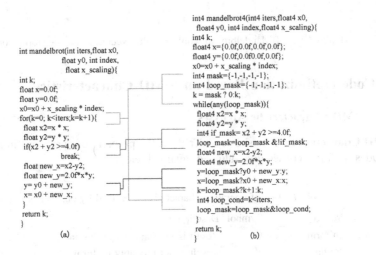

Fig. 1. Visualization of whole function vectorization transformation

After mask generation, the CFG is shown in Fig. 2(a), mexit is the accumulated exit mask of edge c → f, mup is the update operation of that exit mask, and mcomb is the combined exit mask. After select generation, the CFG is shown in Fig. 2(b), rit is the accumulated result vector, rup is the corresponding undate operation. Where r is the final result, blended together from the two values incoming from both exits. Figure 2(c) shows the CFG after partial linearized. After whole function vectorization transformation of Mandelbrot, the vectorized code is shown in Fig. 1(b). As no optimization is implemented when whole function vectorization method is used to generate code in [2], the vectorized code cannot make full use of SIMD extensions. Therefore, we propose an improved code generation method based on SIMD property to achieve better performance.

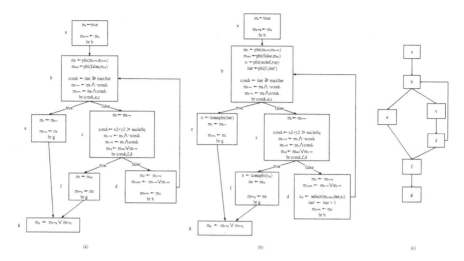

Fig. 2. CFG of Mandelbrot in whole function vectorization transformation

3 Code Optimization Based on SIMD Characteristics

3.1 SIMD Characteristic Analyses

SIMD Characteristics. As shown in Table 1, the property of operations in a program includes uniform, consecutive, aligned, guarded, etc.

Table 1. List of operation characteristics derived by our analyses

Property	Symbol	Description
Uniform	U	Result is the same for all instances
Varying	V	Result is not provably uniform
Consecutive	C	Result consists of consecutive values
Unknown	K	Result values follow no usable pattern
Aligned	A	Result value is aligend to multiple of S
Nonvectorizable	N	Result type has no vector counterpart
Sequential	S	Execute W times sequentially
Guarded	G	Execute sequentially for active instances only

An operation is uniform iff it does not bring any side effects and a single, scalar operation is sufficient to produce all values required for all instances of the executing SIMD group. Otherwise, it is varying. The varying characteristic is a general characteristic that states that a program point is not uniform, i.e., it holds different values for different instances. Characteristics like consecutive, unknown, or guarded describe a varying program point in more detail.

An operation is consecutive iff its results for a SIMD group of instances are natural numbers where each number is by one larger than the number of its predecessor instance, except for the first instance of the group. It is common for vector elements to hold consecutive values. Assume we have a base address and know that the values of the offsets are consecutive for consecutive instance identifiers. Putting the offsets in a vector yields $< n, n + 1, ..., n + W - 1 >$. Using such an offset vector in a gap gives a vector of consecutive addresses. Thus, optimized vector load and store instructions that only operate on consecutive addresses can be used.

An operation is aligned iff its results for a SIMD group of instances are natural numbers and the result of the first instance is a multiple of the SIMD width S. Current SIMD hardware usually provides more efficient vector memory operations to access memory locations that are aligned. Thus, it is also important to know whether an offset vector starts exactly at the SIMD alignment boundary. If, for example, a load operation accesses the array element next to the instance identifier, the vectorized variant would access $< id + 1, id + 2, ..., id + W>$. Without further optimizations, this non-aligned operation would require two vector loads and a shuffle operation.

An operation is non-vectorizable iff its return type or the type of at least one of its operands has no vector equivalent. For example, a call that returns a void pointer is nonvectorizable since there is no vector of pointers in our program representation. More importantly, however, a nonvectorizable value forces operations that use it to be executed sequentially

An operation is sequential iff it has no vector equivalent or if it is nonvectorizable and not uniform. This is mostly relevant for operations with nonvectorizable return type or at least one nonvectorizable operand. It also applies to all those operations that do not have a vector counterpart, for example some SIMD instructions sets do not support a load from non-consecutive memory locations.

An operation is guarded iff it may have side effects and is not executed by all instances. Unknown calls have to be expected to produce side effects and thus require guarded execution. Similarly, store operations must be never executed for inactive instances. Many sources of overhead in vectorized programs are caused by conservative code generation because instances may diverge. If it can be proven that all instances are active whenever an operation is executed, more efficient code can be generated.

As shown in Table 2, the characteristic of block includes by_all, div_causing, blend, rewire. A program point is All-Active Program Point (short for by_all) iff it is not executed by a SIMD group which has an inactive instance. For example, operations that may have side effects have to be executed sequentially and guarded if some instances may be inactive. This is because the side effect must not occur for inactive instances. If the operation is by all, it will be executed by all instances, and no guards

Table 2. List of block and loop characteristics derived by our analyses

Property	Description
by_all	Block is always executed by all instances
div_causing	Block is a divergence-causing block
blend$_v$	Block is join point of instances that divergend at block v
rewire$_v$	Block is a rewire target of a div_causing block v
divergent	Loop that instances may leave at different points

are required. A block b is Divergence-Causing Block (short for div causing) iff not all instances that entered b execute the same successor block.

A loop is divergent iff any two instances of a SIMD group leave the loop over different exit edges and/or in different iterations. It is important to note that this definition of the term "divergence" differs from the usual one that describes a loop that never terminates. A divergent loop in the SIMD context can be left at different points in time (iterations) or space (exit blocks) by different instances. Because of this, the loop has to keep track of which instances are active, which left at which exit, and which values they produced. This involves overhead for the required mask and blend operations. However, if the loop is not divergent, i.e., it is always left by all instances that entered it at once and at the same exit, then no additional tracking of the instances is required. The div causing and rewire characteristics are used to describe the divergence behavior of control flow. Program points marked div causing or rewire correspond to exits and entries of basic blocks, so we simply refer to the characteristics as block characteristics.

Analysis Framework. We give a definition of a framework to describe and derive SIMD characteristics of a program. The framework is built on the ideas of abstract interpretation and closely follows the notation and definitions of Grund.

First, an Operational Semantics (OS) is defined for the program representation. It describes the exact behavior of a program when being executed as transformations between program states. The effects of each operation of the program are described by a transformation rule that modifies the input state.

Second, OS implicitly defines a Collecting Semantics (CS). It abstracts from the OS by combining states to a set of states, which allows to reason about the effects of an operation for all possible input states, e.g. all possible input values of a function.

Third, an Abstract Semantics (AS) is defined. It abstracts from the CS by reasoning about abstract characteristics rather than sets of concrete values. This allows its transformation rules to be implemented as a static program analysis that can be executed during compile time.

Operational Semantics. We consider the scalar function f to be given in a typed, low-level representation. A function is represented as a control flow graph of instructions. Furthermore, we require that f is in SSA form, i.e., every variable has a single static assignment and every use of a variable is dominated by its definition. We will restrict ourselves to a subset of a language that contains only the relevant elements for this thesis. Table 3 shows its types and instructions. Other instructions, such as arithmetic and comparison operators are straightforward and omitted for the sake of brevity.

Let $d = (\rho, \mu, @, \#)$ be a state of the execution at a given node, where ρ: Vars $\rightarrow \tau$ is a mapping of variables to values, $\mu : \pi^* \rightarrow \pi$ is a mapping of memory locations to values, $@ :\gamma \rightarrow$ bool stores for each program point whether it is active or not, and $\# :$ int is an instance identifier. The notation $\rho \oplus \{x \mapsto y\}$ stands for

$$\lambda v. \begin{cases} y & \text{if } x = v \\ \rho(v) & \text{otherwise,} \end{cases}$$

Table 3. Program Representation: types and instructions

Types	Instructions
σ = unit|τ	val: τ
τ = β|ν	tid: unit $\rightarrow$ int
β = bool	arg: int $\rightarrow$ τ
π = ν|π*	phi: $(\tau,\gamma) \times (\tau,\gamma) \rightarrow \tau$
ν = int|float	gep: $\pi^* \times$ int $\rightarrow \pi^*$
γ = program point id	load: $\pi^* \rightarrow \pi$
δ = function id	store: $\pi^* \times \pi \rightarrow \pi$

i.e., the value of x in ρ is updated to y. The evaluation functions (also called the transformer of OS) are defined in Fig. 3. For the sake of brevity, we only show the effects on the updated elements of state d per rule.

$$\Box x \leftarrow c \Box(d) = \rho \oplus \{x \mapsto c\}$$
$$\Box x \leftarrow v \Box(d) = \rho \oplus \{x \mapsto \rho(v)\}$$
$$\Box x \leftarrow tid \Box(d) = \rho \oplus \{x \mapsto \#\}$$
$$\Box x \leftarrow arg(i) \Box(d) = \rho \oplus \{x \mapsto \rho(arg(i))\}$$
$$\Box x \leftarrow phi((v_1,b_1),(v_2,b_2)) \Box(d) = \rho \oplus \left\{x \mapsto \begin{cases} \rho(v_1) & if\, @(b_1) = true \\ \rho(v_2) & if\, @(b_2) = true \end{cases}\right\}$$
$$\Box x \leftarrow load(a) \Box(d) = \rho \oplus \{x \mapsto \mu(\rho(a))\}$$
$$\Box x \leftarrow store(a,v) \Box(d) = \mu \oplus \{\rho(a) \mapsto \rho(v)\}$$
$$\Box x \leftarrow \omega(v_1,v_2) \Box(d) = \rho \oplus \{x \mapsto \omega(\rho(v_1),\rho(v_2))\}$$

Fig. 3. Evaluation functions for the operational semantics

Finally, in order to reason about SIMD programs, we lift OS to operate on vectors instead of scalar values. This is straightforward: # is a vector of instance identifiers now, and every function is evaluated separately for each of these SIMD instances. Instances that are inactive, i.e., their value in @ at the current program point is false, are not updated. Note that this lifting means that phi can blend two incoming vectors if some values in @ are true for either predecessor program point.

In order for this to work, loops iterate until the predicate that describes which instances stay in the loop is entirely false. This means that loops iterate as long as any of the instances needs to iterate. Execution then continues with the loop exit program points, again in topological order. Their predicates are correct since they have been continually updated while iterating the loop.

Collecting Semantics. OS implicitly defines a Collecting Semantics (CS) which combines all possible states of a program point to a set of states. This means it operates on a set of states D which contains all states d of the program at a given program point. The sets ρ, μ, @, and # are lifted to sets of sets. This collection of sets of states allows us to reason about universal characteristics of the states. Most importantly, the alignment of traces that is ensured by post dominator reconvergence prevents cases where we would derive characteristics from values that belong to different loop iterations.

Vectorization Analysis. We now define an Abstract Semantics (AS) that abstracts from the Collection Semantics by reasoning over SIMD characteristics instead of concrete values. In the following, the transfer functions

$$\#: (Vars) \quad (DBAL) \quad (Vars) \quad (DBAL)$$

of AS (the abstract transformer) are defined. They can be computed efficiently by a data-flow analysis we refer to as Vectorization Analysis. The functions D_{Abs}, B, A, and L contain the analysis information. They map variables to elements of the lattices D, B, A, and L (see Figs. 4 and 5).

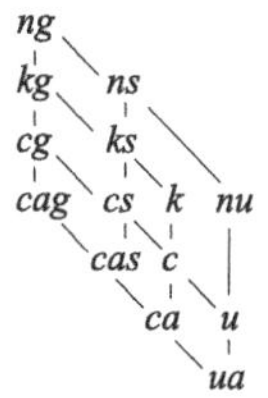

Legend	
Acronym	Property
n	nonvectorizable
g	guarded
s	sequential
k	unknown
c	consecutive
a	aligned
u	uniform

Concrete Value Examples

Element	Shape of Vector	Example
k	$\langle n_0, n_1, \ldots n_{w-1} \rangle$	$\langle 9, 2, 7, 1 \rangle$
c	$\langle n, n+1, \ldots n+w-1 \rangle$	$\langle 3, 4, 5, 6 \rangle$
ca	$\langle m, m+1, \ldots m+w-1 \rangle$	$\langle 0, 1, 2, 3 \rangle$
u	$\langle m+c, m+c, \ldots m+c \rangle$	$\langle 7, 7, 7, 7 \rangle$
ua	$\langle m, m, \ldots m \rangle$	$\langle 4, 4, 4, 4 \rangle$
nu	*type not vectorizable*	*void*, void*, void*, void**
		$n \in$ int, *float*, m=n $\cdot$ W

Fig. 4. Hasse diagram of the lattice D, legend, and value examples

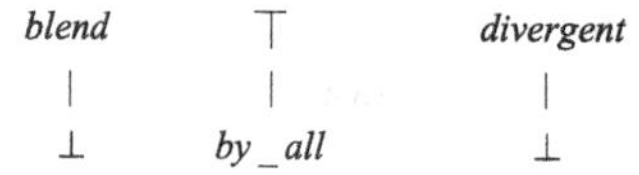

Fig. 5. Hasse diagrams of the lattices B, A and L

Note that, since we consider SSA-form programs, the terms variables and values both refer to program points. A program point also has a concrete operation associated. The presented analyses use join (not meet) lattices and employ the common perspective that instructions reside on edges instead of nodes. Program points thus sit between the instructions (see Fig. 6). This scheme has the advantage that the join and the update of the flow facts are cleanly separated.

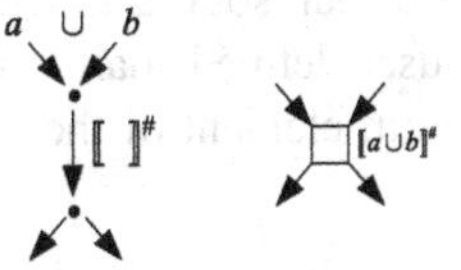

Fig. 6. Left: Analysis setup with separated join and update of flow facts. Right: Classic setup with mixed join/update

As in the Operational Semantics, the notation $D_{Abs} \oplus \{x \to y\}$ stands for

$$\lambda v. \begin{cases} y & \text{if } x = v \\ D_{Abs}(v) & \text{otherwise,} \end{cases}$$

i.e., the value of x in D_{Abs} is updated to y.

The order of characteristics of DAbs is visualized by the Hasse diagram, which is shown in Fig. 5. It describes the precision relation of the characteristics. An element that is lower in the diagram is more precise than an element further up. This relation is expressed by the operators $\subseteq$ and $\supseteq$: The notation a $\subseteq$ b describes the fact that a is at least as precise as b, a $\supseteq$ b means that a is at most as precise as b.

Our analysis tracks the following information for each program point. Is the value the same for all instances that are executed together in a SIMD group (uniform), and is it a multiple of the SIMD width (uniform/aligned) or does it have a type that is not vectorizable (uniform/nonvectorizable)? Such a uniform variable can be kept scalar and broadcast into a vector when needed, or used multiple times in sequential operations if it is nonvectorizable.

Otherwise, a value may hold different values for different instances (varying). If possible, these values are merged into a vector in the SIMD function. We track whether a varying value contains consecutive values for consecutive instances (consecutive) and whether it is aligned to the SIMD width (consecutive/aligned). The latter allows using faster, aligned memory instructions. The former still avoids sequential, scalar execution but needs unaligned memory accesses. If nothing can be said about the shape of a varying value, its analysis value is set to unknown or, if it has a type that cannot be vectorized, nonvectorizable.

Furthermore, we track the information whether a program point is sequential, and whether it is guarded. A sequential operation cannot be executed as a vector operation, but has to be split into W sequential operations. If the program point is guarded, it also requires conditional execution. This means that each sequential operation is guarded by an if statement that evaluates to true only if the mask element of the corresponding instance is set.

Finally, control-flow related information is tracked: Is the program point always executed without inactive instances (by all) Is it a join point of diverged instances of some program point v (blendv) Does it have to be executed if some program point v was executed (rewirev) or can it be skipped under certain circumstances (optional) Finally, we track whether a program point is part of a loop that some instances may leave at different exits or at different points in time (divergent).

Figure 7 shows the transfer functions. The initial information passed to the analysis is twofold: the signatures of the scalar source function f and the vectorized target declaration of fW on one hand, user-defined marks on the other. All other program points are initialized with the bottom element of the lattice, ua (uniform/aligned).

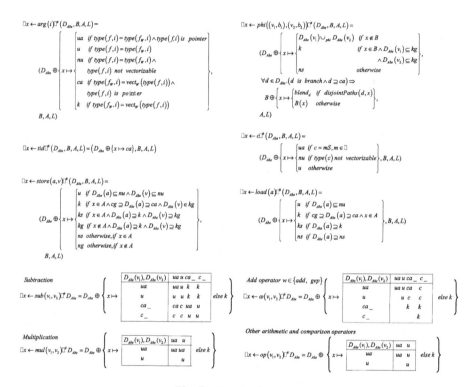

Fig. 7. Transfer functions

3.2 Vectorization Instruction Optimization

The Vectorization Analysis attempts to prove that the address of a memory operation is uniform or consecutive. While an unknown load or store requires W sequential operations, a uniform load or store only requires a single, scalar operation. A consecutive load or store can use a vector operation to access W elements at once.

If the current mask is not true for all instances, memory operations and calls to external, non-native functions have to be split into W guarded scalar operations. This is because we have to conservatively assume them to produce side effects that we do not want to occur for inactive instances. We have to prevent the execution of a store operation for inactive instances. Thus, we have to guard each scalar execution by an if conditional that skips the instruction if the mask of that instance is false. Unfortunately, this involves a lot of extract-, insert-, and branch-operations that sacrifice the overall benefit of vectorization. However, we can optimize such a store operation by generating a load-blend-store sequence of vector operations that is faster than an if-cascade with scalar stores as described above. Another way to circumvent this issue is hardware support of conditional load/store instructions as e.g. AVX will provide.

For linearized regions, an optimization similar to branch-on-superword condition-codes (BOSCC) can be applied. Such a technique reintroduces branches after linearization to skip a basic block or an entire control flow path if the mask of the corresponding entry edge is entirely false at runtime. This way, it trades some performance for the runtime check for a larger gain every time the block can be skipped.

In the setting of WFV, predicates are stored on a per-block basis. In addition, it is easy to maintain information about structured control flow even across the CFG linearization phase. For this variant, only the start and end block of the path have to be stored. After linearization, an edge is introduced that goes directly from the start to the end. If there is a disjoint neighboring path, as in case of an if statement, an additional edge can be introduced.

3.3   Instance Multi-version

If the function in question has complex code that is frequently executed with only one active instance, it may be beneficial to switch back to sequential execution for that part. This optimization first determines the index of the single active instance. On the optimized code path, all required values for that instance are extracted into scalar registers. The rest of that path consists of the scalar code of the original kernel before vectorization. Finally, at the end of the path, the results are packed back into the corresponding vectors. Of course, this transformation is only beneficial if the vectorized code suffers from a higher overhead than the extraction/insertion that is required to execute the scalar code. However, additionally, the scalar code may hold the potential to use an orthogonal vectorization approach such as Superword-Level Parallelism.

The variant can be executed if only a subset of the SIMD group is active. The code transformation for the optimized code path merges independent, isomorphic vector operations into a single vector operation. Intuitively, this can be found as switching the vectorization direction. Where normal WFV vectorizes horizontally (each operation works on combined inputs), SLP vectorizes vertically (different operations with different inputs are combined). Since the original code is already vector code, combining values is more complicated than when transforming scalar code to SLP code. This is because the values of active instances first have to be extracted from their original vectors and then combined to a new one. This means that this variant possibly involves significant amounts of overhead due to the data reorganization.

Consider the example in Fig. 8, which shows a WFV-SLP variant for two out of eight active instances ($W = 8$). Inside the while loop, there are two code paths. The else path is the original vector code that computes products of sums and differences. The results of 3 of these operations that are independent are added up and stored in variable

```
void kernel()
{
    int tid= get_global_id();
    float8 x,y,z,a,b,c,···
    do{
        bool m=x<c8;
        int n=m[0]+···+m[7];
        if(n==2){              //instance multi-version
                               //assume 2 instances active
            int idx0=-1,idx1=-1;
            for(int i=0;i<8;i++){
                if(m[i]==0)continue;
                if(idx0=-1)idx0=i
                    else idx1=i;
            }
            float8 m=(x[idx0],x[idx1],y[idx0],y[idx1]
                ,0,0,0,0);          // instance regroup
            float8 n=(a,a,b,b,0,0,0,0); // instance regroup
            float8 d=m-n;           // switch to SLP
            float8 v=m+n;
            float8 o=d*v;
            r[idx0]=o[0]+o[2];          // extract result
            r[idx1]=o[1]+o[3];          // extract result
        }
        else{          //insist whole function vectorization
            float8 dx=x-a;
            float8 dy=y-b;
            float8 vx=x+a;
            float8 vy=y+b;
            float8 mx=dx*vx;
            float8 my=dy*vy;
            mr=mx+my;
            r=m?mr:r;
        }
    }while(any(m))
    *((float8*)(array+tid))=r;
}
```

Fig. 8. If two out of eight instances are active, values are extracted and SLP is employed

r. This is very efficient if all or most instances are active. However, if only few are active, a lot of computations are wasted because their results are discarded by the blend.

The variant shown in the then part improves on this. It is only executed if two instances are active. First, the indices idx0 and idx1 of the active instances are determined. Then, the input values of the two independent additions and subtractions are combined into two vectors (m, n) instead of six (x, y, a, b) as in the original vector code. Because these input values do not depend on each other, it is safe to use a single vector addition and subtraction instead of two operations of either type. Finally, scalar additions are required per active instance instead of the vector additions used in the original code. However, the variant code does not require an additional blend operation. This is because each result is inserted directly into the result vector at the appropriate index.

3.4   Instance Reorganization

Instance reorganization is only relevant for pumped vectorization with $W = V * S$. Executing more instances in the same function than the number of available SIMD lanes offers an additional optimization opportunity. Instead of executing a single instruction or block V times sequentially, a larger code region is executed V times sequentially, but the instances are reorganized.

Instance Reorganization exploits the fact that there are only two possible decisions where to go for each instance. Instance Reorganization is using to improve control flow coherence. The code makes use of the fact that if the W instances diverge into two sets, they can be reorganized such that there is at most one subset of S instances that do not agree on the direction. As it is shown in Fig. 9, the code is to be executed V − 1 times with uniform control flow after reorganization, and only once the linearized code which accounts for diverged instances. This can be achieved either with code duplication or with a loop with a switch statement that determines which path to execute with which reorganized group.

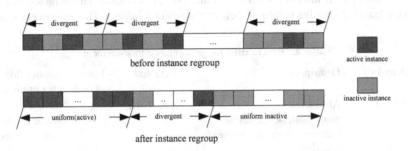

Fig. 9. Instance reorganization

On the flipside, reorganization may impose significant cost if many values are live at the point of reorganization. The code with instance reorganization may be less efficient in many cases than the code obtained by standard WFV or pumped WFV. The reorganization is also only valid for a single varying branch. Only the corresponding control flow of this branch can retain in the V − 1 coherent executions, each nested varying branch requires linearization again.

4 Experimental Evaluation

4.1 Experiment Setup

We implemented the optimization method presented in this paper in the LLVM compiler framework. All experiments are conducted on an Intel Xeon E5520 at 2.26 GHz with 16 GB of RAM running Ubuntu Linux 13.04 64-bit. The vector instruction set is Intel's AVX, yielding a SIMD width of four 64-bit values. The machine run in 64-bit mode, thus 16 vector registers were available.

We propose a set of pragmas for programmers to perform high-level SIMD programming explicitly with a similar look-and-feel of high-level OpenMP parallel programming. Pragma__decl-vec-function denotes a function that can be vectorized. Pragma #pragma simd vec-function denotes a loop that can be vectorized, without regard to calls in the loop. We add a compiling option -opt-wfv = [true/false] in the LLVM to control our optimization method.

4.2   Experiment Result

We choose nine workloads from different application domains to test the performance of our SIMD vector extensions and compiler implementation. These workloads cover a diverse set of real-world problems ranging from sorting algorithms over stock option estimation to physics simulations and computer graphics. Table 4 provides the information on these workloads, including input data size, number of lines of code, control flow characteristics, and a brief description of each workload.

To assess the impact of our analyses and code generation techniques, we compare the performance of code in different configurations that enable or disable certain analyses and optimizations of WFV. If all analyses are disabled, all characteristics are set to the least informative values of their corresponding lattices. This effectively results

Table 4. Benchmarks for performance measurement

Benchmarks	Description	Data size	Lines of code	Control flow characteristic
Mandelbrot	Abstract mathematics-computation of complex quadratic recurrence equation	8 K × 8 K	85	for loop in vector function with early exit
BlackScholes	Stock option estimation	16 K	106	Simple nested for loops
NBody	Simulate the evolution of a dynamic system of N-body or particles under the influence of physical forces	8 K	1294	for loop inside vector function
DwtHaar1D	Compression algorithm	30 K	120	Single branch in vector function
DCT	Compression algorithm	8 K × 8 K	236	Multiple branches in vector functions
Histogram	Image processing	16 K × 16 K	344	Multiple branches in vector functions
MersenneTwister	Random number generator	32 K	80	Single branch in vector function
BitonicSort	Sort algorithm	1 M	120	Single branch in vector function
AoBench	Graphics: shader computes attenuation due to occlusion	1 K × 1 K	78	Multiple branches in vector functions

in W-fold splitting of each statement without using vectors or vector operations. Since this is not very meaningful for a vectorization algorithm, the default mechanism when all analyses are disabled is to initialize all program points with unknown. This means that no optimizations that are based upon uniform values have any effect, but only those operations that have to are split and/or guarded.

Summarized, these are the tested configurations: the scalar configuration does not perform WFV. The naive configuration initializes all values with unknown. This disables all optimizations that exploit uniform values and consecutive memory access operations. Control flow is also fully linearized since the Rewire Target Analysis is disabled. The uniform + consecutive configuration employs uniform, consecutive, and aligned, i.e., it optimizes memory access operations. The all configuration finally enables all analyses, i.e., it now also employs the Rewire Target Analysis to retain structure of the CFG. Figure 10 shows an overview of the performance of the different WFV configurations. The concrete numbers can be obtained from Table 5.

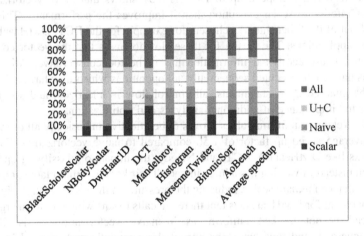

Fig. 10. SIMD performance speedup of workloads

Table 5. The concrete numbers of different configurations

Application	Scalar	Naive	U + C	All	Speed up
BlackScholesScalar	570	177	175	173	1.02 X
NBodyScalar	229	107	63	60	1.78 X
DwtHaar1D	9	23	7	6	3.83 X
DCT	866	2351	960	653	3.60 X
Mandelbrot	1046	774	735	684	1.13 X
Histogram	840	1159	905	762	1.52 X
MersenneTwister	1449	1618	1094	719	2.25 X
BitonicSort	166	156	149	149	1.05 X
AoBench	2123	1344	1358	1350	1.00 X
Average speed up	–	1.30X	1.56X	1.17X	1.91 X

Table 5 Median kernel execution times (in milliseconds) of our method in different configurations for different applications (W = 4). U + C stands for code generation based on SIMD property. The column speedup shows the effect of our optimizations, comparing all to naive. The average speedup denotes the geometric mean of the speed ups compared to the next configuration(e.g. **all** vs. **U + C**).

The overall observation is that performance improves with the addition of analyses and optimizations. In the following paragraphs, we assess the impact of gradually enabling more parts of our analyses. We chose to use an accumulative scheme since the analyses build on top of each other. For example, it does not make sense to classify consecutive memory operations without determining which values are uniform.

Naive Vectorization. The first observation is that even naive vectorization can yield significant speed ups. For example, BlackScholesScalar improves by a factor of 3.22, and NBodyScalar by a factor of 2.14. These benchmarks are dominated by arithmetic operations with only few different control flow paths. This makes them perfect targets for WFV. However, the speed up of DCT is 0.37 shows that naive vectorization is sometimes inferior to scalar execution than it improves the performance. This is the case for 4 out of the 9 benchmarks chosen, for example for the DCT, DwtHaar1D, and Histogram applications. The reason for these slowdowns is that the vector code generated without any additional information has to be too conservative: All memory operations have to be executed sequentially, all operations which may have side effects have to be guarded by if statements, all control flow is linearized, and so on. This highlights the importance of additional analyses and optimizations.

U + C. Retaining uniform values proves to be effective in basically all of the cases with an average speed up factor of 1.56 compared to naive vectorization. For some benchmarks like DwtHaar1D, DCT, and NBody, the impact of classifying operations as uniform instead of varying has a huge impact. Their kernels exhibit large amounts of operations that can remain scalar, reducing the pressure on the vector unit. A big part of the improvement for DwtHaar1D is that there are calls to sqrt with a uniform argument. If this call is not marked uniform, as in naive vectorization, it is lifted to unknown/guarded, and thus has to be executed sequentially and guarded by conditionals. Exploiting consecutive characteristics is especially effective for DCT.

ALL. Finally, our last configuration, all, employs the Rewire Target Analysis. This enables the CFG linearization phase to retain parts of the control flow graph where it could be proven that the instances cannot diverge. These results in less code being executed, less register pressure, and less overhead for mask and blend operations. This also enables further optimization by allowing the Vectorization Analysis to be more precise. In Histogram, for example, there is a store operation inside a loop. Conservatively, it has to be executed sequentially and guarded to account for possibly diverging control flow. However, the loop is proven not to diverge, and furthermore to always be executed by all instances. This allows us to issue a vector store during instruction vectorization.

The effect of the instance multi-version and instance reorganization can be best observed for DCT, Histogram, and MersenneTwister. These benchmarks profit most from the control-flow related improvements, with speedup factors of 1.47, 1.19, and 1.50. As expected, there is no effect on benchmarks that do not have any non-divergent control flow, such as BitonicSort. The average speedup compared to the uniform + consecutive configuration is 1.17. The average speedup compared to the naive configuration is 1.91. The final speedup achieved over scalar execution is on average 1.84 for these benchmarks.

5 Related Work

Currently, there are two major vectorization algorithms. Loop-based algorithm can combine multiple iterations of a loop into a single iteration of vector instructions. Super-word level parallelism (SLP) targets straight-line code.

Loop-based algorithms are based on the notion of data dependence along with several classical loop transformations. Strip-mining, scalar expansion, reduction processing, loop distribution and outer-loop vectorization are major loop transformation techniques used to enhance parallelism [8]. A cost model and a loop transformation framework is proposed to extract SIMD parallelism opportunities and to select an optimal strategy among them, which is based on polyhedral compilation, leveraging its representation of memory access patterns and data dependences as well as its expressiveness in building complex sequences of transformations [9, 10].

SLP has been introduced to taking advance of SIMD extensions for the straight-line code. Larsen and Amarasinghe [11] are the first to present an automatic vectorization technique based on vectorizing parallel scalar instructions with no knowledge of any surrounding loop. A back-end vectorizer in the instruction selection phase based on dynamic programming is introduced by Barik et al. [12]. The approach is different from most of vectorizers as it is close to the code generation stage and can make more informed decisions on the costs involved with the instructions generated. Liu et al. present a vectorization framework that improves SLP by performing a more complete exploration of the instruction selection space while building the SLP tree [13]. Kim et al. propose a technique to detect and exploit more parallelism by dynamically analyzing data dependences at runtime, and thus guiding vectorization [14]. Other straight-line code vectorization techniques that depart from the SLP algorithm have also been proposed in the literature [15–17].

In spite of similarities, there exist several differences between traditional vector machine and SIMD extensions [18]. The most significant differences arise from the weaker memory units of SIMD extensions. In contrast to those of vector processors, the memory units of SIMD extensions usually do not support scatter/gather operations [19]. They only allow to access memory locations that are aligned at vector register length boundaries. Eichenberger et al. proposes a method for vectorizing loops with misaligned stride-one memory references [20]. Ren et al. optimizes a sequence of multiple data reorganization for statically misaligned data [21]. Nuzman and Rosen extends a loop-based vectorization technique to handle computations with non-unit stride accesses to data, where the strides are the powers of 2 [22]. Other memory access optimizations have also been proposed in the literature to enhance SIMD performance [23–25].

Ispc is similar to CUDA and OpenCL in concept, which are both SPMD languages that primarily target GPUs, although some OpenCL implementations also target vectorization for CPUs [26]. Ispc's "SPMD-on-SIMD" approach coupled with its small set of language features enable programmers to easily write a single source code that can efficiently compile to different targets [27]. The ispc language is based upon C and C++. To the best of our knowledge, our work is the first to propose code optimization of whole function vectorization on SIMD extensions.

6 Conclusion

An optimization method based on SIMD property is proposed in this paper. First, the notion and advantage of whole function vectorization are illustrated. Second, we describe the transformation of whole function vectorization based on static single assignment. Third, SIMD property of operations and basic blocks in programs are analyzed. Code optimization of whole function vectorization based on the SIMD properly are implemented, which includes instance multi-version, instance regroup and vectorization instruction optimization. We implement our algorithm in the LLVM compiler infrastructure and evaluate it in nine practical kernels. Compared with un-optimized codes, we show an average speedup of 1.59 times faster for these kernels. An average speedup of 1.91 times faster is achieved by our method compared with scalar code. To exploit function-level vectorization without pragmas will be the further work.

References

1. Gao, W., Zhao, R.C., Han, L., Pang, J., Rui, D.: Research on SIMD auto-vectorization compiling optimization. J. Softw. **26**(6), 1265–1284 (2015)
2. Huo, X., Ren, B., Agrawal, G.: A programming system for xeon phis with runtime SIMD parallelization. In: Proceedings of the 28th ACM International Conference on Supercomputing (ICS), pp. 283–292 (2014)
3. Chen, L., Jiang, P., Agrawal, G.: Exploiting recent SIMD architectural advances for irregular applications. In: Proceedings of the 14th International Symposium on Code Generation and Optimization (CGO) (2016)
4. Lei, R.A., Sierra, H.I.: A SIMD extension for C++. In: Proceedings of the 19th PPOPP Workshop on Programming models for SIMD/Vector processing (WPMVP), pp. 17–24 (2014)
5. Evans, G.C., Abraham, S., Kuhn, B.: Vector seeker: a tool for finding vector potential. In: Proceedings of the 19th PPOPP Workshop on Programming models for SIMD/Vector processing (WPMVP), pp. 41–48 (2014)
6. Wang, Y., Wang, D., Chen, S.: Iteration interleaving–based SIMD lane partition. ACM Trans. Architect. Code Optim. (TACO) **12**(4) (2016)
7. Karrenberg, R., Hack, S.: Whole-Function vectorization. In: Proceedings of the 9th ACM International Symposium on Code Generation and Optimization (CGO), pp. 141–150 (2011)
8. Nuzman, D., Zaks, A.: Outer-loop vectorization-revisited for short SIMD architectures. In: Proceedings of the 2008 International Conference on Parallel Architectures and Compilation Techniques (PACT) (2008)
9. Trifunovic, K., Nuzman, D., Cohen, A., et al.: Polyhedral-model guided loop-nest auto-vectorization. In: Proceedings of the 2009 International Conference on Parallel Architectures and Compilation Techniques (PACT) (2009)

10. Kong, M., Veras, R., Stock, K.: When polyhedral transformations meet SIMD code generation. In: Proceedings of the 2013 Conference on Programming Language Design and Implementation (PLDI) (2013)

11. Larsen, S., Amarasinghe, S.: Exploiting superword level parallelism with multimedia instruction sets. In: Proceedings of the ACM SIGPLAN ·Conference on Programming Language Design and Implementation, pp. 145–156 (2000)

12. Barik, R., Zhao, J., Sarkar, V.: Efficient selection of vector instructions using dynamic programming. In: Proceedings of the 43rd Annual IEEE/ACM International Symposium on Microarchitecture (MICRO) (2010)

13. Liu, J., Zhang, Y., Kandemir, M.: A compiler framework for extracting superword level parallelism. In: Proceedings of the 2012 Conference on Programming Language Design and Implementation (PLDI) (2012)

14. Haque, M., Yi, Q.: Past dependent branches through speculation. In: Proceedings of the 22nd International Conference on Parallel Architecture and Compilation Techniques (PACT). IEEE Computer Society, Washington DC (2013)

15. Porpodas, V., Magni, A., Timothy, M.: PSLP: Padded SLP automatic vectorization. In: Proceedings of the 2015 Annual IEEE/ACM International Symposium on Code Generation and Optimization (CGO) (2015)

16. Porpodas, V., Jones, T.M.: Throttling automatic vectorization: when less is more. In: Proceedings of the 24th IEEE Computer Society International Conference on Parallel Architectures and Compilation Techniques (PACT) (2015)

17. Zhou, H., Xue, J.: Exploiting mixed SIMD parallelism by reducing data reorganization overhead. In: Proceedings of the 14th ACM International Symposium on Code Generation and Optimization (CGO) (2016)

18. Morad, A., Yavits, L., Kvatinsky, S.: Resistive GP-SIMD processing-in-memory. ACM Trans. Architect. Code Optim. (TACO) 12(4) (2016)

19. Chang, H., Sung, W.: Efficient vectorization of SIMD programs with non-aligned and irregular data access hardware. In: Proceedings of the 2008 International Conference on Compilers, Architectures and Synthesis for Embedded Systems (CASES), pp. 167–176 (2008)

20. Eichenberger, A.E., Wu, P., O'Brien, K.: Vectorization for SIMD architectures with alignment constraints. In: Proceeding of the ACM SIGPLAN 2004 Conference on Programming Language Design and Implementation (PLDI), pp. 82–93. ACM Press, New York (2004)

21. Ren, G., Wu, P., Padua, D.A.: Optimizing data permutations for SIMD devices. In: Proceedings of the 2006 ACM SIGPLAN Conference on Programming Language Design and Implementation (PLDI), pp. 118–131 (2006)

22. Nuzman, D., Rosen, I.A.: Auto-vectorization of interleaved data for SIMD. In: Proceedings of the ACM SIGPLAN 2006 Conference on Programming Language Design and Implementation (PLDI), pp. 132–143 (2006)

23. Sharma, N., Panda, P.R., Catthoor, F.: Array interleaving—an energy-efficient data layout transformation. ACM Trans. Design Autom. Electron. Syst. 20(3), 1–26 (2015)

24. Asher, Y.B., Rotem, N.: Hybrid type legalization for a sparse SIMD instruction set. ACM Trans. Architect. Code Optim. (TACO) 10(3), 520–532 (2013)

25. Jie, S.J., Kapre, N.: Comparing soft and hard vector processing in FPGA-based embedded systems. In: Proceedings of the 24th Field Programmable Logic and Applications (FPL) (2014)

26. Pharr, M., Mark, W.R.: ispc: A SPMD compiler for high-performance CPU programming. In Innovative Parallel Computing, pp. 65–74 (2012)

27. Kerr, A., Diamos, G., Yalamanchili, S.: Dynamic compilation of data-parallel kernels for vector processors. In: Proceedings of the 10th International Symposium on Code Generation and Optimization (CGO), pp. 23–32 (2012)

A Stacked Denoising Autoencoders Based Collaborative Approach for Recommender System

Baojun Niu, Dongsheng Zou[(✉)], and Yafeng Niu

Chongqing University, Chongqing 400044, China
dszou@cqu.edu.cn

Abstract. This paper uses an autoencoder neural network as user feature learning component for collaborative filtering task. We propose a stacked denoising autoencoder (SDAE) based model to alleviate the sparseness issues in recommendation system. Our model also extends the scalability of CF-based methods in the Top-N recommendation task. Experiments on MovieLens datasets and the result confirmed the effectiveness and potential of our model.

Keywords: Collaborative filtering · Recommender system · Stacked denoising autoencoder

1 Introduction

Recommender System (RS) is an effective tool to deal with information overload problem [1]. As the main thrust in the field of RS research, collaborative filtering(CF) methods are widely used for its knowledge domain free. CF-based approaches can be divided into two branches: neighborhood-based and model-based [2]. Neighborhood-based approaches focus on spotting the similarity relationships among either users or items. Herlocker et al. [3] carried out a comprehensive research into user-based collaborative algorithms. Badrul et al. [4] presented the item-based collaborative system to pre-compute the item-item similarities.

In contrast to the neighbor-based method, singular value decomposition (SVD) [6] is the most prevalent model-based method for Matrix factorization (MF) [5]. SVD performs well in building accurate models for RS by applying elaborated and succinct algebraic structure. Nevertheless, SVD method is only well-defined when the matrix is complete, still facing the problem of sparseness. Imputation [7] is a technique to get a relatively denser matrix filled with baseline estimations for missing ratings. However, inaccurate imputation might distort the original data considerably.

In recent years, research in the neural network has taken a breakthrough in deep layer architecture. The deep belief networks (DBN) [8] inspire researchers to integrate neural network with CF for patterns learning or useful features extracting [9] uses Restricted Boltzmann Machines (RBM) to perform CF, and [10] extended this work to a noo-iid framework. Autoencoder (AE) as a special neural network has achieved good performance in the text processing, image classification tasks. In this paper, we employ

© Springer Nature Singapore Pte Ltd. 2017
G. Chen et al. (Eds.): PAAP 2017, CCIS 729, pp. 172–181, 2017.
DOI: 10.1007/978-981-10-6442-5_15

AE as a user preference learning component in collaborative filtering and develop a stacked denoising autoencoder (SDAE) based model to alleviate sparseness issue in RS.

2 Traditional Autoencoder

A traditional autoencoder is a feedforward neural network illustrated in Fig. 1. AE can learning representations of the raw data by reconstructing the input data from its output. Usually, it has a three-layered structure, the input and the output has the same number of neurones, and a hidden layer is in the middle.

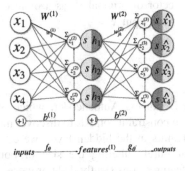

Fig. 1. Traditional autoencoder

Training an autoencoder is to minimise the reconstruction error, solve the following optimisation problem:

$$L(x, \hat{x}) = \frac{1}{2m} \sum_{i=1}^{m} \left\| \hat{x}^i - x^i \right\|^2 + \frac{\lambda}{2} \left(\left\| W^{(1)} \right\|^2 + \left\| W^{(2)} \right\|^2 \right)$$

$$\arg\min_{\theta, \theta'} L(x, \hat{x})$$

$$(1)$$

To get the hyper-parameters $W^{(l)}, b^{(l)}, l = 1, 2$ in (1). Stochastic Gradient Descent (SGD) is an efficient method to train autoencoder. The key steps of SGD is computing the partial derivatives. We use backpropagation algorithm to compute these partial derivatives. Steps as follows:

For each training example x^i

1. Feedforward pass the inputs, we get activations for each layer: $x^i, h^i, \hat{x}^i$.
2. Backpropagate the errors:
   (1) Compute the error of output layer,

$$\delta^{(2)} = (\hat{x}^i - x^i) \cdot \hat{x}^i \cdot (1 - x^i)$$

$$(2)$$

   (2) Backpropagate the output error $\delta^{(2)}$ to the hidden layer,

$$\delta^{(1)} = (W^{(2)})^T \cdot \delta^{(2)} \cdot h^i \cdot (1 - h^i) \tag{3}$$

3. Compute the desired partial derivatives, set as

$$\frac{\partial L}{\partial W^{(1)}} = \delta^{(1)} (x^i)^T, \frac{\partial L}{\partial b^{(1)}} = \delta^{(l)} \tag{4}$$

$$\frac{\partial L}{\partial W^{(2)}} = \delta^{(2)} (h^i)^T, \frac{\partial L}{\partial b^{(2)}} = \delta^{(2)} \tag{5}$$

4. After T rounds iterations when finally the value of (1) tends to be stable, we can take h^i, the high-order vector of original x^i, as a more concentrated and effective feature representation of raw input x^i.

3   Construct SDAE

SDAE stands for Stacked Denoising Autoencoder, which is a variant of traditional autoencoder. We can impose constraints on AE to discover more useful representation. By altering the number of units in the hidden layer, we can get a lossy compressed representation feature. But purely altering the size of bottleneck layers may not guarantee to extract good features consistently. [11] proposed a strategy to corrupt the clean input partially for constructing a denosing autoencoder structure(DAE). There are three commonly used noise models to corrupt the input, the additive Gaussian noise, the salt-and-pepper noise and the Masking noise.

The training process of DAE is the same as an AE, except that it takes corrupted input $\tilde{x}$ generated from a particular noise model and then minimise reconstruction error between its outputs and the original clean inputs. That is, the overall loss function is still the same. Figure 2. shows a schematic for these procedures.

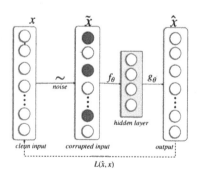

Fig. 2. The schematic of denoising autoencoder

Based on all of the foundations above, learning even higher level representations would be possible by stacking denoising autoencoders. SDAE is a bit deeper neural network constructed with multiple levels of DAEs in which the hidden outputs of each level is accepted as inflow to the successive.

Greedy layer-wise training [12] is an efficient way to obtain parameters for SDAE or SAE. After training the first level autoencoder for corrupted input, its learnt encoding operator f_θ acts on clean input to get the first level hidden layer vector, then use it to train the second level DAE to learn encoder $f_\theta^{(2)}$. From here, repeat the procedure. After training a stack of encoders, perform backpropagation through the whole system to fine-tune the hyper-parameters globally. Figure 3 illustrates the schematic for training the SDAE.

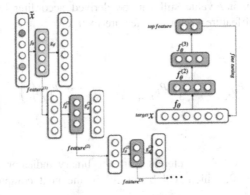

Fig. 3. Schematic for training the SDAE

3.1 SDAE Applied in Collaborative Filtering

In view of the ability of AE to mining inherent structure of data in many domains, we bring SDAE into collaborative filtering, trying to excavate delicate structure of user's preference, and obtain the compressed representation of user's behaviour vector, then apply these high-level features in recommendation task.

We model user's preference behaviour as a vector consisting of ratings over items, denoting $p_{u_i} \in \Re^n$, for example, in a 5-stars scaled rating system for movies. There exist a target user u_t which has a rated items collection $I_{u_t} = \{i_1, i_3, i_4, i_5\}$, in which $r_{t,1} = 5, r_{t,3} = 2, r_{t,4} = 3, r_{t,5} = 4$, then we can get $p_{u_i} = (5, NULL, 2, 3, 4, NULL, \cdots, NULL)^T$.

As shown above, due to the sparseness of rating matrix for RS, the user behaviour preference vector containing a large number of *NULL* value or missing value. There are many reasons for these missing ratings beyond only not liking it, so we can't fill these *NULLs* with "0" and feed this kind of input to autoencoder, otherwise, all these zero ones will be treated as negative preference leading to a strong bias in the system.

To address issue stated above, we extend the user preference vector to a user preference matrix denoting as $p_{u_i} \in \Re^{n \times S}$ where S is the numerical rating scale. In this

$n \times S$ matrix, if exists the rating $r_{i,j}$ on item i_j by user u_i valued with c, then we set the $p^c_{u_i,i_j} = 1$,otherwise 0, and we set the formula as

$$p^c_{u_i,i_j} = \begin{cases} 1, & if \ni r_{i,j} = c, & i_j \in I_{u_i} \\ 0, & if \ c \neq r_{i,j}, & i_j \in I_{u_i} \\ 0, & otherwise \, if & i_j \notin I_{u_i} \end{cases} \tag{6}$$

As for the values of $p^c_{u_i,i_j}$, we illustrate the same example aforementioned, in a 5-starts scaled(S = 5) rating system for movies, as for the partial feature comes from item i_1 in the combined user's preference matrix,we get $p^5_{u_t,i_1} = 1$ and set the rest part as 0, thus we get vector $p_{u_t,i_1} = (1,0,0,0,0)$, the same process as for the rest part, we get $p_{u_t,i_3} = (0,1,0,0,0)$, $p_{u_t,i_4} = (0,0,1,0,0)$, $p_{u_t,i_5} = (0,0,0,1,0)$. For those items $i_{other} \notin I_{u_i}$ called missing value still can be derived according to (6). Finally, we combine each partial feature vector as a feature matrix

$$P_{u_t} = \left(P_{u_t,i_1}, P_{u_t,i_2}, P_{u_t,i_3}, P_{u_t,i_4}, P_{u_t,i_5}, \cdots, P_{u_t,i_n} \right)^T = \begin{pmatrix} 1,0,0,0,0 \\ 0,0,0,0,0 \\ 0,1,0,0,0 \\ 0,0,1,0,0 \\ 0,0,0,1,0 \\ \vdots \\ 0,0,0,0,0 \end{pmatrix}$$

Note that the value "0" is deferent from the binary indicator vector **0**. And we suppose $0 \times NULL = 0$, which will be useful in the next computation process of modified AE.

In this article, we modify the autoencoder to adapt the *NULL*-value (sparseness) scenario illustrated as above. The particular modifications are listed as follows.

1. For a particular user u_i, only activate the input units for the items rated by that user, and then feed those data to SDAE for encoding.

2. In steps of building each level AE of SDAE, when involving decoding the hidden layer in each local level DAE. We also only reconstruct the ratings for the items rated by the user u_i, which also means only backpropagate the error coming from rated items by the current user.c

In this modified model, it seems that every user in the system will have their own personal autoencoder based on their own collection of rated items. Each autoencoder only has a single training case. Weights and biases belonging to one user-specific autoencoder contribute to the entire global neural networks. Thus if the item was co-rated by many users, these users would share the corresponding weights for that item through the path from input to output in their own autoencoder networks. Figure 4 illustrate the modified AE model.

After all above preparation, then we can train the basic DAE and then stack them all to get SDAE. For each user u_i, we substitute p_{u_i} for x^i and carry out the feedforward step illustrated in Sect. 2, the final output $\hat{p}^c_{u_i,i_j} \in [0,1]$ substituted for $\hat{x}^i$ represents the probability of item i_j rated in value c. Here we can regard c as a confidence level for

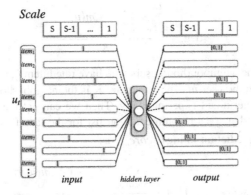

Fig. 4. Modified AE model in collaborative filtering

expressing not only the user rated the item but also indicate the preference extent the user showed when rating it.

By limiting the number of hidden neurons d' to a much smaller but reasonable scale than the total number of items n, the modified AE transform the sparse users behavior vector with null values into a dense feature vector. Thus decreasing the dimensionality of the user behavior vector space from n to d', in which small perturbances inherited from noise in the data are eliminated, leaving only the strongest effects or robust dependency features. Consequently, it decreases the storage and computational requirements for the next collaborative filtering workflow.

3.2 Generate Recommended List and Prediction

After the workflow of SDAE, we can obtain the compact and efficient feature vector learnt from user's original preference behaviour. For a particular user u_i, we regularise the hidden feature vector h^{u_i} to $bicode_{u_i}$, the process of binarization given as follows, where 0.5 is the threshold value to decide if the hidden unit was highly activated by some latent features.

$$bicode_{u_i k} = \begin{cases} 1, & if\ h^{u_i} \geq 0.5 \\ 0, & otherwise \end{cases} 1 \leq k \leq d' \tag{7}$$

We use the modified Hamming distance between these $bicode_{u_i}$ as the similarity metric formulated as

$$sim(u_1, u_2) = 1 - \frac{Hamm(bicode_{u_1}, bicode_{u_2})}{len(hidden\ units\ of\ final\ features)} \tag{8}$$

In the last step, we use the similarity between users to generate prediction ratings and recommendation list. Given the particular user u_i, and a certain item i_j, where $i_j \notin I_{u_i}$, we define deg_{u_i, i_j} to express the degree for item i_j worth of being recommended to user u_i.

$$\deg_{u_i,i_j} = \frac{\sum\limits_{v \in Nbr(u_i,i_j)} sim(u,v)}{K} \tag{9}$$

where $Nbr(u,i)$ are the neighbourhood users who have rated item i_j, and shared the most top-K similarity with the user u_i. The parameter K is the neighborhood size, needed to be optimized and determined experimentally. Then we can get the Top-N items in a descending order of $\deg_{u_i,i_j}$ as recommendation list for the target user.

We also define prediction ratings $score_{u_i,i_j}$ as

$$score_{u_i,i_j} = \bar{r}_{u_i} + \frac{\sum\limits_{v \in Nbr(u_i,i_j)} (r_{v,i_j} - \bar{r}_v) sim(u,v)}{\sum\limits_{v \in Nbr(u_i,i_j)} sim(u,v)} \tag{10}$$

4 Experiments and Discussion

4.1 Dataset and Evaluation

We use MovieLens datasets provided by GroupLens research team to evaluate the performance of our SDAE-CF approach and compare the results with several alternative baseline CF-based methods. To mimic different extent sparseness of the rating matrix, in this paper we experiment on MovieLens 100 k, 1 M, and 10 M datasets with their statistics summarised in Table 1.

Table 1. Statistics for data sets of the MovieLens

Type	Users	Movies	Ratings	Density%
100 k	943	1682	100,000	6.30
1 M	6040	3900	1,000,209	4.26
10 M	71567	10681	10,000,054	1.31

We randomly sample 80% ratings for each user as the training set and the rest 20% is used as the testing set. We generate five independent splits and report the averaged performance in our evaluations. We firstly take MAE as our evaluation metric to report experiments in user preference prediction. Then we use recall as another performance measure particularly for the Top-N recommendation task [13]. The evaluation rules are listed as follows, where $Test_u$ denotes the items having been rated by the user in the testing set. I_r is the set of Top-N recommendation generated from its training set denoted as $Train_u$.

$$MAE = \frac{\sum_{u \in U} \left| r_{i,j} - scrore_{u_i,i_j} \right|}{\sum_{u \in U} |Test_u|}, \; recall = \frac{\sum_{u \in U} |I_r \cap Test_u|}{\sum_{u \in U} |Test_u|}$$

4.2 Results and Discussion

In this section, we present the prediction quality of our SDAE-CF model. The benchmark methods include three conventional CF-based methods, that's the pure user-based method [2], the pure Item-based method [3], and the SVD feature [5].

In order to do a reasonable comparison, pre-settings of the experiments are as follows, according to [2], we set the neighbourhood size with 30 and use person correlation as similarity metric for user-based benchmark algorithm. As for [3], we choose the adjusted cosine as the similarity metric. The third baseline SVD feature method is a latent factor model intuitively having some analogies with our model, which leads us first thought of experiment on a middle sized dataset—MovieLens 1 M, to determine the optimal numbers of hidden units for our model, and simultaneously in line with the number of the latent factor in SVD. The sensitivity of MAE affected by the number of hidden units in a basic DAE is illustrated in Fig. 5, showing that 200 hidden units would be the choice for a better prediction performance and there is no obvious difference when the hidden units go beyond 200. Thus we set the number of latent factors in SVD feature with 200.

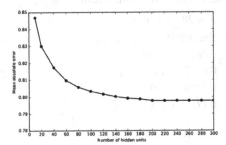
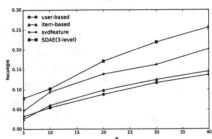

Fig. 5. The sensitivity of the number of hidden units

Fig. 6. Performance comparison by recall

Table 2 presents the MAE results from all three alternative baseline methods and our SDAE based model, in which fundamental structure also be involved. We can express the table from an empirical view that the DAE is relatively more accurate than the basic AE in faced with higher dimensional data. As it shows in Table 2, DAE and SDAE seems underperforming when coping with a lower dimension datasets, but still

Table 2. Performance on MovieLens datasets by MAE

CF-based method	ML-100 k	ML-1 M
Pure user-based	0.7489	0.7592
Pure item-based	0.7293	0.6983
SVD feature	0.7334	0.6944
AE (1 level)	0.7282	0.6965
DAE (1 level)	0.7482	0.6940
SDAE (3 level)	0.7262	0.6825

outperform other alternatives slightly. The purpose of introducing noise in autoencoder is for learning more robust features from the input. However, if the input itself are not strong enough, introducing noise would achieve nothing and may result in unrecoverable injuries to the original input. In general, AE based models are nearly the same level performance with the SVD feature methods by MAE.

The second form of interest prediction is Top-N recommendation task. We carry out experiments on MovieLens 10 M dataset to get the Top-N recommended items just according to (9), requires no further prediction score calculation. And we zoomed the range of N in [5, 10, 40]. Table 3 report the recall results in particular when the length of recommendation list is 20 and 40. Figure 6 illustrates that SDAE-based method outperforms the other three basic methods regarding recall metric, first also followed by the SVD feature. However, there is just a small performance gap between SDAE and SVD. In this respect, the two traditional neighborhood-based methods are in an inferior position, for in a sparser dataset the similarity calculation based on correlation is hindered by lacking enough co-rated items for users.

Table 3. Performance on MovieLens 10 M dataset by recall*

	N = 20	N = 40
CF-based method	Recall	Recall
Pure user-based	0.0877	0.1368
Pure item-based	0.0982	0.1456
SVD feature	0.1386	0.2017
SDAE (3 level)	0.1710	0.2561

5   Conclusion

In this paper, we analysis the problems of traditional collaborative filtering methods, then we proposed SDAE-CF model to alleviate the issues. On the one hand, using SDAE to encode the user preference vector break the limitation of similarity calculation among users depending on the common rated items, thus alleviate the loss of potential information. On the other hand, since the modified similarity based on Hamming distance can be calculated in constant time, decreasing the requirement for storing similarity extensively, which extends the scalability of traditional correlation-based algorithm. Experiments on three type scale MovieLens datasets show that SDAE-CF has potential in dealing with a high-dimension dataset and can achieve relative good performance in score prediction and Top-N recommendation task.

For the future work, we consider integrating with content-based techniques which can provide a more informational data foundation for extracting more useful latent features, and further to confirm the applicability of SDAE-CF model.

Acknowledgment. This work was supported by the National Nature Science Foundation of China (No. 61309013) and Chongqing Basic and frontier research projects (No. CSTC2014J CYJA40042).

References

1. Eppler, M.J., Mengis, J.: The concept of information overload: a review of literature from organization science, accounting, marketing, MIS, and related disciplines. Inf. Soc. **38**(5), 325–344 (2004)
2. Adomavicius, G., Tuzhilin, A.: Towards the next generation of recommender systems: a survey of the state-of-the-art and possible extensions. IEEE Trans. Knowl. Data Eng. **17**, 634–749 (2005)
3. Herlocker, J.L., Konstan, J.A., Borchers, A., Riedl, J.: An algorithmic framework for performing collaborative filtering. In: Proceedings of the 22nd annual international ACM SIGIR conference on Research and development in information retrieval, pp. 230–237. ACM, Berkeley (1999)
4. Badrul S., George K., et al.: Item-based collaborative filtering recommendation algorithms. In: Proceedings of the 10th international conference on World Wide Web (WWW 2001), Hong Kong (2001)
5. Yehuda, K., Robert, B., Chris, V.: Matrix factorization techniques for recommender systems. Comput. J. **42**(8), 30–37 (2009)
6. Chen, T., Zhang, W., Lu, Q., Chen, K., Zheng, Z., Yu, Y.: Svdfeature:a toolkit for feature-based collaborative filtering. JMLR **13**, 3619–3622 (2012)
7. Sarwar B., Karypis, G., Konstan, J., Riedl. J.: Application of dimensionality reduction in recommender systems—a case study. In: Proceedings of the ACM WebKDD Workshop, Boston (2000)
8. Hinton, G.E., Osindero, S., Teh, Y.W.: A fast learning algorithm for deep belief nets. Neural Comput. **18**, 1527–1554 (2006)
9. Salakhutdinov, R., Mnih, A., Hinton, G.: Restricted boltzmann machines for collaborative filtering. In: Proceedings of the 24th International Conference on Machine Learning, pp. 791–798. ACM, Corvalis (2007)
10. Georgiev, K., Nakov, P.: A non-iid framework for collaborative filtering with restricted Boltzmann machines. In: The 30th International Conference on Machine Learning (ICML 2013), pp. 1148–1156, Atlanta (2013)
11. Vincent, P., Larochelle, H., Lajoie, I., Bengio, Y., Manzagol, P.-A.: Stacked denoising autoencoders: learning useful representations in a deep network with a local denoising criterion. JMLR **11**, 3371–3408 (2010)
12. Bengio, Y., Lamblin, P., Popovici, D., Larochelle, H., et al.: Greedy layer-wise training of deep networks. Adv. Neural. Inf. Process. Syst. **19**, 153–160 (2007)
13. Cremonesi, P., Koren, Y., Turrin, R.: Performance of recommender algorithms on top-n recommendation tasks. In: Proceedings of the fourth ACM conference on Recommender systems. ACM, Barcelona (2010)

Research on Adaptive Canny Algorithm Based on Dual-Domain Filtering

Xiajiong Shen[1,2], Xiaoyu Duan[2(✉)], Daojun Han[1,2], and Wanli Yuan[2]

[1] Institute of Data and Knowledge Engineering,
Henan University, Kaifeng Henan 475004, China
[2] School of Computer and Information Engineering,
Henan University, Kaifeng Henan 475004, China
349386172@qq.com

Abstract. In order to overcome the shortcomings of the traditional Canny algorithm, which is easy to lose the edge details of the edge detection, and is prone to a large number of false edges and the threshold needs to be manually determined, an adaptive Canny algorithm based on Dual-domain filtering is proposed. Firstly, the Dual-domain filter is used to instead of the traditional algorithm, which is used to remove the image noise while preserving the image edge information, and more effective than the traditional Canny algorithm; Secondly, the Otsu algorithm is used to calculate the high and low threshold of Canny operator in order to achieve the purpose of high automation. The experimental results show that the improved algorithm can detect more effective edges of the image, but also has a strong adaptability.

Keywords: Canny algorithm · Edge detection · Dual-domain filtering · Gauss filter · Genetic algorithm · Otsu algorithm

1 Introduction

Edge detection is the basis of image recognition, and it is also an indispensable part of image analysis. It can greatly reduce the amount of data, eliminate the irrelevant information and retain the important structural attributes of the image. At present, the operators used in edge detection can be divided into three types: first order differential operator, two order differential operator and optimal operator, while the Canny operator of optimal operator is the most widely used in edge detection.

Canny operator [1, 2] is a classical edge detection operator. The edge detection of the target image is achieved by using the information of edge magnitude and edge direction. Although the Canny operator works better than the others, some inherent deficiencies can't be ignored. The traditional Canny operator uses Gauss filter to smooth the noise, while the Gauss filter can remove the noise, it will lose some edge information. On the other hand, the Canny operator needs to manually determine the threshold, which may lead to inaccurate valuation of threshold, and the inaccurate threshold estimation will lead to the consequence of weak image edges or false edges. Therefore, it is important to improve the filtering method and the way to choose the threshold of Canny operator. Some methods have already be put up to overcome the

© Springer Nature Singapore Pte Ltd. 2017
G. Chen et al. (Eds.): PAAP 2017, CCIS 729, pp. 182–191, 2017.
DOI: 10.1007/978-981-10-6442-5_16

deficiency, such as the article [3] which is using median filtering method, the Gauss filter based on directional wavelet transform is proposed in article [4], article [5] uses Bilateral filtering instead of Gauss filter. Although the above algorithms are improved according to the original one, the effect is not satisfied enough. In the aspect of image de-noising, nowadays, BM3D algorithm [6] has the Best treatment effect, however, the BM3D algorithm is too complex to realize and the computing time is too long to put into practical use. According to the article [7], the effect of Dual-domain filtering de-noising algorithm which based on the spatial domain and transform domain is close to the BM3D algorithm and is feasible in implementation. The Dual-domain filtering algorithm combines the advantages of both spatial domain and transform domain, the high-contrast layer and the low-contrast layer of the original image are separated by image segmentation. Finally, the results of the two layers are added to get the de-noised image. Therefore, this paper decided to uses double domain filter instead of Gauss filter, after the experiment, we can find that the Canny operator with double domain filter de-noising retains the effective and detailed information to the greatest extent and the contrast is clear and with no fuzzy which Gauss filtering would have. Then in the adaptive threshold selection algorithm, Otsu algorithm [8] is a classical nonparametric and unsupervised adaptive threshold selection algorithm. However, the calculation of the Otsu algorithm is way too complex. Refer to some researches [9, 10] about the improvement of the Otsu algorithm, we can conclude that the genetic algorithm is the most suitable method for adaptive threshold selection. Therefore, we use this method to improve the traditional Canny operator, in order to keep the details, smooth noise and be adaptive at the same time.

2 Traditional Canny Operator

In 1986, Canny proposed three optimal evaluation criteria from the perspective of optimal filter: good SNR, good positioning performance and minimum response. An optimal edge detection operator—Canny operator is proposed on this basis. Canny operator can achieve the process through four steps: firstly, to neat the image through Gauss filter, secondly, to calculate the magnitude and direction of gradient by using differential operator, thirdly, Non-maximum suppression and to use double thresholds to determine edge pixels at last.

2.1 Gauss Smoothing Image

First of all, the image $f(x,y)$ was convolution filter and remove noises by Gauss filter, and then, we can get the de-noised image $I(x,y)$, suppose the two-dimensional Gauss function as

$$G(x,y) = \frac{1}{2\pi\sigma^2} exp\left(-\frac{x^2+y^2}{2\sigma^2}\right) \qquad (1)$$

Then the de-noised image is

$$I(x,y) = G(x,y) * f(x,y) \tag{2}$$

The σ is Gauss filter parameters, it can Control the degree of smoothness.

2.2 Calculate the Magnitude and Direction of the Gradient

Canny uses the first order difference operator of 3 * 3 neighborhood to calculate the horizontal and vertical gradient amplitude components, the partial derivative matrix of the X direction and the Y direction partial derivative matrix are

$$P_x[x,y] = (f[x,y+1] - f[x,y] + f[x+1,y+1] - f[x+1,y])/2 \tag{3}$$

$$P_y[x,y] = (f[x,y] - f[x+1,y] + f[x,y+1] - f[x+1,y+1])/2 \tag{4}$$

Thus each pixel gradient amplitude and gradient direction is

$$M[X,Y] = \sqrt{P_x[x,y]^2 + P_y[x,y]^2} \tag{5}$$

$$\theta[x,y] = arctan(P_y[x,y]/P_x[x,y]) \tag{6}$$

2.3 Non-maximum Suppression

In order to ensure the accuracy and uniformity of the image edge location, so non-maximum suppression was performed, it means only keep the point with the biggest change of amplitude. The interpolated pixel is chosen as the center, the interpolation of the adjacent two gradient amplitudes in the current pixel gradient direction is calculated, if the gradient magnitude of the current pixel is greater than the two values, then the point is the edge point and the amplitude is marked as 1, otherwise assigned to a value of 0.

2.4 Double Threshold Edge Pixels

Firstly, using double threshold to determine the edge pixels needs to manually set the threshold, and the value of the high threshold is normally twice the value of the lower one. Secondly, the sub-image $N[X, Y]$, which is classified by non-maximal suppression and gradient histogram, are thresholding to obtain $T_h[x, y]$ and $T_l[x, y]$, where $T_h[x, y]$ is a strong edge, and $T_l[x, y]$ is a weak edge. Finally, track the edges in $T_h[x, y]$, when the edge reaches the end, the edge points are searched in the 8 neighborhood of $T_l[x, y]$ to connect the discontinuity edge in $T_h[x, y]$, recursively search the edges of the low-threshold image until the edges in $T_h[x, y]$ are interrupted.

3 Improved Canny Algorithm

In this paper, the algorithm improves the first step and the fourth step of the traditional Canny algorithm, and replaces the Gaussian filter in the first step into the double domain filtering which can keep the effective image information better for the noise treatment; In the fourth step, the Otsu algorithm is used to automatically select the double threshold. The flow chart of the improved algorithm is:

3.1 Dual-Domain Filtering

Dual-domain filtering is an iterative process, the basic principle is to divide the original image x into two layers: the high-contrast layer and the low-contrast layer, and process these two layers respectively. The process of double-domain filtering is:

(1) Enter the original image, combine the Bilateral filtering to obtain the high-contrast layer $\tilde{s}$, but use the Bilateral filter [11] in the first process.
(2) Eliminate the high-contrast layer from the original image to get the low-contrast layer.
(3) The Short-time Fourier transform and the wavelet shrinkage are performed on the low-contrast layer to obtain a low contrast image $\tilde{S}$.
(4) The high-contrast layer is added to the low-contrast layer which was processed by short-time Fourier transformation wavelet to obtain the de-noised image $\tilde{x}$. The formula can be expressed as

$$\tilde{x} = \tilde{s} + \tilde{S} \tag{7}$$

(5) The de-noised image $\tilde{x}$ is used as the next iteration of the bootstrap. Repeat the above steps, and the image $\tilde{x}$ after the last iteration is the final de-noised image.

Spatial Domain Processing. The processing of the spatial domain is actually the processing of the high-contrast layer, it uses a joint Bilateral filtering algorithm. Firstly, Bilateral filtering is a typical nonlinear filter, which is non-iterative, local and simple. It is similar to the traditional Gaussian filter according to the weighted average definition of pixels, but the Bilateral filtering also uses the change of the intensity to preserve the edge information, while the joint Bilateral filtering algorithm is to simultaneously boot image and noisy image for Bilateral filtering. Supposing that the square neighborhood of the Bilateral filter core is N_P, the center of the neighborhood is P, and the radius is r, then the kernel function of the Bilateral filter is:

$$k_{p,q} = e^{-\frac{|p-q|^2}{2\sigma_s^2}} e^{-\frac{(g_p - g_q)^2}{\gamma_r \sigma^2}} \tag{8}$$

After the Bilateral filtering of the noisy image $\tilde{s}_p$ and the guided image $\tilde{g}_p$ are

$$\tilde{s}_p = \frac{\sum_{q \in N_p} k_{p,q} y_p}{\sum_{q \in N_p} k_{p,q}} \tag{9}$$

$$\tilde{g}_p = \frac{\sum_{q \in N_p} k_{p,q} g_p}{\sum_{q \in N_p} k_{p,q}} \tag{10}$$

where σ_s and y_r are used to adjust the spatial domain and the domain variance (Fig. 1).

Transform Domain Processing. The transform domain is the processing of the low-contrast layer of image. Take the 3 * 3 pixel block as an example, the treatment is to minus the value which corresponding to the original layer from the original layer, as shown in Fig. 2(c).

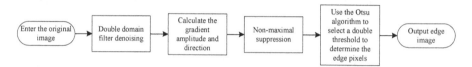

Fig. 1. Improved algorithm flow chart

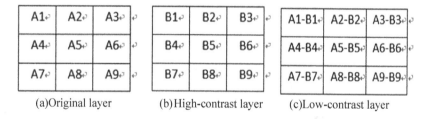

A1	A2	A3		B1	B2	B3		A1-B1	A2-B2	A3-B3
A4	A5	A6		B4	B5	B6		A4-B4	A5-B5	A6-B6
A7	A8	A9		B7	B8	B9		A7-B7	A8-B8	A9-B9

(a)Original layer   (b)High-contrast layer   (c)Low-contrast layer

Fig. 2. Obtain the low-contrast layer

Then the short-time Fourier transform is performed on the low-contrast layer, The Gauss kernel function is chosen as the window function, and the whole process is changed into Gabor transform. Transform the guide map and the low-contrast layer of noise figure respectively, the formula is as follow:

$$G_{p,f} = \sum_{q \in N_p} e^{-\frac{i2\pi(q-p)f}{2r+1}} k_{p,q} (g_q - \tilde{g}_q) \tag{11}$$

$$S_{p,f} = \sum_{q \in N_p} e^{-\frac{i2\pi(q-p)f}{2r+1}} k_{p,q} (y_q - \tilde{S}_q) \tag{12}$$

where $k_{p,q}$ is the Gauss kernel function, the transformation results $G_{p,f}$ and $S_{p,f}$ are functions defined at the frequency f in the window F_p the size of the window F_p is the same as that of the N_p. Assuming that the kernel function $k_{p,q}$ is not affected by noise, the noise variance of the Fourier coefficient of the variation domain is

$$\sigma_{p,f}^2 = \sigma^2 \sum\nolimits_{q \in N_p} k_{p,q}^2 \tag{13}$$

Then, the short-time Fourier transform coefficients $S_{p,f}$ are subjected to wavelet shrinkage. In order to get rid of the noise signal and retain the useful signal, it is necessary to take the inverse of the spectral coefficients, using the guided image spectrum, the shrinkage factor is:

$$K_{p,f} = e^{-\frac{y_f \sigma_{p,f}^2}{|G_{p,f}|^2}} \tag{14}$$

Finally, the spectrum of the details of the signal back to the spatial domain, the formula is

$$\tilde{S}_p = \frac{1}{|F_p|} \sum\nolimits_{f \in F_p} F_{p,f} S_{p,f} \tag{15}$$

The $\tilde{s}$ which obtained by the spatial domain processing and transform domain and the $\tilde{S}$ from the low-contrast layer can be used to get the next bootstrap map, but at the last iteration, the $\tilde{x}$ we get is the de-noised image.

3.2 Adaptive Threshold Selection

N. Otsu proposed the Otsu algorithm in 1979, which is a method to obtain the global optimal solution. The idea is to use the gray histogram in the image to determine the threshold through the maximum of the variance between the target and the background. However, since the Otsu algorithm needs to traverse all the pixels in the gray scale and calculate the variance, it will have a large computational complexity. As a highly efficient parallel global search method, the genetic algorithm and method can be used to solve the shortcomings of Otsu algorithm, so we choose the improved Otsu genetic algorithm as a Canny algorithm adaptive threshold selection method, the Otsu formula is

$$\sigma(t)^2 = w_1(t) * w_2(t) * (u_1(t) - u_2(t))^2 \tag{16}$$

where $\sigma(t)^2$ is the variance, t is the threshold, $w_1(t)$ is the number of gradients whose gradient value is larger than the that of the threshold, $w_2(t)$ is the gradient of the gradient value smaller than the value of threshold, $u_1(t)$ is the average gradient value of the gradient value larger than the threshold value, $u_2(t)$ is the average gradient value of the gradient value smaller than the threshold value. The selection of the threshold is to find the t value that maximizes the $\sigma(t)^2$ value, the obtained one is the high threshold of Canny. From the traditional algorithm, we can see that the high threshold is 2 times higher than the low threshold. Therefore, the low threshold can be obtained by calculation. The specific process can be seen from journal articles [8–10].

4 Experimental Results and Analysis

The processor of the test computer is the AMD Athlon (tm) II * 4 635 Processor 3.10 GHz, 4 G memory, 2804 MB video memory. The test was processed in the Windows environment, with vs2012 + opencv programming. In this paper, the improved part of the algorithm and the overall test results of the algorithm for comparative analysis. Figure 3 is the comparative analysis of Gauss filtering and Bilateral filtering and all the image size are 512 * 512; Table 1 is the comparison of the run time of Lena, Barbara, and Man images through Gaussian filtering, Dual-domain filtering, and BM3D algorithms; Table 2 is the comparison of the PSNR of Lena, Barbara, and Man images through Gaussian filtering, Dual-domain filtering, and BM3D algorithms;

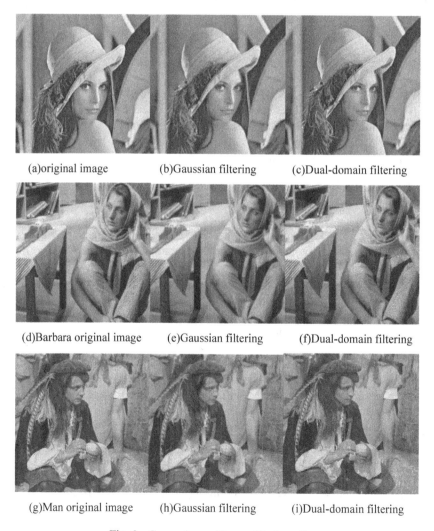

(a)original image   (b)Gaussian filtering   (c)Dual-domain filtering

(d)Barbara original image   (e)Gaussian filtering   (f)Dual-domain filtering

(g)Man original image   (h)Gaussian filtering   (i)Dual-domain filtering

Fig. 3. Comparison of image filtering effects

Fig. 4 is the comparison of the traditional Canny algorithm and the improved algorithm for edge detection of Lena and Baby image.

4.1 Image Denoising Performance Comparison

Firstly, it can be seen from Fig. 3 that compared with the Gauss filter, the Dual-domain filter smoothed the image effectively, and the contrast is clear, while Gauss filter can smooth the image, but the image edge is not clear and have poor contrast. At the same time, the visual effect of the Dual-domain filtering is better than the Gauss filter.

Then we compare the running time of the algorithm from Table 1, we can clearly see that the Gauss filter has the shortest running time while the BM3D has the longest, the running time of the Dual-domain filter is higher than that of the Gauss filter.

Table 1. Running time contrast

Image	Gaussian filtering	Dual-domain filtering	BM3D
Lena	1.47 s	113 s	175 s
Barbara	1.5 s	121 s	173 s
Man	1.45 s	119 s	171 s

Table 2. Peak signal to noise ratio (PSNR/dB) contrast

Image	Gaussian filtering	Dual-domain filtering	BM3D
Lena	28.37	32.17	32.08
Barbara	25.36	30.70	30.72
Man	25.12	29.65	29.62

Finally, Comparison these images. As described in the article [7]. The peak signal to noise ratio (PSNR) of BM3D and the peak signal to noise ratio (PSNR) of Dual-Domain filtering are almost the same, but they are all higher than Gaussian filtering. It can be concluded that the double domain filtering is superior to Gaussian filtering in image de-noising.

In summary, although the running time of the Gauss is short, it performed worse in de-noising. So Gaussian filtering is more suitable for some algorithms that require less de-noising. The image de-noising algorithm based on Canny operator is the basis of subsequent processing, and there are strict requirements for image de-noising. If a large amount of edge information is lost at the time of de-noising, it will cause the final edge detection to be inaccurate. Therefore, it is reasonable to choose the Dual-domain filter instead of Gauss filter to compensate for the shortcomings of traditional Canny algorithm.

4.2 Edge Detection Contrast

First, comparing Fig. 4(c) with Fig. 4(b), we can see from the Fig. 4(c) that the image processing of the hat have better connectivity and fewer pseudo-edge information than

Fig. 4(b). Then compare Fig. 4(f) with Figs. 4(e) and (f) Baby details of the legs have a clear outline and remove a large number of false edges. Finally, with the comparison of the results of the whole treatment, it can be concluded that the image using adaptive thresholding has better connectivity and fewer false edges. Thus, we can draw a conclusion that the improved algorithm can achieve the automatic selection of the threshold and a better detection accuracy at the same time.

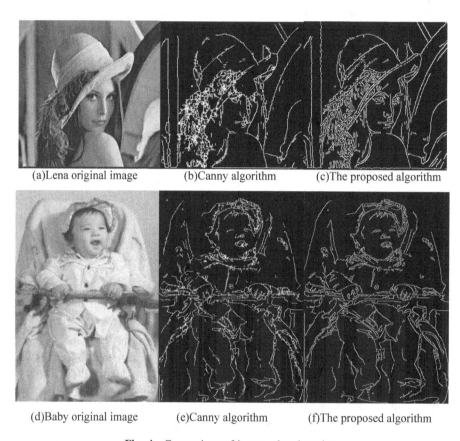

(a)Lena original image   (b)Canny algorithm   (c)The proposed algorithm

(d)Baby original image   (e)Canny algorithm   (f)The proposed algorithm

Fig. 4. Comparison of image edge detection

5   Conclusion

In summary, this algorithm not only maintains the advantages of traditional Canny operator, but also improves the noise suppression ability of the algorithm, and preserves the edge detail information to the greatest degree; On the other hand, the adaptive threshold selection method can automatically detect the edge of the image, so that the ability of automation is greatly improved. However, the time complexity of the short time Fourier transform in the Dual-domain filter is high, which leads to the high time complexity of the whole image processing, and GPU parallel computing is one of the best ways to reduce the running time of the algorithm. Therefore, this algorithm can be continued to be improved by GPU parallel method.

Acknowledgements. We thanks for the support of National Natural Science Foundation of China (61272545, 61402149) and Scientific and technological project of Henan Province (142102210390, 14A520026). We also thank anonymous reviewers for their helpful reports.

References

1. Canny, J.F.: A computational approach to edge detection. IEEE **8**(6): 679–698 (1986)
2. Zhang, F.Z., Wang, S.Y., Xue, T.G.: Digital Image Processing and Machine Vision, 1st edn. Posts and Telecom Press, Beijing (2010)
3. Wang, F.C., Duan, S.Z., Li, T.Y.: An improved Canny edge detection algorithm. Trans. Shenyang Ligong University **26**(6), 20–22 (2007)
4. Han, F.H., Han, S.X.: Edge detection algorithm based on directional wavelet transform. Microelectronics Comput. **29**(7), 55–58 (2012)
5. Li, F.D., Zhao, S.W., Tan, T.H.: A new method of edge detection based on bilateral filtering. Comput. Technol. Develop. **17**(4), 161–163 (2007)
6. Hou, Y., Zhao, F.C., Yang, S.D.: Image denoising by sparse 3-D transform-domain collaborative filtering. IEEE Trans. Image Process. **20**(1), 268–270 (2011)
7. Claude, K.F., Matthias, Z.S.: Dual-domain image denoising. In: IEEE International Conference on Image Processing, pp. 440–444 (2013)
8. Otsu, N.F.: A threshold selection method from gray level histograms. IEEE Trans. Syst. Man Cybern. **9**(1), 62–66 (1979)
9. Dong, F.Z., Jiang, S.L., Wang, T.J.: Modified one-dimensional otsu algorithm based on image complexity. Comput. Sci. **42**(6A), 171–174 (2015)
10. Yang, F.T., Tian, S.H., Liu, T.X.: Research on image segmentation algorithm based on edge detection and otsu. Comput. Eng. **42**(11), 255–265 (2016)
11. Tomasi, C.F., Manduchi, R.S.: Bilateral filtering for gray and color images. In: Sixth International Conference on Computer Vision (IEEE Cat. No. 98CH36271), pp. 839–846 (1998)

A Dynamic Individual Recommendation Method Based on Reinforcement Learning

Daojun Han[1], Xiajiong Shen[1(✉)], Tian Gan[2(✉)], and Ruiqing Cai[3]

[1] Institute of Data and Knowledge Engineering,
Henan University, Kaifeng 475004, Henan, China
shenxj@henu.edu.cn
[2] CITIC Bank Zhengzhou Branch, Zhengzhou 450000, Henan, China
[3] Hikvision Digital Company Limited, Hangzhou 310052, Zhejiang, China

Abstract. As a widely used recommendation method, collaborative filtering can solve the problem of low level of resource utilization which caused by information overload. At present, in order to exhibiting and searching items, we need to use multipole attributes to describe items. Thus request to particularly distinguish every attribute and realize accurate recommendation. While the collaborative filtering method lose sight of the dynamic regulation of items attributes' importance degree, and it cannot interpose the discrimination of attributes. Aiming at this problem, this paper come up with a dynamic individual recommendation method based on reinforcement learning. This method can dig user's attribute tag preference from operant behavior. It can record user's attributes operate path and recall path. Then we build the award-punishment model of attribute tag, and realize the tag weight dynamic regulation. According to the principle that reinforcement learning system always get max award, we make a tag recommend strategy and give user recommendations in accordance with the preferences. The experimental result show that this method can distinguish the validity of user's click, and realize the tag weight dynamic regulation and give user recommendations in accordance with the preferences.

Keywords: Recommendation method · Reinforcement learning · Dynamic

1 Introduction

Information overload means the rapid growth of information on the contrary caused low level of resources utilization. Individual recommendation can solve this problem according to evaluate items take the place of users. Collaborative filtering is a popular individual recommendation method, which is put forward by Goldberg Nicols Oki and Terry in 1992 [1]. This method can predict whether a user like the item by looking for user's neighbor and their item's grade. In this course, similarity must be calculated time and again. When items reach a big order of magnitudes, the expansibility become the choke point of collaborative filtering's performance [2]. In the practical application, we need multiple attributes to describe the item, but collaborative filtering cannot solve the problem of attribute discrimination partition [3]. It lack deep mining and lose sight of the dynamic regulation of items attributes' importance degree. In addition, we found

© Springer Nature Singapore Pte Ltd. 2017
G. Chen et al. (Eds.): PAAP 2017, CCIS 729, pp. 192–200, 2017.
DOI: 10.1007/978-981-10-6442-5_17

that user's click behavior contains personal preference information. How to get these information and apply to individual recommendation, make change as the user on the use of the system. All of these are issues worthy of concern.

The user behavior pattern mining can be a solution of above problems. It can help us understand the regularity characteristics as the user on the use of the system, thus we can improve the system and provide better individual recommendation service to users [4]. The data of user behavior pattern mining include: data of page view, similarity data between pages and users, type of pages and users, relevance between pages, and so on. Technologies of mining these data include: statistical analysis, association rules, sequential patterns, clustering and classification [5]. But these technologies all have their own defects. Association rules technology is easy to lost important patterns, which lead to the user is not interested in excavated model; the sequence pattern shows low coverage when it applied in personalized recommendation [6]. Therefore, the research train is integrated application of different technologies. We hope to import an automatic learning method and dynamic integration different part of user behavior pattern mining. At last realize an automatic, accurate, wide coverage of personalized recommendation.

Based on the above assumptions, this article come up with a dynamic individual recommendation method based on reinforcement learning. This method can record user's click and back tracking operation sequence. Then give reward and punishment to these two kinds of tag nodes combine reinforcement learning and use the reward price as tag's weight. At last recommended for users according to tag's weight. This recommendation method can distinguish the validity of user's click and realize the dynamic change of tag weight.

2 Reinforcement Learning Theory

Depending on the different feedback, machine learning techniques can be divided into three categories: supervised learning, unsupervised learning and reinforcement learning. Reinforcement learning is a special kind of machine learning method which use the environment feedback as input, and can adapt to the environment. The difference between reinforcement learning and supervised learning mainly performance on the teacher's signal. In reinforcement learning, the signal provided by environment is an evaluation of action's stand or fall, but not to tell reinforcement learning system how to do the correct action [7]. The learning technology is divided into two categories. One is searching agent behavior space to find the optimal behavior, which is usually by using technology such as genetic algorithm implementation; Another one is to use statistical techniques and dynamic planning method to estimate utility function value in a certain environment conditions action [8]. Due to reinforcement learning has a good learning performance in a large space and complex nonlinear system, this technology is widely used in the field of four categories: games, control system [9], the robot [10, 11] and scheduling management [12].

The standard agent reinforcement learning framework is shown in Fig. 1, it consists of three modules: state perceptron I, learner L and action selector P.

The perceptron I can sense the environment state s, and map it to the agent internal perception I; Action selector P choice action according to strategy and effects on the

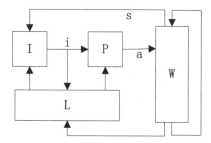

Fig. 1. Reinforcement learning framework

environment W; Learner L develop new strategies according to the internal perception *i* and reward value *r*. The goal of reinforcement learning is to learn a policy *P: S - > A*, action A under the state *S*, and the current action can obtain largest environmental award according to the strategy. Due to the action has its continuity, it is necessary to define a target function to show the reward values of actions sequence in future step h. The functional form is:

$$V^*(s_i) = \sum_{i=0}^{h} r_i \tag{1}$$

Where r_i is the reward value caused by state transition after action. After determine the objective function, we can determine the optimal behavioral strategy based on that reinforcement learning is always obtain the maximum rewards goal:

$$p* = \arg_p \max V * (s) \tag{2}$$

2.1   The Recommendation Model Based on Reinforcement Learning

In the recommendation system, we can regard path nodes as a series of tags described the item. The process of user clicking on the navigation step by step is actually choosing tags which they thought can describe the target item. If the user successfully retrieved the target item along a certain path $<a_1, a_2... a_n>$ (i.e. to browse projects), it means that the node tags on the path can correctly describe the target item and should be reward; If the user did not browse the item after a series of click but back to the higher path, such as there exist $a_i = a_j$ in click sequence $<a_1, a_2...a_n>$, it means back part of the node tags $<a_i...a_j>$ can not correctly describe the target item. It should be punishment. After a series of clicks and back tracking operation, the tags in system have been rewards and punishment for several times, their weight also change from the same initial set to different. When the user selects the tag, the system will ordered all tags in accordance with the weight, and priority recommend tags with big weights to user.

The model's frame is shown in Fig. 2:

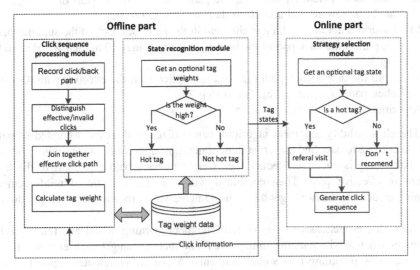

Fig. 2. Dynamic recommended model frame based on reinforcement learning

In Fig. 2, the state recognition module read the label weight data, get the tag status according to the data, and pass the status to a strategy selection module; Strategy selection module choose a strategy to determine the next step of operation according to the label of passed state, the operating effect on click sequence processing module and provide the next click node; Click sequence processing module splice clicked nodes, distinguish the effective click and invalid clicks, and change the label weight. The goal of this model is to constantly adjust the tags weight through the user click on the tags, make it easier for tags to be recommended which is conform to the user preferences, and make those tags gradually sink which do not conform to the user preferences.

In dynamic recommended model based on reinforcement learning, we respectively defined: action sets A, it is used to represent the user's operation of path tag; State sets S, it is used to represent the next tag state; Rewards and punishments value R present the value-added when reward or punishment on target tags, here we set the reward to 10 and the punish to 0; Strategy *Policy* defines which action should be implemented when meet the status s in status set. Specific definition is as follows:

A: {enter (click to enter), cancel (not operating)}

S: {s_0 (current tag weights is high, hot tag), s_1 (the current tag weight is low, not hot tag)}

R: {10 (belong to effective operation), - 10 (the tag located on the back path)}

Policy: {s_0 -> enter, s_1 -> cancel}

The click validity judgment algorithm description:

1. The set C {c_1 and c_2,..., c_n} represent all path nodes in the system, namely tags. Initialized set weight of each tag as 0.

2. Record user's click path of this operation $c_i, \ldots c_m, \ldots c_j$, and record the user's operation of the item.
   (a) If the user operate an item f in the path, each tag in the path of $c_i, \ldots c_m, \ldots c_j$ should be reward and add 10 to their weight. Go to step 4.
   (b) If user did not do anything to the item in the path, but back to the superior path cm, punish the back part of the tags $c_m, \ldots c_j$ and minus 10 of their weight. Go to step 3.
3. To record the user's click path $c_m, \ldots c_n$, and splice with the effective path. Get the new click path of $c_i, \ldots c_m, \ldots c_n$. Go to step 2.
4. Algorithm end.

The click validity judgment algorithm can realize the dynamic weighted of path tags. According to user find the target item or not, this algorithm can determine whether the click of tag is effective. Reduce tags weight of back path and improve the tags weight in effective path. Then combined with the characteristic of reinforcement learning model which always get the highest reward value, recommend high weight tag priority.

Through comprehensive study of reinforcement learning model and related technology in user behavior patterns mining, we found that we can get the user's preference information by recording user's click sequence. Make the preference degree numerically as the reward values in reinforcement learning model. Using the feature of always get maximum reward in reinforcement learning, make recommendations for the user's next click. After a series of click operation, user can find most conforms target. Because of the recommendation is based on the user clicks sequence, so the recommendation is personalized. The user's current click can be recorded as a basis for the future training of reinforcement learning model. As long as using the system, more clicks can be recorded. The reinforcement learning model is more perfect, the coverage goes wide and recommend has higher accuracy.

2.2 Comparison of Three Kinds of Model

There are a lot of differences between the three sequential: Recommended based on reinforcement learning, collaborative filtering recommendation and patterns mining (Table 1).

It can be seen from the chart that of the recommendation model based on reinforcement learning in the previous section is different from the traditional collaborative filtering recommendation. It is because of the collaborative filtering recommendation method is based on a two-dimensional table of graded of user-item. While the recommendations based on user action sequence pattern mining method do its research focus on the combination high frequency sequence sets from past sequence, and using the sequence set to recommend. There is no obvious hierarchy between these sequences. The recommended method is proposed in this paper build on the basis of hierarchical navigation, record user operation path, change the tags weight, and recommended. Because there are differences between recommend object's structures, the algorithm in this paper cannot be compared accuracy and recall rate with the traditional collaborative filtering recommendation methods and recommendation based on user

Table 1. Shows these differences from the processing object, expounds and focus.

	Processing object	Processing mode	Focus	Scope of application
Recommend module based on reinforcement learning	Tags and items in hierarchy navigation	1. Join together click sequence and change the weight 2. Judge tag's weight 3. Select a stratagem	Reward value between click processing	Goods and file resource stored in hierarchy navigation
Collaborative filtering	User-item rate matrix	1. Similarity calculation and find the neighbor 2. Predict the rate of target item based on neighbor	Similarity and grade	Goods and movies recommendation score
Sequential pattern mining	Sequence database	1. Sorting 2. Find the frequent item sets 3. Translate the sequence into frequent item sets 4. Find sequence making use of known frequent item sets	The relationship between items in one business or different businesses	Business of buying goods, web access sequence

behavior patterns mining methods. But as the user clicks the proposed approach can also realize dynamic recommendation. So the experimental analysis in this article focus on the availability of recommendation method based on reinforcement learning.

3 Experimental Analysis

With the increasingly rich of network resources, users can easily download a lot of video and audio files. Then store them separately in different folder path. But how to find these files quickly and accurately is a problem. In this paper, we take the browsing of music files for example, and verify the validity of recommended method based on reinforcement learning by simulating user clicks operation on the folder path.

There exists a music file directory structure as shown in Fig. 3:

Initially set Path label weights to 0, and the tags weight is shown in Table 2.

User's two clicks sequence is shown in Table 3.

From Table 3, the second click path is as follows: "70", "Blues", "happy", "Blues", "dance", "classic", "Japanese", "dance", "pop" and "rock". User operate "Beijing" under the file path. According to the second step b and the third step operation of the

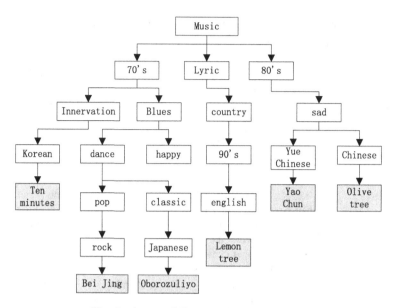

Fig. 3. A part of directory about music files

Table 2. Initial tags weight table

Tag	70's	Lyric	80's	Innervation	Blues	Country	Sad	Korean	Dance
Weight	0	0	0	0	0	0	0	0	0
Tag	Happy	90's	Yue Chinese	Chinese	Pop	Classic	English	rock	Japanese
Weight	0	0	0	0	0	0	0	0	0

Table 3. User's click sequence

First	70's	Blues	Dance	Pop	Rock					
Second	70's	Blues	Happy	Blues	Dance	Classic	Japanese	Dance	Pop	Rock

algorithm in previous section, the user have two backtracking clicks, and the back-tracking tags as "happy" and "classical", "Japanese". These three tags should be punished and their weight minus 10. Stitch tags without backtracking, we can get: "70", "Blues", "dance", "pop", "rock". According to the second step of the algorithm, each tag in the path should be rewarded and their weight plus 10. Tags and weights is shown in Table 4.

Table 4. The tag-weight table after second click

Tags	70's	Lyric	80's	Innervation	Blues	Country	Sad	Korean	Dance
Weight	20	0	0	0	20	0	0	0	20
Tags	Happy	90's	Yue Chinese	Chinese	Pop	Classic	English	Rock	Japanese
Weight	-10	0	0	0	20	−10	0	20	−10

According to the goal of reinforcement learning that it always get the maximum reward and the *Policy*:$\{s_0 \rightarrow$ enter,$s_1 \rightarrow$ cancel$\}$, we recommend tags with high weight to user in priority. When user faced with four choices, each tag, their weight, action, and reward are shown in Table 5.

Table 5. Four choices selection state table

First	Tags	70's	Lyric	80's	Action	Reward value
	Weight	20	0	0	Click 70's	20
Second	Tags	Blues	Innervation		Action	Reward value
	Weight	20	0		Click Blues	40
Third	Tags	Dance	Happy		Action	Reward value
	Weight	20	−10		Click Dance	60
Fourth	Tags	Pop	Classic		Action	Reward value
	Weight	20	−10		Click Pop	80

When the user begin from the root directory "music", three tags and weights are shown in "first time" in Table 5 and recommend "70's" to user.

After click on "70's", users are faced with two tags. Their weights as shown in Table 5 in the "Second Time". According to the policy, we recommend "Blues" to the user.

After click on "Blues", tags and weight are shown in Table 5 in the "Third Time". According to the policy, we recommend "Dance" to user.

After click on "Dance", tags and weight are shown in Table 5 in the "Fourth Time". According to the policy, we recommend "pop" to user.

After click the "pop", the user clicks the "Rock", and operate "Beijing" to. During this process, the reinforcement learning model get total reward $V^* = 20 + 20 + 20 + 20 = 80$. After Verified, we can find that the reward value is the maximum value of the system. It indicate that the principle of reinforcement learning system specified action always get the maximum reward values can be achieved.

After the above case, we found that the recommended model based on reinforcement learning can record the user's clicks and backtracking operations and achieve the reward and punishment to tags. Using the reward value as tag weight. When user clicks the path next time, we priority recommend tags with high weight to the user according to that reinforcement learning model always get the maximum reward value. After multiple recommendation and clicks, users can ultimately find popular files "Beijing . mp3".

4 Conclusion

This article describes some of the issues in the case of information overload during users retrieve. We found that personalized recommendation has an irreplaceable role in solving these problems. By observing existing research we found that collaborative filtering method has defects in similarity calculation complicated. It cannot solved the problem of discrimination division of property and ignore the importance dynamic adjustment of the item's. It cannot take advantage of the user clicks on file path and mining more information. After considering the above issues, this paper proposes a tag recommendations method based on reinforcement learning. The method comprises a reinforcement learning model and a click validity determine algorithm. It can record user's clicks and backtracking operations and assignment different weights to tags. After the user clicks, it will priority recommend tags to the user with high weight. Finally, take the music file as example to verify the validity of this method. It is confirmed that the method distinguishes clicks effectiveness, and achieve a dynamic recommendation. However, the method study and recommend for a single user operating behavior, and at the beginning system initialize tags with undifferentiated average weight. How to achieve multi-user collaborative filtering, and solve the cold start problem is the direction of future research.

Acknowledgements. We thanks for the support of National Natural Science Foundation of China (61272545, 61402149) and Scientific and technological project of Henan Province (142102210390, 14A520026). We also thank anonymous reviewers for their helpful reports.

References

1. Goldberg, D., Nichols, D., Oki, B.M., Terry, D., et al.: Using collaborative filtering to weave an information tapestry. Commun. ACM, 61–70 (1992)
2. Cong, L.: Review of scalability problem in E-commerce collaborative filtering. New Technol. Libr. Inf. Serv. **11**, 7–44 (2010)
3. Li, H., He, K., Wang, J., Peng, Z., Tian, G.: A friends: recommendation algorithm based on formal concept analysis and random walk in social network. J. Sichuan Univ. (Eng. Sci. Edn.) **06**, 131–138 (2015)
4. Chen, T.S., Chou, Y.S., Chen, T.C.: Mining user movement behavior patterns in a mobile service environment. IEEE Trans. Syst. Man Cybern. Part A Syst. Hum. **42**(1), 87–101 (2012)
5. Yang, F., Yan, B.: Research on web usage mining. Microelectron. Comput. **11**, 146–149 (2008)
6. Li, Q., Shi, J., Qin, Z., Liu, T.: Mining user behavior patterns for event detection in email networks. Chin. J. Comput. **05**, 1135–1146 (2014)
7. Zhang, R., Gu, G., Liu Z.: Reinforcement learning theory. Algorithm and its application. Control Theory Appl. **17**(5), 637–642 (2005)
8. Gao, Y., Chen, S., Lu, X.: Reinforcement learning: survey of recent work. Acta Autom. Sin. **30**(1), 86–100 (2004)
9. Bazzan, A.L.C.: Opportunities for multiagent systems and multiagent reinforcement learning in traffic control. Auton. Agents Multi-Agent Syst. **18**(3), 342–375 (2009)
10. Kober, J., Peters, J.: Reinforcement learning in robotics: a survey. Int. J. Robot. Res. **32**(11), 1238–1274 (2014)
11. Liu, Z., Zeng, X., Liu, H., Chu, R.: A Heuristic two-layer reinforcement learning algorithm networks. Comput. Res. Dev. **03**, 579–587 (2015)
12. Jurczyk, P., Xiong, L., Sunderam, V.: Reinforcement learning strategies for A-team solving the resource-constrained project scheduling problem. Neurocomputing **146**(1), 301–307 (2014)

Research on the Pre-distribution Model Based on Seesaw Model

Mingshan Xie[1,2,3], Yanfang Deng[2(✉)], Yong Bai[2,3],
Mengxing Huang[2,3], and Zhuhua Hu[2,3]

[1] College of Network, Haikou College of Economics, Haikou 571127, China
[2] College of Information Science and Technology,
Hainan University, Haikou 570228, China
271190993@qq.com
[3] State Key Laboratory of Marine Resource Utilization in South China Sea,
Haikou 570228, China

Abstract. The relationship among many subsystems in multi-agent complex systems is difficult to quantify and the coordination among multiple processes in multivariate complex processes is hard to clear analysis. In order to solve the problems this paper presents a seesaw model for the basic dual relation of the complex systems and complex processes. The seesaw model can be applied in various fields and in all aspects of people's lives. In this paper, the application of this model is derived, and the pre-distribution model of distribution industry is obtained. We analyze the time factor of the distribution and optimize the distribution process by using the seesaw model. Concorde process makes the dissatisfaction degree of distribution service decreased. Pre-distribution makes that delivery speed and delivery efficiency are improved and ensures that dissatisfaction degree of distribution service is effectively reduced.

Keywords: Pre-distribution · Complex system · The seesaw model

1 Introduction

With the advent of the era of intelligence, the rise of big data, more and more complex computing process and complex computing systems, people face a variety of systems and processes more complex. The application of analytical methods for complex systems is increasing. In [1] the authors apply complex systems analysis methods and theories to national critical infrastructure projects. They analyze the basic concepts of national critical infrastructure and the associated complex system characteristics, and then construct the model of national critical infrastructure based on complex system theory. Huang et al. study the stability of complex network system based on the characteristics and development rules of complex networks in [2]. In [3] the authors point out the main elements for the development of systemic thought from its beginning, through its application in business sciences, to the birth of Complex Systems Theory. Complex systems and complex process analysis methods need to be further improved. The traditional reductionism and holism of complex systems need to be further optimized and integrated. In the process of analyzing complex systems, it is

© Springer Nature Singapore Pte Ltd. 2017
G. Chen et al. (Eds.): PAAP 2017, CCIS 729, pp. 201–213, 2017.
DOI: 10.1007/978-981-10-6442-5_18

necessary to combine the complex process and complex system with the characteristics of complex process.

It is difficult to quantify the relationship between the subsystems in complex systems and the coordination between the various sub processes in the complex process. We can use the aquarium ecosystem as an example. There are fish, fish, microbes and so on in the bathtub. What is the relationship between these creatures? What is the quantitative relationship between these organisms and the environment? What is the interaction between them? How to calculate the degree of mutual influence?

The traditional theory of complex system analysis is based on statistics, and seeks the law of development. However, the studies about the interaction between individuals are not a lot. In complex systems and complex processes, the whole is derived from the parts. We can decompose and refine complex problems. The binary relation is the most basic relationship in complex systems and complex processes. The binary relation is derived to obtain the complex relation, and the complex relation is refined and decomposed to obtain the binary relation. Based on the research of the two basic objects of complex system and complex process, this paper puts forward the seesaw model of the dual relation. The seesaw model is helpful for people to understand the complex phenomena of nature and society, and to reveal the law and the effect of interaction to adjust and control the development of things. In this paper, the application of this model is derived, and the pre-distribution model of distribution industry is obtained.

There are many literatures on the relationship between two random variables. In [4], the authors proposed the Statistical Asynchronous Regression (SAR) method. It determined a relationship between two time varying quantities without simultaneous measurements of both quantities. When seeking the relationship between the two quantities, covariance is commonly used. For example, the papers [5] and [6]. Covariance is based on statistical significance. The seesaw model is more concerned about the state of the two subjects, the relationship between changes in the state, is a micro analysis of the covariance model.

The causal model in complex process is also a common analysis method. The application of causal model is very much. In [7] the authors used causal models to estimate incomplete data. In [8] the authors illustrated evolutionary implications of various kinds of causal mechanisms by means of using causal graph theory. In [9] the authors used testing ecological interventionist causal models to seek to enhance effectiveness of psychological interventions under real-world conditions. In [10] Irvine et al. empirically evaluated a conceptual model developed for a regional aquatic and riparian monitoring program using causal models. In [11] Masa'Deh et al. used causal models to analyze the factors that influence it-business partnership. In the causal model, the two quantities are not equal; their development power is the same. While, the seesaw model focuses on the relationship between variables which have the different development directions and different development powers.

The seesaw model can be applied to the complex fields such as politics, economy, society and so on, because the seesaw model is the basic element for the analysis of complex systems and complex processes. In this paper, it is applied in the field of distribution. The seesaw model makes more comprehensive analysis on the optimization of distribution, and is helpful to seize the main contradiction to be convenient for people to adjust and control the distribution process in time and reasonably, to improve customer satisfaction of the distribution service. This application fully reflects the application value of the seesaw model.

Distribution service satisfaction has been the focus of the field of distribution. The paper constructed the distribution service customer satisfaction evaluation model (DCS) based on different aspects and designed a questionnaire to conduct an empirical research in [12]. In [13] the authors established a TPL distribution service satisfaction model. In [14] the authors sought to propose a conceptual structural equation model to investigate the relationships among city distribution service quality (SQ), supplier-customer relationship (SCR) and customer satisfaction (CS) and to demonstrate their direct and indirect effects on each other. The paper explores the relationship between the quality of distribution center of the direct selling and the distributors' satisfaction in [15]. Commodity distribution network optimization is the process of redesigning the distribution network under the condition of meeting customer require-ments for time and service quality. This requires minimum transportation and inventory costs. Based on the 2 indexes of customer satisfaction and delivery cost, a multi objective distribution model based on customer time satisfaction is established in [16].

In real life, some businesses have been doing pre-delivery. However, there is little research on the systematic analysis and the accurate interpretation and the standard-ization of pre delivery. Seesaw model allows businesses to accurately and scientifically regulate the distribution process, rather than just experience. Experience will be lack of effectiveness and accuracy, and it is difficult to popularize experience. All these make the research of this paper have practical significance.

This paper is organized as follows. Section 2 presents the seesaw model. We introduce the background of the distribution service in Sect. 3. We analyze the condition of the seesaw model for distribution service in Sect. 4. In this section we conclude that we can use the seesaw model to study the distribution sevice. Section 5 analyze the time of pre-distribution. We have an experiment and discuss the advantage of the pre-distribution model. In Sect. 6 we conclude our work and lay out future research.

2 Seesaw Model

The idea of the seesaw model is based on the observation and refining of the two related processes in nature and society and the relationship between the two systems. What is the impact of a system change or the simultaneous change of two systems, as well as how? The degree of influence is beneficial to the two processes or systems or adverse effects, or to a process or system. In order to describe and study the interaction between two processes or systems, we introduce a seesaw model.

For example, if the ambient temperature changes, then the animals have to adjust their functions to adapt the changes of environmental temperature. When the human body fever, the body surface heat dissipation rate can not keep up with the body temperature changes, it will easily burn the body. These related processes or systems have a characteristic: they all have a state. State is changing with time. There is a driving force in state change. The difference between them has an impact on two agents or the third party.

Definition 1: active agent: It is an entity or process with the ability to perform a function autonomously. It can be a system or a process. It can also be a kind of animal or even people, etc. There is a corresponding state at each moment of the subject, that is, the existence of a transient. The active agent has a closed construction, which is referred to as feudalism, namely Feudal Character. The active agent can then include the active agent. The definition of active agent is based on the specific research, from different levels and different angles to define.

Definition 2: the thrust of the agent: In the seesaw model, it is the own power of each active agent which can promote the development and the change. The creator of this power can be human or a process. We can use f to represent the thrust of the agent. The active agent is denoted as:
$$\dot{X} = \left\{ X(Q, f, t); Q \subset R^n, f \in \dot{X}, t \in T, T \subset R^m, X \in R^k, \right\}. \qquad \text{Where} \qquad Q =$$
$\{q_1, q_2, \ldots q_n, n \in N\}$, $q \in R^n$ is the main input parameter. $f \in \dot{X}$ represents the thrust of the agent. Because the agent can also include the agent, this is a nested definition. f $\in \dot{X}$ and $T \in R^m$ denotes the parameters which the agent depend on. Since the thrusts of the most agents are always determined, f is always a constant. The parameter T of many agents has the meaning of time, so T is used to be called time. The function $X(Q, f, T)$ indicates the state of the agent determined by the input parameter vector Q and the time function T. The mapping $X(\cdot, f, \cdot) : R^n \times R^m \to R^k$ is a function of local Lipschitz and satisfies $X(0, f, 0) = 0$. The function of mapping is to quantify the state and get the K state. When the f is determined, we can short agent as:
$$\dot{X} = \left\{ X_t(Q); Q \subset R^n, T \in R^m, X \in R^k \right\}.$$

The relationship between the two agents is the basic element of all complex subject relations, and the seesaw model is based on the study of the relationship between the two agents. In order to explore the mutual relationship and mutual influence between each agent, the concept of the thrust of the agent is put forward. This paper focuses on the coordination between the agent and the other, weakens the relationship among the input parameters of the agent, and emphasizes the closed construction characteristics of the agents. The basis for accurate and precise regulation of complex systems and complex processes is provided.

Definition 3: the saw: It is the requirement which can contact the two agents and fulfill a common function. The correlation coefficient of the two agents is not zero. The existence of the saw is a sufficient condition for the establishment of the seesaw mode. Set up two agents as: $\dot{X}$ and $\dot{X}'$. $R_t < \dot{X}, \dot{X}' >$ represents the saw of two agents at the moment t. $R_t < \dot{X}, \dot{X}' > = \rho_{\dot{X}\dot{X}'}$。

The two agents of the seesaw model need to meet the following conditions:

1. The saw Exists between two agents. Correlation coefficient of two agents $\rho_{\dot{X}\dot{X}'} \neq 0$.
2. The f Exists between two agents. That is, One or two can change. The motive power of the two subjects is not the same. $f_{\dot{X}} \neq f_{\dot{X}'}$
3. Two agents have equality. There is no relation between agents.
4. Two agents are incompatible. They cannot merge. That is, a process or system is transformed into another process or system, or two systems are fused to generate a new system.

That the development of the two agents is not harmonious will cause the seesaw characteristics. The change of two agents is within a certain range, known as the limit of the seesaw. There is no convergence of the two agents' changes, known as the proliferation of seesaw.

Definition 4: The degree of seesaw: It is the degree of disharmony of the two agents' development and change. The concept is concerned with the results of the development of agents, rather than the driving force of development. The seesaw characteristic has a certain impact on either sides or third parties, which is the direct promotion or obstruction or restriction at a certain moment. There are instantaneous effect and delay effect. At t time the degree of the seesaw of two agents $\dot{X}$ and $\dot{X}'$ is denoted as $D_{<\dot{X},\dot{X}'>}(t)$.

The input parameter vector of the agents stores the parameters of different attributes. Each parameter is measured in different units. For example when we study the degree of the seesaw between the two runners in the race, the parameters of the input parameter vector are: the distance between the athletes, the speed of the athletes, the physical strength of the players, the mentality of the players, etc. The degree of influence of these parameters on the agents is different, that is to say, each parameter has its own weight. In the seesaw model the parameters in the input parameter vector of two agents should be one-to-one correspondence. The comparison requires the unit of measurement have to be the same.

The weight distribution law of input parameter vector Q which describes the effect of the seesaw between two agents is: $W\{Q = q_n\} = w_n$. Mathematical expectation of the ratio of the input parameter vectors of the two agents is defined as:

$$E\left(\frac{Q}{Q'}\right) = \sum_n \frac{q_n}{q_n'} w_n \tag{1}$$

The degree of the two agents is:

$$D_{<\dot{X},\dot{X}'>}(t) = C \times \log E\left(\frac{Q}{Q'}\right) \tag{2}$$

where C is the adjustment parameter. By the mathematical expectation of the nature, we can get:

$$D_{<\dot{X},\dot{X}'>}(t) = C \times E\left(\log\left(\frac{Q}{Q'}\right)\right) = E\left(C \times \log\left(\frac{Q}{Q'}\right)\right) \tag{3}$$

The effect of each state of the two agents is constant, which we call parallel. When two agents change over time, the impact of each state presents a fluctuating trend, which we call pull about.

The sign of $D_{<\dot{X},\dot{X}'>}(t)$ reflects the size of input parameters of the two agents.

If $D_{<\dot{X},\dot{X}'>}(t) > 0$, we can get $E(Q_{\dot{X}}(t)) > E(Q_{\dot{X}'}(t))$

If $D_{<\dot{X},\dot{X}'>}(t) < 0$, we can know $E(Q_{\dot{X}}(t)) < E(Q_{\dot{X}'}(t))$

If $D_{<\dot{X},\dot{X}'>}(t) = 0$, we can obtain $E(Q_{\dot{X}}(t)) = E(Q_{\dot{X}'}(t))$

The greater the magnitude of the fluctuation is, the more unstable the relationship between the two agents becomes. This instability needs to be controlled, which we call the process as Concord Process.

Definition 5: The Crack: It means that the degree of seesaw has exceeded the degree of tolerance; the relevant parties will have a qualitative change. We call this the crack of the seesaw. The crack is fuzzy; it should be set according to different application scenarios and requirements. Set the value of the Crack as: $D_{throd} <\dot{X}, \dot{X}' >$.

Definition 6: Concord Process: When the degree of seesaw of two agents has not yet reached $D_{throd} <\dot{X}, \dot{X}' >$, we adjust the input parameters of the agents to promote their harmonious development.

The concept of Crack and Concord is beneficial to the selection of the time points and the setting of the adjustment. This can optimize the complex system adjustment effect. They can be applied in economic, social and natural fields.

The question of concord based on the fuzzy is:

$$\begin{cases} \text{Get: } t \\ \text{Meet: } D_{<\dot{X},\dot{X}'>}(t) \to D_{throd} <\dot{X}, \dot{X}' > \\ t \in T \end{cases} \qquad (4)$$

'$\to$'is fuzzy inequality relation such as' $\leq$ ', ' $\geq$ '. To solve the question of concord based on the fuzzy, the key is to solve (4).

3 Backgrounds

Distribution refers to a series of logistics activities within the scope of economic and reasonable, which is the selection of goods, processing, packaging, segmentation, distribution, delivered to the designated location on time and other operations according to customer requirements. In this paper, the study on the distribution is aimed at the immediate distribution process. It is a new logistics form which should be born by O2O (Online To Offline) such as fast food distribution. It is discrete and sudden. The whole distribution system is composed of the merchant, the deliveryman and customers. The customers are the recipients of the delivery service, after they place an order to buy goods. The merchant is the delivery service provider, when the customer has ordered goods. The deliveryman is the executor of the delivery service. He obtained goods from the merchant and promptly delivered to the customers, in order to complete the delivery service activities.

The specific process of instant delivery service is: Customers order to purchase the required goods, through the network, telephone and other means of communication. Merchant have been waiting for orders. Once some customers order, they immediately prepare goods to customers. The deliveryman arrives at the store where the package will be delivered and loaded into the distribution box to deliver the goods to the customers. After he delivers the goods, the deliveryman returns to the merchant to prepare the next order to start the next round of distribution activities.

4 The Analysis of the Seesaw Model for Distribution Service

In the distribution service, the two agents involved are the order process and the delivery process. The agent $\dot{X}$ indicates the order process of customers. The agent $\dot{X}'$ denotes distribution process of distribution staff. $f_{\dot{x}}$ is the customers. $f_{\dot{x}'}$ is the distribution staff. Obviously $f_{\dot{x}} \neq f_{\dot{x}'}$. The second condition is met. The saw $R_t < \dot{X}, \dot{X}' >$ exists. It indicates the process that merchants need deliveryman to send the goods to the customers after the customers order to buy goods.

The more the orders of the customers are, the greater the distribution task of deliveryman is. It indicates $\rho_{\dot{X}\dot{X}'} \neq 0$. The first condition is met. The distribution process and the customer order process do not contain inclusion relationships and they will not be fusion, so they meet conditions 3 and 4.

In summary, we can use the seesaw model to analyze the changes of the two agents in the distribution process.

Speed is the key factor in Distribution Service. In this paper, the key parameters are considered. $Q_{\dot{X}} = \{q_1\}$, $Q_{\dot{X}'} = \{q_1'\}$. q_1 denotes the speed of customer orders. q_1' denotes the average delivery speed of the distributor.

In the distribution service the degree of seesaw reflects the dissatisfaction level of the distribution system. In the Concord Process, we need to pay attention to the following: The speed of the customers order is what the business is expected to improve, the faster the better. The distribution speed of the distributor is related to many factors, such as weather, the level of proficiency of the distributor, traffic congestion, the distributor's transport equipment for the distribution, the work attitude of the deliveryman and other factors.

In the immediate delivery service, in real life, many merchants are after the user orders, immediately let the distributor to deliver. So there is the time that the distributor's distribution box is not full and empty. This is affecting the distribution speed of the delivery staff. This paper focuses on the problem. This situation is also easy to control, but also easy to ignore. It is necessary to study.

In order to reduce the time that the distributor's distribution box is not full and empty, this paper presents the pre-distribution mechanism.

5 Analysis the Time of Pre-distribution

5.1 Model Hypothesis

For the convenience of research, we need to make the following assumptions:
Hypothesis 1:   The total amount of merchant goods can meet the needs of all customers.

Hypothesis 2:   The customer's demand is less than the maximum delivery quantity for one time. Otherwise, the customer will be delivered separately until the remaining demand is less than the maximum distribution.

Hypothesis 3:   Each delivery service is for a number of customers. All customers need to arrange multiple deliveries. In the actual distribution, it is often delivered to many customers at the same time. However, the purpose of the hypothesis is to make the problem static, which does not affect the applicability of the model.

Hypothesis 4:   The order of each customer must be finished at one time

5.2   Analysis of Delivery Time of Deliveryman

The whole time of the distribution service is decided by the deliveryman. From the point of view of the delivery staff to analyze the delivery time, it can be described as T_{nopre}. It meets the formula as follows:

$$T_{nopre} = T_{stoc} + T_{dtos} + T_{dctoc} + T_{dc} \tag{5}$$

where T_{stoc} is the total amount of time for the delivery of a one-way distance from the merchant to the customer. T_{dtos} is the total time of all one-way distance from the customer to the merchant for the deliveryman, because the clerk must return the merchant to take the goods in order to re distribution. T_{dctoc} is the total time it takes for a distributor to travel from one customer to another, because sometimes the distribution box is equipped with a number of required goods of customers in the distributor's delivery process. T_{dc} is the total time spent by the deliveryman in delivering the goods to each customer. Usually, customers need to sign when they receive the goods.

Merchants prepare goods for customers. For example, in the fast food industry, when the user orders, the merchant to prepare food ingredients, processing materials, packaging and other processes. The time for merchants to prepare goods is set as:

$$T_{prep} = kN_c + b \tag{6}$$

where, k is the average time for merchants to prepare the ordered goods for a single customer. b is the average time for merchants to prepare some ordered goods before the distribution.

If the distribution time is less than the merchant's time to prepare the ordered goods, the situation that the deliveryman has to wait for the store will appear. Otherwise, there will be a situation that the merchant has to wait for the distributor. These two situations will cause the service efficiency of the business is low, so that the customer satisfaction of time decreased. It is the best case that the time of preparing the ordered commodities for merchant is equal to the time of distributing for the deliveryman. The value of the time of best case is the minimum in the distribution model.

The average distance between merchant and customers is described as l. The average distance between each customer is l'. Number of merchants set up 1. There is one deliveryman. The total number of orders for the customer is N_c, The delivery speed of the deliveryman is v. The average value of the quantity of goods delivered at one time corresponding to the number of the customers is m_{av}, t_{dc} is the average time spent by the deliveryman in delivering the goods to each customer, Then:

$$T_{stoc} = T_{dtos} = \frac{1}{v} \times \left\lceil \frac{N_c}{m_{av}} \right\rceil \tag{7}$$

Set the number of times to go out the store and back for deliveryman is a one-to-one correspondence, so $T_{stoc} = T_{dtos}$.

In real life, each delivery is to serve a number of customers. Then:

$$T_{dctoc} = \left\lceil \frac{N_c}{m_{av}} \right\rceil \times (m_{av} - 1) \times \frac{l'}{v} \tag{8}$$

When a distribution is only for one user, there are: m = 1, $T_{dctoc} = 0$.

$$T_{dc} = t_{dc} \times N_c \tag{9}$$

The formula (7), (8) and (9) are substituted into the formula (5), and:

$$\begin{aligned}
T_{nopre} &= 2 \times \frac{1}{v} \times \left\lceil \frac{N_c}{m_{av}} \right\rceil + \left\lceil \frac{N}{m_{av}} \right\rceil \times (m_{av} - 1) \times \frac{l'}{v} + t_{dc} \times N_c \\
&= \left\lceil \frac{N_c}{m_{av}} \right\rceil \times \left(2 \times \frac{1}{v} + (m_{av} - 1) \times \frac{l'}{v} \right) + t_{dc} \times N_c
\end{aligned} \tag{10}$$

5.3   Analysis of Pre-distribution Time

It can be seen from the formula (10) that we can implement the pre-distribution in order to optimize the delivery time. In each distribution, the goods which have a high probability of being ordered and can stand wear and tear in the distribution process can be loaded into the distribution box in advance early. The m_{av} is needed to be the maximum m_{max}, The m_{max} is maximum distribution capacity of deliveryman each time when he is from the merchant to the customer.

The formula (10) is used to get the time formula of the pre-distribution model as follows:

$$T_{pre} = \left\lceil \frac{N_c}{m_{max}} \right\rceil \times \left(2 \times \frac{1}{v} + (m_{max} - 1) \times \frac{l'}{v} \right) + t_{dc} \times N_c \tag{11}$$

Since the m_{max} is much larger than 1, the formula (11) can be approximated as:

$$T_{pre} = \left\lceil \frac{N_c}{m_{max}} \right\rceil \times \left(2 \times \frac{1}{v} + m_{max} \times \frac{l'}{v} \right) + t_{dc} \times N_c \tag{12}$$

5.4   The Delivery Speed of the Distributor

Customer order speed is:

$$q_1 = N_c \tag{13}$$

The distribution speed of on pre-distribution:

$$q_1' = \frac{N_c}{T_{nopre}} = \frac{N_c}{\left\lceil \frac{N_c}{m_{av}} \right\rceil \times \left(2 \times \frac{1}{v} + (m_{av} - 1) \times \frac{1'}{v}\right) + t_{dc} \times N_c} \tag{14}$$

The distribution speed of pre-distribution:

$$q_1' = \frac{N_c}{Tpre} = \frac{N_c}{\left\lceil \frac{N_c}{m_{max}} \right\rceil \times \left(2 \times \frac{1}{v} + m_{max} \times \frac{1'}{v}\right) + t_{dc} \times N_c} \tag{15}$$

Since only 1 parameter is selected, the weight w_1 is about 1. We can set $C = 1$ for comparison. Then the degree of seesaw of no pre-distribution is:

$$D_{<\dot{X},\dot{X}'>}(N_c) = \log\frac{q_1}{q_1'} = T_{nopre} \tag{16}$$

While the degree of seesaw of pre-distribution is:

$$D_{<\dot{X},\dot{X}'>}(N_c) = \log\frac{q_1}{q_1'} = T_{pre} \tag{17}$$

Since the pre-distribution may increase the loss probability of goods in the distribution process, it is not the best to implement the pre-delivery model at the beginning of distribution service. It is necessary to use pre delivery model when there is a large number of orders, that is, a delivery staff is very busy. The time T_{prep} to prepare the goods for merchant has been much less than the time of delivery, otherwise it will occur that the deliveryman has to wait to waste the time, so that the overall satisfaction of the distribution is decreased.

6   Numerical Experiments and Discussion

6.1   Numerical Experiments

The software in the experiment is MatlabR2015a. Set N_c the number of orders per minute increased from 1 to 100. $\frac{1}{v} = 5$, $\frac{1'}{v} = 0.5$, $t_{dc} = 0.1$, $D_{throd}<\dot{X},\dot{X}'>$ is assumed to be 50, the comparison between the pre-distribution model and no pre-distribution model is shown in Fig. 1.

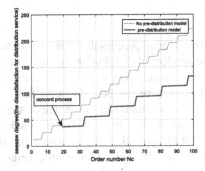

Fig. 1. The comparison between the pre-distribution model and no pre-distribution model

6.2   Discussion

According to the formula (10), m_{av} and v are the key variables that can be adjusted during the total time of delivery. The total time that the delivery clear go comes back from the customer is the time redundancy caused by the case that he goes to the customer from the merchant. To shorten the delivery time, it is necessary to improve the delivery speed, or reduce the time redundancy. It is needed to choose a vehicle with high speed, be familiar with the distribution route, and reduce the unnecessary distance for the deliveryman, in order to improve the speed of delivery. The time redundancy in distribution is related to the distribution capacity each time. As can be seen from Fig. 1, when the current orders increases gradually, that is, the order speed increases gradually, the speed of distribution will change accordingly. With the increase of the number of orders, the degree of seesaw between ordering speed and delivery speed is bigger and bigger. In this paper, the value of the Crack is set to 50. The degree of seesaw is not required to exceed this value. We should adjust the delivery speed when the degree of seesaw is almost over the threshold. As long as the distribution speed is adjusted to improve, the degree of seesaw is not too high. This paper discusses that it is easy to do to increase the distribution capacity for the merchants. In real life, merchants began to distribute, when the customer orders accumulated to a certain amount. It is caused that the distribution boxes are always not full in each distribution process. While the implementation of pre-delivery requires that the distribution box is always full in each distribution. Then this needs merchants to forecast the probability of goods.

When the goods have not been ordered to buy, we can put these goods which have high probability of being purchased and are not easily damaged into the distribution box in advance, so that the distribution boxes are full in each delivery process. If you do not implement the pre- delivery, then, when the order density is large, the risk of dissatisfaction with distribution increases quickly.

7 Conclusions

It can be seen from the experimental process that it is easy to find the problems which are often overlooked in the services of distribution by means of using the seesaw mode. The use of the seesaw model can help people accurately analyze, control or coordinate complex systems. The model can be extended to many fields. The delivery speed is effectively controlled under the constraints of the degree of seesaw and the satisfaction of distribution is guaranteed in the pre-distribution model. If we do not use pre-delivery, the time can not be controlled and the user's dissatisfaction will increase. In the follow-up study, we will use the seesaw model to generate the tug vines structure model for complex systems or complex process, in order to effectively obtain analysis method for complex system.

Acknowledgments. This work was financially supported by the Project of Natural Science Foundation of Hainan Province in China (Grant No. 20166232), the National Natural Science Foundation of China (Grant No. 61561017), Hainan Province Natural Science Foundation of China (Grant No. 617033) and Open Sub-project of State Key Laboratory of Marine Resource Utilization in South China Sea (Grant No. 2016013B).

References

1. Wang, C., Lan, F., Dai, Y.: National critical infrastructure modeling and analysis based on complex system theory. In: 1st International Conference on Instrumentation, Measurement, Computer, Communication and Control. IEEE, pp. 832–836 (2011)
2. Huang, J., Feng, Y., Zhang, S.: Research of complex system theory application on reliability analysis of network system. In: International Conference on Reliability, Maintainability and Safety. IEEE, pp. 1141–1145 (2009)
3. Dominici, G., Levanti, G.: The complex system theory for the analysis of inter-firm networks. A literature overview and theoretic framework. Int. Bus. Res. **4**(2), (2011)
4. O'Brien, T.P., Sornette, D., Mcpherron, R.L.: Statistical asynchronous regression: determining the relationship between two quantities that are not measured simultaneously. J. Geophys. Res. Space Phys. **106**(A7), 13247–13259 (2000)
5. Chamberlain, G.: Analysis of covariance with qualitative data. Rev. Econ. Stud. **47**(1), 225–238 (1980)
6. Byrne, B.M., Shavelson, R.J., Muthén, B.: Testing for the equivalence of factor covariance and mean structures: the issue of partial measurement invariance. Psychol. Bull. **105**(3), 456–466 (1989)
7. Karvanen, J.: Study design in causal models. Scand. J. Stat. **42**(2), 361–377 (2015)
8. Otsuka, J.: Using causal models to integrate proximate and ultimate causation. Biol. Philos. **30**(1), 19–37 (2015)
9. Reininghaus, U., Depp, C.A., Myingermeys, I.: Ecological interventionist causal models in psychosis: targeting psychological mechanisms in daily life. Schizophr. Bull. **42**, 264–269 (2015)
10. Irvine, K.M., Miller, S.W., Al-Chokhachy, R.K., et al.: Empirical evaluation of the conceptual model underpinning a regional aquatic long-term monitoring program using causal modelling. Ecol. Ind. **50**, 8–23 (2015)

11. Masa'Deh, R., Shannak, R.O., Obeidat, B.Y., et al.: Investigating a causal model of it-business partnership and competitive advantage. In: Global Business Transformation through Innovation and Knowledge Management: An Academic Perspective (2016)
12. Xiang, L.I., Liang, W.H.: On customer satisfaction evaluation of the distribtion service under the B2C environment. J. North China Electric Power University (2012)
13. Zhou, X., Guizhou Normal University: Study on TPL distribution service quality evaluation index system based on customer satisfaction. Logist. Technol. (2013)
14. Cui, L., Zhang, H., He, M., et al.: City distribution service quality: a factor impacting on supplier-customer relationship and customer satisfaction. In: International Conference on Logistics Systems and Intelligent Management. IEEE, pp. 813–816 (2010)
15. Wang, H., Wei, F.X.: Evaluation of service quality and customer satisfaction of distribution center-direct selling. J. Yibin University (2007)
16. Jueliang, H., Lihua, W., Han, S., et al.: Apparel distribution model and algorithm based on time satisfaction. J. Text. 31(2), 138–142 (2010)

An Efficient Filtration Method Based on Variable-Length Seeds for Sequence Alignment

Ruidong Guo[1,2], Haoyu Cheng[1,2], and Yun Xu[1,2(✉)]

[1] Computing School of Computer Science,
Key Laboratory on High Performance,
University of Science and Technology of China, Anhui 230027, China
{grd, chhy, xuyun}@ustc.edu.cn
[2] Collaborative Innovation Center of High Performance Computing,
National University of Defense Technology, Changsha 410073, China

Abstract. With the rapid development of next-generation sequencing (NGS) platforms, more than billions of reads are produced quickly. Finding all mapping locations of these reads in the reference genome is not only a bioinformatics issue, but also a large-scale computation issue. Existing all mapping tools are usually divided into the two steps, filtration and verification. Filtration step discards some wrong locations and generates candidates. As for verification step, each candidate is mapped to the reference sequence to determine whether it is a mapping location. Statistics indicated that the verification step is the main part of the whole mapping time. That is to say, less candidates lead to less mapping time. Our strategies improve filtration step to decrease the number of candidates.

We propose a dynamic programming and two heuristic strategies and integrated them into the filtration step. These strategies are applied in the state-of-the-art all-mapper, Bitmapper. Compared with the advanced all-mappers, experiment results show that our method make a significant progress.

Keywords: Read alignment · Bitmapper · Filtration · All mapper

1 Introduction

Nowadays, DNA sequencing has many applications in various fields. The next-generation sequencing (NGS) platforms, such as Illumina system and SOLiD system, can produce reads quickly and cheaply. These platforms boost the development of biological areas and generate a large amount of short reads. The short reads should be quickly mapped to reference genome for further biological studying.

The read mapping problem has been studied for many years. It can be classified into two types: best-mapping and all-mapping. For best-mapping task, it focuses on identifying one or few locations with the best quality, while all-mapping task finds the all matching locations. BWA [1], Bowtie [2] and Bowtie2 [3] are popular best-mappers, which are used in many applications, such as mapping DNA-protein interactions, understanding evolution. However, in some important cases, all-mapping

© Springer Nature Singapore Pte Ltd. 2017
G. Chen et al. (Eds.): PAAP 2017, CCIS 729, pp. 214–223, 2017.
DOI: 10.1007/978-981-10-6442-5_19

problem is received more attention. For instance, in ChIP-seq experiments, it is critical to find all locations and search many binding sites. Those all mapping tools are called all-mappers. Generally, the state-of-the-art all-mappers are mrFAST [4], Yara [5] and Bitmapper [6]. Under the condition that the amount of data is huge and time is limited, most mapping tools could be accelerated by using multi-thread CPU and some could even support SIMD instruction sets and GPU.

Compared with the best-mappers, all-mappers are more complex. The main reason is that all-mappers needs to search all qualified locations, while best-mapping tools only choose the best one. Many reference genome exist many repeat segments, which causes a huge calculation. To solve this problem, all-mappers still need to be optimized.

Most all-mappers are divided into two steps, filtration and verification. In the filtration step, reads are split into shorter segments. These shorter segments are called seeds and then found their locations in the reference genomes. Besides, this step discards some un-mapping locations and remains the possible mapping locations, called candidates. After that, each candidate is verified to determine if it is the truly mapping locations. Statistics indicated that verification step is the dominant part of whole mapping time. It means that reducing verification time significantly accelerates mapping time.

To reduce verification time, decreasing the number of candidates is an efficient strategy. Two classes of seeds selection methods, aimed at decreasing the number of candidates, are proposed in recent years. One class is the methods based on the fixed-length seeds [7, 8]. Those methods are fast but might generate a lot of number of candidates sometimes. The other class is based on the variable-length seeds [9–11]. Compared with fixed-length seeds methods, variable-length methods generate less number of candidates but is much slower. For instance, OSS [10] consumes much time and needs more than 350G memory. It is difficult to be integrated into all-mappers.

In this paper, we focus on filtration step to decrease the number of candidates. The main contributions are as follows.

- We propose a filtration method that can be used in many all-mappers. It is based on a dynamic programming for selecting variable-length seeds. With the two heuristic strategies, limiting the seeds length and lengthening high frequency seeds, it achieves a good balance between the mapping time and total number of candidates.
- We apply these strategies to the state-of-the-art all-mapper, Bitmapper. And we also compared our method with other all-mappers. In addition, like other all-mappers, our method can also be accelerated by multithreads CPU.

2 Preliminaries

2.1 Definition of Read Mapping Problem

First, there is the definition of read mapping.

Definition 1: Given a read and a reference sequence s, finding all locations where hamming distance or edit distance is not greater than threshold k in s of this read.

Two evaluation criteria, hamming distance and edit distance, are commonly used for read mapping. Hamming distance counts the differences which only allow the substitutions between two same length sequences. Different with hamming distance, edit distance does not require the same length. Given two sequence *s1* and *s2*, it just finds the minimum operations for transforming *s1* to *s2*, where the operations allow substitutions, insertions and deletions.

Since edit distance is more suitable for realities, it is more popular used in all-mappers. Therefore, we adopt edit distance as the standard in the remaining parts.

2.2 Seed-and-Extend Strategy

Most all-mappers adopt the seed-and-extend strategy. Therefore, we introduce the procedure of this strategy.

Definition 2: For a read r, we split it into some shorter segments which are called as seeds. A seed, with a length q, can also be expressed as q-grams.

Seed-and-Extend strategy can be divided into two phases: filtration and verification.

1. *Filtration:* The pigeonhole principle is a high-efficient method for filtration. It states that n pigeons take shelter inside $n + 1$ pigeonholes, then one or more pigeonholes are empty. Based on this principle, while the edit distance threshold is k, each read is divided into $k + 1$ non-overlapping seeds. It is easy to know at least one seed can be mapped to the reference sequence without any substitution, intersection or deletion. With the help of index, the locations of $k + 1$ seeds are quickly searched. Similarly, if a read is partitioned into $k + m$ non-overlapping seeds, at least m seeds are exactly mapped. It means that if a location does not appear in m seeds, this location is actually un-mapping and needs to be filtered.
2. *Verification:* After filtration step, the remained locations, called candidates, are verified to the reference sequence by utilizing the edit-distance calculation, such as Smith-Waterman [12], Needleman-Wunsch [13], Gene Myers' bit-vector [14] and so on. Candidates, whose edit distance is no more than k, can pass the verification step and are recorded as truly mapping.

2.3 Index of Seeds

The index is classified as hash table index and FM-index. Hash table index is faster but larger than FM-index. Due to the high demand on time, most of the all-mappers use the hash table index. So we introduce the hash table index in the next.

Figure 1 shows how to construct the *5-gram* hash table index. Each inverted list contains a sorted list of locations where the corresponding hash value appears in the sequence. For instance, at location 201, the hash value of this location is ACGAC, then this location is stored in the ACGAC inverted list. The size of the invert list, called frequency, shows how many times a seed is appeared in the reference sequence. Scanning the whole reference sequence would create the whole hash table index. For a seed, calculating its hash value and searching the correspondingly inverted list can find its locations quickly. The average frequency is lower while the length of seeds is longer. Thus, in order to balance size and frequency, the length of hash table is usually between 10 and 15.

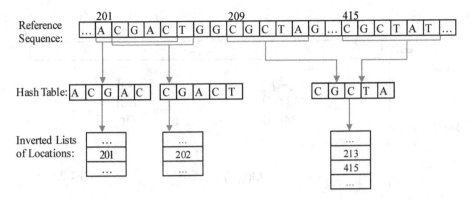

Fig. 1. The construction of hash table index

3   Methods

Firstly, there is a description for the dynamic programming, which is aimed to select $k + 2$ variable-length seeds. Then we show two heuristic strategies for accelerating this algorithm. At last, the index for variable-length seeds is introduced.

3.1   Selecting Seeds for Variable Length

The dynamic programming of selecting $k + 2$ variable length seeds is as follows.

Definition 3: Given a read r, let $|r|$ denote the length of the read and $r[p1, p2]$ denote a sub-segment of the read which begins at position p1 and ends in position p2.

Definition 4: Let $C[p1, p2]$ $(1 \leq p1 < p2 \leq |r|)$ be the minimum frequency of a seed in $r[p1, p2]$, $M(i, j)$ be the minimum sum of frequency for i $(1 \leq i \leq k + 2)$ seeds in $r[1, j]$ $(1 \leq j \leq |r|)$, and $M(k + 2, |r|)$ be the last result that we want to get.
We calculate $C[p1, p2]$ for all sub-segment of r firstly. Assuming the minimum sum of frequency for i seeds in all $r[1, j]$ $(1 \leq j \leq |r|)$ have been stored. To calculate $M(i + 1, j)$ $(1 \leq j \leq |r|)$, $r[1, j]$ is divided into two parts, $r[1, p - 1]$ and $r[p, j]$, at position p $(1 < p \leq j)$. Figure 2 gives the detailed description of this condition. Get $M(i, r[1, p - 1])$ from the first part and $C[p, j]$ from the second part. Calculating all situations in p $(1 < p \leq j)$, the minimum values in those values are the optimal total number of frequency for selecting $i + 1$ seeds in $r[1, j]$ (i.e., $M(i + 1, j)$) Here is the recurrence function in Eq. 1, where $1 \leq i \leq k + 2$ and $1 \leq j \leq |r|$.

$$M(i+1, j) = \min \begin{cases} M(i, 1) + C[2, |r|] \\ M(i, 2) + C[3, |r|] \\ \cdots \\ M(i, |r| - 1) + C[|r|, |r|] \end{cases} \tag{1}$$

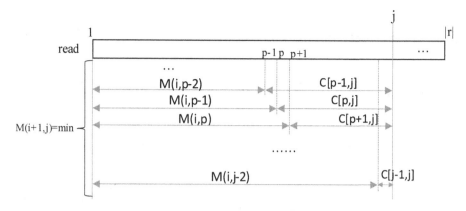

Fig. 2. The process for computing M (i, j)

3.2 Heuristic Strategies

Obviously, the time complexity of this method is $O((k + 2) \times |r|^2)$, and the space complexity is $O((k + 2) \times |r|)$. Although this dynamic programming method has minimum total number of $k + 2$ seeds and generates the least candidates, it is very time-consuming. To solve this problem, we propose two heuristic strategies: bounding seed lengths and lengthening high frequency seeds.

Bounding seed lengths. Bounding seed lengths reduces the time of selecting variable-length seeds. In the previous work, Optimal Prefix Seeds (OPS) [8] proposes a dynamic programming to quickly select fixed-length seeds. In practice, the *11-gram* index achieves best performance in the OPS.

Experiment shows that the index of *11-grams* has a good effect for most of reads. Figure 3 displays the distribution of candidates for 10 k human reads. It shows that the number of candidates of most reads are not high. Statistics results indicate that, for only 20% of the whole reads, their number of candidates are more than 100. Nevertheless, the verification time of those 20% reads accounts for about 80% of the whole verification time. We find that only a few candidates in those 20% reads are truly mapped locations. It means that the candidates of those 20% reads are worth improving.

For our method, the range of variable-length is between 9 and 11. On the one hand, because of the small range, it generates candidates quickly. On the other hand, it can reduce the number of candidates for those 20% reads.

Lengthening seed. After bounding seed lengths, there still exist high frequency seeds. This is because some special reads appear many times in the reference genome. For these reads, only long seeds (e.g. >20) have low frequency. Based on this, we propose a greedy strategy to reduce the frequency. If the frequency of a seed is more than threshold t, we called it as a high frequency seed. After selecting $k + 2$ seeds, we lengthen high frequency seeds as long as possible, as shown in Fig. 4.

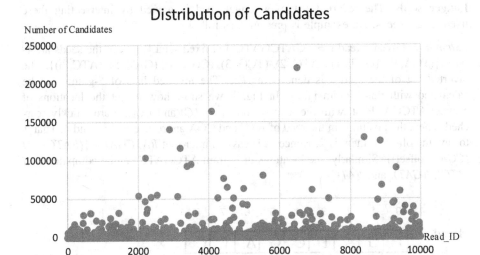

Fig. 3. The distribution of candidates

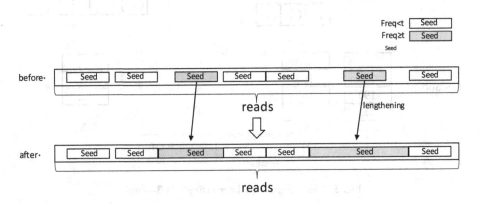

Fig. 4. Lengthening high frequency seeds as long as possible

3.3 Variable-Length Seed Index

The FM-index is widely used in selecting variable-length seeds. However, it is time-consuming. Considering the requirement of high speed, our method adopts hash table index. Although this index is built by the fixed-length seeds, the locations of variable-length seeds can also be got by these fixed-length seeds.

Shorter seeds. As shown in Fig. 1, location 201 belongs to the invert list of ACGAC, but it also belongs to the invert list of *4-gram* ACGA. Consequently, given a *5-gram* hash table index, to get the sorted invert list of *4-gram* ACGA, we just need to union the invert lists of ACGAA, ACGAC, ACGAT, ACGAG and then sort them. In the same way, the invert list of a *3-gram* can be acquired by sixteen *5-grams*. This is the way to compute invert lists of shorter seeds.

Longer seeds. The inverted list of a longer seed is gained by intersecting these fixed-length seeds. An example is presented as follows.

Example 1: Given a read r = TGATCGATC and fixed seed length 3, the seeds are *G (s)* = {(TGA, 0), (GAT, 1), (ATC, 2), (TCG, 3), (CGA, 4), (GAT, 5), (ATC, 6)}. The inverted list of the seed s is denoted as *I(s)*. The inverted lists of 5-grams can be calculated with this 3-gram index. In Fig. 4, we show how to get the locations of *5-gram* ATCGA. First, with the help of index, *I(ATC)* and *I(CGA)* are quickly searched. Then, the positions in the read of ATC and CGA are respectively 2 and 4. That is to say, the offset of them is 2. Hence, it is easy to infer that *I(ATCGA)* = {(*I(ATC)*∩(*I (CGA)* - offset)}. Similarly, the locations of 7-gram ATCGATC can be computed by *I (ATC)*, *I(GAT)*, and *I(ATC)*.

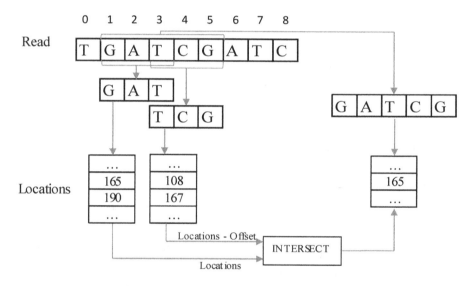

Fig. 5. Get a 5-gram by intersecting two 3-grams

4   Results and Discussion

In the experiments, the human genome HG19 is used as the reference sequence. And the results are gained by mapping 1 million 100 bp real single-end reads which come from specimen HG00096 of the 1000 Genome project [15]. At last, the edit distance (ED) is separately set as threshold k = 5 and k = 6.

4.1   Number of Candidates

As we mentioned before, the length range of seeds is between 9 and 11. According to the statistics, the total number of 11-grams is more than 9-grams and 10-grams. It means that constructing 11-gram hash table index cost minimum calculations, because of the locations of 11-gram are gained directly by this index. In addition, we found that while the threshold t = 14000, lengthening seeds gets the best effect.

We test three advanced seeds selection methods, OPS [8], ASF [9] and CKS [7], with multiple configurations and compare with the best results of those methods in the previous data sets. Table 1 shows the average number of candidates for each read. Compared with OPS, more than 30% of the candidates can be reduced in our method. As for ASF and CKS, our method acquires several times fewer candidates.

Table 1. Results of average number of candidates

Method	Average number of candidates	
	ED = 5	ED = 6
ASF	9849	22698
CKS	6001	9955
OPS	2420	3837
Ours	1120	2666

It is easy to see that the number of candidates is more with higher edit distance threshold k. The reason is that higher threshold k leads to more seeds, which would generate more candidates.

4.2 Running Time Comparison

Our method is compared with three advanced all-mappers, including mrFAST [4], Yara [5] and Bitmapper [6]. Those mappers are used by default configuration and their results are recorded in SAM format output. All mappers run on a machine with an AMD Opteron(TM) Processor 6168 CPU and 24 GB of RAM, running 64-bit Ubuntu OS.

Table 2 lists the mapping time. We can see that our method is about 30% faster than Bitmapper and several times faster than other all-mappers when the ED = 5. While the ED = 6, it is nearly twice as fast as Bitmapper. The reason is due to the time of verification. From the front section, we know that higher edit distance would lead to more candidates. Verifying more candidates causes a larger proportion of the whole mapping time. Meanwhile, our method reduces the number of candidates. Therefore, the effect of accelerating becomes more significant while the ED is larger.

Table 2. Results for 1 million reads of human genomes

All-mappers	Time[min: sec]	
	ED = 5	ED = 6
mrFAST	293:43	666:14
Yara	132:59	818:55
Bitmapper	50:01	151:39
Ours	38:34	78:28

4.3   Memory Usage

For most all-mappers, their memory usage depends on the reference genome and the index. The peak memory of our method and Bitmapper is nearly 15 GB. Human genome is composed of 3.15 billion base pairs which needs 3 GB to store. As for hash table index, the hash table and inverted lists totally occupy approximately 12 GB. Different with the previous methods, mrFAST only requires 7 GB though it uses hash table index. Because it cuts human genome and index into several parts and uses one of them each time. Yara adopts another data structure called FM-index. It only needs small memory size.

5   Conclusion

In this paper, we propose two strategies to improve the filtration. In the experiments, compared with the current state-of-the-art all-mappers, our method significantly accelerates the all-mapping time.

After using the improvements, the number of candidates are greatly reduced. It would decrease the verification time, which is the mainly part of the whole time.

The memory footprint of our method is a little large by comparison with Yara and mrFAST. Considering our fast speed and currently low memory prices, the short-coming of the memory can be bearable sometimes. We prepare to reduce the memory usage by compressing the hash index in the future.

Besides, although the number of candidates is significantly decreased by using our method, it still has a large space to be improved. We plan to find other effective strategy to decrease the number of candidates in further research.

Acknowledgment. This work was supported by the National Nature Science Foundation of China under the grant No. 61672480 and the Program for Excellent Graduate Students in Collaborative Innovation Center of High Performance Computing.

References

1. Li, H., Durbin, R.: Fast and accurate short read alignment with burrows-wheeler transform. Bioinform. **25**(14), 1754–1760 (2009)
2. Langmead, B., Trapnell, C., Pop, M., Salzberg, S.L.: Ultra-fast and memory-efficient alignment of short DNA sequences to the human genome. Genome Biol. **10**(3), R25 (2009)
3. Langmead, B., Salzberg, S.L.: Fast gapped-read alignment with bowtie 2. Nat. Methods **9**(4), 357–359 (2012)
4. Hach, F., Hormozdiari, F., Alkan, C., et al.: mrsfast: a cache-oblivious algorithm for short-read mapping. Nat. Methods **7**(8), 576–577 (2010)
5. Siragusa, E.: Approximate string matching for high-throughput sequencing. Ph.D. Dissertation, Freie University Berlin (2015)
6. Cheng, H., Jiang, H., Yang, J., et al.: Bitmapper: an efficient all-mapper based on bit-vector computing. BMC Bioinform. **16**(1), 192 (2015)
7. Xin, H., Lee, D., Hormozdiari, F., et al.: Accelerating read mapping with fasthash. BMC Bioinform. **14**(1), S13 (2013)

8. Kim, J., Li, C., Xie, X.: Improving read mapping using additional prefix grams. BMC Bioinform. **15**(1), 42 (2014)

9. Marco-Sola, S., Sammeth, M., et al.: The gem mapper: fast, accurate and versatile alignment by filtration. Nat. Methods **9**(12), 1185–1188 (2012)

10. Xin, H., Nahar, S., et al.: Optimal seed solver: optimizing seed selection in read mapping. Bioinform. **32**(11), 1632–1642 (2016)

11. Kim, J., Li, C., Xie, X.: Hobbes3: dynamic generation of variable-length signatures for efficient approximate subsequence mappings. In: IEEE 32nd International Conference on Data Engineering (ICDE). IEEE 2016

12. Smith, T.F., Waterman, M.S.: Identification of common molecular subsequences. J. Mol. Biol. **147**(1), 195–197 (1981)

13. Needleman, S.B., Wunsch, C.D.: A general method applicable to the search for similarities in the amino acid sequence of two proteins. J. Mol. Biol. **48**(3), 443–453 (1970)

14. Myers, G.: A fast bit-vector algorithm for approximate string matching based on dynamic programming. J. ACM (JACM) **46**(3), 395–415 (1999)

15. 1000 Genomes Project Consortium: An integrated map of genetic variation from 1,092 human genomes. Nature, **491**(7422): 56–65 (2012)

An Optimized Fusion Method
for Double-Wearable-Wireless-Band Platform
on Cloud-Health Application

Wenchao Xu[1], Yanbo Liu[1(✉)], Yanqin Yang[1], Xiaoshuang Ning[1],
Tianxing Chu[2], and Hongzhi Song[2]

[1] East China Normal University, Shanghai, China
wchxu@ce.ecnu.edu.cn, 51151214034@ecnu.cn
[2] Texas A&M University-Corpus Christi, Corpus Christi, TX 78412, USA
{tianxing.chu,hongzhi.song}@tamucc.edu

Abstract. This paper presents a stable double-wireless-wearable-band platform that can detect hand gestures. The real-time monitoring and control system utilizes an MCU processor, a wireless transceiver, and a commercial three-axis, digital-output MEMS accelerometer. To detect the user's hand movements, a 3D virtual environment is created via a double-wearable-band controller. Compared with a single wearable band, double wearable bands can identify more gestures with improved stability. Performances in terms of control and detection are discussed in detail. This research development allows the user to specify desired two-hand postures using the multi-sensor information fusion technique for controlling a variety of robotic devices. In the system, the defined two-hand postures also allow the user to add freestyle control to various applications, which bridge the communication gap between humans and the systems. Moreover, the integration of the action recognition algorithm of the combination of two bracelets and the server brings out a real-time approach to analyze and make decisions based on the users' data. Therefore, the system can call for help in a timely manner under critical conditions.

Keywords: Double wearable wireless band platform · Multiple kalman filter fusion · Cloud monitoring health platform

1 Introduction

With the rapid development and evolvement of microelectromechanical systems (MEMS), smart devices, such as smart bands and smart phones, have greatly elevated the quality of human living, and gradually changed our life styles [1]. Smart bands have been extensively used for recognizing and assisting with human activities. Lombardi used a wearable wireless accelerometer for fall detection in ambient assisted living (AAL) applications by communicating with care holders and relatives of the assisted person through an ADSL based gateway [2]. Yao et al. proposed a pre-impact fall early warning system on the waist to alarm and solve the aged falling [3].

Smart handhelds have also been commonly used for building a cloud-based health monitoring platform. Mathavan et al. designed a single wearable appliance gesture

© Springer Nature Singapore Pte Ltd. 2017
G. Chen et al. (Eds.): PAAP 2017, CCIS 729, pp. 224–236, 2017.
DOI: 10.1007/978-981-10-6442-5_20

remote monitoring and fall detection system using wireless intelligent personal communication node [4]. Due to the popularity of handheld devices and the development of web cloud, the company called Xiaomi has built iHealth monitoring system, which is based on handheld terminal and cloud. It can make it easy for the elderly to collect and process body data [5]. When the emergency occurs, the server will dispatch multi-duties dynamic alarm strategy using the cloud-based architecture [6].

Nowadays, however, most applications, as mentioned above, utilize a single handheld device, and recognizing the human gestures sometimes lacks stability or portability. Compared with a single band, using double wearable bands tends to identify more activities with higher stability. Therefore, an optimized fusion method using double wearable wireless bands is proposed in this paper to improve diversity and stability of controlling performance driven by human gestures.

Optimized with the synergic effect between double wearable wireless bands, the proposed platform aims to build an easy-to-use, intuitive, and robust system that compensates the drawback of using a single functional wearable band and broadens its applicability in control design. With cheap hardware equipment, the system uses multi-thread, parallel processing mechanism and monitoring service for continuous, stable and reliable implementation.

The remainder of the paper is structured as follows. Section 2 presents double wearable wireless platform as well as the details of its implementation. Section 3 provides experimental design and results. Summaries are concluded in Sect. 4.

2 Double-Wearable-Wireless-Band Platform

In this section, a double-wearable-wireless-band platform is provided in detail. As shown in Fig. 1, the system integrates multi-terminal and multi-sensor, it is an intelligent perception health management system for caring group such as old-aged people and children. The system adopts a variety of sensors of smart bracelets to collect the physiological features and environmental parameters of the users.

The integration of the action recognition algorithm of the combination of double-wireless-wearable-band platforms and the server brings out a real-time approach to analyze and make decisions based on the users' data, so that the system can call for help under the condition that the user is in the normal state or stays in a dangerous place in a timely manner. The user management of the old-aged care institution or the community center can monitor the healthy activity of the caring group like healthy statistics and location track based on the cloud server community. Also the guardians of caring group can obtain alarm message, location query playback or health campaign test statistics service from the mobile terminal and cloud to help those users protect the safety and quality of the lives of the caring group. The system can detect the physical condition from two hand gestures and transmit the signal via GPRS modules to send message to the cloud server. Through more diversity of action designs, it may be predicted that double-wearable-wireless bands will be used in many communication field.

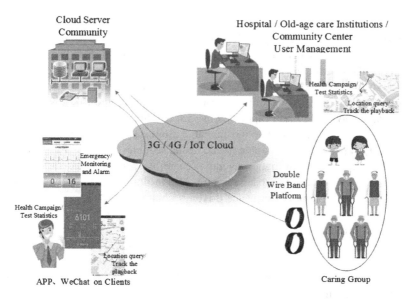

Fig. 1. Block diagram of the whole developed system frame

2.1 The Multi Kalman Filter Fusion

As shown in Fig. 2, after sampling the acceleration information from the band sensor, the data is processed using Kalman filter. Then the system extracts feature and finally makes association and fusion [7]. By multi-sensor fusion processing, the data is sent to the system at the next moment [8]. At the same time, the data from the inertial sensor is sent to the system via wireless module. The inertial sensor data is filtered with a Kalman filter and then fused in a complementary filter. After computing the rotation matrix based on the Euler angles (roll, pitch and heading), we convert different types of data such as acceleration, rotational and linear movement from the frame of human body motion sampling to the frame of space matrix.

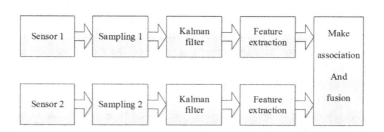

Fig. 2. Method of fusion and association

The process model, as defined in (1), describes the transition of the process state, assuming that the next state is a linear function of the previous state [8].

$$x(k) = Ax(k-1) + Bu(k-1) + w(k-1) \tag{1}$$

The Kalman filter that estimates the state of a discrete-time system tends to a steady value for both stand-alone and data-fusion modes from near constant noise and system parameters [9]. The discrete Kalman filter time update equations include:

$$\begin{cases} \hat{x}(k) = A\hat{x}(k-1) + Bu(k) \\ P(k) = AP(k-1)A^T + Q \end{cases} \tag{2}$$

In (2), $\hat{x}(k)$ stands for the time update state and the measurement update statement. $P(k)$ represents the posteriori estimate error covariance of the uncertainty of the estimation in Kalman filter.

In our system, two wearable wireless devices have the same sampling rates, and these sensors are assumed to have the same noise characteristics. The Kalman filter can fuse sensor data from multiple sources by using a technique to detect and estimate the statement of the system [8]. The approximate discrete-time state model can help to estimate the motion state of human hand continuously as following:

$$x(k) = Ax(k-1) + w(k-1) \tag{3}$$

The hand motion variable is the measurement input. The gain A of the rotation is measured by the accelerometer and computed by the gyroscope. We define the x axis as the direction of the hand motion. After analysis of the hand motion along x axis, we get the following state transition matrix as:

$$A = \begin{bmatrix} 1 & \Delta t & \Delta t^2 \\ & 1 & \Delta t \\ & & 1 \end{bmatrix} \tag{4}$$

In our application, the inertial sensors are sampled at 125 Hz ($\Delta t = 1s/125$).

We apply the continuous Wiener process acceleration model. The acceleration is perturbed with a white noise process whose power spectral density is q [10]. We improve and optimize the Q [7] that matches our system in fusing sensor as following:

$$Q = \begin{bmatrix} 0.070\Delta t^5 & 0.126\Delta t^4 & 0.158\Delta t^3 \\ 0.126\Delta t^4 & 0.158\Delta t^3 & \Delta t^2 \\ 0.158\Delta t^3 & \Delta t^2 & \Delta t \end{bmatrix} q \tag{5}$$

We define the following gesture recognition valid, or otherwise invalid. The double dynamic gesture track recognition range $Di(i \in (1,4))$ is valid including $D1(B \text{ to } S)$, $D2(F \text{ to } S)$, $D3(L \text{ to } S)$ and $D4(R \text{ to } S)$ which will be explained in detail in Sect. 3.

$$Di(k) = Map[x(k)] \tag{6}$$

We define the current hidden state as C(t), and the observed results from the three sub-detectors as *RM(t)* (MEMS), *RP(t)* (pulse), *RB(t)* (blood pressure), where *R* is the output environment type with the highest confidence level from each sub-detector. We also define valid range of *R* that are $RM(t) \in (0.8, 0.9)$, $RP(t) \in (0, 0.1)$ and $RB(t) \in (0, 0.2)$. We set *F* as standard trigger criteria from environment *C1* to *C2* (elaborated as V: valid, I: invalid):

$$F(k) = RM(t)||RP(t)||RB(t)||Di(k) \tag{7}$$

2.2   Hardware Architecture

In this section, we analyze hardware architecture of double wearable wireless devices. Table 1 shows a comparison table of our design and other designs.

Table 1. Compare with different designs

Design	Communication mode	Device
A	Wi-Fi	Smart phone
B	Bluetooth	Smart phone
C	Bluetooth	One wearable device
D	Bluetooth/GPRS	Two wearable devices

Design A used Wi-Fi protocol to transmit control command to a receiver which then controlled the RC car [11]. Design B, which is similar to design A, used Bluetooth protocol as transmission media [12]. Design C from Xiaomi used a wearable device as remote unit. Different from other designs, we choose double wearable wireless bands as hand-gesture recognition devices so as to increase reliability of remote unit control.

In this paper, as can been seen in Fig. 3(a), the wearable device is composed of a micro control unit processor (MCU) module, Sensor modules, a wireless communication module and a GPS module:

- MCU module: MCU is the code of the double-wearable-wireless-band platform, which controls Sensor modules and reads out its digital output. It processes the data with primary Kalman filter.
- Sensor module: in particular, Sensor modules like MPU9255 or GY-251, with fusion function, can be used in gesture commands. It provides a 3-axis gyroscope, 3-axis accelerometer, 3-axis magnetometer and a Digital Motion Processor to interface with multiple non-inertial digital sensors, like pressure sensors, on its auxiliary I2C port.
- GPS module: ATGM332D module from Chinese Micro Technology company can receive signals of BeiDou or GPS.
- Wireless communication module: a wireless communication module contains Bluetooth module and GPRS module.

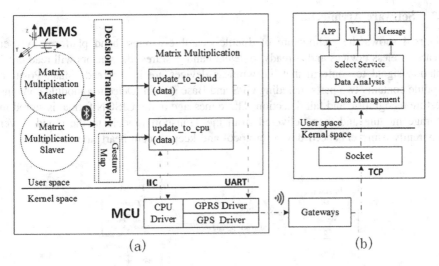

(a)   (b)

Fig. 3. Block diagram of the control frame

The two developed wearable wireless bands are worn on left and right hands separately. One is the master, the other is slaver. After processing (8) and (9), the system gets the matrix operations using Matrix Math Library on matrix multiplication application. Then outputs of matrix operations are delivered to decision framework which contains gesture map which will be explained in detail in Sect. 3. Then wireless module transmits the data to the gateway [12]. In the block diagram of Fig. 3(b), the cloud server mainly contains three functions.

- Firstly, the MCU keeps sending the signal from the wireless module. Proto-threads concept is applied in wireless-transportation to protect the system on the real-time work state [14]. As shown in Fig. 3(b), through the gateways, the cloud server receives message from Fig. 3(a) in the system. We select the slave model for wireless module that pairs with master wireless module in Fig. 3(a).

- Secondly, as shown in Fig. 3(b), the architecture of cloud server contains a fully-functioning Node.js Web APP, using the Express framework, that reads from and writes to a MongoDB database. MCU sends message using wireless module from gateways to the cloud server through TCP socket connection. Then the cloud server manages and analyzes the message to make decision. The information in that message includes the client's PID, the name of the client binary, an optional "tag" message for use in logging events from that client's falling detection, an optional location message of the latitude and longitude, a character that stands for hand gesture recognition and a CRC.

- Thirdly, after making decision from the receiving message, the cloud server selects data to different service platforms including APP, Web and phone message. The monitor listeners or the guardians of caring group can log in different platform view.

2.3   Software Architecture

Figure 4 shows the architecture of double-wearable-wireless-band platform. A user wears wireless devices and is ready to detect hand gestures, the sensors will read out a rotation signal to conform that the wireless devices are in the ready state. Position module contains two parts including GPS and Base station. Detection module contains gesture recognition and fall detection. These message in the CJSON format is stored in Storage module and then delivered from the Transport module to the cloud server coherently using enhanced device protocol and keeps recent heart rate on network.

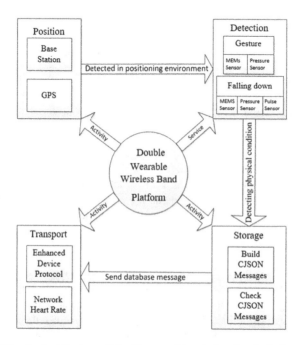

Fig. 4. Architecture of double-wearable-wireless-band platform

As shown in Fig. 5, after initializing the sensor, the system starts thread and obtains acceleration data from different axes. Then the system will preprocess the measured accelerometer signals into Multi Kalman filter fusion, the HMM (Hidden Markov Model) module and DTW (Dynamic Time Warping) module to improve the above detection performance for further analysis [15].

After recognizing gesture, the system will determine the palm rotation angle matched with the correct direction. A fall detection gets not only a fall detection information, but also the blood pressure and pulse of user. The system call the detection results derived from the Sensor modules. The system keeps all four parts continuously monitoring and accordingly updates its return result in a real-time manner.

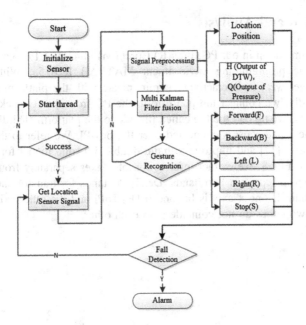

Fig. 5. The frame of hand gesture detection flowchart

Figure 6 shows the frame flowchart of cloud server reading message. ApiKey is the tag message to recognize different names of clients. Periodically, once or twice 3 min or when a threat is detected, the data block of GPS and pulse will be transported to the cloud server. Then the system starts thread and maps different data block.

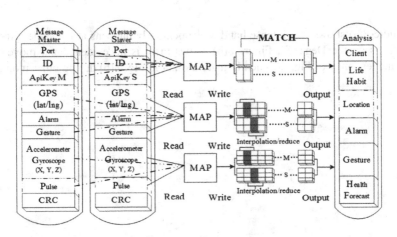

Fig. 6. Cloud server workflow

3 Simulation and Analysis

We make the simulation in our PC: Intel(R) Core(TM) i3-4160, CPU 3.60 Ghz, RAM 8.00 GB on the programming environment MATLAB R2015a. With MATLAB Support Package, we can interactively communicate with the platform over a USB cable [16]. Firstly, we connect the platform to the PC from the IO package of Math Works. Secondly, we upload the file called file.pde to the platform. Thirdly, we access peripheral devices and sensors connected over IIC or SPI. We refer to the code from Ezequiel Gonzalez [17] and improve to build double 3D solid model for recognition.

As shown in Fig. 7, we extract information of x axes separately from two Cubes which are drawn to represent two hands. DegToPi stands for transformation between degree and radian. If one x axis is beyond 0 DegToPi and the other axis is below 0 DegToPi, the two cubes do not coincide with each other.

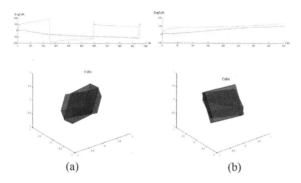

Fig. 7. X axis at different(a)/same(b) direction of two hand gestures

Figure 8(a) shows that intense hand movement only happens at the same direction and effects on the DegToPi. Figure 8(b) shows there is no evident hand movement, while the tendency of Fig. 8(a) nearly matches that in Fig. 8(b) during other moments.

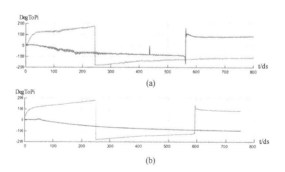

Fig. 8. Data from different Axes between shaking and normal gestures

Figure 9 presents two pairs of signals that map the user's different gestures during the change of states such as "hand waving forward" in Fig. 9(a), "hand stop" in Fig. 9 (b), "hand waving backward" in Fig. 9(c), "hand waving right" in Fig. 9(d) "hand waving left" in Fig. 9(e) and "palm flip 180-degree rotation" in Fig. 9(f). Different cube positions can show the corresponding hand gestures.

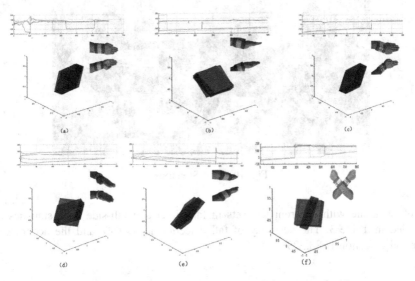

Fig. 9. Relationship between gestures and the sensor signals

Using this principle and method, we identify and classify hand gesture features within different values in Table 2 as following. The output characters of BFRLS represent gesture recognition that stands for the backward, forward, right, left, and stop, respectively.

Table 2. Gesture Map

Gesture	X axis value range	Y axis value range
F(Forward)	−32768–6000	−6000–6000
B(Backward)	6000–32768	−6000–6000
R(Right)	−6000–6000	−32768–6000
L(Left)	−6000–6000	6000–32768
S(Stop)	−6000–6000	−6000–6000

4   Experiment and Result

The proposed double-wearable-wireless-band platform applies a gesture recognition system and a fall detection system which bridge the communication gap between humans and the systems. We have trained our platform with different types of gestures.

In order to evaluate the performance of the proposed system and method, we implement it on the double-wearable-wireless-band platform and test its performance around the ECNU sports ground as shown in Fig. 10.

Fig. 10. Test Scenarios

We evaluate with different test sets in fall-direct and fall-side. The statistics are provided in Table 3. The accuracy of fall detection is 84.16% and the accuracy of none-fall detection is 83.33%.

Table 3. Fall Test

Data	Fall		None-fall		Fall detection	None-fall detection
	Test 1	Test 2	Test 3	Test 4		
R(Right)	30	30	3	3	50	5
L(Left)	30	30	3	3	51	5
S(Stop)	60	60	6	6	101	10

The cloud server can effectively take the statement of user regularly. As shown in Fig. 11, when the user falls down or has other accidents, the cloud server will give an alarm in real time on the web screen. Also it is proposed to identify activities of caring group in their daily life using playback function in SDK of Baidu. These will help decrease the potential risk of falling down for aged people and will also help them receive health care in a timely manner.

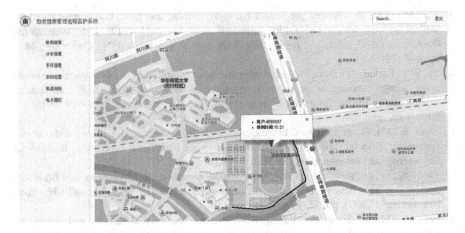

Fig. 11. Alarm and Playback of test result delivered to the cloud server

5 Conclusions

In this work, we have implemented a system from a practical hardware platform and the relevant embedded firmware. This system provides accurate hand gesture recognition by using a multi-Kalman filter fusion scheme. We set up a strong real-time, multithread wireless platform with wireless duplex data channel. We selected fusion algorithm and optimized it to deal with data information.

In the simulation, we build the double 3D control model of the multiple sensors for the inversion of the algorithm by analyzing collected data. Experimental test results indicate that double-wearable-wireless-band platform in the gesture recognition system can be adopted for caring group guarantee monitoring based on cloud computing remotely. The platform can detect falling down events of aged people and help relieve potential risks. The cloud can make health management provided by the system to track the daily movement and analyze the medical motion data in order to make reasonable recommendations for the caring group users.

Acknowledgment. The research work was supported by the National Natural Science Foundation of China (61300043, 61373156 and 91438121), and the Science and Technology Commission of Shanghai Municipality (14DZ2260800).

References

1. Noda, K., et al.: MEMS on robot applications. In: Transducers 2009–009 International Solid-State Sensors, Actuators and Microsystems Conference, pp. 2176–2181 (2009)
2. Lombardi, A., Ferri, M., Rescio, G., Grassi, M., Malcovati, P.: Wearable wireless accelerometer with embedded fall-detection logic for multi-sensor ambient assisted living applications. In: Sensors, 2009, pp. 1967–1970. IEEE (2009)

3. Yao, M., et al.: A wearable pre-impact fall early warning and protection system based on MEMS inertial sensor and GPRS communication. In: 2015 IEEE 12th International Conference on Wearable and Implantable Body Sensor Networks (BSN), Cambridge, MA, pp. 1–6 (2015)

4. Yi, W.J., Sarkar, O.: Design flow of wearable heart monitoring and fall detection system using wireless intelligent personal communication node. In: 2015 IEEE International Conference on Electro/Information Technology (EIT), Dekalb, IL, pp. 314–319 (2015)

5. http://www.miui.com/

6. Zhou, B., Ma, Q., et al.: Cloud-based dynamic electrocardiogram monitoring and analysis system. In: 2016 9th International Congress on Image and Signal Processing, BioMedical Engineering and Informatics (CISP-BMEI), Datong, pp. 1737–1741 (2016)

7. Tsuda, K., et al.: Proposal for a seamless connection method for remotely located Bluetooth devices. In: 2014 Seventh International Conference on Mobile Computing and Ubiquitous Networking (ICMU), pp. 78–79 (2014)

8. Feng, S., Murray-Smith, R.: Fusing Kinect sensor and inertial sensors with multi-rate Kalman filter. In: IET Conference on Data Fusion and Target Tracking 2014: Algorithms and Applications (DF&TT 2014), Liverpool, UK, pp. 1–8 (2014)

9. Yadav, A., Naik, N., Ananthasayanam, M.R., Gaur, A., Singh, Y.N.: A constant gain Kalman filter approach to target tracking in wireless sensor networks. In: 2012 IEEE 7th International Conference on Industrial and Information Systems (ICIIS), pp. 1–7 (2012)

10. Lawrence, P.J., Berarducci, M.P.: Navigation sensor, filter, and failure mode simulation results using the distributed Kalman filter simulator (DKFSIM). In: Position Location and Navigation Symposium, 1996, pp. 697–710. IEEE (1996)

11. Jing, Y., Zhang, L.: AndroRC: an Android remote control car unit for search missions. In: Systems, Applications and Technology Conference (LISAT), 2014, Long Island, pp. 22–27. IEEE (2014)

12. Mora, A., et al.: Speed digital control for scale car via Bluetooth and Android. In: 2015 CHILEAN Conference on Electrical, Electronics Engineering, Information and Communication Technologies (CHILECON), pp. 129–135 (2015)

13. Min, S., Kim, J.: Inertial sensor based inverse dynamics analysis of human motions. In: 2015 IEEE International Conference on Advanced Intelligent Mechatronics (AIM), pp. 177–182 (2015)

14. Klein, I., Rusnak, I.: Joint Kalman Filter for formation moving with wiener process acceleration. In: 2014 IEEE 28th Convention of Electrical and Electronics Engineers in Israel (IEEEI), pp. 1–4 (2014)

15. Tra, K., Pham, T.V.: Human fall detection based on adaptive background mixture model and HMM. In: 2013 International Conference on Advanced Technologies for Communications (ATC 2013), Ho Chi Minh City, pp. 95–100 (2013)

16. https://github.com/ezgode

17. http://www.mathworks.com/academia/arduino-software/arduino-matlab.html

Research on Concept Drift Detection
for Decision Tree Algorithm in the Stream
of Big Data

Shangdong Liu[1], Lili Lu[2], Yongpan Zhang[1], Tong Xin[1],
Yimu Ji[1,3,4(✉)], and Ruchuan Wang[1,3]

[1] Jiangsu High Technology Research Key Laboratory for Wireless Sensor
Networks, Nanjing University of Posts and Telecommunications,
Nanjing 210023, China
jiym@njupt.edu.cn
[2] College of Software, Nanjing College of Information Technology,
Nanjing 210046, China
[3] Key Laboratory of Computer Network and Information Integration,
Southeast University, Nanjing 211189, China
[4] Key Laboratory of Intelligent Perception and Systems for High-Dimensional
Information of Ministry of Education, Nanjing University of Science and
Technology, Nanjing 210094, China

Abstract. With the rapid development of information technology, various
industries have to deal with an increasing number of data. Compared with the
traditional static data, stream data under big data environment was rapid, continuous and always changed with time. At the same time, the implicit distribution of data stream brought about the concept drift. A stream data concept
drift detection algorithm named ADDS (Anti-concept Drift Detection Algorithm) was put forward, which is mainly used to detect and process the hidden
concept drift of unsteady data stream, under big data environment. The ADDS
was focused on the improvements of traditional classification algorithms with
incremental way to adapt to the demand of streaming data processing. The
experimental results showed that the ADDS had a better concept drift detection
effect.

Keywords: Stream computing · Decision tree · Classification algorithms ·
Bigdata

1 Introduction

With the rapid development of Internet of Things (IOT) and all kinds of information
technologies, various industries have to process an increasing number of data, such as
telecom operators, securities finance banks and Internet terminals [1]. In light of
massive data grow explosively over a long period, it is required to explore how to
extract more values from big data [2]. The growth of big data in terms of volume not

© Springer Nature Singapore Pte Ltd. 2017
G. Chen et al. (Eds.): PAAP 2017, CCIS 729, pp. 237–246, 2017.
DOI: 10.1007/978-981-10-6442-5_21

only facilitates people's lives, but also unearths many challenges for big data mining. Classification data mining algorithm belongs to supervised learning algorithm in the field of machine learning, the classification methods for which include decision tree, Bayes algorithm, neural network, K-nearest neighbor, support vector machine, etc. [3]. Classification mining algorithm has been widely applied in the fields including operator/user behavior analysis, sensor network, forecast of traffic data stream, securities risk evaluation, etc. [4]. Stream classification mining algorithm based on big data environment should be able to timely adapt to the accuracy decrease brought by the sudden change of hidden data distribution, and adjust algorithm model quickly when concept drift is detected. Most of the current data mining frameworks and platforms only analyze massive offline data, instead of being able to update models in real time to make them better adaptive to real-time data stream mining scenario.

Comparing with traditional classification algorithms, the decision tree classification method for stream data is more difficult in terms of real-time restriction of storage space and classification. Traditional data classification method can hardly meet the requirements for data stream classification in most cases; therefore, traditional classification model needs to be readjusted. In the references [5], the VFDT (Very Fast Decision Tree) brought up by Domingos et al. is a classification algorithm under stream data environment. VFDT establishes decision tree based on Hoeffding inequality, and makes a new sample transverse the decision tree from upward to downward every time when it flows in, until it reaches the leaf nodes of the tree at last. Gama et al. [6, 7] brought up VFDTc algorithm based on VFDT, so that it can process continuous data. However, VFDTc algorithm calculates the information entropy of all possible values, which aggravates the calculation burden under the environment of high-speed data stream environment. Anagnostopoulo et al. [8] brought up a stream classification algorithm based on probability estimation method, but the estimation result obtained with this algorithm shows large deviation affected by data themselves. The references [9] brought up CVFDT (Concept-Drift Very Fast Decision Tree), which was the extension of VFDT. It not only remains the speed and accuracy of VFDT, but also ensures the real-time capability and accuracy of model in a better manner when the hidden information of data stream changes constantly. As the hidden distribution of data stream changes constantly, model accuracy and stability cannot be ensured, which leads to concept drift [10]. The references [11] classified the existing processing concept drift strategies with time sequence analysis method, and described the adaptive learning process. In the references [12], Gama et al. brought up time sliding window mechanism which improved the real-time capability of model training. Kuncheva et al. [13] carried out the in-depth study on stream data classification model, only to find that model accuracy is not only related to data quality and the concept drift degree, but also the size of sliding window. Therefore, Kuncheva et al. brought up the stream data classification method based on variable sliding window. With this method, sliding window can adjust window size adaptively when concept drift occurs to data stream, so that classification model can resist concept drift more effectively and the real-time capability and accuracy of classification model can be ensured. Liang et al. [14]

improved and expanded CVFDT based on the study of CVFDT, and brought up puuCVFDT algorithm, which can process the data without label or with undetermined attribute value. Similarly, Xu et al. [15], on the basis of improving CVFDT algorithm, brought up CFDT, which introduced a new tree type indexed structure. This algorithm pre-processes data with clustering method first, and then scans clusters and extracts hidden information to adjust tree structure, making CFDT tree be able to improve the real-time capability of model, and ensuring high classification accuracy.

This Paper puts forward the concept drift detection algorithm for stream data under big data environment, ADDS, to detect and process the hidden concept drift of unsteady stream data. Section 2 of this Paper introduces Hoeffding theorem, Sect. 3 puts forward ADDS, Sect. 4 analyzes experiment result, and Sect. 5 summarizes the whole work.

2 Hoeffding Theorem and Lemma

Hoeffding inequality [16] describes the difference between probability of the trueness of a certain event and the frequency observed from m different Bernoulli trials. When machine learning algorithm was adopted, we built up an abstract model to simulate the laws implied in true data, in hope that we could obtain assumptions that fitted training data to the highest degree, which means we sought for the lowest error rate. However, we could hardly find a perfect classifier the error rate of which is 0. With the theory of probability statistics, we could only endeavor to minimize the error rate of model in training data set with the greatest efforts.

Theorem 1 (Hoeffding Theorem): Suppose $X_1, X_2 \ldots X_n$ are n independent random variables within the range [0,1], and the empirical mean of these variables is $\overline{X} = \frac{1}{n}(X_1 + \cdots X_n)$. Hoeffding Theorem 1 is:

$$P(\overline{X} - E[\overline{X}] \geq t) \leq e^{-2nt^2} \tag{1}$$

Theorem 2 (Hoeffding Lemma): If X is a random variable the mean value of which is 0, and $P(X \in [a, b]) = 1$, then for all $\lambda \in R$:

$$E[e^{\lambda X}] \leq \exp\left(\frac{\lambda^2(b-a)^2}{8}\right) \tag{2}$$

Based on the Hoeffding lemma and the Law of Markov, the demonstration of Hoeffding inequality can be obtained: suppose there are n independent random variables, $X_1, X_2 \ldots X_n$, and $P(X_i \in [a_i, b_i]) = 1, 1 \leq i \leq n.$, and make $S_n = X_1 + \ldots + X_n$, then for any s, t $\geq$ 0. It can be then learned from Markov inequality and the Law of Hoeffding that:

$$P(S_n - E[S_n] \geq t) = P\left(e^{s(S_n - E[S_n])} \geq e^{st}\right) \leq e^{-st} E\left[e^{s(S_n - E[S_n])}\right]$$

$$= e^{-st} \prod_{i=1}^{n} E\left[e^{s(X_i - E[X_i])}\right] = e^{-st} \prod_{i=1}^{n} e^{\frac{s^2(b_i - a_i)^2}{8}} = \exp\left(-st + \frac{1}{8}s^2 \sum_{i=1}^{n} (b_i - a_i)^2\right) \quad (3)$$

To calculate the most accurate upper limit of Hoeffding means calculating the minimum value at the right side of the equality. There is:

$$s = \frac{4t}{\sum_{i=1}^{n} (b_i - a_i)^2} \quad (4)$$

Substitute s into the equality, it can be obtained that:

$$P(S_n - E[S_n] \geq t) \leq \exp\left(-\frac{2t^2}{\sum_{i=1}^{n} (b_i - a_i)^2}\right) \quad (5)$$

During the machine learning of decision tree algorithm, the minimum quantity of samples needed by the internal node split of decision tree within the given confidence interval can be obtained by solving the inequality in Theorem 1, when Hoeffding inequality splits at the nodes of decision tree.

3 ADDS: Anti-Concept Drift Detection Algorithm

ADDS brought up in this Paper detects concept drift of data streams, the distribution of which changes constantly, with the thought of Hoeffding inequality, calculates the concept drift occurrence threshold with Hoeffding boundary, and tests the classification result of drift data stream that changes constantly by setting up an n-sized window on classification model. Then, we inserted maker bit True into the window when the classification result obtained by classification model is consistent with the label of marked sample, and marker bit False into the window when it is not. When the total number of True and False marker bits was up to n along with the data streams keeping flowing into the model, the real-time classification accuracy, P_{real}, in sliding window at a certain time point, t, was obtained by dividing the quantity of True marker bits by n-sized window. At the same time, we defined the optimal classification accuracy, $P_{optimal}$, so as to record the historical maximum value of window sample accuracy.

With the constant flowing in of data stream, the degree of fitting of decision tree model became increasingly higher. At this moment, the classification accuracy of decision tree became more stable, and the difference between the classification accuracy, P_{real}, in sliding window and the optimal classification accuracy, $P_{optimal}$ also became smaller gradually. We defined $\Delta P = P_{optimal} - P_{real}$ to detect the error between the optimal classification accuracy and the real-time classification accuracy in sliding window. Suppose we give the confidence level, δ, and confidence interval, h, for accuracy error, it can be obtained from the inequality in Hoeffding theorem 1

$$P\left(\overline{X} - \mathrm{E}[\overline{X}] \geq \mathrm{h}\right) \leq e^{-2nh^2} \tag{6}$$

Suppose $\overline{X}$ the estimate accuracy, while $\mathrm{E}[\overline{X}]$ is the real accuracy, then meaning of above inequality is that the probability that the estimation accuracy is deviating from the real accuracy more than h is not higher than e^{-2nh^2}.

According to symmetry, there is:

$$P\left(-\overline{X} + \mathrm{E}[\overline{X}] \geq \mathrm{h}\right) \leq e^{-2nh^2} \tag{7}$$

It can be obtained from (6) + (7) that:

$$P\left(\left|\overline{X} - \mathrm{E}[\overline{X}]\right| \geq \mathrm{h}\right) \leq 2e^{-2nh^2} \tag{8}$$

For given confidence level δ, within the confidence interval where the width of the expected accuracy $\mathrm{E}[\overline{X}]$ is 2 h, there is

$$\delta \leq 2e^{-2nh^2} \tag{9}$$

Since $n \geq -\frac{\log(\delta/2)}{2h^2}$ was obtained by solving the inequality, at least $-\frac{\log(\delta/2)}{2h^2}$ samples were needed in order to meet the confidence level $1 - \delta$ within the given confidence interval $\mathrm{E}[\overline{X}] \pm \mathrm{h}$. According to the conclusion obtained with Hoeffding inequality, we learned that the scope of accuracy error should be within $\sqrt{\frac{1}{2n} \ln \frac{1}{\delta}}$ for the size, n, of the given sliding window and the confidence level δ. Therefore, we obtained that the concept drift occurrence threshold of data stream is $\varepsilon = \sqrt{\frac{1}{2n} \ln \frac{1}{\delta}}$, which means when $\Delta P = P_{\text{optimal}} - P_{\text{real}} \geq \sqrt{\frac{1}{2n} \ln \frac{1}{\delta}}$, ADDS will prompt that the occurrence of concept drift is detected.

According to ADDS, the real-time accuracy will be calculated when the total number of True and False marker bits in accuracy detection window reaches n, and at the same time, updating will also be activated when the window threshold n is reached. The label that enters the marker bit queue the earliest will be abandoned and the latest sample classification result will be added to the queue constantly. Moreover, the optimal classification accuracy, P_{optimal}, in ADDS will also be updated constantly along with time series update. When the classification accuracy, P_{real}, in the window is larger than P_{optimal}, P_{optimal} will be updated.

See Algorithm 1 for the pseudo-code of concept drift detection algorithm ADDS:

Algorithm 1 Concept drift detection algorithm ADDS
Input: decision tree model M;
    E, forecasted sample stream (x, y);
    n, window size;
    δ, the preset confidence level;
Output: Whether concept drift occurs.

---

1 Initialization window N = empty queue, $P_{optimal} = 0$;
2 **for** every forecast sample (x, y) in sample stream E
3   Sample (x, y) forecasts through classification model M to obtain decision value
    d;
4   **if** d = y, **then**
5   Insert marker bit True into window queue;
6   **else**
7   Insert window queue into the marker bit False;
8   **end if**
9 **end for**
10 **if** the markers bits in the window are unfilled, **then**
11   Add new marker bits;
12 **else**
13   Calculate the proportion, P_{real}, of marker bits in the window;
14   Delete the oldest marker bit and insert the latest marker bit;
15   **if** $P_{optimal} - P_{real} < 0$ **then**
16     $P_{optimal} = P_{real}$;
17   **else**
18   **if** $P_{optimal} - P_{real} > \sqrt{\dfrac{1}{2n} \ln \dfrac{1}{\delta}}$ **then**
19   It is promoted that concept drift occurs;
20   Clear the window N, $P_{optimal} = 0$;
21   **end if**
22   **end if**
23 **end if**
24 **end**

---

4 Test Result and Analysis

Three commonly used synthetic data sets: MIXED dataset, SINE1 dataset, and CIR-CLES data set were used in this Paper to analyze the classification accuracy of ADDS. During test, we added 10% noise to all data sets to test its effect to algorithm accuracy. The description of the 3 data sets is as below:

(1) MIXED data set: MIXED data set has 2 numerical attributes, x and y, which are uniformly distributed within the [0, 1] range, and there are also 2 Boolean type attributes, v and w. When at least 2 out of the 3 variables, v, w and y, are ensured

smaller than $0.5 + 0.3 * \sin(2\pi x)$, a sample will be divided into positive class. Classification rules will be reversed when concept drift occurs; therefore, we set a mutant concept drift for every 20,000 samples.

(2) SINE1 data set: SINE1 data set contains 2 attributes, x and y, which are uniformly distributed within the range [0, 1], and the classifier function is a sine function $y = \sin(x)$. Sudden concept drift is also hidden in data distribution. We set a sudden concept drift for every 20,000 samples cyclically. Before the first concept drift occurs, all samples under the sine function curve will be classified as positive, and the rest samples will be classified as negative.

(3) CIRCLES data set: Gradual concept drift is hidden in the distribution of CIRCLES data set. This data set contains 2 attributes, x and y, which are uniformly distributed within the range [0, 1]. Classifier function takes a circular ring $(x - x_c)^2 + (y - y_c)^2 = r_c^2$ as its standard, with (x_c, y_c) as the circle center and r_c as the radius. The 4 circles <(0.2, 0.5), 0.15>, <(0.4, 0.5), 0.2>, <(0.6, 0.5), 0.25>, <(0.8, 0.5), 0.3>, classify samples in sequence. Samples inside the circular ring are classified as positive, while those outside it will be classified as negative. Concept drift occurs at 25,000 samples.

Confusion matrix is an important evaluation method for supervised learning algorithm, by which, the classification result of classifiers can be obtained by testing and comparing labeled samples. Confusion matrix is as shown in Table 1:

As shown in Table 1, all samples can be classified into one of the following 4 types in confusion matrix:

(1) If a sample is positive and it is classified as positive by the classifier, then it is True Positive (TP);
(2) If a sample is positive but it is classified as negative by the classifier, then it is False Negative (FN);
(3) If a sample is negative but it is classified as positive by the classifier, then it is False Positive (FP).
(4) If a sample is negative and it is classified into negative by the classifier, then it is True Negative (TN).

Table 1. Confusion matrix

		Actual value		Total
		p	q	
Forecast values	p'	True Positive (TP)	False Negative (FP)	Actual Positive (TP + FN)
	q'	False Positive (FN)	True Negative (TN)	Actual Negative (FP + TN)
Total		Predicted Positive (TP + FP)	Predicted Negative (FN + TN)	TP + FN + FP + TN

In general, a good anti-concept drift classifier should be with the most TF samples, the least FP samples and the least false samples. For the evaluation of concept drift detection methods mentioned in this Section, we defined an acceptable delay range, h, and users set the value of h to identify the error between the result tested by ADDS and the true concept drift position. If the detector detects concept drift at moment t, but concept drift does really exist within the range [t − h, t + h] indeed, then this concept drift will be deemed TP. If the detector prompts that there is no true concept drift existing within the range around h at the position where concept drift occurs, then this concept drift will be deemed FP. Similarly, if concept drift occurs to true data, but the detector does not prompt that there is concept drift within the range around h at the position where concept drift occurs, then this concept drift will be deemed FN.

We used decision tree as the classifier to test the average detection delay of ADDS, DDM and ADWIN in MIXED data set, SINE1 data set and CIRCLES data set respectively, the quantity of TP, FP and FN concept drifts, and the classification of decision tree. On MIXED data set and SINE1 data set, we set the acceptable delay range h as 250, and the detection window size as 25. Since concept drift in CIRCLES data set was of gradual type, and it took longer to adapt to a new concept, its acceptable delay range h was set as 1,000 and the test window size was set as 100. Each algorithm was used for 50 tests to obtain the average value. Comparison test was carried out and the performance of concept drift detection algorithm was as shown in Tables 2, 3 and 4:

As shown in Table 2, the test result in MIXED data set showed that the average TP of ADDS and ADWIN is the highest and there was no FN. At the same time, ADDS has the least FP, while the FP of ADWIN and DDM is relatively high. From the aspect of average delay, ADDS is far lower than DDM, and slightly lower than ADWIN. In terms of accuracy, the classification accuracy of the 3 concept drift detection algorithms is within 81–82%, and the accuracy of ADDS is slightly higher than that of ADWIN and DDM.

Table 2. Evaluation indexes of algorithm in MIXED data set

Detector	Average delay	TP	FP	FN	Accuracy
ADDS	16.5 ± 1.4	4.0	0.02 ± 0.15	0.0	87.37% ± 0.16
ADWIN	21.40 ± 1.48	4.0	21.81 ± 5.69	0.0	86.75% ± 0.18
DDM	137.15 ± 23.76	3.57 ± 0.69	2.93 ± 1.88	0.56 ± 0.73	86.03% ± 1.09

Table 3. Evaluation indexes in SINE1 data set

Detector	Average delay	TP	FP	FN	Accuracy
ADDS	16.0 ± 1.2	4.0	0.21 ± 0.49	0.0	82.60% ± 0.1
ADWIN	23.80 ± 1.67	4.0	18.90 ± 5.25	0.0	81.92% ± 0.13
DDM	170.02 ± 23.10	2.68 ± 0.88	2.94 ± 1.98	1.33 ± 0.96	81.68% ± 1.79

Table 3 shows that ADDS has the lowest FP and the lowest FN. ADWIN has the same TP and FN with ADDS, and they are respectively 4.0 and 0.0. However, the FP of ADWIN is far higher than that of ADDS, which shows that it is easy for ADWIN to prompt wrong concept drift. The average delay of DDM in SINE1 data set is still far higher than that of ADDS and DDM.

Table 4. Evaluation indexes of algorithm in CIRCLES data set

Detector	Average delay	TP	FP	FN	Accuracy
ADDS	94.12 ± 27.23	3.0	0.04 ± 0.25	0.0	88.71% ± 0.13
ADWIN	242.70 ± 158.67	2.57 ± 0.60	18.01 ± 5.57	0.44 ± 0.57	86.29% ± 0.21
DDM	612.3 ± 123.16	2.31 ± 0.84	1.68 ± 1.52	0.73 ± 0.77	84.97% ± 0.86

Table 4 shows that the average delay of the 3 algorithms in CIRCLES data set is all increased; however, the average delay of ADDS is far lower than that of ADWIN and DDM. At the same time, ADDS has the highest TP and the lowest FP and FN, which shows that ADDS can detect the concept drifts that truly exist accurately, and the probability of wrong prompts of ADDS is also low when data stream is steady and there is no concept drift. Therefore, it can be learned that the accuracy of ADDS is the highest.

From the several tests of the detector with sudden concept drifts and gradual concept drifts in the above three data sets, we can learn that ADDS, comparing to ADWIN and DDM, can achieve better concept drift detection result; therefore, the classification accuracy is higher when ADDS is adopted.

5 Conclusion

By putting forward the concept drift data stream classification method based on decision tree, testing the accuracy maintenance sliding window for decision tree classification algorithm, and deriving the threshold of concept drift in sliding window with Hoeffding inequality theorem and lemma, this Paper brings up ADDS, which, by experiments, is demonstrated to be applicable to fast decision tree algorithm, with better detection rate and lower false detection rate. What we will do next is to parallelize ADDS and use it in big data platform for the purpose of realizing real-time stream processing.

Acknowledgement. This paper was supported in part by project on the National Key Research and Development Program of China (2017YFB0202200); Program of National Natural Science Foundation of China (61373017, 61572261, 61170065); Outstanding Young Fund Project of Jiangsu Natural Science Foundation of China (BK20170100); Jiangsu Key Research and Development Program (BE2017166); Open-End Fund of Jiangsu High Technology Research Key Laboratory for Wireless Sensor Networks (WSNLBZY201514) and Research Project of Nanjing University of Posts and Telecommunications (NY214067).

References

1. Reed, D.A., Dongarra, J.: Exascale computing and big data. Commun. ACM **58**(7), 56–68 (2015)
2. Assunção, M.D., Calheiros, R.N., Bianchi, S., et al.: Big data computing and clouds: Trends and future directions. J. Parallel Distrib. Comput. **75**(5), 3–15 (2014)
3. Gaber, M.M., Zaslavsky, A., Krishnaswamy, S.: Data stream mining. In: Data Mining and Knowledge Discovery Handbook, pp. 759–787. Springer, Berlin (2009)
4. Lu, S., Xie, G., Chen, Z., et al.: The management of application of big data in internet of thing in environmental protection in China. In: IEEE First International Conference on Big Data Computing Service and Applications (BigDataService), pp. 218–222. IEEE (2015)
5. Domingos, P., Hulten, G.: Mining high-speed data streams. In: Proceedings of the 6th ACM SIGKDD International Conference on Knowledge Discovery and Data Mining, pp. 71–80. ACM, New York (2002)
6. Gama, J., Rocha, R., Medas, P.: Accurate decision trees for mining high-speed data streams. In: Proceedings of the 9th ACM SIGKDD International Conference on Knowledge Discovery and Data Mining, pp. 523–528. ACM, New York (2003)
7. Gama, J., Fernandes, R., Rocha, R.: Decision trees for mining data streams. Intell. Data Anal. **10**(1), 23–45 (2006)
8. Anagnostopoulos, C., Tasoulis, D.K., Adams, N.M., et al.: Temporally adaptive estimation of logistic classifiers on data streams. Adv. Data Anal. Classif. **3**(3), 243–261 (2009)
9. Hulten, G., Spencer, L., Domingos, P.: Mining time-changing data streams. In: Proceedings of the ACM SIGKDD International Conference on Knowledge Discovery and Data Mining, pp. 97–106. ACM, New York (2001)
10. Suzuki, Y., Kido, K.: Big-data streaming applications scheduling with online learning and concept drift detection. In: Proceedings of the Design, Automation & Test in Europe, pp. 1547–1550. IEEE, Piscataway (2015)
11. Kuncheva, L.I.: Classifier ensembles for changing environments. In: Roli, F., Kittler, J., Windeatt, T. (eds.) MCS 2004. LNCS, vol. 3077, pp. 1–15. Springer, Heidelberg (2004). doi:10.1007/978-3-540-25966-4_1
12. Gama, J.: A survey on learning from data streams: current and future trends. Prog. Artif. Intell. **1**(1), 45–55 (2012)
13. Chunquan, L., Yang, Z., Peng, S., et al.: Learning very fast decision tree from uncertain data streams with positive and unlabeled samples. Inf. Sci. **213**(23), 50–67 (2012)
14. Wenhua, Z.: Constructing decision trees for mining high-speed data streams. Chin. J. Electron. **21**(2), 215–220 (2012)
15. Hoeffding, W.: Probability inequalities for sums of bounded random variables. Am. Stat. Assoc. **58**(301), 13–30 (1963)

Review of Various Strategies for Gateway Discovery Mechanisms for Integrating Internet-MANET

Lin Yang, Zhijie Han[✉], Rui Wang, and YongHang Yan

Hennan University, Kaifeng 475001, Henan Province, China
hanzhijie@126.com

Abstract. Mobile Ad hoc network (MANET) is a short range network which enables various mobile nodes to communicate with each other. The range of mobile node for communication is limited in the MANET so two nodes of different network which are placed in long distance are unable to communicate between them. Utility of MANET can be enhanced by providing Internet connectivity to the mobile nodes. MANET is infrastructureless and Internet is infrastructure type network so they require an interface to connect with each other. So in this situation, Gateway, a device which acts as an interface and a router, provides the MANET nodes connectivity to fixed networks. A mobile node in MANET has to find a route to a gateway first to communicate with the Internet host. It requires an efficient gateway discovery mechanism for improve the overall performance of network. For this process various approaches have been proposed in the past. The popular approaches are reactive, proactive and hybrid gateway discovery mechanisms. This paper focuses on various issues for MANET-Internet integration and their proposed technical solutions by various proposals for different Internet gateway discovery mechanisms to improve the performance.

Keywords: Gateway discovery mechanism · MANET · Routing

1 Introduction

A mobile ad hoc network (MANET) defines a collection of mobile nodes that can communicate with one another without using any fixed networking infrastructure. Due to their free mobility, the network's topology is dynamic and unpredictable. It is useful for communication among mobile nodes located at a short range such as official meetings, classroom environments, military and controlling of crowds. Each mobile node acts as a router as well as host. Wireless communication is successful in other scenarios, provoked the extension of this technology for conferences, visiting parks or recreation areas. PDA, laptop, cellular phones are different types of element that can coexist for these situations. Among all the nodes for successful communication within MANET range is not always sufficient. If mobile nodes in MANET want to communicate with other networks, it needs to connect to the Internet, there should have a special gateway between them as a bridge for connecting MANET into the internet. Whenever a mobile node wants to connect with internet, it has to search for the

© Springer Nature Singapore Pte Ltd. 2017
G. Chen et al. (Eds.): PAAP 2017, CCIS 729, pp. 247–257, 2017.
DOI: 10.1007/978-981-10-6442-5_22

gateways which are called as gateway discovery. Gateway plays the role of Access Router. Two main tasks is executed by gateway, first routing the packet from Internet to mobile nodes and second generating Modified Router Advertisement (MRA) messages for informing the configuration parameter. Upon getting the MRA message by the mobile nodes, the mobile nodes create, update or optimize the route to the gateway which is sending the packet to the Internet hosts. Discovery processes are classified in to three main categories: proactive, reactive, and hybrid schemes [1–3]. A proactive approach [4] can achieves good connection and low latency, but requires considerable overheads. In contrast, the method of the reactive [5] achieves low routing overhead at the expense of increased latency. Previous strategies proactive and reactive approaches are used to discover the gateways, in hybrid approach [1, 3, 6], it uses a proactive approach within a gateway's advertisement range, while it uses a reactive approach outside the coverage.

In this paper, we analyze various strategies of gateway discovery and mechanisms involved for the advancement in technical issues for MANET-Internet integration and review their advantages and disadvantages.

2 Analysis of Different Proposals for Gateway Discovery Approach

Gateway Discovery is a method which allows a MANET node to discover an Internet gateway (IGW) via which traffic for the Internet can be delivered, and from which traffic returned from the Internet can be received. Existing methods can be divided into three methods: proactive methods, reactive methods and hybrid methods.

2.1 Analysis of Different Proposals Using Proactive Gateway Discovery Approach

In a proactive scheme, gateway periodically broadcast a gateway advertisement message (GWADV). These GWADV message throughout ad hoc network. When a node receives the advertisement message, the node stores the route information to the gateway or updates its route table if it had any previous entry for the gateway. And then forwards it to other nodes until this message is flooded in the entire network. There are many approaches implemented by using proactive gateway discovery.

Khan et al. [7] proposed a proactive gateway discovery scheme integrating Efficient DSDV (Eff-DSDV) protocol and mobile IP that allows routing table for each mobile node to store the number of hops, sequence number for all destinations. The routing table updates periodically. DSDV protocol used in this approach since in DSDV packet delivery ratio is very slow because of stale route due to broken links. Therefore, the packet can be forwarded through other neighbor nodes that may have a route to the destination. Whenever a link is broken, DSDV create a temporary link through the neighbor having a valid route to the destination. To create a temporary link, a one-hop request ROUTE_ACK is sent. The advantage is that it does not require the flooding of the gateway advertisements for registration of mobile nodes with Mobile IG

(MIG). It also overcomes the drawback of conventional DSDV by reducing the packet loss due to broken links.

Lei et al. [8] integrated the routed (modified RIP) Ad Hoc routing protocol with Mobile IP routing Protocol which resulted in a combined routing table and also enabled foreign agents (FAs) to participate in the Ad Hoc network routing.

2.2 Analysis of Different Proposals Using Reactive Gateway Discovery Approach

The reactive gateway discovery is initiated by a mobile node that is to create or update a route entry to a gateway. If MANET is using reactive routing protocol, When a node has the need to connect to the network, it may send the Route Request (RREQ) packet containing the gateway solicitation information, mobile node broadcast a RREQ message with an "I" flag (RREQ_I) to the IP address for the group of all gateways in a MANET (ALL_MANET_GW_MULTICAST Address). Therefore, only the gateways are addressed by this message and only they process it, so intermediate mobile nodes just rebroadcast it. When a gateway receive a RREQ_I, it unicasts back a RREP_I packet containing IP address of the gateway to the source node. After receiving the RREP_I, source node can creates a route path to gateway, then selects one of the hop counts and forwards the data packets to the selected gateways.

Ammari et al. [9] proposed scheme is based on three-layer approach using Mobile protocol and DSDV ad hoc routing protocol [10] (Fig. 1). The first layer contains Mobile IP foreign agents, the second layer contains mobile gateways and mobile Internet nodes having one hop distance from Mobile IP foreign agents, the third layer includes all MANET nodes and also visiting mobile Internet nodes that are at least one hop away from gateways. This framework considers using some border MANET nodes (referred to as mobile gateways (MGs)) to connect the rest of the MANET nodes to the Internet. Second layer provides Internet connectivity to mobile nodes. Based on the distance and the load criteria, MG selects a closest and/or a least loaded FA. MANET nodes select a closest and/or least loaded MG. Mobile gateways are designed to use both Mobile IP and DSDV protocol for routing packets in MANET.

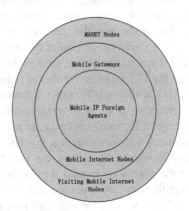

Fig. 1. Architecture based on three layer mobile gateway [10]

A modified routing protocol AODV is proposed by Nilsson et al. [11], in which I-RREQ and I-RREP messages are used to discover the gateway. This solution relies on the signaling of AODV to find an access providing Internet Gateway that can distribute a globally routable prefix for the ad hoc network.

2.3   Analysis of Different Proposals Using Hybrid Gateway Discovery Approach

A hybrid gateway discovery scheme is a hybrid combination of the proactive and the reactive algorithm. In hybrid gateway approach, All the mobile nodes in a certain range around a gateway use the proactive gateway discovery scheme while the mobile node residing outside that range use reactive gateway discovery to communicate with gateway.

In previous Hybrid method, this limitation of area supposed to have uncertainty of good outputs. So it cannot be said fully advantageous method of gateway operation.

The paper [12] presents a new solution to this kind of problems in previous works and adopts a novel Hybrid gateway discovery method in which the similar hop count is used to determine the area of gateway advertisement but in a broader, an adaptive way will overcomes the limit of the hop count range. Computation of the new radius of n-hops is based on an algorithm of average of route requester nodes according to hop count under gateway supervision. Whenever the gateway receives a request packet from a request node, we use the hop count information from the packet. We increase the hop count values of all the request packets received from the requestors during each advertisement interval. At the end of the interval we find the arithmetic mean of these hop count values by dividing the total hop count by the total requestor or users. These hop count value will be used as the advertisement zone for the next gateway advertisement message.

There are some further classification of hybrid gateway discovery approach are used by researchers. Some of these schemes are described here:

Strategy 1:

In paper [2, 13] covers only one aspect of dynamisms which is adjusting range of GWADV messages dynamically. The strategy is called maximal source coverage (MSC) as it covers the active source at the farthest distance from the gateway. For sending the data packet, mobile node uses the gateway either proactive or reactive scheme called the active sources. Using the IP header of the packet, gateway finds out the number of hops at which each of the active sources located. After locating the farthest active node it changes the TTL value to number of hops of this active source. In next advertisement, the GWADV with new TTL value is broadcasted. Further this node drifts from reactive zone to proactive zone. MSC strategy not only limits the range of the active region with the average TTL value, and in order to control the high overhead of the reactive gateway discovery scheme, provides a gateway request message sent only in Proactive Zone, they broadcast throughout the Reactive Zone. The gateway reply message is returned by the edge node located at proactive zone, not by a gateway reply. It is a very simple heuristic with very less implementation complexity. The main advantage of this heuristic is that low overhead and high packet delivery ratio as close as possible. But the disadvantage is that even if a single active source is located at a far distance the proactive zone will be increased to the distance of this node thus increasing the packet overhead. Since the MANET is dynamic, the TTL value sometimes can be close to optimal as nodes at more number of hops from gateway may not be requiring Internet connectivity.

Strategy 2:

This paper [14] also dynamically adjusts the range of GW_ADV message. The TTL value is decided in a distributed way, not by the gateway. The GW_ADV message is sent to the active sources which are the part of some route. All nodes in the original network are overridden in the reaction area, so the nodes which desire connectivity with gateway initiate the communication. Nodes send GW_SOL messages and in response gateway broadcasts GWADV messages with TTL = 1. As the intermediate node of the part of the route to the requesting node sets their TTL = 1 and broadcasts the message. And the nodes which are not the part of this requesting route set the TTL = 0 and discard this message. In this way, only nodes which desire internet connectivity receive the GWADV message. Overhead is less in this approach compared to other approaches.

Strategy 3:

This strategy is discussed in [15] that introduce maximal benefit coverage (MBC) strategy. The MBC policy refers to the maximum amount of GWADV messages broadcast by the gateway to the source node, that is, the Maximization range of proactive zone. This will ensure a large number of source nodes that have Internet access needs can receive GWADV messages and find the gateway, only a small amount of the source node that are not in this range to detected by passive expanding ring search to find the available gateway. Gateway set the TTL value dynamically for GW_ADV messages those gives the maximal benefit in Maximal Benefit Coverage. Maximal benefit is found as: A ratio between the costs of flooding the whole ad hoc network to the cost of providing internet connectivity in which not any GW_ADV messages sent. This ratio is used to find the most appropriate TTL value for GW_ADV message. Here, N is the message of flooding cost in the network and S(s) is the function and this is represent the number of active sources for the gateway and distance is less than are equal to s. The advantage of these strategies is that higher packet delivery.

The gateway will maximize the savings by selecting the TTL (hop count) of the broadcast gateway advertisement message. Maximal benefit is found as: A ratio between the costs of flooding the whole ad hoc network to the cost of providing internet connectivity in which not any GW_ADV messages sent.

This ratio is used to find the most appropriate TTL value for GW_ADV message.

$$\partial(n) = N * \frac{S(n)}{n(n+3)} \tag{1}$$

Where, N (the number of nodes in the network) indicates the cost of flooding the entire network; Function represents the number of active source nodes that are less than or equal to the N hop distance from the gateway, which can be obtained from the routing table "TTL". $n(n+3)/2$ is the number of nodes that are within the range of N hops from the gateway, where 2 is omitted. That is to the ratio of the number of active source nodes and the number of nodes in the range of N hops from the gateway is multiplied by the cost of the flooding of the whole network N, when the ratio is 1, the cost is N. The goal of the MBC strategy is to find the maximum TTL value of n. Then, the next gateway advertisement message flooding TTL: when $n \in [1 \ldots n_{\max}]$, $\partial(n) = \max_{1 \leq x \leq n_{\max}} \partial(x)$. In which $n_{\max}$ is the TTL value of the source node farthest from the gateway.

MBC strategy to maximize $\partial(n)$ to avoid source node that have Internet access necessary to find the available gateway through the internet flooding that caused a large number of overhead. Here, the flooding mechanism used by the source node refers to the expanded ring search method of reactive gateway discovery strategy.

Strategy 4:

Paper [16] has described the complete adaptive gateway discovery approach. A GW_ADV messages are sent periodically at large intervals and the periodicity is adapted with mobility detection in the MANET. Maximal Benefit Coverage algorithm is used for maintain the TTL value of the GW_ADV message in dynamic way. To decide whether to adapt TTL periodicity, a heuristic function is used. At regular interval, regulated mobility factor (RMF) is calculated by the each gateway, which estimates the amount of mobility of the source nodes which has registration with the gateway. Periodicity of the GW_ADV message whether to bring adapted is determine by RMF parameter which is used as threshold value. Maximal Benefit coverage algorithm is used for the demand of adaption. If there is a requirement of adaptation, it is done according to the maximal benefit coverage algorithm. Following equation, calculate the total AF nodes:

$$TAFN = \sum_{1}^{nextTTL > 1} Ms(TTL - 1) \qquad (2)$$

Ns = number of sources communicating with the internet Maximal Benefit Coverage algorithm decides the value of next TTL. Following Expression, is used when next TTL value is greater than 1. Adaptive advertisement is always beneficial if value of next TTL is one because the distance between source nodes and internet gateway is one hop. Regulated Mobility Factor is calculated by Eq. 3 for compute the performing an adaptive advertisement benefit.

$$RMF = \frac{MS}{TAFN} \qquad (3)$$

RMF is calculated by gateway for every X second where X is minimum advertisement period.

Strategy 5:

This paper [17] uses the Ant Colony optimization technique. Mobile agents behaved as ants are used to explore the entire MANET and collect routing data. Mobile agents are basically packets. The gateway contains two routing tables, one for updating the information in the local MANET and the other for updating information from the Internet. In order to discover flooding gateway in the network ACO messages. There are two types of ACO messages: FRMA and BRMA. FRMA messages are used to establish paths to the gateways, which use pheromone values and pheromone decay times to determine the stability of the link. The gateway that received the FRMA transmits the required data to the FRMA packet and converts it to BRMA. This BRMA is unicasted to the requesting source node through the path that FRMA is used to get to the gateway. The decision process is that the greater the pheromone value, the shorter the time to reach the gateway, and vice versa. Use the entries created by the ACO mechanism (such as hops, next hop, pheromone and pheromone decay time and balance index) to forward the packets to the Internet destination to the gateway.

The main advantage of this strategy is to set the link-adaptive parameters based on the current topology of the link through which the mobile agent passes over a certain period of time.

Strategy 6:

In paper [18] adaptive distributed gateway discovery (ADD) algorithm is discussed, which is a fully distributed gateway discovery method. The goal of ADD is only those nodes which need the gateway and other nodes hampered with periodic gateway advertisement. In ADD algorithm, the active source node initial discover the gateway by in reactive manner. Once the route to the gateway is known, source node sending the data by using this route. By finding the IP header of the packet received by the gateway, the gateway knows the number of hops to reach the active source node. Based on this information, a source table is constructed by the gateway. Based on this information, all the intermediate nodes between the active source node and the gateway also capture this information. The GW_ADV message is sent with TTL = 1 by gateway during the gateway advertisement periodically. The node that receives this GW_ADV message then proves whether it is an intermediate node. If the node finds that it has sent the message to the gateway, then the node is the intermediate node. If find the intermediate node then it forwards further the GW_ADV message with TTL = 1 and other than intermediate node does nothing. All intermediate nodes forward this GW_ADV message. Then path towards the gateway is hop by hop for communicating with gateway. Between gateway and active sources an active region is formed and each active region has at least one active source for communication. From this distributed approach other than intermediate node not involve for gateway discovery, therefore significantly decrease the overhead. However, if high mobility condition exit intermediate node help the active sources. Intermediate node also change if active sources is moves, from this situation active region is also moves. To this end, active region is always intact with active source node, and active region adapts itself based on the active source movement. If an active sources down the route to the gateway then this active node send the query to the any node of active region and then update the route to the gateway. Finally advantage of this approach without generating more overhead in dynamic network this is because update the route to the gateway by active source node in very less time.

3 Proposed Scheme Comparison

The comparison of different proposed schemes is illustrated in Table 1.

Table 1. The comparison of different proposed schemes

Gateway discovery scheme	Proposed scheme	Multiple gateways support	Ad hoc routing protocol	Mobile IP support
Proactive discovery approach	Khan et al. [7]	YES	DSDV	YES
	Lei et al. [8]	YES	Modified version of RIP	YES
Reactive discovery approach	Ammari et al. [9]	YES	DSDV	YES
Hybrid discovery approach	Ruiz et al. [13]	NO	AODV	NO
	Rashid et al. [14]	YES	AODV	YES
	Ruiz et al. [15]	YES	AODV	NO
	Bin et al. [16]	YES	AODV	NO
	Velmurugan et al. [17]	NO	AODV	YES
	UsmanJaved [18]	NO	AODV+	NO

4 Gateway Discovery Overhead Calculation by Analytic Model

Our analysis model assumes that new traffic generated by the hosts connected to mobile nodes follows Poisson distribution and is generated independently of each other. All hosts have the same traffic generation pattern.

The network overhead caused by reactive gateway discovery is, when a node tries to discover a route towards IG (internet gateway), it uses the method of ring expansion to broadcast RREQ packets to the network, Each IGWs node that receives the packet replies to a RREP to mobile requestors (MRs) node. The source wants to reactively discover an IG there is an overhead which includes the IGRQ broadcast messages, plus IGRP reply messages from every IG to the source. The overhead of the reactive IG discovery by one source can be computed as follows:

$$\theta_R = M \left(M_s \cdot \partial_{IGRQ} \cdot \Delta_t(R) + H_{Rh}(\Delta_t(R)) + \frac{\beta_L h}{\omega} \right) \tag{4}$$

Where θ_R is the overhead by the reactive approach, M_s is the number of source nodes communicating with a host in the Internet, ∂_{IGRQ} is the sum of route requests and replies during the time interval $\Delta_t(R)$ for reactive requests, $H_{Rh}(\Delta_t(R))$ is the number of hello packets emitted by a AMN for $\Delta_t(R)$ second, $\frac{\beta_L h}{\omega}$ is route maintenance overhead, that is called ρ_R, where β_L is the number of active links and h is a hop count. If link layer is used to detect link failures, H_{Rh} is 0. Route lifetime follows an exponential distribution with a mean route lifetime of $\frac{\omega}{h}$. The average rate of route failures is given by $\frac{h}{\omega}$. The discovery overhead of the reactive approach is proportional to the number of active routes in the network. Therefore, reactive overhead increases with the number of sources and destinations in the network.

In a proactive approach, IGs will periodically broadcast IGAM messages to the entire network. Therefore, the overhead of proactive schemes includes hello messages for route update plus the messages sent out by IGs themselves. Total overhead required by the proactive approach can be expressed as follows:

$$\theta_p = M\left(M_{IG} \cdot \partial_{IGAM} \cdot \Delta_t + H_{Rh}(\Delta_t) + \frac{\alpha_N}{\omega}\right) \tag{5}$$

Where θ_p is the overhead of proactive approach, M is the number of nodes, M_{IG} is the number of IGs, ∂_{IGAM} is the rate at which IGAM messages are emitted by IGs, $H_{Rh}(\Delta_t)$ is the number of the hellos packets by a AMN per a time interval, $\frac{\alpha_N}{\omega}$ is a route maintenance cost which is called ρ_p, where ω is average communication link lifetime and α_N is the number of active neighbor nodes.

By a hybrid/adaptive approach, IGs periodically send IGAM messages within a certain range which is determined by a proactive area. Thus, the overhead of the hybrid/adaptive approach is computed as follows:

$$\theta_H = M_{TTL}^{IG}(M_{IG} \cdot \partial_{IGAM} \cdot \Delta_t + H_{ph}(\Delta_t) + \rho_p) + \\ M_{N-TTL}(M_s \cdot \partial_{IGRQ} \cdot \Delta_t(R) + H_{Rh}(\Delta_t(R)) + \rho_R) \tag{6}$$

Where θ_H is the overhead of hybrid/adaptive approach, M_{TTL}^{IG} is the number of nodes in the TTL range from an IG, and M_{N-TTL} is the number of nodes for each source outside the proactive area.

5 Performance Analysis of Gateway Discovery

According to the mechanism of the three gateways, it is known that with the increase of the number of mobile nodes and the increase of the moving speed of the mobile nodes, the overall control overhead and the average packet transmission delay will increase. In the control overhead, the overhead of proactive gateway discovery method is significantly higher than that of reactive and hybrid methods, but with the increase of node mobility, the global connectivity of the active method is better than the other two. Although the reactive gateway discovery can get the lowest cost, the packet delivery rate and the average packet transmission delay are relatively large. When the network size is large (60 mobile nodes), the hybrid method has better adaptability. By means of the gateway notification message in the vicinity of the gateway, the mobile node can quickly find the gateway closest to it and get better handoff performance, which means the lower average packet transmission delay. The mobile nodes, which are far from the gateway, can quickly obtain the required global connection information by proactively sending out the request message, and ensure that the global connection information needed by the mobile node can be received at regular intervals. On the whole, the hybrid gateway discovery method can obtain the packet delivery rate of the proactive scheme in the near reactive cost, and the comprehensive performance is better in different scenarios.

6 Conclusion

This paper throws light on the various issues and solutions proposed by various researchers for gateway discovery for internet connectivity in MANETs. After studying a number of gateway discovery strategies, it becomes quite clear that a lot of work has been done in this field with lots of efficiency and all the strategies do offer some advantages over the others. By these gateway discovery schemes we can conclude that pure proactive and reactive approaches are easy to implement but they may increase overload, delay and congestion problems. However hybrid approach includes the advantages of both proactive and reactive schemes in addition with some new metrics. In case of connectivity, proactive gives better connection than reactive and hybrid approaches. But in case of overhead proactive have the highest and hybrid approaches gives lesser overhead. Thus we can conclude that hybrid has the best stability, less overhead and better packet delivery.

References

1. Ghassemian, M., Hofmann, P., Prehofer, C., Friderikos, V., Aghvami, A.H.: Performance analysis of internet gateway discovery protocols in ad hoc networks. In: WCNC2004 (2004)
2. Ruiz, P.M., Gomez-Skarmeta, A.F.: Maximal source coverage adaptive gateway discovery for hybrid ad hoc networks. In: Nikolaidis, I., Barbeau, M., Kranakis, E. (eds.) ADHOC-NOW 2004. LNCS, vol. 3158, pp. 28–41. Springer, Heidelberg (2004). doi:10. 1007/978-3-540-28634-9_3
3. Ratanchandani, P., Kravets, R.: A hybrid approach to internet connectivity for mobile ad hoc networks. In: Proceedings of WCNC 2003, March 2003
4. Jonsson, U., Alriksson, F., Larsson, T., Johnasson, P., Maguire, G.Q.: MIPMANET: Mobile IP for mobile ad-hoc networks. In: Mobihoc, August 2000
5. Wakikawa, R., Maline, J.T., Perkins, C.E., Nilsson, A., Tuominen, A.H.: Global connectivity for IPv6 mobile ad-hoc networks. In: IEFT Internet-Draft, draft-wakikawa-manet-globalv6-03.txt, October 2003
6. Hamidian, A.: A study of internet connectivity for mobile ad hoc networks in NS 2. Master's thesis, Department of Communication Systems, Lund Institute of Technology, Lund University, January 2003
7. Khan, K.U.R., Reddy, A.V., Zaman, R.U.: An efficient integrated routing protocol for interconnecting mobile ad hoc network and the internet. International connecting mobile ad hoc network and the internet. Int. J. Comput. Electr. Eng. 1(1), 1793–8198 (2009)
8. Lei, H., Perkins, C.: Ad hoc networking with mobile IP. In: Proceedings of 2nd European Personal Mobile Communication Conference (1997)
9. Ammari, H., El-Rewini, H.: Integration of mobile ad hoc networks and the internet using mobile gateways. In: Proceedings of the 18th IPDPS (2004)
10. Bansal, S., Gaur, P.K., Marwaha, A.: Review of various strategies for gateway discovery mechanisms for integrating internet-WANET. Ph.D.
11. Bin, S., Bingxin, S., Bo, L., Zhonggong, H., Li, Z.: Adaptive gateway discovery scheme for connecting mobile ad hoc networks to the internet. In: Proceedings of International Conference on Wireless Communication, Networking and Mobile Computing, vol. 2, pp. 795–799 (2005)

12. Nilsson, A., Perkins, C., Tuominen, A., Wakikawa, R., Malinen, J.: AODV and IPv6 internet access for ad hoc networks. SIGMOBILE Mob. Comput. Comm. Rev. **6**(3), 102–103 (2002)

13. Rajesh, H., Davda1, Noor Mohammed: Text detection, removal and region filling using image inpainting. Int. J. Futur. Sci. Eng. Technol. **1**(2), 2320–4486 (2013). ISSN

14. Javaid, U., Rasheed, T.M., Meddour, D., Ahmed, T.: Adaptive distributed gateway discovery in hybrid wireless networks. In: WCNC-2008, pp. 2735–2740 (2008)

15. Ruiz, P.M., Gomez-Skarmeta, A.F.: Enhanced internet connectivity for hybrid ad-hoc networks through adaptive gateway discover. In: 29th Annual IEEE International Conference on Local Computer Networks, pp. 370–377 (2004)

16. Bin, S., Haiyan, K., Zhonggong, H.: Adaptive mechanisms to enhance internet connectivity for mobile ad hoc networks. In: International Conference on Wireless Communications, Networking and Mobile Computing, WiCOM 2006, pp. 1–4, 22–24 September 2006

17. Bin, S., Bingxin, S., Bo, L., Zhonggong, H., Li, Z.: Adaptive gateway discovery scheme for connecting mobile ad hoc networks to the internet. In: Proceedings of International Conference on Wireless Communications, Networking and Mobile Computing, vol. 2, pp. 795–799 (2005)

18. Zhuang, L., Liu, Y., Liu, K., Zhai, L., Yang, M.: An adaptive algorithm for connecting mobile ad hoc network to Internet with unidirectional links supported. J. China Univ. Posts Telecommun. **17**(1), 44–49 (2010)

19. Ayyadurai, V., Ramasamy, R.: Internet connectivity for mobile ad hoc networks using hybrid adaptive mobile agent protocol. Int. Arab J. Inf. Technol. **5**(1), January 2008

20. Yuste, J., Trujillo, F.D., Trivino, A., Casilari, E.: An adaptive gateway discovery for mobile ad hoc networks. In: 5th ACM International Workshop on Mobility Management and Wireless Access, pp. 159–162 (2007)

21. Jiang, H., Jin, S.: Design and analysis of adaptive strategies for locating Internet—based servers in MANETs. J. Perform. Eval., 464–479 (2006). Elsevier

22. Prakash, J., Kumar, P.: A Survey on Adaptive Gateway Discovery Strategy in MANET (2016)

23. Prakash, J., Verma, R.: Gateway Discovery and Gateway Selection Schemes for Connecting MANET and Internet: A Review (2016)

24. Pandey, A., Bajpai, A., Singh, D., Kumar, R.: Survey of Adaptive Gateway Discovery Mechanisms (2015)

25. Xie, F., Du, L., Bai, Y., Chen, L.: Adaptive gateway discovery scheme for mobile ubiquitous networks. In: WCNC 2008, pp. 2916–2920 (2008)

Research on Extraction Algorithm of Palm ROI Based on Maximum Intrinsic Circle

Gang Liu and Jing Zhang$^{(\boxtimes)}$

School of Computer Science and Software, Tianjin Polytechnic University,
Tianjin 300387, China
1912830275@qq.com, china_ximeng@sohu.com

Abstract. Aiming at the problem that using maximum inscribed circle algorithm to extract the region of interest (ROI) in palm vein image is retrieved and recorded too much, an improved maximum inscribed circle algorithm is proposed. First of all, the palm vein image was done by background separation and smoothing filter preprocessing. Secondly, the grids were added for the preprocessed image, the selection range of the center was determined by the assistant of grids, the initial radius and the variable value of radius were defined. Then, changed the radius in variable size, recorded all inscribed circles. Finally, compared and obtained the maximum inscribed circle of the palm vein image. The circle was the ROI which was needed. The experimental results show that the execution time of the new algorithm is reduced by 10.7 ms, 10.2 ms, 11.3 ms and 10.8 ms under four groups of samples respectively. Therefore, the new algorithm effectively improves the efficiency of the maximum inscribed circle algorithm.

Keywords: Palm vein · Region of interest · Extract · Maximum inscribed circle

1 Introduction

The demand of reliability and security in identity authentication is being raised in the ever-evolving society. The traditional way, such as private documents, passwords, passwords, etc., has some threats such as loss, forgetting, theft and forgery. In this context, identity authentication with some qualities such as unique, permanent, characteristic that can't be stolen has a wide range of market prospects. Biometric identification technology [1], according to the physiological and behavioral characteristics of human beings, has become a popular identification of the way today.

Palm vein recognition technology is a new bio-identification technology. Through the near infrared emission of near infrared equipment, scan the palm of your hand palm vein then distribution map. According to the special alignment algorithm on the palm vein distribution map to extract and store as a template image. Finally, compare the target image with the template image according to the specific feature matching algorithm [2].

ROI (region of interest) extraction is the key step of palm vein recognition technique. At present, the commonly used ROI extraction algorithm are the method based

G. Chen et al. (Eds.): PAAP 2017, CCIS 729, pp. 258–267, 2017.
DOI: 10.1007/978-981-10-6442-5_23

on the rectangular image block segmentation and the method based on the maximum inscribed circle [3]. Method based on the rectangular image block segmentation is trying to find two key points in the image, through the key points to establish a Cartesian coordinate system, draw a specific size of the rectangular block as ROI. The basic idea of the maximum inscribed circle method is that intercept the largest inscribed circle in the hand region as ROI. As the palm vein recognition technology requires to extract the ROI contains a lot of blood vessels in the hand palm, so the maximum inscribed circle method is more suitable for extracting the palm vein to get the ROI.

The way of maximum inscribed circle algorithm [4] need to extract a large amount of vein information in the palm area, and eliminate finger interference. The idea of the largest inscribed circle algorithm is that select the center of the circle in the palm position, define the initial radius, set the radius increment to change the radius, loop the whole process. The algorithm performs too much traversal in finding the center of the circle determining the radius which result in inefficient and unexpected maximum inscribed circle. In this paper, based on the maximum inscribed circle algorithm, preprocess the original image, import the grid as a reference, and determine the center and radius more efficient.

2 Pretreatment

Flow through the veins of blood, hemoglobin in red blood cells can absorb near 700–1000 nm wavelength infrared ray. As a result, the transmission of the near infrared rays in the vein portion is greatly reduced. Due to the absorption of hemoglobin, when the near infrared ray is transmitted, the vein is highlighted on the image sensor senses. On the contrary, palm muscles, bones and other parts are weakened. Therefore, through the irradiation of near infrared devices, you can get a clear palm vein image.

The original image captured by near-infrared equipment has a lot of noise, it will affect the image quality of the extracted ROI. Therefore, the original image is first subjected to get on background separation [5] and smoothing preprocessing [6].

2.1 Background Separation

Binarization, setting the gray value of the pixel on the image to 0 or 255, is a method of presenting the entire image with only black and white visual effects which is in order to achieve background separation. First, according to formula (1) to transform the palm vein image into an 8-bit grayscale image.

$$\text{Gray} = 0.3R + 0.59G + 0.11B \tag{1}$$

Niblack algorithm [7] is the way that the threshold is calculated of the pixel point of the center neighborhood area. Formula (2) is the way of calculating threshold, k is a coefficient. Formula (3) is the calculation formula for each pixel threshold, m is the variance, and s is the standard deviation.

$$T_{Niblack} = m + k * s \tag{2}$$

$$
\begin{aligned}
T_{\text{Nibalck}} &= m + k\sqrt{\frac{1}{NP}\sum (p_i - m)^2} \\
&= m + k\sqrt{\frac{\sum p_i^2}{NP} - m^2} \\
&= m + k\sqrt{B}
\end{aligned}
\tag{3}
$$

After Niblack binarization algorithm processing, the image is achieved by the separation of the palm region and the non-palm region. Background separation image is shown in Fig. 1:

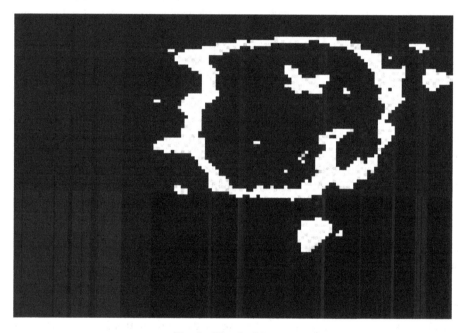

Fig. 1. Binarized image

2.2 Smoothing Filter

The image is used to reduce the noise of the image by Gaussian smoothing filter [8]. Through the Gaussian kernel function to generate Gaussian filter, formula (4) is the two-dimensional spatial distribution equation.

$$G(x,y) = \frac{1}{2\pi\sigma^2} e^{\frac{-(x^2-y^2)}{2\sigma^2}} \tag{4}$$

Set the standard variance as 0.625, the contour generated by the formula is a normal distribution of concentric circles which is beginning from the center. The image processing image quality can be ensured by calculate matrix only. The 5 * 5 matrix which is generated by the Gaussian distribution is listed as follow.

$$\begin{bmatrix} 1 & 2 & 3 & 2 & 1 \\ 2 & 7 & 11 & 7 & 2 \\ 3 & 11 & 17 & 11 & 3 \\ 2 & 7 & 11 & 7 & 2 \\ 1 & 2 & 3 & 2 & 1 \end{bmatrix}$$

After Gaussian filter processing, a smooth image is shown in Fig. 2:

Fig. 2. Image of smoothing the filtered

In summary, the preprocessing process consists of background separation and smoothing filtering. The preprocessing steps are shown in Fig. 3.

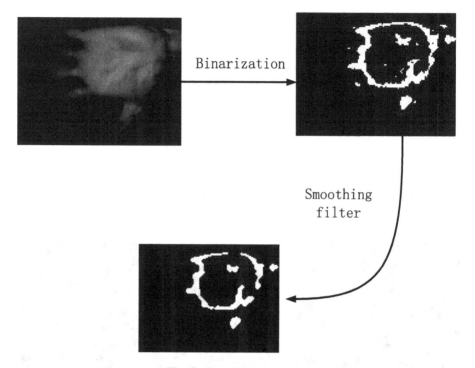

Fig. 3. Pretreatment process

3   ROI Extraction

The purpose of ROI extraction is extracting invariant features in the palm vein. The idea of ROI extraction is extracting areas with higher venous information intensity and contrast.

The basic steps of the maximum inscribed circle algorithm can be listed as follow. First, retrieve each pixels point on the binarized palm vein image. If the point is in the palm, change the radius of the circle with a certain unit value. Second, when the edge of the circle is tangent to the edge of the palm, stop the action of searching for the radius, record the value of radius. Then, continue to change the center of the circle, change the radius, determine inscribed circle. By looping this process above, Traversal all points. Record all radius, find the longest one. The circle with the largest radius in all records is the largest inscribed circle. The shortcoming of the algorithm is the "blindness" of determining the center and radius, which will result in excessive execution of the algorithm loop and may even cause the stack overflow of program to exit abnormally.

ROI extraction of the maximum inscribed circle should be done by extracting the ROI as much as possible throughout the palm area, avoiding contact with the palm border and finger area. Therefore, the center of the circle should be defined in the geometric center of the palm region. The palm image is an irregular geometric figure that is difficult to quantify the center point.

In order to optimize the maximum inscribed circle extraction algorithm, add grids to the binarized image of the palm region, and obtain the binarized image with grids. The image with grids is shown in Fig. 4.

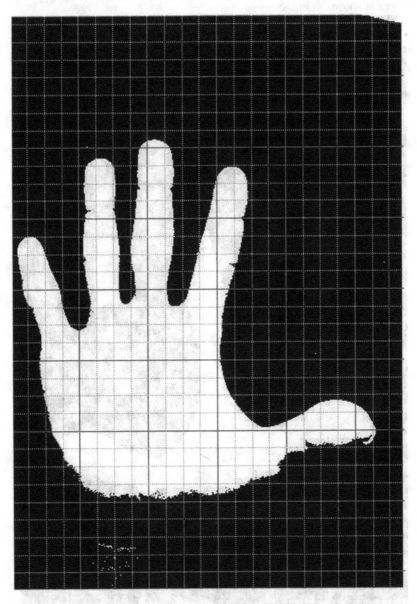

Fig. 4. Image with grids

Set each grid be 1 unit, so that the existence of grids can quantify the palm area, calculate the number of vertical direction and horizontal direction grids. The total number of grids in the two directions is divided by two, and the general grid of center area can be obtained. The center grid is the center of the inscribed circle. The image of center area is shown in Fig. 5.

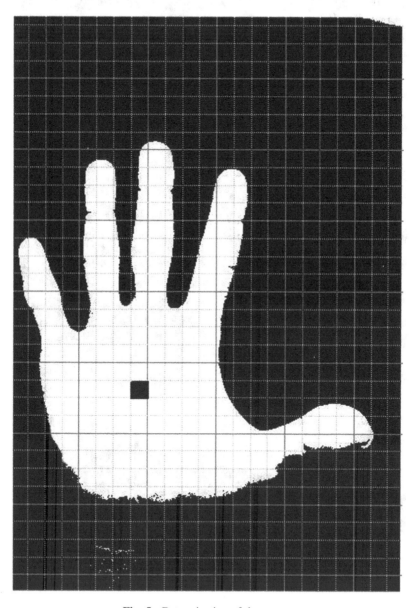

Fig. 5. Determination of the center

Take the center area grid, set the intersection point of the lines of the opposite sides' midpoints as the initial center. Set 4 units as the initial radius to determine the initial circle. Change the radius with 1/3 grid units every time, retrieve the inscribed circle, and record the radius. Loop these steps in the center area. The inscribed circle with longest radius in the record is the largest inscribed circle. It is the ROI of palm. The effect of algorithm execution is shown in Fig. 6.

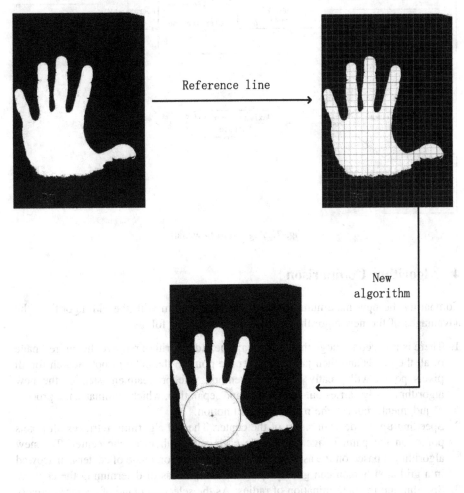

Fig. 6. Algorithm execution effect

The flow chart of the complete algorithm is shown in Fig. 7.

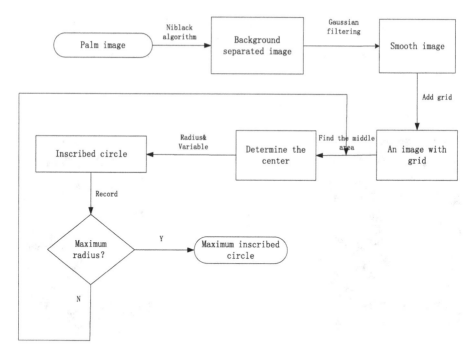

Fig. 7. Algorithm flowchart

4 Algorithm Comparison

Comparing the new maximum inscribed circle algorithm with the old algorithm, the advantages of the new algorithm can be summarized as follow:

1. There is no need to judge the palm area: The old algorithm retrieve the entire image of all the pixels and then judge whether the point in the palm or not. Search for all pixels points will greatly consume memory. The pretreatment step of the new algorithm firstly carries out the background separation, which eliminates the process of judgment, reduces the memory consumption.
2. Speeding up the determination of the center: The old algorithm retrieves all pixels points on the palm image and define each one of them as the center. The new algorithm is based on the assistant of grids, the selection range of center is narrowed in a grid area, which can greatly speed up the process of determining the center.
3. Speeding up the determination of radius: As the selection range of center is greatly reduced, with assistant of grids, calculate the number of grids. Define a larger initial radius that close to the maximum radius. Define the independent variable which is defined the value by a grid. Relative to the old algorithm, the new algorithm speeds up the radius determination process.
4. Reducing the number of cycles: With the assistant of grids, the new algorithm narrow the selection range of center, speed up the determination of radius. Thus it greatly reduce the number of cycles to improve the efficiency of the algorithm.

Four groups of palm image samples were selected, with each group containing 100 images, use the old and new algorithm to extract the ROI respectively. The average extraction time of each group is shown in Table 1.

Table 1. Comparison of algorithm execution time (unit: ms)

	New algorithm	Old algorithm
Sample1	18.6	29.3
Sample2	19.5	29.7
Sample3	17.1	28.4
Sample4	18.3	29.1

According to the data given above, the execution time of the new algorithm is reduced by 10.7 ms, 10.2 ms, 11.3 ms and 10.8 ms under the four groups of samples respectively. Therefore, the new algorithm effectively improves the efficiency of the maximum inscribed circle algorithm.

References

1. Jain, A.K., Ross, A., Pankanti, S.: Biometrics: a tool for information security. IEEE Trans. Inf. Forensics Secur. **1**(2), 125–143 (2006)
2. Lee, J.C.: A novel biometric system based on palm vein image. Pattern Recogn. Lett. **33**(12), 1520–1528 (2012)
3. Dahmouche, R., Andreff, N., Mezouar, Y., et al.: Dynamic visual servoing from sequential regions of interest acquisition. Int. J. Robot. Res. **31**(4), 520–537 (2016)
4. Yan, X., Kang, W., Deng, F., et al.: Palm vein recognition based on multi-sampling and feature-level fusion. Neurocomputing **151**(151), 798–807 (2015)
5. Samorodova, O.A., Samorodov, A.V.: Fast implementation of the niblack binarization algorithm for microscope image segmentation. Pattern Recogn. Image Anal. **26**(3), 548–551 (2016)
6. Psiaki, M.L.: Gaussian mixture nonlinear filtering with resampling for mixand narrowing. IEEE Trans. Sig. Process. **64**(21), 5499–5512 (2016)
7. Sen, S., Ai-hua, L., Liang, Y., et al.: Gun code binary image algorithm based on wavelet packet and niblack method. Guangzi Xuebao/Acta Photonica Sin **42**(3), 354–358 (2013)
8. Ohtani, T., Kanai, Y., Kantartzis, N.V.: A 4-D subgrid scheme for the NS-FDTD technique using the CNS-FDTD algorithm with the shepard method and a Gaussian smoothing filter. IEEE Trans. Magn. **51**(3), 1–4 (2015)

Power Adaptive Routing Scheme with Energy Hole Avoidance for Underwater Acoustic Networks

Xian-yi Chen[1] and Guo-lan Lin[2(✉)]

[1] School of Information Science Technology, Hainan University, Haikou, China
[2] College of Applied Science and Technology,
Hainan University, Haikou, China
lglgl3128@163.com

Abstract. Underwater Wireless Sensor Networks (UWSNs) have grown rapidly in recent decade. As acoustic communication consumes much more power and underwater deployment environments are much harsher than that of the terrestrial wireless sensor networks, energy efficiency is much more critical for UWSNs. Therefore, it is essential to deal with the energy hole problem and balance the power consumption in underwater acoustic network. In this article, we propose a Power Adaptive and Energy Balance routing scheme (PAEB) to avoid the energy holes taken place owing to the unbalanced energy consumption in UWSNs. In order to balance the energy consumption, a Binary Exponential Transmission (BET) scheme is introduced. In BET scheme, sensor node located in different layer may have different transmission distance according to its residual energy. When sensor node sends packet to the sink, it selects the optimal transmission radius and adjusts the transmission power based on its current power level. The much energy is residual, the further the transmission radius is selected. The experimental results show that PAEB performs better than the existing representative protocols in terms of network lifetime and end-to-end delay.

Keywords: Underwater acoustic network · Energy hole avoidance · Energy efficient routing · Energy consumption balance

1 Introduction

The growing use of underwater acoustic sensor networks (UASNs) is driven by the desire to provide support for many activities, such as tsunami warning, mine reconnaissance and offshore exploration. As RF radio and optical communication technology are not suitable for UASNs due to their poor propagation in the acoustic channel, acoustic communication technology is typically adopted in the underwater environment [1, 2].

Compared to the radio signals used in the terrestrial sensor networks, there are many challenges of acoustic signals used in the underwater sensor networks as the following reasons.

© Springer Nature Singapore Pte Ltd. 2017
G. Chen et al. (Eds.): PAAP 2017, CCIS 729, pp. 268–278, 2017.
DOI: 10.1007/978-981-10-6442-5_24

- The available bandwidth of acoustic channel is severely limited than radio channel, which is normal in hundreds of KHz.
- As the speed of acoustic signals is 1500 m/s while radio signals run at a speed of 3×10^8 m/s, the propagation delay in the UASNs is substantially larger than that in the terrestrial sensor networks.
- The bit error rate of the underwater acoustic sensor networks is very high due to the multipath fading, the path loss and the transmission loss of the acoustic channel.

Similar to the terrestrial sensor network, the energy hole problem still occurs in UASNs. Moreover, the energy of the nodes in UASNs is severely limited, and the nodes will not work if there are no powers. Therefore, how to deal with the energy hole problem and lower the power consumption of the nodes in UASNs is the most key issue. In recent decades, many researchers have studied widely and deeply on the energy hole problem of the terrestrial networks, and have proposed lots of solution to overcome the energy hole problem. However, those solutions available for terrestrial wireless sensor networks are not suitable for UASNs due to the aforementioned reasons. Therefore, we focus on the energy hole problem for UASNs. To balance and decrease the energy consumption of underwater acoustic sensor networks efficiently, we propose a Power Adaptive and Energy Balance routing scheme (PAEB), which can avoid the energy holes taken place owing to the unbalanced energy consumption in UWSNs.

The rest of the paper is organized as follows. In Sect. 2, we give the related literatures. The detailed operation of PAEB is presented in Sect. 3. Section 4 shows the performance evaluation of our method and comparison with related algorithms. Finally, we draw the main conclusions in Sect. 5.

2 Related Work

The energy hole created due to unbalanced energy consumption is one of the key issues, which have attracted lots of attentions over the world. In this section, we discuss some existing protocols related to energy balance in UWSNs.

In [3], Nominal Range Forwarding (NRF) scheme is used for the nodes to transmit data to the sink with one hop transmission range, which results in unbalance load and decrease the network life time. Unlike NRF scheme, Balanced Routing (BR) scheme [4] uses two level transmission ranges, where the node divides the data into two fractions: small and large. BR send the small fraction data to its one hop neighbors with a low transmission range, while sending the large fraction data to its two hop neighbors with another transmission range.

A hybrid routing mechanism named Balanced Transmission Mechanism (BTM) is proposed in [5], where each node divides its energy into different levels and balances the network load among all sensor nodes. In BTM, the node will send data directly to the sink in case of excessive energy consumption in multi-hop transmission, which results in high energy consumption when the network range is very large. To improve the performance of BTM, the Balanced Energy consumption Technique (EEBET) is propose in [6]. EEBET does not directly transmit data to the sink if the distance between the node and the sink is very long, which can save more energy consumption and increase the network life time.

A balanced routing scheme is proposed in [7], where each sensor has two available transmission ranges. Moreover, the optimal load weight for each available range has been discussed, and the network energy consumption is balanced. In [8], a depth adaptive routing protocol (DARP) is proposed, which the sound speed as well as the depth and the distance to the sink are taken into consideration.

Anupama et al. [9] propose a location-based clustering algorithm for data gathering (LCAD) in UASNs. In LCAD, the sensor nodes are deployed at different depths statically and grouped into some clusters. The sensor nodes in the same cluster send data to their cluster head, and the cluster head preprocess the data and relays the data to the sink. To prolong the network life time, a Distributed Underwater Clustering Scheme (DUCS) is proposed in [10]. Similar to [9], the cluster member nodes send the data to their cluster head in a single hop, and then the cluster head forwards the data to the sink in a multi-hop. In [11], the monitor area is divided into several SCs, and these SCs are regarded as clusters. As a result, the cluster head can be selected in a SC and the power efficient can be improved.

3 Power Adaptive and Energy Balance Routing Scheme (PAEB)

In this section, we first setup the network model and energy consumption model. Cluster-head node selection, binary exponential transmission scheme and the energy balanced inter cluster dead routing scheme are described in detail.

3.1 Network Model

The architecture of underwater wireless sensor network for PAEB is shown in Fig. 1, where the following assumptions have been hold:

(1) There are N sensor nodes deployed in the monitoring area, which is a cylinder with a radius of R and a height of H.

(2) Base station is deployed on water surface statically, and is equipped with radio and acoustic modems, which can communicate with the satellite and sensor nodes respectively.

(3) As sensor nodes are attached to anchors or buoys, they can be deployed statically in the different depth of the cylinder monitoring region. Each sensor node can obtain the geographical location and calculate the distances to its neighboring nodes by ToA.

(4) All the sensor nodes have the same initial energy and the maximum transmission power. Each nodes can adjusted its band and power to reach a certain level according to the communication distance. The power of the base station is unconstrained.

(5) The cylinder monitoring region is divided into some clusters, and each sensor node in the same cluster can be elected as cluster head. It takes Eagg (n J/bit) energy for cluster head to aggregate the collected data.

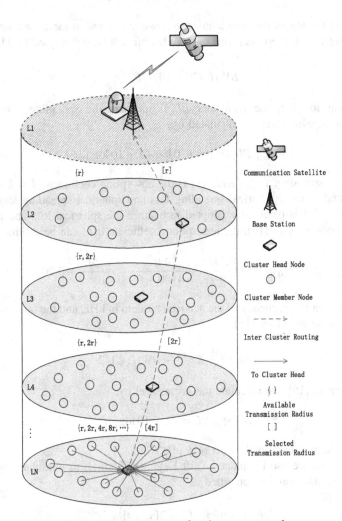

Fig. 1. The architecture of underwater network

3.2 Energy Consumption Model

For an underwater acoustic channel, if the distance of the transmitter and receiver is (km), then the required transmit power (dB re 1 μPa2) can be calculated as [13, 14]:

$$T_x(\ell) = BW(\ell) + PL(\ell) + NL(\ell) + R_x + \alpha \tag{1}$$

where $T_x(\ell)$ is the sound level of transmitter in dB re 1 μPa2, $BW(\ell)$ is the optimal transmit bandwidth, $PL(\ell)$ is the path loss, $NL(\ell)$ is the noise level, R_x is the signal-to-noise ratio (SNR) required for receiver to decode the received signal, and α is the penalty factor for the receiver.

Assuming the transmitter can adjust the power and band to meet the capacity C at the given distance, the optimal transmit bandwidth will be determined by [15]:

$$BW(\ell, C) = 10^{\frac{\alpha_1(C)}{10}} \ell^{\alpha_2(C)} \qquad (2)$$

According to [13], the path loss $PL(\ell)$ includes the spreading loss and the absorption loss, which can be calculated by:

$$10 \log_{10} PL(\ell, f) = k 10 \log_{10} \ell + 10 \log_{10} a(f) \qquad (3)$$

where k is the spreading factor and $a(f)$ is the absorption coefficient. $k = 1$ and $k = 2$ are correspond to the cylindrical spreading loss and spherical spreading loss, respectively, and $k = 1.5$ is practically adopted to balance the spherical loss and the cylindrical loss. Following to [16], the absorption coefficient $a(f)$ can be estimated by:

$$10 \log_{10} a(f) = \frac{0.11 f^2}{1 + f^2} + \frac{44 f^2}{4100 + f^2} + \frac{2.75 f^2}{10^4} + 0.003 \qquad (4)$$

where f is the central frequency of the acoustic modem in kHz, and the optimal transmit frequency can be found by [15]:

$$f = 10^{\frac{\alpha_3(C)}{10}} \ell^{\alpha_4(C)} - BW(\ell, C)/2 \qquad (5)$$

According to [16], we calculate the noise level by:

$$10 \log_{10} NL = \delta_0 - 10 \log_{10} f \qquad (6)$$

where δ_0 is the constant level in dB re 1 $\mu Pa^2/Hz$.

In this paper, we use Frequency Shift Keying (FSK/FH) to modulate signal. The signal-to-noise ratio can be computed as

$$R_x = (\text{erfcinv}(2 - 2(1 - \varepsilon)^{1/PS})^2 \qquad (7)$$

where ε is a given packet error rate, and PS is the packet length.

As the unit of acoustic power $T_x(\ell)$ in Eq. (1) is dB re 1 μPa^2, we use the empirical relation proposed in [13] to translate $T_x(\ell)$ into electrical power in Watt as

$$E_{Tx}(\ell) = T_x(\ell) \times 10^{-17.1} \qquad (8)$$

Therefore, the energy consumption of transmitting a packet over the distance ℓ can be calculated by

$$E_{Tx}(\ell) = \frac{T_x(\ell) \times PS}{10^{17.1} \times C} \qquad (9)$$

Compare to the energy used for packet transmission, the energy consumption of receiving a packet can be computed simply by

$$E_{Rx} = P_{Rx} \times PS/C \qquad (10)$$

where P_{Rx} is the reception power of the acoustic modem in Watt.

3.3 Binary Exponential Transmission Scheme

The nodes located in different layer may have the different transmission distance. We introduce a binary exponential transmission radius to set the transmission distance of the nodes in the different layer. The set of the available transmission distance of the nodes in the layer L is defined as

$$d_L \in \{ r, 2r, \ldots, 2^{(k_L-1)}r \} \qquad (11)$$

where k_L is the number of the available transmission radius of the nodes in the layer L, and k_L is determined by Eq. (12). d_L^i denotes the ith element of the set d_L, for example, $d_L^1 = r$, $d_L^2 = 2r$, $d_L^3 = 4r$.

$$k_L = \lfloor \log_2(L-1) \rfloor + 1 \quad L \geq 2 \qquad (12)$$

When node transmit packet toward the sink, it select the transmission radius in the set d_L according to its residual energy. The much energy is residual, the further the transmission radius is selected. Considering a node in the 5th layer, it has 3 available transmission radiuses according to Eq. (12). Firstly, it uses the transmission radius of $4r$ to transmit packet. When the residual energy is less than a given value, it transmit packet using the $2r$ transmission radius. Finally, the r transmission radius is used to transmit packet to the sink if the node's energy is lower than another threshold.

Let W_L^i denote the ith threshold of the residual energy in percent of the initial energy for the nodes in the layer L, and W_L^i will be set by:

$$\sum_{i=1}^{k_L} W_L^i = 1 \qquad (13)$$

Therefore, the number of the packets transmitted in the corresponding transmission radius in the layer L can be calculated by:

$$N_L^i = W_L^i \times E_{ini}/E_{Tx}(d_L^i) \qquad (14)$$

Following Eq. (14), it is simple to compute the number of total packets transmitted in the layer L by:

$$S_L = \sum_{i=1}^{k_L} N_L^i \tag{15}$$

It is notable that the value of S_L is not the same in the different layer because the transmission packet includes the packets created in its own layer and packets received from the lower layer which need to forward to the sink except for the bottom layer.

Assuming there are S packets created in each layer in the entire network lifetime, S_L can be denoted as:

$$S_L = \begin{cases} S & L = LN \\ S + R_L & 2 \le L \le LN \end{cases} \tag{16}$$

where R_L is the number of packets received from the lower layer, it can be calculated by

$$R_L = N_{L+1}^1 + N_{L+2}^2 + \cdots + N_{L+2^{(k_{LN}-1)}}^{k_{LN}} \quad L + 2^{(k_{LN}-1)} \le LN \tag{17}$$

To balance the energy consumed in the different layer and maximize the network lifetime, the energy of the nodes in the network should be used up at the same time, as well as the Eq. (16) must be hold. Therefore, we have:

$$\begin{cases} \sum E_{Tx} = E_{ini} & L = LN \\ \sum E_{Tx} + \sum E_{Rx} = E_{ini} & 2 \le L \le LN \end{cases} \tag{18}$$

Obviously, Eq. (18) can be rewritten as:

$$\begin{cases} \sum_{i=1}^{k_L} (N_L^i \times E_{Tx}(d_L^i)) = E_{ini} & L = LN \\ \sum_{i=1}^{k_L} (N_L^i \times E_{Tx}(d_L^i)) + R_L \times E_{Rx} = E_{ini} & 2 \le L \le LN \end{cases} \tag{19}$$

By putting the values of RL in Eq. (19) from Eq. (16), we have:

$$\begin{cases} \sum_{i=1}^{k_L} (N_L^i \times E_{Tx}(d_L^i)) = E_{ini} & L = LN \\ \sum_{i=1}^{k_L} (N_L^i \times (E_{Tx}(d_L^i) + E_{Rx})) - S \times E_{Rx} = E_{ini} & 2 \le L < LN \end{cases} \tag{20}$$

Our goal is to find the optimal value of W_L^i to balance the energy consumption. From Eqs. (14), (16) and (20), we can acquire the Eq. (21), which W_L^i must be hold.

$$
\begin{cases}
\sum_{i=1}^{k_L} (W_L^i/E_{Tx}(d_L^i) + W_L^i \times E_{Rx}/E_{Tx}(d_L^i)) \\
\quad - \sum_{i=1}^{k_{LN}} (W_{LN}^i \times E_{Rx}/E_{Tx}(d_{LN}^i)) = 1 \\
\sum_{i=1}^{k_L} (W_L^i/E_{Tx}(d_L^i)) = \sum_{i=1}^{k_{LN}} (W_{LN}^i/E_{Tx}(d_{LN}^i)) \\
\quad + W_{L+1}^1/E_{Tx}(d_{L+1}^i) + \cdots + W_{L+2^{k_L-1}}^{K_L-1}/E_{Tx}(d_{L+2^{k_L-1}}^i) \\
\text{subject to} \\
\sum_{i=1}^{k_L} W_L^i = 1, \forall L, 2 \le L \le LN
\end{cases}
\tag{21}
$$

4 Performance Evaluation

In this section, we present the performance evaluation of our proposed PAEB protocol. The performance of PAEB was compared to NRF [3] and BR [4].

The simulation is done by Matlab and the simulation settings are shown in Table 1.

Table 1. System parameters in the simulations.

Parameter	Value
Width	3000 m
Depth	2000 m
Location of BS	(0, 0, 0)
Frequency	30 kHz
Capacity	30 kbps
Data packet size	1024 bits
Initial energy	50 J
Packet error rate	0.001
Spreading factor	1.5

4.1 Performance Metrics

The following metrics are used to evaluate the performance of our proposed protocol:

- Network life-time is defined as the time when 1% of total nodes, 10% of total nodes and 20% of total nodes are died respectively in the network due to the energy exhaustion.
- Energy consumption is the total energy consumption in the network until the data packets are transmitted successfully.
- End-to-end delay is the time taken by a data packet transmitted from a source node to the sink node.

4.2   Simulation Results and Analysis

The network life time is shown in Fig. 2, from which it can find that the average network life time of PAEB is 109 rounds, which is higher 31.8% and 18.4% than that of NRF and BR, respectively. Therefore, the network life time of PAEB is much better than of NRF and BR. The end-to-end delay is shown in Fig. 3, from which it can see that the average end-to-end delay of PAEB is 789 ms, which is lower 23.5% and 13.6% than that of NRF and BR, respectively. Therefore, PAEB has the lower end-to-end delay compared to NRF and BR. The reason is that sensor nodes located in different layer have different transmission distance according to its residual energy. When sensor node sends packet to the sink, PAEB selects the optimal transmission radius and adjusts the transmission power based on its current power level. Hence, the power consumption is balanced among the sensor nodes, which improves the network life time and decreases the end-to-end delay.

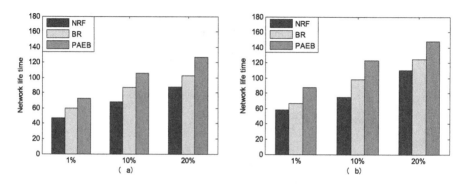

Fig. 2. The network life time

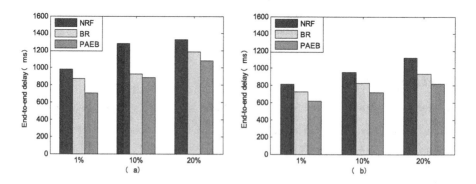

Fig. 3. The end-to-end delay

5   Conclusion

In this paper, we proposed an energy efficient routing protocol named PAEB for UWSNs. PAEB uses the Binary Exponential Transmission (BET) scheme to determine the transmission radius based on the residual energy of the node located in the different layer. Furthermore, in order to improve the network life time, an energy balancing is employed using the residual energy information of the sensor nodes.

The energy hole problem of underwater acoustic networks will become more interesting when the nodes are not static. In the future, we are going to deal with such circumstance.

Acknowledgment. This work was supported by Natural Science Foundation of Hainan Province of China (Grant No. 61706), Scientific Research Foundation for the Ph.D. Scholars of Hainan University (Grant No. kyqd1651), the Key Research Project of Hainan Province of China (Grant No. ZDYF2016153), and the National Natural Science Foundation of China (Grant No. 61363071).

References

1. Akyildiz, I., Pompili, D., Melodia, T.: Underwater acoustic sensor networks: Research challenges. Ad Hoc Netw. **3**(3), 257–279 (2005)
2. Sozer, E.M., Stojanovic, M., Proakis, J.G.: Underwater acoustic network. IEEE J. Oceanic Eng. **25**(1), 72–83 (2000)
3. Ammari, H.M., Das, S.K.: Promoting heterogeneity, mobility, and energy-aware voronoi diagram in wireless sensor networks. In: IEEE Transactions on Parallel and Distributed Systems (2008)
4. Zidi1, C., Bouabdallah, F., Boutaba, R.: Routing design avoiding energy holes in underwater acoustic sensor networks. Wirel. Commun. Mob. Comput. **16**, 2035–2051 (2016)
5. Cao, J., Dou, J., Dong, S.: Balance transmission mechanism in underwater acoustic sensor networks. Int. J. Distrib. Sensor Netw. **11**(3), 1–12 (2015)
6. Javaid, N., Shah, M., Ahmad, A., Imran, M., Khan, M.I., Vasilakos, A.V.: An Enhanced Energy Balanced Data Transmission Protocol for Underwater Acoustic Sensor Networks. Sensor (2016)
7. Jiang, P., Liu. J, Ruan B, Jiang L, Wu F. A New Node Deployment and Location Dispatch Algorithm for Underwater Sensor Networks. Sensors (2016)
8. Chen, Y.D., Lien, C.Y., Wang, C.H., et al.: DARP: a depth adaptive routing protocol for large-scale underwater acoustic sensor networks. In: Oceans. IEEE, pp. 1–6 (2012)
9. Anupama, K.R., Sasidharan, A., Vadlamani, S.: A location-based clustering algorithm for data gathering in 3D underwater wireless sensor networks. In: Proceedings of International Symposium on Telecommunications, pp. 343–348 (2008)
10. Domingo, M.C., Prior, R.: Design and analysis of a GPS-free routing protocol for underwater wireless sensor networks in deep water. In: International Conference on Sensor Strategies and Applications, Valencia, pp. 215–220 (2007)
11. Wang, K., Gao, H., Xu, X.L., Jiang, J.F., Yue, D.: An energy-efficient reliable data transmission scheme for complex environmental monitoring in underwater acoustic sensor networks. IEEE Sensors J. **16**(11), 4051–4062 (2016)
12. Urick, R.: Principles of Underwater Sound. McGraw-Hill, New York (1983)

13. Zorzi, M., Casari, P., Baldo, N., Harris, A.F.: Energy-efficient routing schemes for underwater acoustic networks. IEEE J. Selected Areas Commun. **26**(9), 1754–1766 (2008)
14. Stojanovic, M.: On the relationship between capacity and distance in an underwater acoustic communication channel. In: Proceedings of the International Workshop on Under Water Networks (WUWNet), Los Angeles, USA (2006)
15. Berkhovskikh, L., Lysanov, Y.: Fundamentals of Ocean Acoustics. Springer, Heidelberg (1982)
16. Ahuja, R.K., Mehlhorn, K., Orlin, J.B., Tarjan, R.E.: Faster algorithms for the shortest path problem. J. ACM **37**, 213–223 (1990)

A Lightweight Algorithm for Computing BWT from Suffix Array in Disk

Jing Yi Xie[1], Bin Lao[1], and Ge Nong[1,2(✉)]

[1] Department of Computer Science, Sun Yat-sen University, Guangzhou, China
issng@mail.sysu.edu.cn
[2] SYSU-CMU Shunde International Joint Research Institute, Shunde, China

Abstract. The Burrows-Wheeler transform (BWT) and the suffix array (SA) of an input string are important data structures widely used in modern bioinformatics researches such as full-text search, alignment etc. In this paper, we present a lightweight external memory algorithm for computing the BWT from a given suffix array and the input string. The algorithm has a linear I/O complexity $O(n)$ and a workspace of at most $n/2$ integers. An experiment study is conducted to evaluate the time and space performance of the proposed algorithm on a number of realistic datasets. The experimental results are consistent with the theoretical complexities of the algorithm.

Keywords: External memory algorithm · Induced sorting · BWT

1 Introduction

The Burrows-Wheeler transform (BWT) was proposed by Burrows and Wheeler in 1994 [2] for data compression. Given an input string, its BWT is a permutation of the characters of this string, which can be computed by lexicographically sorting all the suffixes, then the list composed of the preceding character of each sorted suffix is the BWT. The same characters often span consecutively in a BWT, so it is easier to be compressed than the original string [6], e.g. the bzip2 was designed in this way [1]. Besides data compression, BWT was later applied to other applications such as genome alignment, construction of compressed full-text index [7] and etc. As a result, it has become an important data structure in modern bioinformatics. The algorithms for computing the BWT can be directly or indirectly, in terms of that, the former directly computes the BWT from the input string, and the latter first computes an interim data structure such as suffix array (SA) [5] and then from which computes the BWT. The SA can play as a succinct alternative for replacing suffix trees in many applications [10, 11]. This paper presents a lightweight external memory algorithm to compute BWT from SA, where "lightweight" means a small required space.

The input T is a size-n string with the characters from a constant alphabet Σ of size $O(1)$. Let A and B be the suffix array and the BWT of T, respectively, where A is a size-n integer array with the pointers to all the suffixes of T in their increasing lexicographical order, and $B[i] = T[A[i] - 1]$ for $A[i] > 0$ or else $B[i] = T[n - 1]$. Using internal memory, B can be easily computed from A and T in linear time $O(n)$ and workspace $O(1)$, where the workspace excludes the input and the output. In external

© Springer Nature Singapore Pte Ltd. 2017
G. Chen et al. (Eds.): PAAP 2017, CCIS 729, pp. 279–289, 2017.
DOI: 10.1007/978-981-10-6442-5_25

memory, B can be computed from A in a workspace of $2n$ integers by a straightforward method as follows: (1) compute the inverse suffix array of T from A, called A^{-1}, which is a size-n integer array satisfying that $A^{-1}[i] = j$ if and only if $A[j] = i$; (2) scan A^{-1} to get the preceding character of each suffix; (3) sort the preceding character of each suffix by the key as the suffix position in A^{-1}. This method is rather space consuming and becomes the space bottleneck in some existing applications such as a suffix array checker.

In addition to be constructed in a deterministic way, a suffix array can also be constructed by a probabilistic method [4]. In this case, the resulting suffix array might be incorrect with a very small probability. Hence, it has to be checked to ensure the correctness. In addition, a suffix array is usually checked after its construction to avoid errors caused by program bugs. A suffix array checker is commonly provided in the open source software for SA construction nowadays. A key step in an SA checker is computing the BWT from the given SA and T, which uses the aforementioned straightforward method and hence constitutes the space bottleneck of the whole checking process. A lightweight approach with less space consumption is demanded for overcoming this issue, which motivates our work reported here.

To the best of our knowledge, in the existing literature, there is no study devoted to designing an external memory algorithm for computing BWT from SA. In this paper, we propose a lightweight and linear I/O complexity external memory algorithm called SA2BWT for computing BWT from SA, with a workspace of at most $n/2$ integers. SA2BWT employs a divide-and-conquer method, which first splits the original problem into a number of sub-problems that each can be solved in internal memory, then solves in internal memory each sub-problem using induced sorting [9] and merges in external memory the sub-solutions to get the final result.

The rest of this paper is organized as follows. Section 2 gives the algorithm with a complexity analysis. Section 3 presents the algorithm with a running example for illustration. Section 4 conducts an experiment to evaluate the algorithm's time and space performance and discusses the optimization of algorithm. Section 5 gives the summary.

2 Our Solution

2.1 Basic Notations

The problem to be solved is that given a size-n input string T and its suffix array, use an I/O complexity $O(n)$ and a workspace of at most $n/2$ integers to compute in external memory the BWT of T, where $\|\Sigma\| = O(1)$. Some basic notations are recapitulated from [9] for presentation convenience.

Let $T[0, n) = T[0]T[1]...T[n - 1]$, where $T[n - 1]$ is a sentinel \$ that is the unique lexicographically smallest character in T but not in Σ. A suffix $suf(T, i)$ starts from T $[i]$ to the sentinel, it is classified as S- or L-type as follows: S-type if (1) $i = n - 1$ or (2) $T[i] < T[i + 1]$, or (3) $T[i] = T[i + 1]$ and $suf(T, i + 1)$ is S-type; or else L-type. T $[i]$ is S-type if $suf(T, i)$ is S-type, or else L-type. Moreover, if $T[i]$ is S-type and $T[i - 1]$ is L-type, $T[i]$ is an LMS (leftmost S-type) character; if $T[i]$ is L-type and $T[i + 1]$ is

S-type, $T[i]$ is a RML (rightmost L-type) character. For a given substring in T, it is an LMS/RML-substring if it starts at an LMS/RML character and ends at the first LMS/RML character on the right hand, respectively.

In the suffix array A of T, the suffixes of a same head character ch occupy a consecutive range called a bucket, denoted by $bkt(ch)$. In a bucket, the L- and S-type suffixes are clustered at the left and right ends, constituting two sub-buckets called the L- and S-bucket denoted by $bkt\_l(ch)$ and $bkt\_s(ch)$, respectively.

2.2   Algorithm Framework

The algorithm's underlying idea is sketched first. Let n_L and n_S be the number of L- and S-type characters in T, respectively. Without loss of generality, unless specified otherwise, suppose $n_L < n_S$ in the rest of this article.

1. Scan T to divide T and its suffixes into logical blocks.
2. Scan the A from left to right, for each scanned suffix, detecting its type and sequentially putting the L-type suffixes into their block buffers in disk.
3. Compute in internal memory the B of each block by induced sorting.
4. Scan the A to merge the B of each block into the final B.

In step 1, T is scanned to divide its characters and suffixes into blocks, in a way such that the B of each block can be computed in internal memory.

In step 2, we need to detect the type of each scanned suffix. For $\|\Sigma\| = O(1)$, we can scan T to collect the size of each L- and S-bucket in the A. Given the size of each sub-bucket, we can on-the-fly determine the type of the sub-bucket that a scanned suffix resides at, and hence know the type of a scanned suffix.

In step 3, given all the sorted L-type suffix of a block, we can induce the order of all the suffixes in the block, from which the B of the block is obtained.

In step 4, given the B of each block, we scan the A to sequentially retrieving the preceding character of each scanned suffix from the B of the block that the suffix belongs to. Notice that for any two suffixes in a block, their relative order computed by step 3 is identical to that in the suffix array. As a result, this step produces the final B.

Following this idea, we proceed to design the detailed algorithm SA2BWT in the rest of this section. Figure 1 outlines the algorithm framework. Lines 1–6 detect the type of each character to divide all the suffixes of T into blocks. Lines 7–8 scan the A to sequentially put each suffix to its block, and compute in internal memory the B of each block. Line 9 merges in external memory the B of each block to get the final B. The I/O complexity of each step is $O(n)$, while the peak disk usage occurs in Line 7.

2.3   Dividing T into Blocks

Given $n_L < n_S$, RML characters are used to split T, otherwise, LMS characters are used instead. The rules for dividing T into $n_1 = n/m$ blocks $\{T\_b_i \mid i \in [0, n_1-1]\}$ are as follows [8], where m is the block size:

1. Each block starts with $T[0]$ or a RML character and ends with another RML character or $T[n - 1]$.

Algorithm 1 The algorithm framework for computing BWT from SA

Input: T, A
Output: B
 1: Scan T once to detect the type of each character;
 2: **if** n_L is less **then**
 3:     Scan T once to split T using RML-substrings;
 4: **else**
 5:     Scan T once to split T using LMS-substrings;
 6: **end if**
 7: Put the sorted L- or S-type suffixes into each block;
 8: For each block, induced sort the suffixes to compute the B in memory;
 9: Merge the B of each block to get the final B;
10: **return**

Fig. 1. The algorithm framework for computing BWT from SA.

2. Any pair of neighboring blocks overlaps on a common RML character.
3. A block $T[g, h]$, $0 \leq g < h \leq n - 1$, can have more than m characters only if there is no RML character in $T[g + 1, h - 1]$.

Initialize $i = n_1 - 1$ and T_b_i as empty. While scanning T from right to left, each RML-substring will be added to T_b_i if the addition won't cause $\|T\_b_i\| > m$. Otherwise, i is decreased by 1 and this substring is added into a new block. If the length of a substring is larger than m, it will create a single block. For determining the position of each block in T, record their lengths.

Splitting T in this way, the lengths of blocks may not be equal, we can determine the location information of each block in T using the boundary characters as follows [8]. For $T[j]$, where $i * m \leq j \leq (i + 1) * m$, and $i \in [0, n_1 - 1]$, it must belong to (1) the block containing $T[i * m]$, or (2) the block containing $T[(i + 1) * m]$, or (3) the block between them. In this way, given a suffix, the target block can be located in $O(1)$ time and $O(n_1)$ space.

2.4   Computing and Merging the BWT of Each Block

After splitting T into blocks, the sorted L-type suffixes are selected for inducing the suffix array of each block, while the type of each suffix can be determined on-the-fly when scanning A. Given the alphabet size $O(1)$, we can scan T to get the location and size information of each L- or S-bucket in A, then scan the A to determine the type of each suffix by the sub-bucket information, and put each suffix to its block sequentially in the external memory, where each block i is denoted by TS_b_i.

The key idea of SA2BWT is to compute the A and B of each block in internal memory by the induced sorting method, then the whole B of T is completed by merging the B of each block. To compute the B of each block in internal memory, we use a block data structure BLOCKBUF to provide information below:

- *SA_b*: an array storing all suffixes belonging to this block.
- *ch*: an array consists of all characters belonging to this block.

- *type*: an array storing the type of each character in *ch*.
- *start_pos*: the start position of this block.
- *end_pos*: the end position of this block.

With the required data structure, the approach for induced sorting A from the L-type suffixes is given below:

1. Find the head of each L-type bucket in SA_b, and scan TS_b_i from head to end. For each suffix $suf(T, i)$, put it to the current head of the L-type bucket for $ch[suf(T, i) - start\_pos]$ and shift right the current head one step.
2. Find the end of each S-type bucket in SA_b, and scan SA_b once from end to head. For each item $SA\_b[i]$, if $ch[SA\_b[i] - 1 - start\_pos]$ is S-type, put $SA\_b[i] - 1$ to the current end of the S-type bucket for $ch[SA\_b[i] - 1 - start\_pos]$ and shift left the current end one step.

Analogously, if n_S is less, the algorithm for induced sorting the A from the S-type suffixes is modified as follows:

1. Find the end of each S-type bucket in SA_b, and scan TS_b_i from head to end. For each suffix $suf(T, i)$, put it to the current end of the S-type bucket for $ch[suf(T, i) - start\_pos]$ and shift left the current end one step.
2. Find the head of each L-type bucket in SA_b, and scan SA_b once from head to end. For each item $SA\_b[i]$, if $ch[SA\_b[i] - 1 - start\_pos]$ is L-type, put $SA\_b[i] - 1$ to the current head of the L-type bucket for $ch[SA\_b[i] - 1 - start\_pos]$ and shift right the current head one step.

The correctness of step 2 is established by Lemma 3.9 in [9]:

Lemma 1 [9]. *Given all the sorted L-type (or S-type) suffixes of T, the order of all the suffixes of T can be induced in $O(n)$ time.*

After sorting all the suffixes in a block, the B of this block can be immediately computed from the sorted suffixes and input characters for this block. Then, we scan A once to merge the B of each block, by sequentially retrieving the B of each block to produce the overall B.

2.5   Complexity Analysis

Theorem 1 *(I/O and Workspace Complexities). Given a size-n input string T of a constant alphabet, the I/O and workspace complexities for SA2BWT is $O(n)$ and $n/2$ integers, respectively, where an integer is log n bits.*

Proof. Each step in SA2BWT only scans A and/or T a number of $O(1)$ passes, hence the total I/O complexity is $O(n)$. Because $\min(n_L, n_S)/n \leq 1/2$, the workspace is at most $n/2$ integers, which is caused by storing the selected L- or S-type suffixes for each block.

3   A Running Example

In Fig. 2, a running example of SA2BWT is illustrated for a sample string T = im-miissiissiip$, where $ is the sentinel. Assuming the block size of 8 bytes, the algorithm starts scanning T from right to left to detect the type of each character. Line 2 lists the type of each character, where the L-type characters are less. Then, the algorithm continues the following steps:

```
00   Index   :   00  01  02  03  04  05  06  07  08  09  10  11  12  13  14
01      T    :   i   m   m   i   i   s   s   i   i   s   s   i   i   p   $
02      t    :   S   L   L   S   S   L   L   S   S   L   L   S   S   L   S
03   RML     :           *               *               *           *
04   step 1 split string:
05   block   :   |_____||_____||_____|
06   step 2 determine the positions of boundary characters:
07   bound_ch:   *                               *
08      A    :   14  11  07  03  00  12  08  04  02  01  13  10  06  09  05
09   step 3 compute BWT of each block then merge them (take TS_b0 for example):
10   TS_b0   :   02  01  06  05
11   TS_b1   :   10  06  09
12   TS_b2   :   13  10
13   bkt     :           i           m           s
14   SA_b        {-1  -1  -1} {-1  -1} {-1  -1}
15                ^           ^       ^
16               {-1  -1  -1} {02  -1} {-1  -1}
17                ^               ^   ^
18               {-1  -1  -1} {02  01} {-1  -1}
19                ^               ^ ^
20               {-1  -1  -1} {02  01} {06  -1}
21                ^               ^   ^
22               {-1  -1  -1} {02  01} {06  05}
23                ^               ^       ^
24   SA_b        {-1  -1  -1} {02  01} {06  05}
25                        ^       ^   ^@
26               {-1  -1  04} {02  01} {06  05}
27                    ^           ^ @ ^
28               {-1  -1  04} {02  01} {06  05}
29                    ^           ^@  ^
30               {-1  00  04} {02  01} {06  05}
31                ^           @   ^       ^
32               {-1  00  04} {02  01} {06  05}
33                ^           @       ^       ^
34               {03  00  04} {02  01} {06  05}
35                ^       @           ^       ^
36               {03  00  04} {02  01} {06  05}
37                ^ @                 ^       ^
38   TB_b0   :   m   p   i   m   i   s   i
39   TB_b1   :   s   i   s   null i
40   TB_b2   :   null s   i   i   null
41      B    :   s   s   m   p   i   i   i   m   i   i   s   s   i   i
```

Fig. 2. A running example of SA2BWT.

Step 1: Scan T from right to left to split T according to the RML characters. Notice that all the RML characters are marked by '*' at Line 3. Thus T is split into three blocks, which contains "immiiss", "siiss", and "siip\$" with size of 7, 5 and 5, respectively.

Step 2: Determine the positions of boundary characters. The boundary characters in T is $T[0]$ and $T[8]$, since the block size is set as 8 bytes. They belong to block_0 and block_1, respectively.

Step 3: Compute the B of each block in internal memory, then merge them for the final B. Line 8 lists all the suffixes in A by their start positions in T. As the number of L-type suffixes is less, we scan A to find the L-type suffixes and put them into their blocks. For example, the first scanned suffix starts at $A[0] = 14 = n - 1$, it is S-type and belongs to the last block. The second one starts at $A[1] = 11$, which is also S-type as $bkt\_l(i) = 0$. Until $A[8] = 2$ is scanned, it is detected as a L-type suffix. Furthermore, position 2 is within the start and end positions of block_0, so this suffix belongs to block_0. After scanning T, Lines 10–12 show the final TS_b, which are $TS\_b_0 = \{2, 1, 6, 5\}$, $TS\_b_1 = \{10, 6, 9\}$ and $TS\_b_2 = \{13, 10\}$.

Now the algorithm induced sorts the suffixes in each block. Take induced sorting A in block_0 as an example. There are 3 buckets 'i', 'm' and 's' in SA_b. Firstly, the suffixes in block_0 are initialized as -1. As the number of L-type is less, Line 14 finds the head of each bucket and marks them by '^'. Then Lines 14–23 scan TS_b_0 rightwards. The first suffix is $TS\_b_0[0] = 2$ with an initial character 'm', so it is put into the head position of bucket 'm', and the current head of the bucket is increased by one. Notice that the initial characters are retrieved from ch in BLOCKBUF, so as to avoid random access to disk. The next suffix is $TS\_b_0[1] = 1$ with the same initial character 'm', and it is put into the bucket 'm' at the position pointed by '^' now. Similarly, the next two suffixes are appended to the bucket 's'. The result is shown in Line 22 with pointers at the next positions after the end of bucket 'm' and 's'. Afterwards, Line 24 finds and marks the end of each bucket with '^' and scans SA_b leftwards. Herein, the current item being visited is marked by '@'. When $SA\_b[6] = 5$ is visited in Line 24, its type in BLOCKBUF is not L-type, thus the preceding suffix $suf(T, 5\text{-}1 = 4)$ is added to the bucket 'i' and the current end of the bucket is decreased by one. The next suffix being visited is $SA\_b[5] = 6$ and it is L-type, so it is skipped and the symbol '@' moves leftwards by one. Scanning SA_b in this way, all the suffixes in block_0 are finally sorted as Line 36. The same procedure is applied to the other blocks to produce the resulting B of each block shown in Lines 38–40, i.e. TB_b_i.

Finally, scan A from left to right to determine the corresponding block for $A[i]$ and retrieve $B[i]$ from TB_b, where $A[i] = n - 1$ is skipped as it is for the sentinel. The final result of $B[0..: n - 2] = \{s, s, m, p, i, i, i, m, i, i, s, s, i, i\}$ is shown in Line 41.

4 Experiments

For performance evaluation, we implement SA2BWT using the STXXL library [3], which is an implementation of the C++ standard template library STL for external memory computations. The experiments were performed on Linux (version 3.13.0-86-generic) with CPU (Intel Core i3 3.20 GHz) and 4.00 GiB RAM. The

Table 1. Datasets for experiments, n in Gi, one byte per character.

Dataset	n	$\|\Sigma\|$	Description
english	1	237	The English text files selected from Gutenberg collections etext02 to etext05, at http://pizzachili.dcc.uchile.cl/texts.html
proteins	1.1	27	The newline-separated protein sequences from the Swissprot database, at http://pizzachili.dcc.uchile.cl/texts.html
uniprot	2.81	96	The Uniprot knowledgebase release 4.0, at ftp://ftp.expasy.org/databases/uniprot
android	10	256	The source files of project Android, at https://source.android.com/source/downloading
guten	22.44	256	The Gutenberg collection guten1209, at http://algo2.iti.kit.edu/bingmann/esais-corpus/gutenberg-201209.24090588160.xz
enwiki	48.44	256	The English Wikipedia enwiki1503, at https://meta.wikimedia.org/wiki/Data_dump_torrents#enwiki

program was implemented in C/C++ and compiled by gcc/g++ of version 4.8.4. The datasets chosen for experiments are listed in Table 1, which are of varying sizes, alphabets and redundancy.

The performance measures are the runtime, peak disk usage and I/O volume. To smooth the performance fluctuation, each result is the mean of three runs. The peak disk usage and I/O volume are returned by the STXXL library, the time (μs/char) and peak disk usage (bytes/char) are normalized by n.

The experimental results of runtime, peak disk usages and I/O volumes for different datasets are shown in Figs. 3, 4 and 5, where enwiki_1g, enwiki_2g and enwiki_4g are the 1, 2 and 4 GiB prefixes of enwiki, and guten_4g and guten_8g are the 4 and 8 GiB prefixes of guten, respectively. In addition, Fig. 6 shows the $\lambda = \min(n_L, n_S)/n$ of each dataset, which can affect the algorithm's performance. The horizontal axis in each figure are the datasets ordered in non-decreasing n.

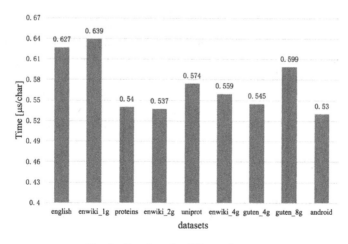

Fig. 3. Runtime for different datasets.

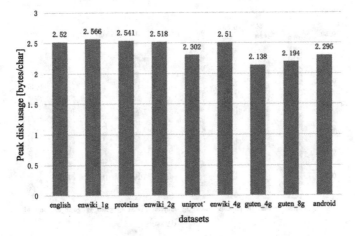

Fig. 4. Peak disk usages for different datasets.

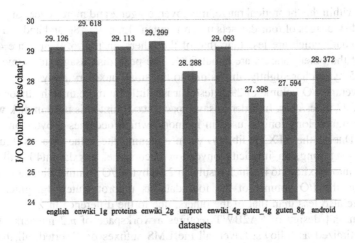

Fig. 5. I/O volumes for different datasets.

The time bottleneck of an external algorithm is mostly due to the I/O volume. To test the machine's I/O throughput, we use the shell command "time cp" to evaluate the time for copying enwiki_1g. The mean time for three runs is 22.54 s. The I/O volume for this copy is 2 GiB, hence the I/O throughput of this machine is about $22.54/(2 * 1024) = 0.011$ μs per byte. The I/O throughput of our program is about $0.55/29 = 0.0189$ μs per byte, where 0.55 is the average runtime per character in Fig. 3 and 29 is the average I/O volume in bytes per character observed from Fig. 5. So the I/O throughput of our program is $0.011/0.0189 = 58\%$ of the machine's saturated I/O throughput.

The peak disk usage of the algorithm is at most $n/2$ integers, i.e., $2.5n$ bytes for 40-bit integers in this experiment. As shown in Fig. 4, the peak disk usage of each

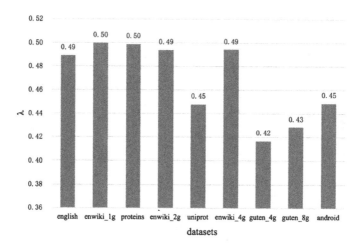

Fig. 6. λ for different datasets.

dataset is within the theoretical range, moreover, bigger λ and more space. In particular, the peak disk usages of four datasets uniprot, guten_4g, guten_8g and android are 2.1–2.3 bytes/char, which are less than that of the others. In Fig. 6, their λ are 42–45%, while λ of the other datasets are about 50%. The peak disk usages for some datasets exceed 2.5 bytes/char slightly, this is due to the small auxiliary arrays.

The average I/O volume of 29 bytes/char is a little bit more than the theoretical I/O volume of 27 bytes/char. In order to save space, stxxl::uint40 is used in disk while the type of unsigned long long is used in memory, which occupies 8 bytes on a 64-bits machine. Due to the STXXL library, when assigning the value of a stxxl::uint40 to unsigned long long, an implicit conversion occurs, i.e., stxxl::uint40 will be first converted into stxxl::uint64, which results in $2.5n$ bytes I/O volume. Similar to the peak disk usage, the I/O volume of the four datasets uniprot, guten_4g, guten_8g and android are also less than that of the others, due to the difference in λ.

The current design of SA2BWT requires a workspace of $n/2$ integers. It can be further optimized as follows. Given all the LMS suffixes of T sorted, all the L-type suffixes can be sorted in $O(n)$ time [9]. So it is feasible to induced sort the SA from the LMS suffixes, which are the suffixes starting from LMS characters. Also from [9], given the probabilities for each character to be S- or L- type as 1/2, the mean size of a non-sentinel LMS substring is 4, i.e., the number of LMS substring is at most $n/3$, resulting in that the expecting space complexities for improving the algorithm to use LMS suffixes is $n/3$ integers. The key is to detect the LMS suffixes. Now we only need to judge the type of a scanned suffix, but if we use the LMS suffixes, not only the character $T[A[i]]$ needs to be S-type, but also the preceding $T[A[i]-1]$ needs to be L-type. An approach for scanning A to detect each LMS suffix starting at $A[i]$ without retrieving $T[A[i]-1]$ is demanded and will be addressed in an extended version of this paper.

5 Summary

A lightweight external memory algorithm is proposed for computing the BWT from the suffix array and the input string. The algorithm requires a linear I/O complexity $O(n)$ and a workspace of at most $n/2$ integers. Currently, computing the BWT constitutes the space bottleneck in an external memory suffix array checker. This algorithm can be employed for computing the BWT, and hence reduce the peak disk usage for checking a big suffix array in external memory. As a next step, we are developing efficient methods to further improve the time and space performance of the algorithm.

Acknowledgments. The work of G. Nong was supported by the Guangzhou Science and Technology Program grant 201707010165 and the Project of DEGP grant 2014KTSCX007.

References

1. Beller, T., Zwerger, M., Gog, S., Ohlebusch, E.: Space-efficient construction of the burrows-wheeler transform. In: Kurland, O., Lewenstein, M., Porat, E. (eds.) SPIRE 2013. LNCS, vol. 8214, pp. 5–16. Springer, Cham (2013). doi:10.1007/978-3-319-02432-5_5
2. Burrows, M., Wheeler, D.J.: A block-sorting lossless data compression algorithm (1994)
3. Dementiev, R., Kettner, L., Sanders, P.: STXXL: standard template library for XXL data sets. Softw. Pract. Exp. **38**(6), 589–638 (2008)
4. Kärkkäinen, J., Kempa, D., et al.: Faster sparse suffix sorting. In: LIPIcs-Leibniz International Proceedings in Informatics, vol. 25. Schloss Dagstuhl-Leibniz-Zentrum fuer Informatik (2014)
5. Manber, U., Myers, G.: Suffix arrays: a new method for on-line string searches. SIAM J. Comput. **22**(5), 935–948 (1993)
6. Manzini, G.: An analysis of the Burrows-Wheeler transform. J. ACM **48**(3), 407–430 (2001)
7. Navarro, G., Mäkinen, V.: Compressed full-text indexes. ACM Comput. Surv. **39**(1), 2 (2007)
8. Nong, G., Chan, W.H., Hu, S.Q., Wu, Y.: Induced sorting suffixes in external memory. ACM Trans. Inf. Syst. **33**(3), 12 (2015)
9. Nong, G., Zhang, S., Chan, W.H.: Two efficient algorithms for linear time suffix array construction. IEEE Trans. Comput. **60**(10), 1471–1484 (2011)
10. Weiner, P.: Linear pattern matching algorithms. In: IEEE Conference Record of 14th Annual Symposium on Switching and Automata Theory, 1973. SWAT'08, pp. 1–11. IEEE (1973)
11. Wu, Y., Nong, G., Chan, W.H., Han, L.B.: Checking big suffix and lcp arrays by probabilistic methods. IEEE Trans. Comput. **1**, 1 (2017)

H_∞ Filtering in Mobile Sensor Networks with Missing Measurements and Quantization Effects

Xueming Qian[1,2(✉)] and Baotong Cui[1,2]

[1] Key Laboratory of Advanced Process Control for Light Industry
(Ministry of Education), Jiangnan University, Wuxi 214122, China
xmqian81@gmail.com
[2] School of Internet of Things Engineering,
Jiangnan University, Wuxi 214122, People's Republic of China

Abstract. This paper is concerned with the H_∞ filtering problem for an array of 2D distributed parameter systems over lossy mobile sensor networks. The mobile sensor network suffers from missing measurements as well as quantization effects that are presented in a new framework. Bernoulli distribution is introduced to govern the data missing. A new H_∞ filtering technique is proposed for the addressed 2D semi-linear parabolic systems. Sufficient conditions are established in terms of some inequalities and the velocity law of each mobile sensor, such that the filtering error system is globally asymptotically stable in the mean square and has a guaranteed prescribed disturbance attenuation level γ for all nonzero noises. Finally, a numerical example is exploited to show the effectiveness of the proposed filtering scheme.

Keywords: Mobile sensor networks · H_∞ filtering · 2D distributed parameter systems · Data missing · Quantization effects

1 Introduction

A sensor network is typically consisted of a large number of sensing devices that can collect information and coordinate their implement tasks via wireless communication. The development of sensor networks was motivated by industrial automation, managing inventory, environmental monitoring and intelligent buildings [1]. To improve the flexibility of sensing devices, mobile sensor networks can be constructed by mobile robots. In [2], a decentralized output feedback control scheme and its dual problem have been developed for a linear parabolic system using mobile actuator-sensor networks. And the optimized criteria of mobile actuators and sensors such that the performance of spatial distributed system were enhanced. Similarly, a group of mobile sensors has been employed in the estimation problem of a spatial distributed system in [3]. Compared to the fixed sensors, the state estimator which designed by mobile sensors has the better performance as can be seen from [3]. The related works that the use of mobile sensors for the state estimation of air traffic flow modeled as a modified Lighthill-Whitham-Richards partial differential equation in [4] has been considered.

© Springer Nature Singapore Pte Ltd. 2017
G. Chen et al. (Eds.): PAAP 2017, CCIS 729, pp. 290–300, 2017.
DOI: 10.1007/978-981-10-6442-5_26

In practical engineering, missing measurement phenomenon usually occurs in networked environment owing to sensor temporal failure or network transmission delay. In [5], missing measurements which described by Bernoulli distribution was first presented to deal with the filtering problem, and has then been investigated in [6, 7] for various filtering problem of networked control systems with probabilistic missing measurements. Also, incomplete information of measured output from fixed sensors was assumed in the studies above. However, so far, missing measurements has little been taken into account in measured output which using mobile sensor networks.

As discussed above, H_∞ filtering in mobile sensor networks with missing measurements and quantization effects has never been tackled. This situation motivates us, in this paper, to solve the problem of H_∞ filtering for an array of 2D distributed parameter systems over lossy mobile sensor networks. We aim to estimate the state of 2D distributed parameter systems through available output with missing measurements and quantization effects such that the filtering error converges to zero asymptotically stable in the mean square and a specified H_∞ disturbance rejection attenuation level is guaranteed.

2 Problem Formulation

The target plant under consideration is described by the following an array of 2D distributed parameter systems.

$$
\begin{cases}
\frac{\partial x(t,\xi)}{\partial t} = \sum_{k=1}^{2} \frac{\partial}{\partial \xi_k}\left(a_k(\xi_k)\frac{\partial x(t,\xi)}{\partial \xi_k}\right) - \phi(t,\xi,x(t,\xi))x(t,\xi) \\
z(t) = \int\int_S b(\xi)x(t,\xi)d\xi
\end{cases}
\tag{1}
$$

where $\xi = [\xi_1,\xi_2]^T \in S \subset R^2$ is a 2D spatial vector and $t \in [0,+\infty)$ is the time variable. S is a compact set with smooth boundary ∂S and mes$S > 0$ in R^2. $x(t,\xi)$ denotes the state of the 2D system at time t and at point ξ. The diffusion operators $a_k(\xi_k) \geq \bar{a}_k > 0, k = 1$, Nonlinear function $\phi(t,\xi,x(t,\xi))$ satisfy $\phi_m \leq \phi(t,\xi,x(t,\xi)) \leq \phi_M$, where ϕ_m and ϕ_M are known bounds. $b(\xi)$ denotes the spatial distribution of the output.

The 2D distributed parameter system (1) subject to the initial condition

$$
x(0,\xi) = x_0(\xi)
\tag{2}
$$

and having Dirichlet boundary conditions

$$
x(t,L_0) = 0, \ x(t,L) = 0, \ \forall t \geq 0,
\tag{3}
$$

where $L_0 = [0,0]^T$ and $L = [l_1,l_2]^T$.

The measurement output from ith mobile sensor is given by

$$
y_i(t) = \alpha_i(t)\int\int_S c_i(\xi;\xi_i(t))x(t,\xi)d\xi + d_iw(t), \quad i = 1,2,\cdots,N
\tag{4}
$$

where $c_i(\xi; \xi_i(t)), i = 1, 2, \cdots, N$ are nonnegative bounded functions, and denote the spatial distributions of mobile sensors. Without loss of generality, the spatial distribution of each mobile sensor which centroid at $\xi_i(t)$, is given by

$$c_i(\xi; \xi_i(t)) = \begin{cases} \frac{1}{\varepsilon} & \text{if } \xi \in U(\xi_i, \varepsilon) \\ 0 & \text{otherwise}, \end{cases} \qquad (5)$$

where $U(\xi_i, \varepsilon) = \{\xi| \, |\xi - \xi_i| \le \varepsilon\}$ is a bounded closed region and denotes a ε_i neighborhood of ξ_i. The spatial point $\xi_i(t) = [\xi_{i1}(t), \xi_{i2}(t)]^T$ denotes the time-dependent position of ith mobile sensor in the spatial domain S. Also, $\xi_i(t)$ describe the time-varying trajectory of ith mobile sensor within $[0, T]$. For every i, the random variable $\alpha_i(t) \in \mathbf{R}, i = 1, 2, \cdots, N$ is a Bernoulli distributed white sequence, takes values of 1 and 0 with

$$\begin{cases} \text{Prob}\{\alpha_i(t) = 1\} = \bar{\alpha}_i \\ \text{Prob}\{\alpha_i(t) = 0\} = 1 - \bar{\alpha}_i, \end{cases} \qquad (6)$$

where $\bar{\alpha}_i \in [0, 1], i = 1, 2, \cdots, N$ are known positive constants, and denoting $\bar{\Lambda}_\alpha = \text{diag}(\bar{\alpha}_1, \bar{\alpha}_2, \cdots, \bar{\alpha}_N)$. It is assumed that the stochastic variables $\alpha_i(t)$ is independent of both $w(t)$ and the initial state of system. d_i is the ith intensity of the disturbance. $w(t)$ denotes the external disturbance, and satisfy the following assumptions

$$\int_0^{+\infty} w^T(t)w(t)dt < +\infty.$$

In addition, the quantization effects are considered in this paper. The quantizer $q(\cdot)$ is defined as

$$\bar{y}(t) = q(y(t)) = [q_1(y_1(t)), q_2(y_2(t)), \cdots, q_N(y_N(t))]^T$$

where $\bar{y}(t) \in \mathbf{R}^N$ is the signals which after quantization. Then, $\bar{y}(t)$ would be transmitted into the filter. Here, for each $q_i(\cdot)$, the set of quantization level is given by

$$\mathcal{U}_i = \left\{ \pm u_l^{(i)}, \ u_l^{(i)} = \rho_i^{(i)} u_0^{(i)}, \ l = 0, \pm1, \pm2, \cdots \right\} \cup \{0\}, \ 0 \le \rho_l \le 1, \ u_0^{(i)} > 0.$$

where $\rho^{(i)}$ is a given constant and denotes the quantization density. The logarithmic type quantizer $q(\cdot)$ is given by

$$q_i(y_i(t)) = \begin{cases} \chi_l^{(i)}, & \frac{1}{1+\delta_i} \chi_l^{(i)} < y_i(t) \le \frac{1}{1-\delta_i} \chi_l^{(i)} \\ 0, & y_i(t) = 0 \\ -q(-y_i(t)), & y_i(t) < 0 \end{cases}$$

where $\delta_i = (1 - \rho_i)/(1 + \rho_i)$. It follows from that $q(y_i(t)) = (1 + \Delta_i(t))y_i(t)$ such that $|\Delta_i(t)| \le \delta_i$. Denoting $\Delta(t) = \text{diag}(\Delta_1(t), \Delta_2(t), \cdots, \Delta_N(t))$, the measurements after quantization can be described as

$$\bar{y}(t) = (I + \Delta(t))y(t). \tag{7}$$

Accordingly, the quantizing effects have been transformed into bounded uncertainties [8].

To investigate the H_∞ filtering problem using Lyapunov direct method, it is need to rewritten the 2D distributed parameter system (1) in an abstract form.

Let $\mathcal{H}$ be a Hilbert space with inner product $\langle \cdot, \cdot \rangle$ and corresponding induced norm $|\cdot|$. Let $\mathcal{V}$ be a reflexive Banach space with norm denoted by $\| \cdot \|$, and assume that $\mathcal{V}$ is embedded densely and continuously in $\mathcal{H}$. Let $\mathcal{V}^*$ denote the conjugate dual of $\mathcal{V}$ with induced norm $\| \cdot \|_*$. It follows $\mathcal{V} \hookrightarrow \mathcal{H} \hookrightarrow \mathcal{V}^*$ with both embedding dense and continuously, and as a consequence we have

$$|\varphi| \leq b \| \varphi \|, \quad \varphi \in \mathcal{V}, \tag{8}$$

for some positive constant b [9].

The system operator $\mathcal{A} : \mathcal{V} \to \mathcal{V}^*$ is a linear operator, satisfying the following assumptions:

(A1) $\mathcal{A}$ is bounded, that is

$$|\langle \varphi, \mathcal{A}\psi \rangle| \leq \alpha_0 \| \varphi \| \| \psi \|,$$

for $\varphi, \psi \in \mathcal{V}$ and constant $\alpha_0 > 0$.

(A2) $-\mathcal{A}$ is coercive, that is

$$\langle \varphi, -\mathcal{A}\varphi \rangle \geq \beta_0 \| \varphi \|^2,$$

for $\varphi \in \mathcal{V}$ and constant $\beta_0 > 0$.

The operator $\mathcal{B} : \mathbf{R}^2 \to \mathcal{V}^*$ is similarly provided by

$$\langle \mathcal{B}\varphi, \psi \rangle = \int\int_S b(\xi)\varphi(\xi)\psi(\xi)d\xi,$$

having assumption as follows:

(A3) $\mathcal{B}$ is bounded, that is

$$\langle \varphi, \mathcal{B}\varphi \rangle \leq \sigma_b \langle \varphi, \varphi \rangle.$$

Hence, the 2D distributed parameter system ([model]) can be expressed in the following abstract form:

$$\begin{cases} \dot{x}(t) = \mathcal{A}x(t) \\ z(t) = \mathcal{B}x(t) \end{cases} \tag{9}$$

In this case, the state space is $\mathcal{H} = L_2(\Omega)$, the infinitesimal operator $\mathcal{A}$ generates a strongly continuous semigroup $T(t), t \geq 0$ and the domain $\mathcal{D}(\mathcal{A})$ of the operator $\mathcal{A}$ is dense in $\mathcal{H}$.

Similarly, the measurement output (4) can be expressed as

$$y(t) = \Lambda_\alpha(t)\mathcal{C}(\xi(t))x(t) + \mathcal{D}w(t) \tag{10}$$

where $y(t) = [y_1(t), y_2(t), \cdots, y_N(t)]^T$, $\Lambda_\alpha(t) = \text{diag}[\alpha_1(t), \alpha_2(t), \cdots, \alpha_N(t)]$. The output operator $\mathcal{C}(\xi(t))$ is given by

$$\langle \mathcal{C}(\xi(t))\varphi, \psi \rangle = \begin{bmatrix} \iint\limits_S c_1(\xi; \xi_1(t))\varphi(\xi)\psi(\xi)d\xi \\ \iint\limits_S c_2(\xi; \xi_2(t))\varphi(\xi)\psi(\xi)d\xi \\ \vdots \\ \iint\limits_S c_N(\xi; \xi_N(t))\varphi(\xi)\psi(\xi)d\xi \end{bmatrix},$$

where the vector of sensor spatial location parameterized by $\xi(t) = [\xi_1(t), \xi_2(t), \cdots, \xi_N(t)]^T$. Indeed, since $c_i(\xi; \xi_i(t))(i = 1, 2, \cdots, N)$ is nonnegative and bounded, then $\mathcal{C}(\xi(t)) : \mathcal{V} \to \underbrace{R^2 \times R^2 \times \cdots \times R^2}_{N}$ satisfying the following assumption:

(A4) $\mathcal{C}(\xi^s(t))$ is bounded, that is

$$\langle \varphi, \mathcal{C}(\xi^s(t))\varphi \rangle \leq \sigma_c \langle \varphi, \varphi \rangle.$$

Additionally, the disturbance operator $\mathcal{D} = [d_1, d_2, \cdots, d_N]^T$ and satisfying the following assumption:
(A5) $\mathcal{D}$ is bounded, that is

$$\langle \varphi, \mathcal{D}\varphi \rangle \leq d \langle \varphi, \varphi \rangle.$$

Therefore, the measurement output after quantization is given by

$$\bar{y}(t) = (I + \Delta(t))\Lambda_\alpha(t)\mathcal{C}(\xi(t))x(t) + \mathcal{D}w(t) \tag{11}$$

In this paper, we focus on H_∞ filtering in mobile sensor networks with missing measurements and quantization effects. Here, the following H_∞ filter structure is adopted for system (9):

$$\begin{cases} \dot{\hat{x}}(t) = \mathcal{A}\hat{x}(t) + \mathcal{C}^*(\xi(t))\Gamma[\bar{y}(t) - (I + \bar{\Delta})\bar{\Lambda}_\alpha \mathcal{C}(\xi(t))\hat{x}(t)] \\ \hat{z}(t) = \mathcal{B}\hat{x}(t) \end{cases}. \tag{12}$$

where $\hat{x}(t)$ is the state estimate and $\hat{z}(t)$ is the estimated output. $\Gamma = \Gamma^T > 0$ is the observer gains. Moreover, $\hat{x}(0) = \hat{x}_0 \neq x(0)$. $\bar{\Delta} = \text{diag}(\delta_1, \delta_2, \cdots, \delta_N)$.

Letting $e(t) = x(t) - \hat{x}(t)$ and $\tilde{z}(t) = z(t) - \hat{z}(t)$, the dynamics of the filtering error system can be obtained from (9), (11) and (12) in the following:

$$e(t) = A_c(\xi(t))e(t) - C^*(\xi(t))\Gamma[(I + \Delta(t))\Lambda_\alpha - (I + \Delta)\bar{\Lambda}_\alpha]C(\xi(t))x(t) + Dw(t)z(t)$$
$$= Be(t)$$

$$(13)$$

where $A_c(\xi(t)) = A - C^*(\xi(t))\Gamma(I + \bar{\Delta})\bar{\Lambda}_\alpha C(\xi(t))$, along with $e(0) = x(0) - \hat{x}(0) \neq 0$.

3 Main Results and Proofs

Our objective of this paper is to design a H_∞ filter in the form (12) of such that filtering error output $\tilde{z}(t)$ satisfies the H_∞ performance constraint, namely

$$|\tilde{z}(t)|^2 \leq \gamma^2 |w(t)|^2 \qquad (14)$$

for the given disturbance attenuation level $\gamma > 0$.

3.1 Stability Analysis of Filtering Error Systems

Theorem 1. Under the filter (12) and the disturbance attenuation level $\gamma > 0$ be given, the zero solution of filtering error system (13) with $w(t) = 0$ is globally asymptotically stable in the mean square, if there exit two positive constants p and q such that the following inequalities:

$$q \geq \frac{1}{2}, \quad p \geq \frac{\gamma^2(1 + \delta_i)^2 \bar{\alpha}_i(1 - \bar{\alpha})\sigma_c^4 Nb^2}{4a_0}, \qquad (15)$$

and the velocity law of mobile sensor as follow:

$$\dot{\xi}_{i1}(t) = -\mu_{i1}\gamma_i(1 + \delta_i)\bar{\alpha}_i\Xi_1(t), \qquad (16)$$

$$\dot{\xi}_{i2}(t) = -\mu_{i2}\gamma_i(1 + \delta_i)\bar{\alpha}_i\Xi_2(t), \qquad (17)$$

hold, where

$$\Xi_1(t) = \int_{\xi_{i1}-\varepsilon}^{\xi_{i1}+\varepsilon} e^2\left(t, \xi_1, \xi_{i2} - \sqrt{\varepsilon^2 - (\xi_1 - \xi_{i1})^2}\right)$$
$$- e^2\left(t, \xi_1, \xi_{i2} + \sqrt{\varepsilon^2 - (\xi_1 - \xi_{i1})^2}\right)d\xi_1$$

$$\Xi_2(t) = \int_{\xi_{i2}-\varepsilon}^{\xi_{i2}+\varepsilon} e^2\left(t, \xi_{i1} - \sqrt{\varepsilon^2 - (\xi_2 - \xi_{i2})^2}, \xi_2\right)$$
$$- e^2\left(t, \xi_{i1} + \sqrt{\varepsilon^2 - (\xi_2 - \xi_{i2})^2}, \xi_2\right)d\xi_2$$

with $\mu_{i1} > 0$ and $\mu_{i2} > 0$ are velocity gain of each mobile sensor. The guidance strategy for mobile sensors enhances the filter performance in the sense that the filtering error $e(t)$ converges to zero faster.

Proof: It can be inferred from (A1) and (A2) that closed-loop operator $\mathcal{A}_c(\xi(t))$ is bounded and coercive.

Consider the following parameter-dependent Lyapunov functional

$$V(t) = -\langle e(t), \mathcal{A}_c(\xi^s(t))e(t)\rangle + \langle x(t), px(t)\rangle. \tag{18}$$

The infinitesimal operator $\mathcal{L}$ along the solution of the filtering error system (13), we obtain that

$$
\begin{aligned}
\mathcal{L}V(t) = & -\mathbb{E}\langle \dot{e}(t), \mathcal{A}_c(\xi(t))e(t)\rangle - \mathbb{E}\langle e(t), \mathcal{A}_c(\xi(t))\dot{e}(t)\rangle \\
& - \mathbb{E}\left\langle e(t), \frac{d\mathcal{A}_c(\xi(t))}{dt}e(t)\right\rangle + \langle \dot{x}(t), px(t)\rangle + \langle x(t), p\dot{x}(t)\rangle
\end{aligned}
\tag{19}
$$

By considering the facts of $\mathbb{E}\{\alpha_i(t) - \bar{\alpha}_i\} = 0$, $\mathbb{E}\{(\alpha_i(t) - \bar{\alpha}_i)^2\} = \bar{\alpha}_i(1 - \bar{\alpha}_i)$ and $\mathbb{E}\{(\alpha_i(t) - \bar{\alpha}_i)(\alpha_j(t) - \bar{\alpha}_j)\} = 0 (i \neq j)$. Also, noticing the fact that the operator $\mathcal{A}_c(\xi_i^s(t))$ is self-adjoint, the following result deduced easily.

$$
\begin{aligned}
-\mathbb{E}\langle \dot{e}(t), & \mathcal{A}_c(\xi(t))e(t)\rangle - \mathbb{E}\langle e(t), \mathcal{A}_c(\xi(t))\dot{e}(t)\rangle \\
& \leq (-2 + q^{-1})\mathbb{E}\langle \mathcal{A}_c(\xi(t))e(t), \mathcal{A}_c(\xi(t))e(t)\rangle \\
& + q\sum_{i=1}^{N}\gamma^2(1 + \delta_i)^2\bar{\alpha}_i(1 - \bar{\alpha})\sigma_c^4|x(t)|^2
\end{aligned}
\tag{20}
$$

Also, we obtain

$$
\begin{aligned}
& -\mathbb{E}\langle e(t), \frac{d\mathcal{A}_c(\xi(t))}{dt}e(t)\rangle \\
& = 2\sum_{i=1}^{N}\gamma_i(1 + \delta_i)\bar{\alpha}_i \iint_S c_i(\xi; \xi_i(t))\dot{\xi}_i(t)\frac{dc_i(\xi; \xi_i(t))}{d\xi}e^2(t, \xi)d\xi \\
& = \frac{2}{\varepsilon^2}\sum_{i=1}^{N}\gamma_i(1 + \delta_i)\bar{\alpha}_i\left(\dot{\xi}_{i1}(t)\Xi_1(t) + \dot{\xi}_{i2}\Xi_2(t)\right)
\end{aligned}
\tag{21}
$$

where $\Xi_1(t)$ and $\Xi_2(t)$ are defined in Theorem 1.

The choice

$$\dot{\xi}_{i1}(t) = -\mu_{i1}\gamma_i(1 + \delta_i)\bar{\alpha}_i\Xi_1(t), \tag{22}$$

$$\dot{\xi}_{i2}(t) = -\mu_{i2}\gamma_i(1 + \delta_i)\bar{\alpha}_i\Xi_2(t), \tag{23}$$

deduces (21) negative definite. Notice that $\dot{\xi}_i(t) = [\dot{\xi}_{i1}(t), \dot{\xi}_{i2}(t)]^T$ is the velocity of each mobile sensor.

$$\langle \dot{x}(t), px(t) \rangle + \langle x(t), p\dot{x}(t) \rangle \\ \leq - 2pa_0 \parallel x(t) \parallel^2 \leq -\tfrac{2pa_0}{b^2} |x(t)|^2. \tag{24}$$

Substituting (20)–(24) into (19), we obtain

$$\mathcal{L}V(t) = (-2 + q^{-1})\mathbb{E}|\mathcal{A}_c(\xi(t))e(t)|^2 \\ + \sum_{i=1}^{N}\left(q\gamma^2(1+\delta_i)^2\bar{\alpha}_i(1-\bar{\alpha})\sigma_c^4 - \frac{2pa_0}{Nb^2}\right)|x(t)|^2 \\ - \frac{2}{\varepsilon^2}\sum_{i=1}^{N}\gamma_i^2\left(1+\delta_i)^2\bar{\alpha}_i^2(\mu_{i1}\Xi_1^2(t) + \mu_{i2}\Xi_2^2(t)\right) \tag{25}$$

The feasibility of the inequalities (25) implies $\mathcal{L}V(t) \leq 0$.

Therefore, it follows readily from (25) and the embedding, that

$$\mathcal{L}V(t) \leq -\beta\mathbb{E}|e(t)|^2, \tag{26}$$

where $\beta > 0$ is a function of the embedding constant b and coercivity constant β_0. In view of Itô formula and inequality (26), we have

$$\mathbb{E}V(t) \leq \mathbb{E}V(0) - \beta\int_0^t \mathbb{E}|e(t)|^2 ds \tag{27}$$

which implies that

$$\int_0^t \mathbb{E}|e(t)|^2 ds \leq \frac{1}{\beta}V(0).$$

In addition, it can verify that $\mathbb{E}|e(t)|^2$ is uniformly continuous on $[0, +\infty)$. Therefore, from Barbalat's Lemma [10], it follows that

$$\lim_{t \to +\infty} \mathbb{E}|e(t)|^2 = 0.$$

Therefore, the filtering error system (13) with $w(t) = 0$ is globally asymptotically stable in the mean square, which completes the proof.

4 H_∞ Performance Analysis

Theorem 2. Let the filter gain Γ, and the disturbance attenuation level $\gamma > 0$ be given. Then, the zero solution of filtering error system (13) with $w(t) = 0$ is globally asymptotically stable in the mean square, and $\tilde{z}$ satisfies the H_∞ performance constraint (14) under the zero initial condition for all nonzero $w(t)$, if under the Assumption (A1)–(A5) such that the following matrix inequality holds:

$$\Psi = \begin{bmatrix} -\beta + \sigma_b^2 & d\varrho_0 \\ d\varrho_0 & -\gamma^2 \end{bmatrix} < 0.$$

Proof: According to Theorem 1, the filtering error system (13) is globally asymptotically stable in the mean square. Now, let us focus our attention on the H_∞ performance of the filtering error system. Construct the same Lyapunov functional candidate $V(t)$ as in Theorem 1. The similar line calculation as in Theorem 1 leads to

$$\mathcal{L}V(t) \leq \mathbb{E}\langle \eta(t), \tilde{\Psi}\eta(t) \rangle \tag{28}$$

where $\eta(t) = [e(t), w(t)]^T$ and $\tilde{\Psi} = [array * 20c - \beta d\varrho_0 d\varrho_0 0]$.
Under the zero initial condition, for all non-zero external disturbance $w(t)$, we have

$$\begin{aligned} J &= E \int_0^{T_f} |\tilde{z}_i(t)|^2 - \gamma^2 |w(t)|^2 dt \\ &\leq \int_0^{T_f} E\langle \eta(t), \Psi\eta(t) \rangle dt \end{aligned} \tag{29}$$

where Ψ is defined in Theorem 2.

Along the same line as in the proof of Theorem 1, we can derive that $J < 0$. Letting $T_f \to \infty$, we obtain that

$$|\tilde{z}_i(t)|^2 < \gamma^2 |w(t)|^2$$

which completes the proof of the Theorem 2.

5   A Numerical Example

Consider an array of 2D spatially distributed processes (1) with Dirichlet boundary conditions, having initial condition $x(0, \xi) = \sin(2\pi\xi)e^{-6\xi^2}, \xi = [0, 1] \times [0, 1]$. The diffusion operator is $a_0 = 0.006$. Bounded nonlinear function $\phi(x(t, \xi)) = 2.6\sin(0.6x(t, \xi))$. Three mobile sensors are considered in this system, and their initial positions are chosen as $(\xi_{11}(0), \xi_{12}(0)) = (0.15, 0.28)$, $(\xi_{21}(0), \xi_{22}(0)) = (0.50, 0.68)$ and $(\xi_{31}(0), \xi_{32}(0)) = (0.75, 0.88)$. The spatial distribution of each mobile sensor which centroid at $(\xi_{i1}(t), \xi_{i2}(t))$, given by

$$c(\xi; \xi_i) = \begin{cases} \frac{1}{0.03} & \text{if } \xi \in U(\xi_i, 0.03) \\ 0 & \text{otherwise.} \end{cases}$$

The probabilities are taken as $\bar{\alpha}_1 = 0.9, \bar{\alpha}_2 = 0.85$ and $\bar{\alpha}_3 = 0.8$. For the H_∞ filter, initial conditions are chosen as $\hat{x}_{11}(0, \xi_1) = \hat{x}_{21}(0, \xi_1) = \hat{x}_{31}(0, \xi_1) = \hat{x}_{12}(0, \xi_2) = \hat{x}_{22}(0, \xi_2) = \hat{x}_{32}(0, \xi_2) = 0$. The filter gains are given by $\gamma_1 = 80, \gamma_2 = 85$ and $\gamma_3 = 90$. The output estimation errors of filter i, $i = 1, 2, 3$ in mobile sensors are presented in Fig. 1.

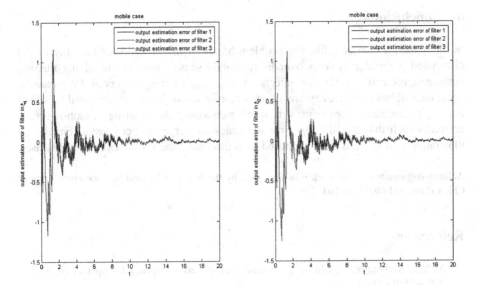

Fig. 1. The output estimation error of filter i, $i = 1, 2, 3$.

As a comparison, we take into account three fixed-in-space sensors which fixed at $(\xi_{11}, \xi_{12}) = (0.15, 0.28)$, $(\xi_{21}, \xi_{22}) = (0.50, 0.68)$ and $(\xi_{31}, \xi_{32}) = (0.75, 0.88)$. The trajectories of three sensors for the fixed and mobile cases are depicted in Fig. 2.

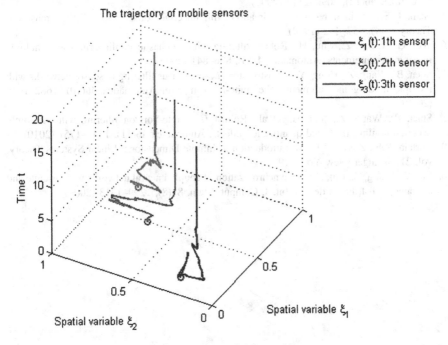

Fig. 2. The trajectories of three mobile sensors.

6 Conclusions

In this paper, a new H_∞ filtering problem has been investigated for an array of 2D distributed parameter systems over lossy mobile sensor networks involving missing measurements and quantization effects. A new H_∞ filtering technique by means of some inequalities and velocity law of each mobile sensor has been presented to satisfy the H_∞ performance constraint. It is worth mentioning that filtering in mobile sensor networks such that the filtering error dynamics converges to zero faster. Finally, an illustrative example has been proposed to demonstrate the usefulness of the results.

Acknowledgements. This work was supported by the National Natural Science Foundation of China (Nos. 61174021, 61104155).

References

1. Yick, J., Mukherjee, B., Ghosal, D.: Wireless sensor network survey. Comput. Netw. **52**(12), 2292–2330 (2008)
2. Demetriou, M.A.: Hussein. I.I.: Estimation of spatially distributed processes using mobile spatially distributed sensor network. SIAM J. Control Optim. **48**(1), 266–291 (2009)
3. Demetriou, M.A.: Guidance of mobile actuator-plus-sensor networks for improved control and estimation of distributed parameter systems. IEEE Trans. Autom. Control **55**(7), 1570–1584 (2009)
4. Work, D.B., Bayen, A.M.: Convex formulations of air traffic flow optimization problems. Proc. IEEE **96**(12), 2096–2112 (2008)
5. Nahi, N.E.: Optimal recursive estimation with uncertain observation. IEEE Trans. Inf. Theory **15**(4), 457–462 (1969)
6. Wei, G., Wang, Z., Shu, H.: Robust filtering with stochastic nonlinearities and multiple missing measurements. Automatica **45**(3), 836–841 (2009)
7. Shen, B., Wang, Z., Hung, Y.S.: Distributed H_∞-consensus filtering in sensor networks with multiple missing measurements: the finite-horizon case. Automatica **46**(10), 1682–1688 (2010)
8. Shen, B., Wang, Z., Shu, H., et al.: Robust H_∞ finite-horizon filtering with randomly occurred nonlinearities and quantization effects. Automatica **46**(11), 1743–1751 (2010)
9. Curtain, R.F., Zwart, H.J.: An Introduction to Infinite Dimensional Linear Systems Theory, vol. 21. Springer, New York (1995)
10. Liu, Y., Wang, Z., Liu, X.: On synchronization of coupled neural networks with discrete and unbounded distributed delays. Int. J. Comput. Math. **85**(8), 1299–1313 (2008)

A Cost-Effective Wide-Sense Nonblocking k-Fold Multicast Network

Gang Liu, Qiuming Luo$^{(\boxtimes)}$, Cunhuang Ye, and Rui Mao

Guangdong Key Laboratory of Popular High Performance Computers,
College of Computer Science and Software Engineering, Shenzhen University,
Shenzhen 518060, People's Republic of China
{gliu,lqm}@szu.edu.cn

Abstract. Multicast is one of the most dense communication patterns. Any destination node of a k-fold multicast network can be involved in up to k simultaneous multicast connection. The hardware cost of traditional k-fold switching network for wide-sense nonblocking multicast is typically very high. In this paper, we propose a new wide-sense nonblocking k-fold multicast network and multicast routing algorithm. The k-fold design has significantly lower network cost than that of k copies of 1-fold multicast networks. The time complexity of the corresponding routing algorithm is no higher than that of previous works.

Keywords: Multistage interconnection network (MIN) · Multicast · Wide-sense nonblocking · k-Fold network

1 Introduction

Multicast communication involves simultaneous transmitting data from a single source to different destinations, and is one of the most dense communication patterns. Multicast is highly demanded in various interactive applications, such as multimedia, teleconferencing, web servers and electronic commerce on the Internet, as well as communication-intensive applications in parallel and distributed systems, such as distributed database updates and cache coherence protocols [1]. Most of these applications also demands foreseeable communications performance, such as guaranteed multicast latency and bandwidth, which is known as quality of services (QoS) along with multicast capability.

Multistage interconnection network (MIN) is a promising solution to construct a large dimension multicast switch fabric by cascading small unicast and multicast switches. There has been much work in the literature on MIN with multicast capability [1–18]. The multistage multicast network can be categorized as strictly nonblocking (SNB), wide-sense nonblocking (WSNB) and rearrangeable nonblocking (RNB). SNB network guarantees there is always possible to establish an available path for a new connection request, independent of existing connections and path routing algorithms. RNB network guarantees there is always possible to establish a new connection by rearranging the paths existing connections. In WSNB network, it is always possible to set up a new connection by suitably choosing from the free links under certain

© Springer Nature Singapore Pte Ltd. 2017
G. Chen et al. (Eds.): PAAP 2017, CCIS 729, pp. 301–310, 2017.
DOI: 10.1007/978-981-10-6442-5_27

strategies without affecting any existing connection. SNB generally has prohibitively high hardware cost and RNB may disrupt existing communication which can cause quite long latency and packet out-of-order problem. WSNB is a trade-off for non-blocking capability and network hardware Cost between WSNB networks and RNB networks. Wide-sense nonblocking multicast switching network have received more and more attention [4–7].

Nonblocking multicast capability of MIN have been extensively studied under different types of nonblocking capabilities, most of which focus on the Clos network [8]. Clos network is one type of traditional three-stage interconnection network with good flexibility, realizing permutation and multicast with different hardware costs. A three-stage $N \times N$ Clos network denoted as $v(n, m, r)$ has r switch modules of size $n \times m$ in input-stage, m switch modules of size $r \times r$ in middle-stage, and r switch modules of size $m \times n$ in output-stage. The parameters n and r vary only in a small range, so the number of middle-stage switches m becomes the key factor in deciding the whole cost. When $m \geq n$, Clos network can realize rearrangeable permutation [9]. When $m \geq \lfloor (2 - 1/F_{2r-1})n \rfloor$ (F_{2r-1} is the Fibonacci number), Clos network can realize wide-sense nonblocking permutation by using packing Strategy [10]. Masson and Jordan [11] and Hwang [12] gave the sufficient conditions on strictly nonblocking and rearrangeable nonblocking multicast Clos networks. The necessary and sufficient condition for wide nonblocking multicast in Clos network is that the number of middle-stage switches m satisfies $m \geq 3(n - 1) \log r / \log \log r$ [13, 14].

In real world, overlapping among destination of different multicast connections may be possible. Yang proposed a wide-sense nonblocking k-fold multicast network in which each destination can be involved in up to k different multicast connections at a time [15]. In a k-fold multicast network, fold number indicates the number of request coming from different sources to a particular destination.

If the number of middle stage switches satisfies $m \geq n(2 + k)\frac{\log r}{\log \log r}$, the network is nonblocking for any k-fold multicast assignments. The design in [13, 14] can be consider as a special case of the new design when k = 1.

The hardware cost of multistage interconnection network is referred to as the number of cross-points. The hardware cost $O(N^{3/2}\log r / \log \log r)$ [13, 14] for Clos network to realize wide-sense nonblocking multicast is much higher than $O(N^{3/2})$ for only nonblocking permutation. The hardware cost of the k-fold nonblocking multicast network is $O(kN^{3/2}\log r / \log \log r)$, which is also very high [15]. As it can be seen, there is a considerable gap between multicast network and permutation network.

Yang and Wang [16] proposed a wide-sense nonblocking four-stage multicast network. The hardware cost of the network is fully analyzed as $16.114N^{3/2}$ [17]. Hasan [18] evaluated a model for fine tuning the value of a k in a k-fold multicast network under different traffic loads under Poisson traffic with finite queue at each node. In this paper, we will extend the switching network to a new wide-sense nonblocking k-fold multicast network named k-PMPM which can simultaneously support all possible k-fold multicast connections. The hardware cost of the design is $H_k \approx (2k + 6\sqrt{k} + 4)N^{3/2}$, which is significantly lower than that of k copies of 1-fold multicast switching networks. Compared to previous works, the hardware cost of the new design can also be decreased significantly. An efficient routing algorithm will also be proposed, the time complexity of the routing algorithm is no higher than previous works.

The rest of this paper is organized as follows. In Sect. 2, the structure and definitions of k-PMPM network is presented. Section 3 presents the multicast routing algorithm and analyzes the time complexity. In Sect. 4, the sufficient condition for realizing wide-sense nonblocking multicast assignments in k-PMPM using the proposed routing algorithm will be given and proved. In Sect. 5 we analyze the hardware cost and compare both hardware and routing time complexity with that of other networks. Section 6 concludes the paper.

2 Construction of k-Fold Multicast Network

In this section, we consider a four-stage interconnection network which can realize any k-fold multicast assignments.

As shown in Fig. 1, the four-stage $N \times N$ Clos type network (with N inputs and N outputs) consists of four stages: r switch modules of size $n \times m_1$ in the first stage (or the input stage), m_1 switch modules of size $r \times m_2$ in the second stage, m_2 switch modules of size $m_1 \times r$ in the third stage, and r switch modules of size $m_2 \times n$ in the fourth stage (or the output stage), where $N = n \times r$. There is exactly one link between every two switches in two consecutive stages. This k-fold multicast network is denoted as k-PMPM (k, n, m_1, m_2, r).

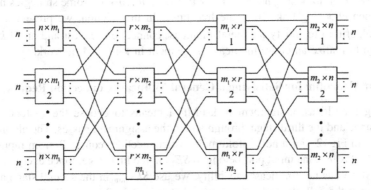

Fig. 1. Four-stage multicast network k-PMPM (k, n, m_1, m_2, r)

Let $SE_{i,j}$ denotes the jth switch module in the ith stage. Each switch in the second stage and the fourth stage is required to have multicast capability (fan-out capability), while those switches in the first stage and the third stage are not. In addition, each switch in the fourth stage is assumed to also have fan-in capability which accommodates k-fold connections at each network output. The two implementations, such as dedicated sublinks and input/output buffers, was illustrated in [15].

In this multicast network, each destination port can be involved in up to k different multicast connections. Since every switch in the fourth stage has multicast capability, we can use the set of the third stage switches in where the assignment's destination ports are located to denote a multicast assignment's destination, instead of a set of the destination port.

Let $R_i = \{x_1, x_2, \ldots, x_j\}$ $(1 \leq i \leq nr, \ 1 \leq j \leq r)$ denotes a multicast connection request from input port i, and $|R_i|$ is referred to as the fan-out of the multicast connection. Since every switch in the third stage has multicast capability, R_i is actually the subset of the switch modules in the output stage. For the ith switch in the second stage, the sum of the fan-out of all multicast connections that use this switch is referred to as FOL_i (fan-out load).

Let $O(1, i) = a_{m_1} a_{m_1-1} \ldots a_2 a_1$ denote the status of the output links of the ith switch in the first stage: $a_k = 1$ indicates the link between $SE_{1,i}$ and $SE_{2,k}$ is idle, while $a_k = 0$ indicates the link is busy. Let $O(2, i) = a_{m_2} a_{m_2-1} \ldots a_2 a_1$ denote the status of the output links of the ith switch in the second stage: $a_k = 1$ indicates the link between $SE_{2,i}$ and $SE_{3,k}$ is idle, while $a_k = 0$ indicates the link is busy. Let $I(4, i) = b_{m_2} b_{m_2-1} \ldots b_2 b_1$ denote the status of the input links of the ith switch in the fourth stage: $b_k = 1$ indicates the link between $SE_{4,i}$ and $SE_{3,k}$ is idle, while $b_k = 0$ indicates the link is busy.

Then, a k-fold multicast assignment in such a network is a set of multicast connections between network inputs and network outputs and can be characterized by a set.

3 Routing Algorithm for a k-Fold Multicast Network

In wide-sense nonblocking network k-PMPM (k, n, m_1, m_2, r), some strategies must be taken when choosing links from the free links. In this section, we present a $O(kN)$ routing algorithm to satisfy a general multicast connection request and an $O(k^2 N^{3/2})$ algorithm to realize an entire multicast assignment in k-PMPM network.

3.1 An Algorithm for Satisfying a General Multicast Connection Request

Realizing a multicast assignment in k-PMPM means to choose the switches in the second stage and the third stage through which the assignment passes. The algorithm is described in Fig. 2. For a new k-fold multicast connection request R_i from input port i with fan-out is $|R_i|$, the input port belong to $SE_{1,j}$ in the first stage. Since every switch in the first stage has not multicast capability, we use $SE_{2,\min}$ in the second stage and $S_{3,i}$ referred to as the collection of the chosen switches in stage 3 to describe this connection path. There are four steps in this algorithm:

Step 1: Make sure the new connection request R_i is an illegal request.

Step 2: Select one switch $SE_{2,\min}$ in the second stage, of which the $FOL_c (1 \leq c \leq m_1)$ is the smallest of all available switches in the second stage.

Step 3: Since the switches in the third stage are not required to have multicast capability, select $|R_i|$ available switches in the third stage.

Step 4: Connect R_i through the input stage switch $SE_{1,j}$, $SE_{2,\min}$ in the second stage, $|R_i|$ switches in $S_{3,i}$, and $|R_i|$ switches in the fourth stage.

MRA_PMPM (Multicast Routing Algorithm in k-PMPM (k, n, m_1, m_2, r))

Input: A new connection request $R_i = \{x_1, x_2, ..., x_f\}$ with fan-out is $|R_i|$, which belongs to $SE_{1,j}$

Output: $SE_{2,\min}$ and $S_{3,i}$ (The set of selected switch elements in stage3)

Begin

Step1: if any destination in R_i is currently involved in more than k multicast connection

then exit (R_i is an illegal request) ;

otherwisegotoStep2;

Step2: $min\_fol=1$;

for($c=1; c \le m_1; c++$)

if($O(1,j)_c = 1 \,\&\&\, FOL_c < FOL_{\min\_fol}$) then $min\_fol = c$;

$SE_{2,\min} = min\_fol$;

$O(1,j)_{min\_fol} = 0$; //The link is marked as busy

Step3: for($c=1; c \le |R_i|; c++$) {//select$|R_i|$ available switches in stage3

$d = 1$;

while ($O(2, min\_fol)_d \,\&\&\, I(4, x_c)_d = false$)$d++$;

$S_{3,i} = S_{3,i} \cup \{d\}$;

$O(2, min\_fol)_d = I(4, x_c)_d = 0$; //The link is marked as busy

}

Step4: connect R_i through the switch $SE_{1,j}$, $SE_{2,\min}$, $|R_i|$ switches in $S_{3,i}$, and $|R_i|$ switches in the fourth stage.

End.

Fig. 2. The description of multicast routing algorithm in k-PMPM (k, n, m_1, m_2, r)

The time complexity is analyzed as follows. It is oblivious that the time complexity of step 2 is $T_{step1} = m_1$; In step 3 of the algorithm, for each destination of the request, the time complexity is no more than m_2, so the time complexity of step 3 is $T_{step2} \le m_2 f_i$. Then the total time complexity is $T = T_{step1} + T_{step2} = m_1 + m_2 f_i \le m_1 + m_2 r$. It will be show in Sects. 4 and 5 that $m_1 = O(\sqrt{N})$, $m_2 = O(k\sqrt{N})$, so the time complexity of our algorithm for realizing a general multicast assignment is $O(kN)$.

3.2 An Algorithm for Realizing an Entire Multicast Assignment

If we proceed to route the connection request until every output port has been connected to some input port, then the total time will be $O(k^2 N^{3/2})$. To prove this, note that there are at most N connection request and the sum of the fan-out of all connection request is no more than kN. The time to satisfy connection request R_i is $m_1 + m_2 f_i$, thus the total time to satisfy all the connection requests is:

$$\sum_{i=1}^{N}(m_1 + m_1f_i) = Nm_1 + m_2\sum_{i=1}^{N}f_i = Nm_1 + kNm_2$$

It will be show in Sects. 4 and 5 that $m_1 = O(\sqrt{N})$, $m_2 = O(k\sqrt{N})$, so the time complexity of our algorithm for realizing the entire multicast assignment is $O(k^2 N^{3/2})$.

4 Sufficient Condition for Wide-Sense Nonblocking Multicast

In this section, the sufficient condition for realizing wide-sense nonblocking multicast assignments in k-PMPM under the routing algorithm will be given and proved.

Theorem 1. If $m_1 = n - 1 + x$, $x > 0$, then $FOL_i \leq (kN - n)/x$, $1 \leq i \leq m_1$.

Proof: We use reduction to absurdity to prove this theorem. Suppose that there is one switch in stage 2, say, $SE_{2,k}$ whose fan-out load $FOL_{min} > (kN - n)/x$. Then let's consider the last connection request that select $SE_{2,k}$: if this connection request is R_i, the fan-out of the request is f_i and the input-port i belongs to the switch element j of stage1. Because there are n input-ports in $SE_{1,j}$, there are at least x available switch elements in the second stage for R_i. The routing algorithm MRA-PMPM always select a switch elements in the second stage, of which the FOL_{min} is the smallest of all available switches in the second stage. ($1 \leq k \leq m_1$). So after R_i is routed, there are still at least $x - 1$ switch elements whose $FOL_{min} > (N - n)/x$ besides $SE(2, min)$, as we suppose $FOL_l > (kN - n)/x$, then one can derive that $\sum_{k=1}^{m_1} FOL_k > x \cdot (N - n)/x + n - 1 + |I_i| = kN + (|I_i| - 1) \geq kN$. This contradicts with $\sum_{k=1}^{m_1} FOL_k \leq kN$. Therefore, the supposition is false, and the theorem is proved.

Theorem 2. In step 3 of the routing algorithm MRA_PMPM, for any one destination of the request, if there are still r available switch in the third stage, a switch elements of stage3 can be selected to satisfy the destination without any blocking.

Proof: Because there are r output-ports in each switch of the fourth stage, any switch in the fourth stage can be involved up to r connection requests. Before the destination is satisfied, if there are still r available switch in the third stage, each of them has a link connected to the destination switch element. There is at least one link that is idle, then the switches in the third stage can be selected to satisfy the destination without any blocking.

Theorem 3. If $m_1 \geq n - 1 + x$ and $m_2 \geq (kN - n)/x + r$, $x > 0$, then k-PMPM (k, n, m_1, m_2, r) is wide-sense nonblocking for multicast by using the routing algorithm MRA_PMPM.

Proof: We consider any multicast connection request, say R_i, the fan-out of the request is f_i and input-port i belong to the switch elements j of the first stage. Because there are n input-ports in $SE_{1,j}$ and $m_1 \geq n - 1 + x$, there are at least x available switch elements in the second stage for R_i. The step 2 in the routing algorithm MRA_PMPM can always be performed without any blocking or any rearrangement. Suppose that $SE_{2,a}$ is selected in step 1 of the routing algorithm. For the step 2, we have $FOL_{min} \leq (N - n)/x + r$. Because the switches in the third stage have not multicast capability, for the last destination to be routed of R_i, there are at most $(N - r - (n - 1))/x + r - 1$ switches in the third stage that are not available for $SE_{2,k}$. Because $m_2 \geq (kN - n - r + 1)/x + r + n - 1$, there are at least n available switches the third stage for $SE_{2,k}$. From theorem 2, we derive that step 3 of the routing algorithm can be performed without any blocking. So using the routing algorithm MRA_PMPM, the k-PMPM is wide-sense nonblocking for k-fold multicast.

5 Analysis and Comparison

In this section, the hardware cost of k-PMPM network is further analyzed. Both the hardware cost and the time complexity of the routing algorithm are compared with those of previous works.

5.1 Comparison of Network Cost

The hardware cost of k-PMPM is $H = rnm_1 + m_1rm_2 + m_2m_1r + rm_2n = N(m_1 + m_2) + 2rm_1m_2$. Consider $m_1 \geq n - 1 + x$, $m_2 \geq (kN - n)/x + r$. To simplify the problem, we assume $n = r = \sqrt{N}$. Then we obtain:

$$\min_x\{H_k\} \approx (2k + 6\sqrt{k} + 4)N^{\frac{3}{2}}, \text{ when } x = \sqrt{\tfrac{3k}{5}n}$$

When k = 1, the hardware cost is $H_1 \approx 12N^{\frac{3}{2}}$, the hardware cost of k copies of 1-PMPM is $H'_k \approx 12kN^{3/2}$. It is obvious that the k-fold design has significantly lower network cost than that of k copies of 1-PMPM. Figure 3 shows the comparison between the hardware cost of k-PMPM and k copies of 1-PMPM, when k = 8,16.

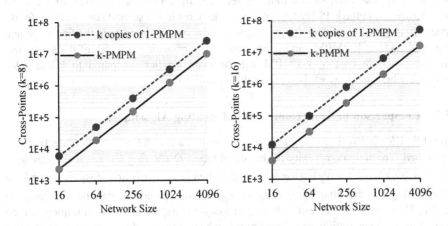

Fig. 3. Comparison between the hardware cost of k-PMPM and k copies of 1-PMPM

The necessary and sufficient condition for k-fold wide nonblocking multicast in Clos network is that the number of middle-stage switches m satisfies $m \geq (k + 2)n \log r / \log \log r$ [15]. If we also assume that $n = r = \sqrt{N}$, then the hardware cost of the network is

$$H_{clos} = 3(k+2)N^{\frac{3}{2}}\frac{\log r}{\log \log r}$$

Figure 4 shows the comparison between the hardware cost of k-PMPM and wide-sense nonblocking k-fold Clos network. As it can be seen, the hardware cost of k-PMPM is much lower than that of the k-fold Clos network.

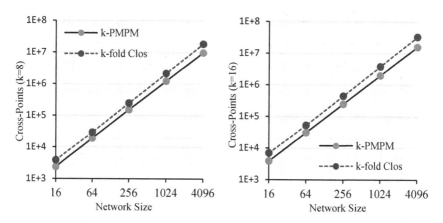

Fig. 4. Comparison between the hardware cost of k-PMPM and k-fold Clos network

Each switch in the Clos network is required to have multicast capability [13, 14], which means that each switch must be replaced by a multicast network of the same size. However, in the k-PMPM, only switch elements in the second stage and the fourth stage are required to have multicast capability; the switch elements in the first stage and the third stage of k-PMPM can be replaced by a permutation network of the same size. So the hardware cost of k-PMPM will be even smaller compared to k-fold Clos network when the network is recursively constructed.

5.2   Comparison of Time Complexity of Routing Algorithms

In k-fold Clos network, routing algorithm to realize a general multicast connection was presented. The time complexity of the algorithm is $O(kN\sqrt{\log N})$, when the hardware cost is $O(kN^{3/2} \cdot \log N/\log \log N)$ [15]; when the hardware cost increased to $O(kN^{3/2} \log N)$, the time complexity of the routing algorithm can be decreased to $O(kN)$ [15]. To realize an entire multicast assignments, the connection request can be routed in sequence until every output port has been connected to some input port, then the total time will be no less than $O(k^2N^{3/2})$.

In this paper, we have presented a $O(kN)$ routing algorithm to satisfy a general multicast connection request and an $O(k^2N^{3/2})$ algorithm to realize an entire multicast assignment in the wide-sense nonblocking k-PMPM network. It is obvious that the time complexity of the new algorithms is lower than or as good as the corresponding routing algorithms in k-fold Clos network [15].

6 Conclusion

In this paper, we proposed a new wide-sense nonblocking k-fold multicast network named k-PMPM together with its multicast routing algorithm. The k-fold design has significantly lower network cost than that of k copies of 1-fold multicast switching networks. Compared to previous works, the hardware cost of the new design can also be significant decreased. The time complexity of the corresponding routing algorithm is no higher than that of previous works.

Acknowledgement. The research was jointly supported by project grant from Shenzhen Sci. &Tech.Foundation: JCYJ20150930105133185/JCYJ20150324140036842, and National Natural Science Foundation of China: NSF/GDU1301252.

References

1. Zhang, Z., Yang, Y.: Performance analysis of k-fold multicast networks. IEEE Trans. Commun. **53**(2), 308–314 (2005)
2. Hwang, F.K., Wang, Y., Tan, J.: Strictly nonblocking f-cast networks $\log_d(N, m, p)$. IEEE Trans. Commun. **55**(5), 981–986 (2007)
3. Jiang, X., Pattavina, A., Horiguchi, S.: Strictly nonblocking design of f-cast photonic multi-log2N networks with crosstalk constraints. In: IEEE Workshop High Performance Switching Routing, pp. 1–6, May 2007
4. Yan, F., et al.: Nonblocking four-stage multicast network for multicast capable optical cross connects. J. Lightwave Technol. **27**(17), 3923–3932 (2009)
5. Ye, T., Lee, T.T., Hu, W.: AWG-based non-blocking Clos networks. IEEE/ACM Trans. Netw. **23**(2), 491–504 (2015)
6. Ge, M., Ye, T., et al.: Multicast routing and wavelength assignment in AWG-based Clos networks. IEEE/ACM Trans. Netw. **99**, 1–18 (2017)
7. Wan, Y.: Nonblocking multicast Clos networks. In: 19th Annual Wireless and Optical Communications Conference, pp. 1–5, May 2010
8. Clos, C.: A study of non-blocking switch networks. Bell Syst. Tech. J. **32**(2), 406–424 (1953)
9. Benes, V.E.: On rearrangement three-stage connecting networks. Bell Syst. Tech. J. **41**(5), 1481–1492 (1962)
10. Yang, Y., Wang, J.: Wide-sense no blocking Clos networks under packing strategy. IEEE Trans. Computers **48**(3), 265–284 (1999)
11. Masson, G.M., Jordan, B.W.: Generalized multi-stage connection networks. Networks **2**, 191–209 (1972)
12. Hwang, F.K.: The Mathematical Theory of Nonblocking Switching Networks, pp. 99–108. World Scientific, Singapore (2004)
13. Yang, Y., Masson, G.M.: The necessary condition for Clos-type nonblocking multicast networks. IEEE Trans. Comput. **48**, 1214–1227 (1999)

14. Yang, Y., Masson, G.M.: No blocking broadcast switching networks. IEEE Trans. Comput. **40**(9), 1005–1015 (1991)
15. Yang, Y., Wang, J.: Nonblocking k-fold multicast networks. IEEE Trans. Parallel Distrib. Syst. **14**(2), 131–141 (2003)
16. Yang, Y., Wang, J.: A new design for wide-sense nonblocking multicast switching networks. IEEE Trans. Commun. **53**(3), 497–504 (2005)
17. Wang, J., Yang, Y.: Four-stage multicast switching networks: nonblocking conditions and cost analysis. J. Lightwave Technol. **30**(3), 290–297 (2012)
18. Hasan, M.M.: Fine-tuning of k in a K-fold multicast network with finite queue using markovian model. Int. J. Comput. Netw. Commun. **5**(2), 195–204 (2013)

The Design of General Course-Choosing System in Colleges and Universities

Chunmin Qiu[✉], Shaojie Du, and Bailu Zhao

Binzhou Polytechnic, Binzhou, Shandong, China
bzqicm@163.com

Abstract. The college credit system allows students to take their own courses according to their own interests and development plans, but since the time for course-choosing is limited and the huge corresponding data requires a large amount of human resources, it is necessary to use computer software for course-choosing. The proposed system follows the complete design flow of computer software. System requirement analysis was done around the core function of the system which is course selection. Based on the system requirement analysis, the system function and the data structure were determined. And then, the system information flow, the code generation algorithm, the back-end database, the course setting algorithm, and the course-choosing algorithm were designed. In addition, a new algorithm for student ID generation is proposed. The student ID generated by this algorithm has simple structure, and also follows the principle of unique, reasonableness and extensibility. Therefore, this course-choosing system could be applied in every college and university.

Keywords: Requirement analysis · System design · Algorithms

1 Introduction

1980s, a climax of the credit system reform appeared in China's colleges and universities. The credit system makes the setting of specialized courses more flexible. Students can choose their own courses according to their interests and development plans. High level education is more diversified [1]. After applying the credit system, since the time for course-choosing is limited and the data needed to be processed is big, it is difficult to manage the data manually at the beginning of each semester.

As the continuous development of information and network technology, using the computer technology to manage the student course-choosing information has become an indispensable part of the teaching management work in colleges and universities. Through online course-choosing systems, students can get the information about courses, lecturers and teaching arrangements in a short time. Thus students could avoid the blindness during course-choosing, and save a lot of time and effort. Also, for the college, a lot of human and material resources could be saved. Moreover, the data could be processed in a short time [2].

© Springer Nature Singapore Pte Ltd. 2017
G. Chen et al. (Eds.): PAAP 2017, CCIS 729, pp. 311–320, 2017.
DOI: 10.1007/978-981-10-6442-5_28

At present, some colleges and universities have developed their own course-choosing systems. But it is difficult to promote these systems because of their limitations. This paper introduces the design of a Web-based general course-choosing system, which is suitable for all colleges and universities.

2 Requirement Analysis

2.1 Functions Analysis

During the management of teaching in colleges and universities, each department is asked to submit the courses which are open in this semester to the office of teaching affairs. Also, the required credits and courses for each student are submitted to the office of teaching affairs in accordance with the requirements of the training objectives. And then, for these courses, the office of teaching affairs coordinately assigns their schedules, lecturers and classrooms. Next, students are organized to select their courses. After course-choosing, the teaching process would be supervised and managed by the office of teaching affairs. At the end of each semester, the office of teaching affairs organizes all the final exams, and records and archives the grade of all students.

It can be seen from the teaching process that the office of teaching affairs focuses on the arrangement of teaching resources like courses and teachers and serves the students of each department. What students are supposed to do is completing the course-choosing, study, and examination based on training objectives. According to the work process that the office of teaching affairs organizes students to choose courses in each semester, the functions that a course-choosing system should have were analyzed as shown in the following figure. In the course-choosing system, the function on the student side is choosing course and the management jobs of the office of teaching affairs are managing departments, specialties, classes, students, teachers, courses and other information, setting up semester courses and inputting the test scores (see Fig. 1).

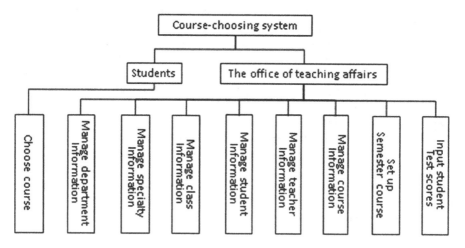

Fig. 1. The course-choosing system function diagram

2.2 Data Analysis

The course-choosing for students is the core function of the course-choosing system and other -functions support this core function. Specifically, the data that is generated and required by the course-choosing function must be provided and managed by other functions. For students who log in this system, the information about the courses that could be taken, classrooms, schedules and lecturers and their profiles could be provided by this system. And the students' course-choosing situations would be saved. Besides, back-end management functions must input the information about departments, specialties, classes, students, courses, teachers and other entities. Compared to the number of all the courses, the courses that could be selected in each term are only a small part. These courses must be set up and assigned with lecturers and schedules and the students who are allowed to choose these courses are also needed to be determined. this part must be set up, designated teacher, teaching time, classes allowed to choose courses. After final exams, the back-end management function must be capable of inputting the marks of students [3]. System E-R diagram is shown as Fig. 2.

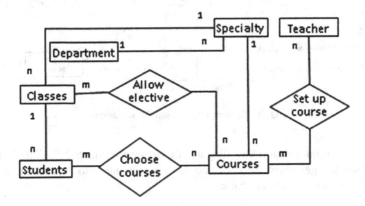

Fig. 2. System E-R diagram

3 System Design

3.1 Information Flow Design

In a school's organizational structure, from bottom to top, the hierarchy consists of students, classes, specialties and departments. Besides, teachers are subordinate to specialties and specialties decide which courses are open in each semester. Thus, there is an order for the input and management of the data when the system is running. Firstly, the information about students, classes, specialties and departments is inputted by back-end data administrators in certain. Secondly, the information about courses and lecturers is inputted respectively. Next, courses would be setted up and lecturers would be assigned to each course. And then, students would choose courses. Finally, the grades of students would be inputted by administers [4]. System information flow diagram is shown as Fig. 3.

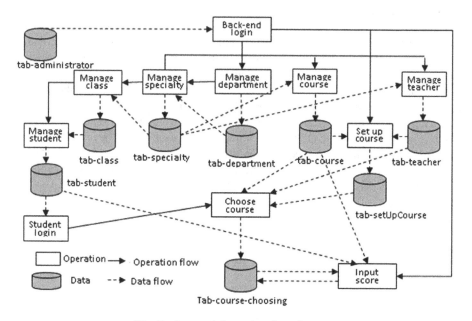

Fig. 3. System information flow diagram

In the system information flow diagram, the arrow pointing to "data" indicates that the operation has generated the data, and the arrow pointing to "operation" indicates that data is provided to the operation.

3.2   Code Design

Code design is related to the ease of use and persistence of the system. In 1999, the world paid a great deal to solve the problem of computer millennium [5], which explains the importance of code design from a different point of view.

The course-choosing system includes six entities which are departments, specialties, classes, students, courses, and teachers. The design of the entity code is unique, reasonable, and extensible [6].

As for the uniqueness, there are two meanings which are horizontal uniqueness and vertical uniqueness. The student ID is unique for each student and it is not possible for two students to use the same student ID, which is the horizontal uniqueness. A student ID cannot be changed from beginning to end which is the vertical uniqueness.

As for the code rationality, the student ID is not only used to distinguish each student, but also used as a user name of some management systems. So the design of the student ID not only need to ensure the uniqueness, but also need to be easy to remember. And each part of the student ID should be designed with certain meaning. For example, a student ID could be designed as "201701001", among them, "2017" is the grade, "01" is the specialty, and "001" is a serial number. This design of student ID is neatly and easy to remember and also looks comfortable.

The extensibility of code is often subjected to rationality. Specifically, for a student ID like "201701001", the grade code can be unlimitedly expanded, but the specialty code and the serial number are subjected to a significant increase in the number of restrictions. In addition, if a student switches to another specialty, his or her student ID must be changed which would destroy the vertical uniqueness.

By comprehensively considering the uniqueness, rationality and extensibility of the entity code designed in the general course-choosing system, the code of the five entities which are departments, specialties, classes, courses, teachers, is designed with an integer number, and as for the student ID, unlike the routine, it is designed as positive integers, the first four digits denotes the grade, followed by a variable long serial number. According to this design, the first registered student in the year 2017, his or her student ID is "20171", and the 999th registered student, his or her student ID is "2017999" and so on. This design is suitable for schools of various sizes.

Algorithm for generating new student ID is:

(1) Input the grade N.
(2) Find the maximum existing student ID denoted by M (If not found, then $M = N * 10$).
(3) Calculate the digit number of M, denoted by L.
(4) Calculate the grade of M, denoted by N_{max}.

$$N_{max} = (int)M/10^{L-4}$$

(5) Calculate the new student's serial number denoted by K.

$$K = \begin{cases} M\%10^{L-4} + 1, & N_{max} = N \\ 1, & N_{max} < N \end{cases}$$

(6) Calculate the digit number in the sequence number K, denoted by J.
(7) Generate a new student ID, denoted by P.

$$P = N*10^{max(J,L-4)} + K$$

3.3   Database Design

In the relational database, there must be a certain relationship among the data. There are "many to one" and "one to many" relationships. How to deal with these relationships will be related to whether the database generates redundant and abnormal data. The relational database normalization theory can handle these issues. The data of the general course-choosing system is divided into six entities which are departments, specialties, classes, students, courses, teachers and three relationships. And nine tables are created according to these relationships respectively. Each attribute value in each table is an indecomposable data item. Each non-master attribute in each table has a full function that depends on any of its candidate codes. All non-master attributes do not pass functions that depend on any candidate code. So these nine tables are in third normal form [7].

A distinguished ID was assigned for each table. And the constraints for the primary key were established. For the course table, the Course ID was regarded as the constraint of its primary key. For the table of the classes that are allowed to choose courses, the course ID and the class ID constitute its composite primary key constraints. And for the Course table, the Course ID and the student ID constitute its composite primary key constraints. By defining the primary key of the table, the uniqueness of the data, i.e., the integrity of the data is ensured. Why not set the course ID or the teacher ID as the primary key for the course table? The reason is that there are some special situations like many teachers teaching one course and one teacher teaching a number of courses.

At the same time, by defining the foreign key constraints of the table, the referential integrity was ensured. The department ID column in the specialty table is the foreign key that points to the department table. The specialty ID column in the class table is the foreign key that points to the specialty table. The class ID column in the student table is the foreign key that points to the class table. The specialty ID column in the course table is the foreign key that points to the specialty table. The specialty ID column in the teacher table is the foreign key that points to the specialty table. The course ID column in the Course table for the open courses is the foreign key that points to the course table for all courses. The open course ID and the class ID column in the tab-allowElective table are the foreign keys that respectively points to the open Course table and the course table. The open course ID and student ID column in the course selection table are the foreign keys that respectively points to the open course table and the student table.

Database tables and the relationship between tables are shown as Fig. 4.

3.4 Detailed Design

The design of the system function is described in detail so that the programmer can encode and debug according to the detailed design report.

When courses are setted up, some courses are only for the students of a specific department and some courses are only for the students in certain specialties and some courses are only the students of certain grades. These restrictions relying on administrative means are often ineffective. Only the usage of technical means could ensure the courses that are shown to the students who log in the system are the courses that can be chosen. This requires a clear definition of the range of students who are allowed to choose for each course when the course is setted up. In order to make it convenient for the students who has logged on the system to query the course that they can choose, the class ID of classes that are allowed to take a course need to be stored when set up a course.

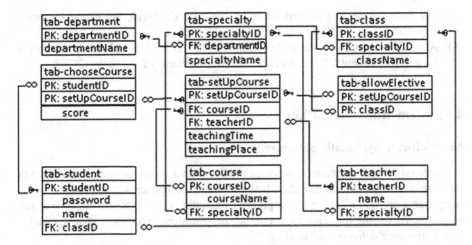

Fig. 4. The relationship between tables

The algorithm of setting up course is:

(1) Add the course to the setting up course list.
(2) Assign teachers for these courses.
(3) Input the teaching place and time.
(4) By default, allow all students to choose all courses.
(5) If the right of a grade to choose the course is canceled, all students in that grade will not be able to choose this course.
(6) If the right of a department to choose the course is canceled, all students in that department will not be able to choose this course.
(7) If the right of a specialty to choose the course is canceled, all students in that specialty will not be able to choose this course.
(8) If the right of a class to choose the course is canceled, all students in that class will not be able to choose this course.
(9) For each course, record the grades that are allowed to choose this course.
(10) Traverse each class, if the class allows choose the course, and then add a tuple in the data table, the set up course ID. And the class ID was stored in the two attributes of the tuple.

Students who log in the system with the student ID could browse the list of open courses and choose courses.

The algorithm of student course-choosing function is:

(1) Student login.
(2) Take the first four digits of the student ID as the student's grade.
(3) Query the class where the student is located.
(4) Query for courses that allow the class and the grade that the student is located to be elective.

(5) List each course that is allowed to be chosen and provide the course-choosing function.

(6) After choosing course, the tuple is added to the data table, and the set up course ID and the student ID are stored in the two attributes of the tuple.

4 System Implementation

4.1 Software Systematic Structure

Software systematic structure defines the partial and total computational components of software and also the relationship between these components. Specifically, these components are some software components including servers, clients, database, and programs. As for the relationship, it consists of procedure invocation, shared variable access, message delivery and so on.

The client and server structure, also known as C/S structure, would reasonably assign the client and the server to different tasks. Specifically, the job of the client is to submit the requirement of users to the server and then show the results coming from the server to users in a certain way. As for the server, its job is to receive the requirement of users, do some corresponding process, and send back the result to the client. The advantages of the C/S structure are as follows: first, the workload of the server is not heavy; second, the storage and management functions of data are transparent at certain extent; also, the response speed on the side of the client is fast. However, this structure also has some drawbacks. Specifically, some specific programs are required on the client. And there are some limitations for the operation system of the client. Besides, the expense of investment and maintenance is high.

The browser and server structure, also known as B/S structure, is also a widespread structure. The WEB browser is the most important program on the client. The core function of the system implementation is applied on the server. And the browser exchanges data with the database through the WEB server. This kind of structure could simplify development, maintenance, and appliance of the system [8].

Nowadays, every undergraduate student has at least one smartphone, which makes it very convenient to connect to the Internet. And the internet speed is satisfactory. Therefore, if the B/S structure is used to develop a course-choosing system, it would be quite convenient for the students.

4.2 Program Code

The system due to the use of extensible student ID generation algorithm, could be applied in every college and university. The following is the server-side C# code that generates the new student ID.

```
protected SqlConnection DBconn(){
  // The method of connecting to the database
  string strDBconn ="Data Source=.\\SQLEXPRESS;";
  strDBconn+="AttachDbFilename=E:\\data\\";
  strDBconn+="ChooseCourse.mdf;";
  strDBconn+="Integrated Security=True;";
  strDBconn+="Connect Timeout=30;";
  strDBconn+="User Instance=True";
  SqlConnection conn = new SqlConnection(strDBconn);
  return conn;
}
protected String executeQuery(String strSql){
// The method of executing the T-SQL statement
  SqlConnection conn = DBconn();
  conn.Open();
  SqlCommand com = new SqlCommand(strSql, conn);
  return com.ExecuteScalar().ToString();
}
protected int calculateDigitNumber(int m){
// The method of calculating the integer number of bits
  int digitNumber = 0;
  do
  {
    digitNumber++;
    m = m / 10;
  } while (m != 0);
  return digitNumber;
}
protected int generateStudentID(int N){
// The method of generating a new student ID
// N indicates the new student's grade, from the client
  SqlConnection conn = DBconn();
  string strSql="select max(studentID) "
  strSql+="from [tab-student]";
  string strMaxStudentID = executeQuery(strSql);
  // Find the largest student ID
  int M=N*10;// Store the largest student ID
  if (strMaxStudentID != null && strMaxStudentID != "")
    M = int.Parse(strMaxStudentID);
  int L = calculateDigitNumber(M);
  // Calculate the number of bits in M
  int Nmax = M / (int)Math.Pow(10, L - 4);
  // Calculate the grade of student ID M
  int K = 1;// Store the serial number of the new student
```

```
if(N==Nmax)
  K = M % (int)Math.Pow(10, L - 4) + 1;
  // Calculate the serial number of the new student
  int J = calculateDigitNumber(K);
  // Calculate the number of bits
  //in the serial number of the new student
  int P = N * (int)Math.Pow(10,Math.Max(J,L - 4)) + K;
  // Generate a new student's student ID
  return P;
}
```

5 Summary

The general course-choosing system follows the complete design flow of the software system. Around the student course-choosing, the core function of the system, the system needs analysis and the system design is finished. The system uses B/S software structure and advanced JSP Web application development technology. It also takes SQL Server 2008 as the backend database. And it uses MVC three-tier architecture for development. The system portability is satisfactory and it is easy to be expanded in terms of functions. After the test, the desired goal was achieved.

References

1. Yan, W., Wu, M.: The theoretical basis and system guarantee for the reform of university academic credit system. Educ. Res. **22**(7), 57–63 (2015)
2. Liu, Z., Han, X., Zhang, W.: Design and implementation of sinatra-based course-choosing system. Comput. Knowl. Technol. **12**(8), 76–78 (2016)
3. Wang, S., Ma, Y., Liu, Y.: Information system requirement analysis process and method. Des. Tech. Posts Telecomm. **22**(12), 6–11 (2015)
4. Wang, X.: In the design of software system some difficult problem in the study. Public Commun. Sci. Technol. **4**(1), 171, 184 (2012)
5. Wei, Z.: The year-2000 computer problem. J. Shanxi Finac. Econ. Univ. **21**(S1), 97–98 (1999)
6. Gao, Y.: Systems analysis and design of management information system. Electron. Test **21**(9), 87–89 (2014)
7. Yang, M.: The shallow on the standardization of relations in relational database. Inf. Technol. Inf. **21**(11), 172–173 (2014)
8. Shen, J., Nie, H.: Introduction to software architecture. Comput. Dev. Appl. **15**(2), 49–51, 54 (2008)

Research on Vectorization Technology for Irregular Data Access

Wang Qi[(⊠)], Han Lin, Yao Jinyang, and Liu Hui

State Key Laboratory of Mathematical Engineering and Advanced Computing,
Zhengzhou 450001, China
wangqi19920406@163.com

Abstract. Current program vectorization methods support continuous memory access forms. There are few researches on vectorization of irregular data access. In order to improve the program execution efficiency, vectorization technology of irregular data access is researched. A vectorization method for non-continuous access and indirect array indexes is proposed. This paper designs a calculation method of vectorization performance gains and analyzes the different performance gains of different non-continuous access vectorization methods. Finally the experimental results show that this method can vectorize irregular data access effectively and improve the program execution efficiency.

Keywords: Vectorization · Irregular data access · Non-continuous data access · Indirect array index · Benefit analysis

1 Introduction

With the continuous enhancement of personal computers, people's expectations for the computers are getting higher and higher. The faster speed of computer calculation is required. At present, most general processors have integrated SIMD (Single Instruction Multiple Data) expansion components [1]. SIMD extension components are widely used in high-performance computing. At the same time more and more research uses SIMD expansion components to accelerate the program [2, 3]. Most of the compiler also integrates automatic vectorization module in order to facilitate the effective use of SIMD expansion components [4]. There are many vectorization analysis methods to realize the vectorization of more programs [5–7]. The basic principle of these methods is to find parallel statements based on data dependencies, and then parallel statements can be executed in parallel with the SIMD extension components.

Vectorization is a kind of fine-grained parallelization. It not only needs the dependence to meet the needs of the parallel, but also has strict requirements for access modes. For continuous data access, the automatic vectorization module of compiler makes it easy to mine its vector execution section, vectorizes it and obtains a certain performance benefit. But for irregular data access, such as non-continuous data, indirect array index, the process of loading valid data into registers is complicated because of its complex situation. The compiler abandons vectorization directly for some complex irregular data access. In fact, the AVX512 instruction set [8] introduced by Intel is very rich on the vector processing. Its use of mask operations, selection instructions, stride access, data

© Springer Nature Singapore Pte Ltd. 2017
G. Chen et al. (Eds.): PAAP 2017, CCIS 729, pp. 321–334, 2017.
DOI: 10.1007/978-981-10-6442-5_29

sorting instructions and so on can deal with the problem of irregular data access vectorization. It can handle eight double-precision floating-point numbers or sixteen single-precision floating-point numbers in one operation. But it should be noted that, GCC does not support the vectorization of stride access actually. ICC supports for stride access, but it doesn't vectorize the program for some complex irregular data access.

There is some research on the vectorization of irregular data access currently. Li et al. used SIMD ADD operation to calculate the sum of the base address and the offset address in the indirect array index [9]. It gets the address of the data, and then extracts the data according to the address. Its calculation of data address is a vector operation, but the data access is a serial operation. Xu et al. did some research on vectorization of irregular data access, but it didn't solve the problem of indirect array index [10]. Kim et al. studied the indirect array index, but the introduction of packaging instructions has additional costs [11].

The above research on the vectorization technology for irregular data access made progress in the program performance. With the introduction of the Intel AVX512 instruction set, its rich data sorting instructions, stride access instructions and mask instructions for irregular data access provide a lot of better and more flexible methods to improve the program performance effectively. There is a lot of irregular data access in mathematical matrix operations, physics simulation applications and spec programs [12]. So this paper researches on the vectorization technology for irregular data access based on second-generation MIC architecture Intel Xeon Phi processor Knights Landing with the AVX512 instruction set and provides a more efficient solution for the vectorization of irregular data access [13]. The main contributions of this paper are as follows:

(1) Analyze the vectorization scheme of non-continuous data access based on AVX512 instruction set.
(2) Provide the vectorization scheme for the indirect array index based on the AVX512 instruction set.
(3) Analyze the vectorization benefit of non-continuous data access based on the AVX512 instruction set.

The structure of this paper is as follows: Sect. 2 describes the reason and type of irregular data access. Section 3 presents the vectorization scheme of non-continuous data access and indirect array index. Section 4 analyzes the vectorization benefit of irregular data access. Section 5 tests and analyzes the experimental results of the irregular data access vectorization scheme with the applications. And at last make a conclusion.

2 Irregular Data Access

2.1 Non-continuous Data Access

There are many reasons why data access is not continuous. For example, the value of the loop index variable is not one, if conditional statement causes that data access is not continuous and the data access in a basic block is not continuous as shown in Fig. 1.

After analyzing the dependence, these non-continuous operations can be executed in parallel. So how to choose a vectorization scheme whose execution efficiency is higher is needed to be studied.

```
for(i=0;i<N;i=i+2)
{
    c[i] = a[i]+b[i];
}
```

(1)

```
for(i=0;i<N;i=i++)
{
    if(i%8 != 7)
        c[i] = a[i]+b[i];
}
```

(2)

```
for(i=0;i<N;i=i++)
{
    c[4i] = a[4i]+b[4i];
    c[4i+1] = a[4i+1]+b[4i+1];
    c[4i+3] = a[4i+3]+b[4i+3];
}
```

(3)

Fig. 1. An example program of non-continuous data access

2.2 Indirect Array Index

The indirect array index is that the value of the array index is accessed from another array. Actually it is similar to a secondary index as shown in Fig. 2. Index value of the

```
for(i=0;i<N;i=i+1)
{
    j=index[i];
    c[3*i] = a[j] + b[j];
    c[3*i+1] = a[j+1]+b[j+1];
    c[3*i+2] = a[j+2]+b[j+2];
}
```

Fig. 2. An example program of indirect array index

indirect array access is read from another array, so the order of data access is not regular. Most of the time, it is difficult for the compiler to form the data into a vector. And that the dependency is more complex because its index values are uncertain. For the sake of conservatism, the compiler does not vectorize the program of indirect array access.

3 Vectorization Technology for Irregular Data Access

3.1 Vectorization Realization of Non-continuous Data Access

As shown in Fig. 3, the data access in the sample program is not continuous and its stride is 2. The program can be vectorized in several schemes with the AVX512 instruction set.

```
for(i=0;i<N;i=i+2)
{
    c[i] = a[i]+b[i];
}
```

Fig. 3. An example program

Vectorization Realization with Mask Instructions

The AVX512 instruction set provides a very rich mask instruction, including access, mathematical calculations, logical calculations, data sorting and so on. All instructions have their corresponding mask instruction versions. So we can use the mask instruction to calculate valid data, and then write effective results back to the memory. The vectorization of non-continuous data access is realized as shown in Fig. 4.

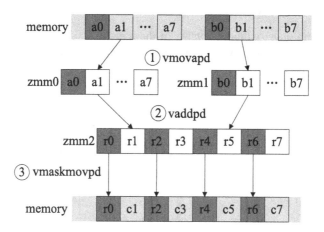

Fig. 4. Vectorization realization of non-continuous data access with mask instructions

In the Fig. 4, ① means that data a0–a7 and b0–b7 in the memory is loaded into the vector registers zmm0 and zmm1 with two vector loading instructions. ② Means vector addition. ③ Means that the valid data in the vector register is written back to memory according to the mask with the vmaskmovpd instruction.

The use of mask instructions can be more flexible to solve the problem of non-continuous access. However, there is redundant data which does not need to be written back to the memory in the vector register used by the mask operation. This makes that the use of vector registers is not sufficient. It will affect the efficiency of the program to further enhance.

Vectorization Realization with Gather/Scatter Instructions

The AVX512 instruction set provides the gather/scatter instruction which can realize the vectorization of non-continuous data access directly in order to use the vector register more sufficiently. It can get the offset address of each data with the index vector and then achieve the stride loading vectorization of non-continuous data. Finally it gathers the discrete data from the memory into one vector register with one instruction. It is similar for writing the data in the vector register back to the memory. It gets the offset address of each data with the index vector and then achieves the stride storing vectorization of non-continuous data. Finally it scatters the discrete data from the vector register into the memory with one instruction as shown in Fig. 5.

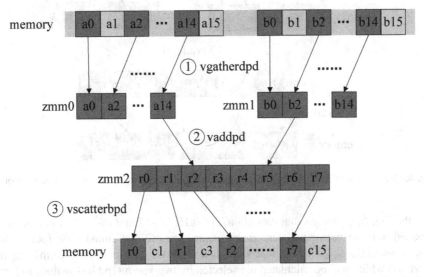

Fig. 5. Vectorization realization of non-continuous data access with gather/scatter instructions

In the Fig. 5, ① means that the discrete data a0, a2, to a14 and b0, b2 to b14 in the memory is loaded into the vector registers zmm0 and zmm1 with two vgatherdpd instructions. ② Means vector addition. ③ Means that data in vector register zmm2 is stored into the corresponding position of the memory with the vscatterdpd instruction and the corresponding index vector.

Vectorization Realization with Data Sorting Instructions

Data in several vector registers can be sorted by data sorting instructions in the AVX512 instruction set such as permute instructions, blend instructions, pack and unpack instructions and so on. It can sort the data of multiple vectors and it can deal with the data of different data types and different vector length. The valid data is loaded into one vector register. All these instructions operate on vector registers so the cost is little. When the stride of array access is relatively small, the use of data sorting instructions will be better to accelerate the program. At the same time, it can increase the reuse of vector registers and reduce the number of times to load data from the memory. Vectorization realization with data sorting instructions is as shown in Fig. 6.

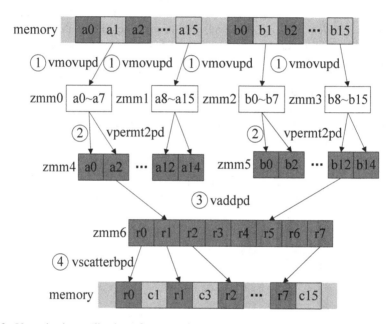

Fig. 6. Vectorization realization of non-continuous data access with data sorting instructions

In the Fig. 6, ① means that data a0–a7, a8–a15, b0–b7 and b8–b15 in the memory is loaded into the vector registers zmm0, zmm1, zmm2 and zmm3 with four vector loading instructions. ② Means that valid data in vector register zmm0, zmm1, zmm2 and zmm3 which will be calculated are selected by two vpermt2pd instructions to form a vector. The result of selection is returned to the vector registers zmm4 and zmm5. ③ Means vector addition. ④ Means that data in vector register zmm6 is stored into the corresponding position of the memory with the vscatterdpd instruction and the corresponding index vector.

3.2   Vectorization Realization of Indirect Array Index

Vectorization Read Operation of Indirect Array Index
Indirect array access is like a[index[i]]. The base address of array a[] is determinate. Its offset address is the value of index[i]. The offset address is indeterminate at compiling time. However the offset address in array index[i] can be loaded into the vector register and an index vector is gotten. Then the vectorization read operation of indirect array index for array a[] can be realized with the gather instruction and the index vector through the AVX512 instruction set as shown in Fig. 7.

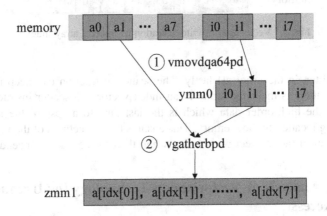

Fig. 7. Vectorization read operation of indirect array index

In the Fig. 7, ① means that the integer index data is loaded into the 256-bit vector register ymm0 with a vector loading instruction. ② Means that the discrete data of indirect array index in the memory is loaded into the vector registers zmm1 with a vgatherdpd instruction and the index vector ymm0.

Vectorization Write Operation of Indirect Array Index
In fact, the vectorization write operation of the indirect array index is actually similar to the vectorization read operation of the indirect array index. For vectorization write operation of the array a[index [i]], the offset address which is gotten in array index[i] is also loaded into the vector register. Then the vectorization write operation of indirect array index for array a[] can be realized with the scatter instruction and the index vector through the AVX512 instruction set as shown in Fig. 8.

In the Fig. 8, ① means that the integer index data is loaded into the 256-bit vector register ymm0 with a vector loading instruction. ② Means that discrete data of indirect array index in vector register zmm1 is stored into the corresponding position of the memory with the vscatterdpd instruction and the corresponding index vector.

However, there is something different between vectorization read and write operation of indirect array index. Since the value of the index array index[i] may be the same in several iterations before and after, there may be output dependence between the iterations for the write operation of array a[]. So the final value of array a[] should be

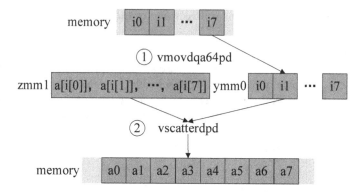

Fig. 8. Vectorization write operation of indirect array index

the value written in the array a[] lastly. The scatter instruction can keep this dependence. When there is the same stride in an index vector, the scatter instruction eventually writes the high-order data which is the last data to access in the loop to the corresponding location in the memory. This ensures the correctness of the vectorization write operation of the indirect array index when there is the output dependence.

4   Benefit Analysis of Vectorization Technology for Irregular Data Access

4.1   Benefit Calculation of Vectorization

Section 3 presents some vectorization schemes of non-continuous data access and indirect array index. So it is important to choose the best one to make the program execution efficiency highest. Benefit calculation of vectorization is necessary. Vectorization benefit is the performance improvement achieved after the vectorization of the program. It can be calculated by comparing the cost performed by the scalar execution with the cost of the vector execution of a loop.

The cost performed by the scalar execution is the time required to calculate the entire loop by scalar execution. It can be calculated by multiplying the scalar instruction overhead of each iteration by the number of scalar loop iterations. The cost performed by the vector execution is the time required to calculate the entire loop by vector execution. It can be calculated by multiplying the vector instruction overhead of each iteration by the number of vector loop iterations.

The total number of instructions in the loop will be less after vectorization, but the introduction of the vector data sorting instructions, access instructions and so on will have a relatively large overhead. We can analyze the benefit that the program vectorization can earn by analyzing and comparing the different cost performed by the scalar execution and vector execution to choose whether to vectorize the serial program or which vectorization scheme is selected for non-continuous data access.

4.2   Benefit Analysis of Vectorization for Non-continuous Data Access

There are many schemes to realize the vectorization of non-continuous data access. The three schemes presented by Sect. 3.1 can all achieve the vectorization operations of non-continuous data access. But for each scheme, their instruction costs are different; their suitable situations are different, therefore, the gained performance benefits are also different. The three programs as shown in Fig. 9 are tested by the three schemes in Sect. 3.1 and N = 4800.

```
for(i=0;i<N;i=i+2)
    c[i] = sin(a[i]+b[i]);
```

(1)

```
for(i=0;i<N;i=i+3)
    c[i] = sin(a[i]+b[i]);
```

(2)

```
for(i=0;i<N;i=i++)
{
    if(i%4 != 3)
        c[i] = sin(a[i]+b[i]);
}
```

(3)

Fig. 9. Example programs to test the performance benefits of different vectorization schemes

At last, the vectorization speedup of the three programs tested by the three schemes is as shown in Table 1.

Table 1. Speedup comparison of three non-continuous data access vectorization schemes

	Realization with mask instructions	Realization with gather/scatter instructions	Realization with data sorting instructions
Program 1	6.87	9.90	10.8
Program 2	4.56	9.91	8.83
Program 3	10.38	7.29	9.90

First tested speedup is greater than the theoretical speedup 8 times. Because the call of sin mathematical function is converted to the call of short vector math library after vectorization and the short vector math library does some optimization in mathematical function. Through the test results, we can find that the gained performance benefits are different facing to different situations.

Cost of mask access instruction is smaller. While because of the existence of the mask, there is redundant data in the vector register. The number of valid data for a mask vector operation is smaller and this will affect the speedup of the program. Therefore, the mask operation is suitable for the IF conditional statement. The control flow is converted into data flow. Then the program is vectorized. At the same time the valid data to be executed in one vector register should be sufficient.

Cost of gather/scatter instruction is bigger. But it can load the discrete data whose stride maybe large in the memory into a vector register through one instruction and the data in the register is all valid. Multiple data can be operated in one time. Finally, the data in the vector register is written to the corresponding position in the memory. Therefore the gather/scatter instruction is suitable for the introduction of non-contiguous arrays vectorization and the implementation process is simple.

For vectorization realization with data sorting instructions, the referenced data is non-continuous, the valid data is loaded into multiple vector registers with multiple continuous vector loading instructions. And then the valid data from multiple vector registers is organized into one vector register with data sorting instructions. There are many extra instructions in vectorization realization with data sorting instructions, but the cost of each instruction is relatively small. It is often efficient for some situations where the stride is small for example the stride is 2. But when the program stride is relatively large or irregular, the implementation process is often more complex, the number of total instructions is large and the total cost will be large.

Therefore, the three vectorization schemes have their own advantages and disadvantages. While generating the corresponding vectorization instructions, the cost model should be used to calculate the vectorized benefits so that the acceleration of the program is optimal.

5   Experimental Test Results

5.1   Experimental Environment

The experimental platform used in this paper is the new generation Intel Xeon Phi processor Knights Landing released in June 2016. The number of cores is more than 60. Its technological design is upgrade to 14 nm. It owns 16 GB MCDRAM which is the high-bandwidth memory integrated on-package. It offers over three TFLOP/s of double precision and six TFLOP/s of single precision peak floating point performance.

The experimental hardware environment used in this paper is that the processer is Intel Xeon Phi Processor 7210, the basic frequency is 1.30 GHz, the turo frequency is 1.50 GHz, the L2 cache is 32 MB (Each two cores make up a tile and share 1 MB cache), instruction set extension is Intel AVX-512, memory is 8 × 16 GB DDR4-2133 ECC REG memory and the 16 GB MCDRAM is set to cache mode. The experimental software environment is as shown in Table 2.

5.2   Vectorization Testing of Non-continuous Data Access

434.zeusmp, 456.hmmer, 462.libquantum and 482.sphinx3 programs in spec2006 are selected to test the vectorization scheme of non-continuous data access proposed by this paper. There is non-contiguous data access in the core loop of these programs. The

Table 2. Experimental software environment

Software environment	Software version
Operating system	CentOS Linux release 7.2.1511
Linux Kernel	Linux version 3.10.0-327.el7.x86_64
Compiler	icc 17.0.0 Beta

test, train and ref scale for these spec2006 programs are tested respectively. Serial execution time of the program is divided by the vectorization execution time of the program to calculate the program vectorization speedup. The results are as shown in Table 3, and the comparison of the speedup is as shown in Fig. 10.

Table 3. Vectorization speedup for the spec2006 programs of different scale

	Test	Train	Ref
434	2.22	2.24	2.55
456	2.23	2.45	2.54
462	1.36	1.27	1.23
482	1.69	1.71	1.71

Through the test we can find that 434.zeusmp, 456.hmmer, and 482.sphinx3 programs in spec2006 are accelerated better. Acceleration effect of the programs after vectorization is obvious. The speedup of 462.libquantum program is slightly lower because its non-continuous data access is caused by the array of the structure, the situation is relatively complex.

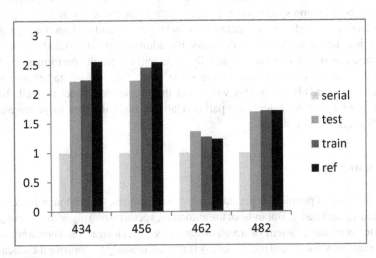

Fig. 10. Vectorization speedup comparison for the spec2006 programs of different scale

5.3 Vectorization Testing of Indirect Array Index

CFV (Compact high order finite volume method on unstructured grids) program, 416. gamess and 435.gromacs programs in spec2006 are selected to test the vectorization scheme of indirect array index proposed by this paper. There are indirect array indexes in the core loops of these programs. The core loops of these programs are tested separately. The vectorization speedup of core loops in these programs is calculated. The results are as shown in Fig. 11.

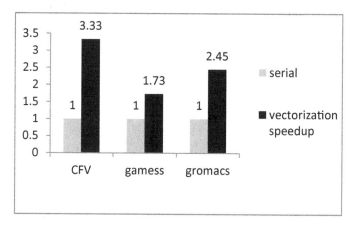

Fig. 11. Vectorization speedup of indirect array index for core loops

Through the dependence analysis of CFV, there is no data dependence in the indirect array index, so the indirect array index can be vectorized in the program. The theoretical vectorization speedup is 8. However, there are reduction addition operations in the indirect array index write operations of 416.gamess and 435.gromacs programs. Therefore it is necessary to divest the reduction addition operations and the reduction addition operations will be executed serially. This will reduce the performance benefits of the programs to some extent. At the same time, the cost of gather and scatter instructions is relatively large, this will affect the accelerated effect as well. But the acceleration of the vector calculation part is relatively high, the core loops have certain acceleration effects as a whole.

6 Conclusion

With the increase of people's expectations for the computers, the computer is required to calculate faster and faster. But most of the automatic vectorization algorithm is not a good solution to solve the problem of irregular data access vectorization. Therefore, vectorization technology for irregular data access is needed to study to improve the calculation speed of the program. Based on this, this paper divides irregular data access into non-continuous data access and indirect array index and corresponding vectorization schemes for the above two cases are presented. The choice of vectorization schemes for non-continuous data access is more, so the performance benefits and its suitable situations of each vectorization scheme is analyzed aim at different computational features. Finally, the scientific calculation program and the programs in spec2006 are selected to verify the validity of the method proposed in this paper.

This paper mainly researches on the problem of irregular data access from two aspects: non-continuous data access and indirect array index. It doesn't introduce other features such as the non-continuous data caused by the array of the structure. At present, the vectorization for the array of the structure can be carried out by data reorganization. The array of the structure is converted into the structure of the array. Realization of data reorganization is complex and it needs extra cost. Therefore, if the vectorization of non-continuous data access the data caused by the array of the structures can be achieved, the cost of data reorganization can be reduced. This needs further study. Moreover, methods proposed in this paper can't be fully applied to the internalization of the basic block. For example the scalar data in the basic block is loaded into one vector register at a time. Therefore, subsequent work will focus on the problem of irregular data vectorization within the basic block. At last, this paper analyzes the vectorization benefits of non-continuous data access for different schemes. Subsequent work will research on the performance benefits of vectorization deeply and make the cost model of the vectorization benefit analysis universal.

References

1. Wei, G., Cai, Z.R., Lin, H., et al.: Research on SIMD Auto-vectorization compiling optimization. J. Softw. **26**(6), 1265–1284 (2015)
2. Song, D., Chen, S.: Exploiting SIMD for complex numerical predicates. In: IEEE, International Conference on Data Engineering Workshops, pp. 143–149. IEEE Computer Society (2016)
3. Tian, X., Saito, H., Preis, S.V., et al.: Effective SIMD vectorization for intel Xeon Phi coprocessors. Sci. Program. **2015**, 1–14 (2015)
4. Diken, E.E., Jordans, R.R., O'Riordan, M.: MoviCompile: an LLVM based compiler for heterogeneous SIMD code generation (2015)
5. Kansal, R., Kumar, S.: A vectorization framework for constant and linear gradient filled regions. Vis. Comput. **31**(5), 717–732 (2015)
6. Chen, J.Z., Lei, Q., Miao, Y.W., et al.: Vectorization of line drawing image based on junction analysis. Sci. China Inf. Sci. **58**(7), 1–14 (2015)
7. Sui, Y., Fan, X., Zhou, H., et al.: Loop-oriented array- and field-sensitive pointer analysis for automatic SIMD vectorization. ACM SIGPLAN Not. **51**(5), 41–51 (2016)
8. Jeffers, J., Reinders, J., Sodani, A.: Chapter 12 – Vectorization with AVX-512 intrinsics. In: Intel Xeon Phi Processor High Performance Programming, pp. 269–296 (2016)
9. Li, P., Zhao, R., Zhang, Q., et al.: An SIMD Code Generation Technology For Indirect Array. **8**(3), 218–222 (2016)
10. Jinlong, X.U., Zhao, R., Liu, P., et al.: Research on irregular memory access problem for programs vectorization. Comput. Eng. **41**(12), 86–90 (2015)
11. Kim, S., Han, H.: Efficient SIMD code generation for irregular kernels. ACM SIGPLAN Not. **47**(8), 55–64 (2012)

12. Wang, Q., Ren, Y.X., Li, W.: Compact high order finite volume method on unstructured grids II: extension to two-dimensional Euler equations. J. Comput. Phys. **314**, 883–908 (2016)
13. Sodani, A., Gramunt, R., Corbal, J., et al.: Knights landing: second-generation Intel Xeon Phi product. IEEE Micro **36**(2), 34–46 (2016)

A New Simple Algorithm for Scrambling

Xing Zeng, Xiulai Li, Yali Luo, and Mingrui Chen[✉]

Hainan University, Haikou 570228, China
mrchen@hainu.edu.cn

Abstract. The paper proposes a new algorithm which use simple divisible and xor operation to change the values of pixels and the site of pixels in digital images based on logistic system. This algorithm encrypted images successfully. Experiments show that the algorithm is simple and easy to do. It has a large secret-key space and higher-security. Furthermore, the algorithm not only has preferable practicability, but also can resists various stacks.

Keywords: Logistic · Divisible operation · Xor operation · Chaotic scrambling

1 Introduction

With the development of information industry, digital image is an important part of people's life. Therefore, the security of images is also the focus of many researchers. Image scrambling, however, is an important means of encrypting images. The current scrambling algorithm mainly achieves the purpose of encryption by changing the location of image pixels or changing the image pixel value. Such as the Arnold transform and the magic transformation, but these algorithms are only a single pixel of image position or pixel value scrambling algorithm, some periodic, easy to break.

But for the initial value, the sequence generated by its sensitivity is irregular, and chaotic system becomes the new favorite of scrambling. Using chaotic system scrambling can improve the security of scrambling system, thus ensuring image security. On the basis of zai'z'g, a new algorithm is proposed to change the image pixel position and a Logistic mapping sequence to change the pixel size by using the chaotic sequence pairs generated by 2 coupled chaotic maps. It is proved that the algorithm is simple and easy to implement, and has good encryption effect. It can resist many kinds of attacks, and the image restoration effect is excellent [1].

2 Image Scrambling Encryption Algorithm

Select a digital image to be encrypted matrix J, for each pixel value to do divisible 4 and 4 to take the other operations, the specific calculation is as follows:

$$divi(i,j) = fix\left(\frac{J(i,j)}{4}\right) \tag{1}$$

$$residue(i,j) = mod(I(i,j), 4) \tag{2}$$

© Springer Nature Singapore Pte Ltd. 2017
G. Chen et al. (Eds.): PAAP 2017, CCIS 729, pp. 335–342, 2017.
DOI: 10.1007/978-981-10-6442-5_30

Take the initial value as the initial value of the random sequence, using the formula (2) to iterate, thus generating chaotic sequences and sequences. The chaotic sequence is sorted from small to large way to get a new sequence, and save the new array of elements in the original array of the location, get arrX sequence and arrY sequence.

Use the permutation sequence of the generated double-coupled chaotic sequences to rearrange the resulting matrix of the shuffled 4 of the stored image, the replacement can be used to complete the storage Divide 4 results to complete the matrix scrambling.

Take the initial value, use the formula (1) to generate a one-dimensional logistic sequence of length m, in the sequence from small to large order to get a new sequence, save the new array of elements in the original array of the location, ArrZ, and will arrZ the number of 64 to take over, let his range in the interval [0, 63] to update the arrZ specific operation as follows:

$$arrZ(1,j) = \mod(arrZ(1,j), 64) \tag{3}$$

Step 5: will be different from the XOR, XOR result multiplied by 4, plus, get the final scrambling matrix, the specific operation is as follows:

$$Result(I,j) = (divi(I,j) \oplus arrZ(1,j)) * 4 + residue(i,j) \tag{4}$$

In the process of image encryption, we introduce four secret keys, the secret search space is large, and each key change will have a great impact on the image, thus ensuring the security of the algorithm [4].

3   Logistic Generation of Chaotic Sequences

3.1   Logistic Chaotic

Chaos is defined as random and irregular phenomena in deterministic systems. The behavior of a chaotic system is uncertain, non repeatable and unpredictable. Chaos is the inherent characteristic of nonlinear dynamical systems. It has the characteristics of simple structure, sensitive to initial value and white noise. It improves the security of the encryption system and improves the difficulty of the crack [2]. Chaotic systems play an excellent role in encryption.

Logistic mapping was first proposed by the 1976 mathematical ecologist R. May in the UK's Nature magazine, which originated from the population model. The Logistic map definition sequence is:

$$x_{k+1} = ux_k(1 - x_k) \tag{5}$$

$x_k \in (0, 1)$, u is the branch parameter. When u is in the interval (3.5699456,4), the sequence it generates is very complex and does not follow the rules. At this point, the logistics system is chaotic and produces chaotic sequences. When the U is not within this range, the value will continue to approach the fixed data.

3.2 Logistic Chaotic Mapping

In order to improve the anti – decipherability of the random sequence and increase the secret key space, two random logistic maps are combined to control the security and randomness of the sequence by interleaving the branching parameters of the other [3]. The coupled chaotic map is as follows:

$$\begin{cases} x_{k+1} = u_x x_k (1 - x_k) \\ y_{k+1} = u_y y_k (1 - y_k) \end{cases} \tag{6}$$

3.3 Image Gray Value and '4' Relationship

Grayscale images have a total of 256 gray levels, from 0–255 different represent a different series. When the value of 0–255 on the 4 to be divisible, the business range between 0–63, while the remainder between 0–3. Calculate the maximum value of its quotient multiplied by 4 plus the maximum of the remainder, and its equal to 255 just falls within the gray scale range.

In addition, through the ergodic calculation we can find 0–63 between the value of different or 0–63 any value, the results still fall between 0–63. This gives us a very convenient and simple way to change the image gray value.

4 Scrambling Algorithm

The decryption algorithm is just the opposite of the encryption algorithm. First, the corresponding logistic sequence is computed, and the sequence is sorted to small and large, and the corresponding position sequence is obtained. The number of sequences is accomplished by taking the remainder of 64. Then the ciphertext digital matrix H to 4 to take the operation, to obtain the corresponding remainder matrix. The ciphertext matrix minus the remainder matrix, and then divisible 4, the divisors on the same or on the arrangement of arrZ', the specific formula is as follows:

$$\text{divi}'(i,j) = \text{fix}\left(\frac{H(i,j) - \text{residue}'(i,j)}{4}\right) \oplus \text{arrZ}''(1,j) \tag{7}$$

And then according to the calculation of the corresponding chaotic sequence, to get the corresponding sequence. Will be replaced with. Restore the original position, and finally multiplied by 4 plus the previously obtained the remainder to find the original image, the specific operation is as follows:

$$I(i,j) = \text{divi}'(i,j) * 4 + \text{residue}'(i,j) \tag{8}$$

Generate the I instantly scrambled before the original image. After comparison, after the chaos decrypted the image and the original encrypted image is not the slightest difference, the reduction effect is excellent [5].

5   Simulation Experiment and Result Analysis

5.1   Histogram Analysis

Gray level histogram is a function of gray level, which represents the number of pixels in gray, and can directly reflect the statistical characteristics of the image. Using histograms, you can get a lot of image features.

In the simulation experiment, this paper uses Matlab R2016a on the liftingbody image to scrambling encryption, encryption before and after the histogram, such as Fig. 1 is the original image of the histogram, Fig. 2 is scrambled encrypted histogram.

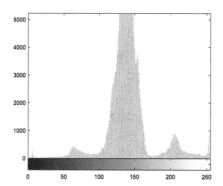

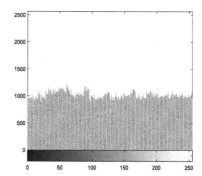

Fig. 1. The original image of the histogram   **Fig. 2.** Scrambled encrypted histogram

It can be seen that the encrypted histogram is more uniform than the histogram before encryption, and the pixel information of the original image is successfully shielded. The encryption effect is good, and it can resist the anti plane attack and statistical analysis effectively [6].

5.2   Secret Key Sensitivity Analysis

$x_0 = 0.666, y_0 = 0.777, z_0 = 0.444, u_z = 3.7$, the liftingbody image is scrambled and encrypted as an initial key set. Its encryption effect as shown below, Fig. 3 for the original image, Fig. 4 for the scrambling image.

From the encrypted image, we almost do not see any information about the original image, we can see that the algorithm encryption effect is good. Now the initial secret key to make minor changes,

$x_0 = 0.677, y_0 = 0.777, z_0 = 0.444, u_z = 3.7, x_0 = 0.666, y_0 = 0.777, z_0 = 0.466$, $u_z = 3.7$ and $x_0 = 0.666, y_0 = 0.777, z_0 = 0.444, u_z = 3.799$, The decrypted images are as follows, Figs. 5, 6 and 7.

As can be seen from the diagram above, even if we use the key different from the original key to decrypt the scrambled image, it can not decrypt the correct image or image contour. The image still has very good scrambling effect. Therefore, the ciphertext search space of this method is very large, and it can resist some exhaustive attacks effectively [7].

Fig. 3. The original image

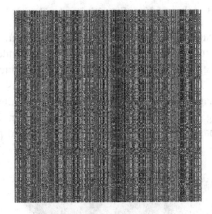

Fig. 4. The scrambling image

Fig. 5. The X changing decrypted images

Fig. 6. The Z changing decrypted images

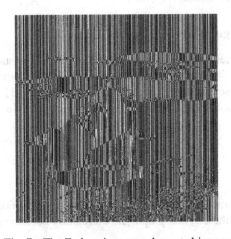

Fig. 7. The Z changing array decrypted images

5.3   Secret Key Sensitivity Analysis

Image collection, processing, storage, transmission and a series of processes often introduce image noise, which has an indelible impact on the quality of the image. Therefore, the noise has a certain resistance is very important.

In the simulation experiment, we use different noise pollution after scrambling images, and observe the image effect after the restoration. As shown in the figure, Fig. 8 uses the salt and pepper noise, Fig. 9 is used Poisson noise. In the contaminated image after the decryption, the decrypted picture is as follows:

Fig. 8. The salt and pepper noise   **Fig. 9.** The used Poisson noise

As can be seen from the figure above, the scrambling method can still retain the basic contours of image and information better after different chaotic noise attacks. It can be seen that the method is more robust against attack. It can resist general noise pollution and is a better scrambling effect.

5.4   Relevant Calculation of Scarcity Evaluation Index

The correlation coefficients of the two images are related to the statistics of the degree of image correlation, which is the statistical parameter which reflects whether the two pairs of images are closely related. The value is in the range [0, 1], the closer the value is to 1, the closer the linearity of the image is, the closer the value is to 0, the less the linear correlation is:

$$r_{xy} = \frac{cov(x, y)}{\sqrt{D(x)}\sqrt{D(y)}} \tag{11}$$

According to the above formula, we calculate the correlation coefficient of the scrambling graph and the original graph, the correlation coefficient of the scrambling graph and the decrypted graph, the correlation coefficient of the decrypted graph and the original image. The calculated values are as follows (Table 1):

Table 1. The correlation coefficient between images.

Image	The original image	Encrypted image	Restore image
The original image		0.0046	1
Encrypted image	*0.0046*		0.0046
Restore image	*1*	0.0046	

It can be seen from the above chart that the correlation coefficient between the encrypted image and the original image is very small, and there is almost no linear correlation between the two, indicating that the correlation between the original image and the scrambled image is very weak, and the effect of the scrambling method is very good. In addition, the restored image and the original picture correlation is 1, indicating that the decrypted image and the original picture no difference [8].

6 Conclusions

In this paper, we propose a scrambling method based on 4-rounding and 4-ary operations. This method introduces a coupled chaotic sequence and a one-dimensional logistic sequence. It uses the coupling sequence to change the pixel position, and uses or operates to change the pixel size. This algorithm introduces four keys, the key search space and the key sensitivity are very high, and it is easy to be cracked. In addition, the algorithm can still recover images before noise. Simulation results show that the algorithm is simple and easy to implement, and can resist all kinds of pollution.

References

1. Pareek, N.K., Patida, V., Sud, K.K.: Image encryption using chaotic logistic map. Image Vis. Comput. **24**(9), 926–934 (2006)
2. Mathews, R., Goel, A., Saxena, P.: Image encryption based on explosive inter-pixel displacement of the RGB attributes of a pixel. In: Proceeding of the World Congress on Engineering and Computer Science, pp. 19–21 (2011)
3. Lee, Y.S., Lee, Y.J.: Performance improvement of power analysis on AES with encryption-related signals. IEICE Trans. Fundam. Electron. Commun. Comput. Sci. pp. 1091–1094 (2012)
4. Li, S., Li, C., Chen, G., et al.: A general quantitative cryptanalysis of permutation-only multimedia ciphers against plaintext attacks. Image Commun. **23**(3), 212–223 (2008)
5. Hong K., Jung K.: Partial encryption of digital contents using face detection algorithm, pp. 632–640, Berling: Springer (2006)
6. Kwok, H.S., Tang, K.S.: A fast image encryption system based on chaotic maps with finite precision representation. Chaos, Solitons Fractals **32**(1), 1518–1529 (2007)

7. Chen, Y., Zhang, W.: A new image scrambling technology based on noncommutative wavelets domain. In: International Symposium on Computer Science and Society, pp. 20–22 (2011)
8. Wang, Z., Bovik, C., Sheikh, H., Simoncelli, E.: From error visibility to structural similarity. IEEE Trans. Image Process. **13**(4), 600–612 (2004)

The Framework of Relative Density-Based Clustering

Zelin Cui[1] and Hong Shen[2,3(✉)]

[1] School of Computer and Information Technology,
Beijing Jiaotong University, Beijing, China
14120379@bjtu.edu.cn
[2] School of Data and Computer Science,
Sun Yat-sen University, Guangzhou, China
hongsh@gmail.com
[3] School of Computer Science, University of Adelaide, Adelaide, Australia

Abstract. Density-based clustering, using two-phase scheme which consists of an online component and an offline component, is an effective framework for data stream clustering, it can find arbitrarily shaped clusters and capture the evolving characteristic of real-time data streams accurately. However, the clustering has some deficiencies on offline component. Most algorithm don't adapt to the unevenly distributed data streams or the multi density distribution of the data streams. Moreover, they only consider the density and centroid to connect the adjacent grid and ignore similarity of attribute value between adjacent grids. In this paper, we calculate the similarity of neighboring grids and take the similarity as a weight that affects the connection of the neighboring grids and propose the relative density-based clustering that cluster the grids based on relative difference model that considers the density, centroid and the weight of similarity between adjacent grids, simply, we connect neighboring grids which are the relative small difference to form clusters on offline component. The experimental results have shown that our algorithm apply to the unevenly distributed data streams and has better clustering quality.

Keywords: Density-based clustering · Relative density-based clustering · Similarity of the neighboring grids

1 Introduction

With the rapid improvement of information systems, modern computers can collect large amounts of data easily. Data mining as an effective technique is applied to analyze the information in decision support systems and plays an important role in practical applications. There are a lot of meaningful and valuable information hidden behind large scale data, such as the statistics of weather information supplied by satellite remote sensors and the forecast of market information of financial and stock markets [6], in such cases, these data are data streams that are real-time, rapid, massive and potentially infinite [4]. How to mine valuable knowledge from data streams is challenging as traditional data mining methods cannot be directly adopted to this new type of data. In

© Springer Nature Singapore Pte Ltd. 2017
G. Chen et al. (Eds.): PAAP 2017, CCIS 729, pp. 343–352, 2017.
DOI: 10.1007/978-981-10-6442-5_31

the research on data stream clustering, the dominant existing algorithms can be divided into two types: single-phase model and two-phase scheme [1] algorithm [3, 7, 13].

The single-phase model divides data streams into many segments as static data and discovers the clusters in these segments using k-means algorithm in a limited space. Actually, it treats the data stream as a successive version of static data [5]. The two-phase scheme [1] is composed of an online component to deal with the original data stream and store statistical information and an offline component to calculate the treated statistics of the data streams and then produce clusters. This model leads to the data stream clustering analysis framework CluStream [1] which is superior to the single-phase model algorithms. However, it still has some deficiencies which is based on the k-means algorithm, be unable to find clusters with arbitrary shapes, is unapplicable to large-scale data stream and be unable to detect outliers and noise effectively.

Density-based clustering which is major clustering algorithm for data stream [11] has been long proposed, it can finds clusters with arbitrarily shaped and detects the noise efficiently, and it is an one-scan algorithm that only examines the original data once. Moreover, it does not need a prior knowledge of the number of the clusters k such as the k-means algorithm does. Then numerous algorithms based on density-based clustering have been put forward, such as D-Stream I [3], DD-Stream [7], D-Stream II [3] and PKS-Stream [10], to overcome the limitations of CluStream. These algorithms map each original data record into a corresponding grids and compute the grid density on online component, and they cluster the neighboring grids based on the density on offline component. These algorithms use a density decaying skill to capture the dynamically changing data streams and exploit the anfractuous factors between data density, the decay factor and cluster shape, these algorithms can generate and adjust the clusters efficiently and effectively. From the Fig. 1, it gives a framework of density-based clustering that how to cluster. But, we discover disadvantages on offline component of the existing algorithms. First, the determination of dense or sparse grid is based on a fixed threshold that do not adapt to the unevenly distributed or the multi density data streams that the density of some clusters is very sparse and others is dense, it is tricky to determine the appropriate threshold of density status. Second, most algorithm only consider the density and centroid to cluster the adjacent grid and ignore similarity of attribute value between adjacent grids. Third, most algorithm don't consider boundary points in sparse grids around a cluster which may be belong to the cluster.

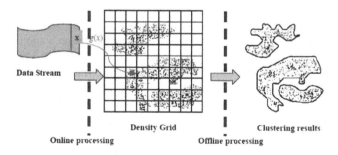

Fig. 1. The density-based clustering framework [3].

Many algorithms based on density-based clustering are usually able to get better clustering results only in the case of a uniform distribution of data sets, but can't achieve satisfactory results for the multi density distribution of the data set or the unevenly distributed data streams. At present, the clustering algorithms used in multi density datasets are Chameleon [8] and SNN [2]. Chameleon and SNN algorithm cluster the multi density data sets effectively, but these algorithms are sensitive to outliers, that lead to a decrease in clustering accuracy, and the time efficiency is high when cluster the large data sets.

In this paper, according to the above three disadvantages, we present improved algorithm to optimize the deficiency and the main contributions of this paper can be summarized as follows:

- We calculate the similarity of neighboring grids and take the similarity as a weight that affects the connection of the neighboring grids.
- We propose the relative density-based clustering that we connect the grids based on relative difference model that considers the density, centroid and weight of similarity between the adjacent grids, we connect the relative small difference neighboring grids to form clusters.

The rest of the paper is organized as follows. In Sect. 2, we introduce the related basic concepts of density-based clustering framework and our algorithms. In Sect. 3, we describe the details of our improved algorithm. In Sect. 4, we show experimental results of our algorithm in comparison with D-Stream and F-Stream on real-world data sets from clustering accuracy and clustering time. We conclude the paper in Sect. 5.

2 Preliminaries

In this section, we introduce the related basic concepts of density-based clustering framework and our algorithms, assume that the original data streams has d dimensions and is defined within d-dimensional data space $S = S_1 \times S_2 \times \ldots \times S_d$ where S_i is the definition space for the ith dimension. Like other density-based clustering algorithms, we partition the data space S into many grid cells.

Definition 3.1 (Grid Cell): For data space $S = S_1 \times S_2 \times \ldots \times S_d$, each $S_i(1 \leqslant i \leqslant d)$ is divided into p_i parts evenly, we defined the intersection of $S_i(1 \leqslant i \leqslant d)$ as the grid cell g. Namely, $g_{j_1 j_2 \cdots j_d} = S_{1,j_1} \cap S_{2,j_2} \cdots \cap S_{d,j_d}, 1 \leqslant j_t \leqslant p_t, 1 \leqslant t \leqslant d$.

So, when a data record $X = (x_1, x_2, \ldots, x_d)$ arrived, it can be mapped to a grid cell $g(x)$ as follows: $g(x) = (j_1, j_2, \cdots, j_d)$ where $x_i \in S_{i,ji}, 1 \leqslant i \leqslant d, 1 \leqslant j_t \leqslant pt, 1 \leqslant t \leqslant d$; The grid density is not a simple sum of data in a grid, in order to show the time characteristics of data stream, for every data record X, we use decay factor coefficient $\lambda \in (0, 1)$ [3]. The worth of the old data decays and the grid density decreases with time.

Definition 3.2 (Grid Density): The grid density is the sum of the current density of each data points in the grid cell. Namely, the density of g at t is $D(g,t) = \sum_{X \in g} D(X,t)$, where $D(X,t) = \lambda^{t-t_c}$ is density coefficient of X, t is now time and t_c is arrival time of X.

Although grid density is different with time, it is unnecessary to calculate grid density at every time step. In truth, only when a new data record arrives at that grid, it is necessary to calculate new grid density proven in [3]. Suppose the grid receives a new data at t_n and receives the last data record is $t_l(t_n > t_l)$, namely the density of g at t_n can be renewed as follows:

$$D(g, t_n) = \lambda^{t_n - t_l} D(g, t_l) + 1 \tag{1}$$

In this paper, we no longer calculate the status of grid density that is dense, sparse or transitional, and derive clusters not just using the density information, but based on relative difference model that considers the density, centroid and similarity of attribute value detailed in Sect. 3.2. In order to advance the accuracy, we use the weighted centroid [9] for grid cell. The weighted centroid of the grid cell is not only related to the coordinates of the data points, but also related to the weight of the data points.

Definition 3.3 (Weighted Centroid): For a grid cell g, at a given time t, namely, the weighted centroid of g at t is $X_{cen}(g, t) = \sum\limits_{X \in g} D(X, t)X/D(g, t)$. So for the grid cell g and h, at a given time t, the distance between the weighted centroid of g and h is $dist(g, h, t) = \sqrt{(X_{cen}(g, t) - X_{cen}(h, t))^2}$. In order to get better clustering, we define a grid characteristic vector for each non-null grid cell, our relative density-based clustering is based on the information of it detailed in Sect. 3.2.

Definition 3.4 (Grid Characteristic Vector): The characteristic vector of non-null grid g is defined as $(t, D(g,t), X_{cen}(g,t), label, snapshot)$, where t is the last time when g receive data, $D(g,t)$ is the last updated density of g, $X_{cen}(g,t)$ is weighted centroid, *label* is the class of g and *snapshot* contains specific mapped data points.

Intuitively, a cluster is made up of more close density grids and more similar attribute value than the surrounding grids. Note, we all the way try to merge various shapes clusters whenever possible, so the resulting clusters are encompassed by relative remote, sparse, and large differences grids. When the clustering is finished at each gap, we clean the snapshot and save other vector values in grid characteristic vector that be used for next gap.

3   Relative Density-Based Clustering and Boundary Detection for Data Stream

The whole algorithm of relative density-based clustering is described in Algorithm 1. For a continuous data stream, our algorithm map the data record into a corresponding discretized density grid and renew the characteristic vector of the grid cell on the online component. Our algorithm dynamically adjusts the clusters by our relative density-based clustering described in detail in Sect. 3.2 at each *gap* time steps, that *gap* is an parameter.

Algorithm 1 The Framework of Relative Density-Based Clustering

---

Input: data stream S
Output: clusters
(1) time = 0;
(2) create an empty gird − list;
(3) **while**(read record X =< X1, X2, · · · , Xd > from S)
(4)   map X into the corresponding discretized density grid g;
(5)   **if**(g not in grid − list)
(6)   add g into grid − list;
(7)   **end if**
(8)   renew the grid characteristic vector of g;
(9)   **if**(time/gap == 0)
(10)   clusters = call **relative density-based clustering**(gird − list);
(11)   **end if**
(12)   time = time+1;
(13) **end while**

3.1 Weight of Similarity Between the Adjacent Grid

As we all know, the similarity between two data-sets can be regarded as a numerical measure that the more the same attributes value are, the closer the similarity is. For the d dimensions data stream, each grid cell has maximum 3d − 1 adjacent grid cells, and we note that the adjacent non-null grid cells of g is $adj(g)$, so $|adj(g)| \leq 3^d - 1$. For example of two-dimensional space in Fig. 2, we can see that the g has 8 adjacent grid cells. In fact, the grid cells B, D, E, H and the g is connected by the public edge which contain the same one dimensional attribute value, and the grid cells A, C, F, I and the g is connected by a common vertex which have not the same attribute value. So we can't treat these adjacent grid cells equally, the similarity between the grid cells B, D, E, H and the g should be more close than the grid cells A, C, F, I and the g apparently. According to such case, we propose weight of similarity between the adjacent grid. Suppose the weight of similarity between grid B and g is w_1, and the weight of similarity between grid A and g is w_2, so $w_1 > w_2$.

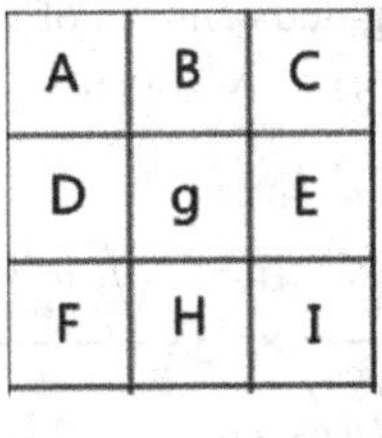

A	B	C
D	g	E
F	H	I

Fig. 2. The adjacent grid cells of g.

Definition 4.1 (The Weight of Similarity): For data space $S = S_1 \times S_2 \times \ldots \times S_d$, the grid cell g and one of the $adj(g)_i$, the Weight of Similarity between g and $adj(g)_i$ is:

$$\text{WOF}(g, \text{adj}(g)_i) = \beta^{d-1-\text{attr}(g,\text{adj}(g)_i)} \tag{2}$$

where $\beta \in (0,1)$ is coefficient, $1 \le i \le 3d - 1$ and $attr(g, adj(g)_i)$ is the number of the same dimension value between g and $adj(g)_i$.

3.2   Relative Density-Based Clustering

For the multi density data streams, for example of two-dimensional space in Fig. 3, we can see the density of cluster A is very dense and the density of cluster B is very sparse. If the threshold of density status is set inaccurate, as in D-Stream [3], the cluster B is considered as outliers. In practice, it is tricky to determine the appropriate threshold of density status for unknown data set. In this sub-section, we propose a relative density-based clustering algorithm on the offline component. First, we find the gird cell g which has the highest density in all gird cells without new label ($D(g,t) > Min_{density}$, where $Min_{density}$ is threshold) and make a new label for g. Second, we calculate the distance between g and each adj(g) using the formula 3 that the distance is the relative difference between two grids. And if $distance(g, adj(g)_i) \le \theta$ (θ is threshold for relative distance), we put the $adj(g)_i$ into a list named satisfy-list, and make $adj(g)_i$ the same new label with g. Last, repeat the Second work based on each point in satisfy-list replace g, if it has not satisfying gird cell that satisfy-list is empty, one cluster with new label is connected and repeat the all above work to connect another cluster. The algorithm of relative density-based clustering is described in Algorithm 2.

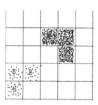

Fig. 3. The unevenly distributed data streams.

Definition 3.2 (Distance between Two Adjacent Grids): For two adjacent grids g_1, g_2, and the density of g_1 is $D(g_1,t)$, weighted centroid of g_1 is $X_{cen}(g_1,t)$, and the density of g_2 is $D(g_2,t)$, weighted centroid of g_2 is $X_{cen}(g_2,t)$.

Namely, the distance(g_1,g_2) is:

$$\text{distance}(g_1, g_2) = \delta_1 \cdot \frac{\dfrac{1}{WOF(g_1,g_2)}}{\dfrac{|D(g_1,t) - D(g_2,t)|}{D(g_1,t)}} \times \delta_2 \cdot \frac{\sqrt{(X_{cen}(g_1,t) - X_{cen}(g_2,t))^2}}{2\sqrt{dlen}} \times \delta_3 \tag{3}$$

where $\delta_1, \delta_2, \delta_3 \in (0, 1)$ is weight factor coefficient, $\delta_1 + \delta_2 + \delta_3 = 1$, *len* is the size of gird, t is now time.

Algorithm 2 The Relative Density-Based Clustering Algorithm

Input: grids G=<g1,g2,...,gn>, D(gi,t)>Mindensity

Output: clusters

(1) **while**(G is not empty)

(2)   Basic point g = find the grid with the highest density and without new label in G;

(3)   make new label for g and put the g into a list satisfy-list;

(4) **end while**

(5) **while** satisfy-list is not empty

(6)   g = satisfy-list.pop();

(7)   **for** adj(g)i to adj(g) do

(8)   **if** distance(g,adj(g)i) ≤ θ and adj(g)i is without label

(9)   make adj(g)i the same new label with g and put adj(g) into clusters;

(10)   put adj(g)i into satisfy-list;

(11)   **end if**

(12) **end for**

(13) remove g from G;

(14) **end while**

(15) **return** clusters;

---

4 Experimental Results

We evaluate the precision, recall and efficiency of our algorithm and compare it with D-Stream [3] and F-Stream [9]. The experiment environment: Intel Core i3-3220 3.3 GHz CPU and 4G memory in Windows 7. We test the real data set KDD CUP-99 that is network intrusion detection data set and has been cited by many articles of data stream clustering. It collected 9 weeks of TCPdump (*) network connection and system audit data by the MIT Lincoln laboratory which contains the simulation of various types of users, a variety of network traffic and attack means, and it like a real network environment and the network intrusion detection data stream. It contains a total of 41 dimensional properties, of which 34 are continuous attributes. Each data stream of network connection is marked as normal or abnormal, and the abnormal type is sub-divided into 4 main categories that are DOS, R2L, U2R and PROBING. In each of the experiments, we standardize all the attributes value of the data to [0,1].

In order to compare with other algorithms, we use the evaluation indexes which are precision, recall, comprehensive index F1 and time consumption. If a is the data set belong to the class 1, b is the data set that our algorithm label the class 1 and c is the data set that are correctly classified. The precision and recall are defined as

$$\text{Precision} = \frac{c}{b} \text{ and Recall} = \frac{c}{a} \tag{3}$$

The comprehensive index F1 is defined as

$$F1 = \frac{2 \times \text{Precision} \times \text{Recall}}{\text{Precision} + \text{Recall}} \tag{4}$$

4.1 Result Comparison Using the Real Data

In this subsection, we test our algorithm on the KDD CUP-99 data set described above that use $\beta = 0.9, \delta_1 = 0.1, \delta_2 = 0.1, \delta_3 = 0.8, \theta = 0.85, len = 0.5, \gamma = 0.5, \lambda = 0.998$, and we compare the precision, recall and F1 of the results with D-Stream and F-Stream. From the result in Fig. 4, We can see the precision and the recall is better than D-Stream and F-Stream at various times, So the average the precision, recall and F1 is much better than D-Stream and F-Stream.

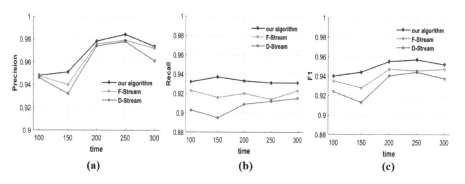

Fig. 4. The precision, recall and F1 comparison on boundary point sets with time.

4.2 Time Performance Comparison

First, we test the processing speed of our algorithm, D-Stream and F-Stream on different sizes of data set that have the some 10-dimensional attributes. From the result of

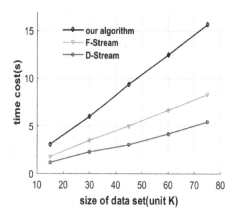

Fig. 5. Efficiency comparison with different sizes of data set.

the Fig. 5, we can see the efficiency of our algorithm increases linearly with the sizes of data set, and is little lower than others efficiency. Next, we test the processing speed of our algorithm, D-Stream and F-Stream on different dimensions of data set whose size is 100 K. From the result of the Fig. 6, we can see the efficiency of our algorithm increases nonlinearly with the dimensions of data set, and is lower than others efficiency varying the dimensionality in the range of 2 to 40.

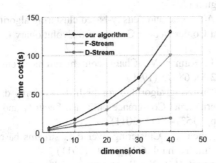

Fig. 6. Efficiency comparison with different dimensions of data set.

5   Conclusion

In this paper, according to defects of density-based clustering using two-phase scheme, we present improved algorithm to optimize the deficiency. First we calculate the similarity of neighboring grids and take the similarity as a weight that affects the connection of the neighboring grids. Second, we propose the relative density-based clustering that we cluster the grids based on relative difference model that considers the density, centroid and similarity of attribute value, and we connect the relative small difference neighboring grids to form clusters. Compared with D-Stream and F-Stream, although we are less efficiency than them, our algorithm has better clustering quality from experimental results.

Acknowledgment. This research is sustained by Research Initiative Grant of Sun Yat-sen University under Project 985 and Australian Research Council Discovery Projects funding DP150104871.

References

1. Aggarwal, C.C., Yu, P.S., Han, J., Wang, J.: A framework for clustering evolving data streams. In: International Conference on Very Large Data Bases, pp. 81–92 (2003)
2. Chen, X., Liu, S., Chen, T., Zhang, Z., Zhang, H.: An improved semi-supervised clustering algorithm for multidensity datasets with fewer constraints. Procedia Eng. **29**(4), 4325–4329 (2012)

3. Chen, Y., Tu, L.: Density-based clustering for real-time stream data. In: Proceedings of the 13th ACM SIGKDD International Conference on Knowledge Discovery and Data Mining, pp. 133–142 (2007)

4. Guha, S., Mishra, N., Motwani, R., O'Callaghan, L.: Clustering data streams. In: Symposium on Foundations of Computer Science, pp. 359–366 (2000)

5. Guha, S., Meyerson, A., Mishra, N., Motwani, R., O'Callaghan, L.: Clustering data streams: theory and practice. IEEE Trans. Knowl. Data Eng. **15**(3), 515–528 (2003)

6. Han, J., Kamber, M., Pei, J.: Data Mining: Concepts and Techniques. Morgan Kaufmann Publishers Inc., Burlington (2011)

7. Jia, C., Tan, C., Yong, A.: A grid and density-based clustering algorithm for processing data stream. In: International Conference on Genetic and Evolutionary Computing, pp. 517–521 (2008)

8. Karypis, G., Han, E.H., Kumar, V.: Chameleon: hierarchical clustering using dynamic modeling. Computer **32**(8), 68–75 (1999)

9. Liu, W., Ouyang, J.: Clustering algorithm for high dimensional data stream over sliding windows. In: IEEE International Conference on Trust, Security and Privacy in Computing and Communications, pp. 1537–1542 (2011)

10. Ren, J., Cai, B., Changzhen, H.: Clustering over data streams based on grid density and index tree. J. Converg. Inf. Technol. **6**(1), 83–93 (2011)

11. Sander, J., Ester, M., Kriegel, H.P., Xu, X.: Density based clustering in spatial databases: the algorithm gdbscan and its applications. Data Min. Knowl. Discov. **2**(2), 169–194 (1998)

Efficient Algorithms for VM Placement in Cloud Data Center

Jiahuai Wu[1] and Hong Shen[1,2(✉)]

[1] School of Data and Computer Science,
Sun Yat-sen University, Guangzhou, China
849197033@qq.com, hongsh01@gmail.com
[2] School of Computer Science, University of Adelaide, Adelaide, Australia

Abstract. Virtual machine (VM) placement problem is a major issue in cloud data center. With the rapid development of cloud computing, efficient algorithms are needed to reduce the power consumption and save energy in data centers. Many models and algorithms are designed with an objective to minimize the number of physical machines (PMs) used in cloud data center. In this paper, we take into account the execution time of the PM, and formulate a new optimization problem of VM placement, which aims to minimize the total execution time of the PMs. We discuss the NP-hardness of the problem, and present heuristic algorithms to solve it under both offline and online scenario. Furthermore, we conduct experiments to evaluate the performance of the proposed algorithms and the result show that our methods are able to perform better than other commonly used algorithms.

Keywords: Cloud data center · Virtual machine placement · Bin packing · Heuristic algorithm

1 Introduction

Cloud computing has become an emerging technology that transforms the IT industry and affects people's lives in recent years. Today in most modern cloud data centers, such as Amazon EC2 [16] and Google data center, there are a large number of physical machines (PMs), also called servers or hosts, and the total number of the PMs in each data center can reach hundreds of thousands. However, due to the uneven resource demand of the applications, a good few of the PMs have very low utilization most of the time. An unnecessarily great number of PMs have to be opened with high cost for the management and maintenance, which results in a serious waste of resources. Hence, virtualization technology [2] is applied to settle these issues.

With the tremendous benefits of virtualization, applications are running on the VMs, not directly on the PMs. Furthermore, a single PM can accommodate multiple VMs as long as their resource demands are satisfied. In another word, the applications may be able to share the resources on the PMs in an isolated way. And in the majority of cases the load of a VM has almost no effect on the performance of the co-located VMs [15]. In some cloud case, particularly in the Infrastructure-as-a-Service (IaaS), the VMs are provided directly to the customers. Typically, the VM placement problem is a serious

© Springer Nature Singapore Pte Ltd. 2017
G. Chen et al. (Eds.): PAAP 2017, CCIS 729, pp. 353–365, 2017.
DOI: 10.1007/978-981-10-6442-5_32

challenge for the data centers. Usually, the customers submit their resource requirements in terms of the basic resource, including CPU, memory, network, etc., to the cloud system, and the cloud system needs to decide the resource allocation. As the cost caused by PMs is proportional to the number of running PMs [10], lots of research aim to use the virtualization technology to consolidate the VMs onto a smaller number of PMs so that the saved PMs can be switched to a low power mode or shut down. Therefore, the objective is to minimizing the number of the PMs that hosts the requested VMs.

To study the VM placement problem for PM cost minimization. We explain that the cost caused by PMs is proportional to not only the number but also the execution time of the PMs. We further assume that the VM requests from the tenants contain both resources demand and the running time of VMs, and the PM cost is mainly determined by the total execution time of all the used PMs accordingly. For example, in Fig. 1, there are three VMs submitted to the cloud system with the same resource demand. Their running time are 3, 4 and 5 respectively, Fig. 1a and b are two different VM placements. As a result, the placement in Fig. 1b has fewer PM cost since the total execution time of the PMs are fewer.

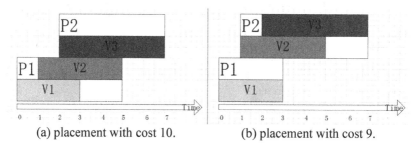

| (a) placement with cost 10. | (b) placement with cost 9. |

Fig. 1. Two placements for the same request (V1, V2, V3). All the three VMs are of half size of the PM and different in the running time.

Obviously, packing VMs onto a number of PMs and minimizing the total execution time of the PMs is an effective way to reduce the cost in cloud data center. In this paper, we formulate the VM placement problem for minimizing the total execution time of the used PMs under both off-line and on-line scenarios. Due to the NP-hardness of this problem, we propose heuristic algorithms to give an efficient method.

The remainder of this paper is organized as follows. First, we make a summary of the related work in Sect. 2. Then we introduce the problem of minimizing the total execution time of the PMs for VM placement in Sect. 3. In Sect. 4, we propose heuristic algorithms for both off-line and on-line scenarios. Experiments and performance evaluation follows in Sect. 5, and Sect. 6 concludes this paper finally.

2 Related Work

Virtual machine placement problem is one of the major issues in cloud data center. An ideal approach can be greatly beneficial to both cloud users and service providers. Recently, a great deal of algorithms and models have been proposed. There are various objectives of these works, which play important roles in the data centers, the VMs and their executions [13]. Due to the features of cloud computing, there have been lots of new models and algorithms with different constraints and objects, for example, availability [7], fairness [17], energy [3] and the communication cost between VMs [4, 11]. In addition, as the result in [5] shows that the physical machines will consume about forty-five percent overall cost in cloud data center, the PM cost is another important objective being considered in the literature, and the goal is to minimizing the number of the used PMs.

Many VM placement schemes consider the VM placement problem as the classical bin packing problem, which is the common way to deal with this problem. In [6], the authors design a new model for saving the power in real-life cloud computing, and they present three different bin packing algorithms with different aspects of the cloud as constraints, it is concluded that simple bin packing algorithms like 2D and 3D bin packing algorithms can reduce power consumption. Due to hardness of bin packing problem, other approach should be applied to solve it with an acceptable complexity, so evolutionary algorithms are very common. In [1], the authors use the modeling of multiple knapsack problem for VM placement problem, they give an algorithm based on ant colony and compare the performance with other solutions. And in [8], a hybrid genetic algorithm using Best Fit Decreasing is designed to deal with infeasible solution due to the bin-used representation, they also conduct experiments to show better performance of their algorithm.

When considering the online scenario, other heuristic methods are proposed. In [18], the authors propose a new energy-aware approach based on the online bin packing algorithm to improve the energy efficiency and resource utilization in cloud data center. In order to deal with the varying resource demands from users, they present an over-provision method. Another online-bin-packing-based research is discussed in [14]. In their work, the VMs, regarded as items in bin packing problem, are divided into four types based on their sizes, and the PMs (called the bins) can be also divided into different types. Their main idea is to keep the gap of most bins within 1/3, and the algorithm can achieve an approximation ratio of 3/2 while the number of movements of the primitive operations, insert and change, are at most 6, which dynamically based on application demands and support green computing. [9] extends the previous one's work and achieve a better approximation ratio of 4/3 with an even finer division of the item types, the proposed approach supports all operations that [14] has constructed but moves at most ten items per operation, which may be a little worse than the 3/2 approximation one. Considering different factors of the cloud environment, there can be different aspects that make VM placement more complex than bin packing. Besides the above works, there are other variants of bin packing problem for modeling VM placement problem [12]. When PM capacities and VM sizes account for different resource types, the problem becomes the vector bin packing problem, when PMs are characterized in different sizes which is motivated by its heterogeneity, the problem becomes the variable sized bin packing, when the VM requests change during the time, the problem becomes the dynamic bin packing problem.

As a result, the goal in most of the work is to minimize the number of the PMs, taking into account the resource demands of the VM, the capacity of the PMs and other factors in cloud computing environment. None of them consider the execution time. In this paper, we aim to reduce the number of PMs and minimize the total execution time of all the PMs.

3 Problem Description

In this section, we study the VM placement problem under the offline scenario, and there are some differences between offline and online VM placement. Under the offline scenario, we know all the information about the requested VMs set a priori, such as the number of VMs and the resource requirement of each VM, while under online scenario, the VM requests are coming one by one without knowing the information beforehand.

Suppose that there are infinite PMs in the cloud system, where the capacity of each PM is C. And n VMs are waiting to be allocated to the PMs. Different from most of the exist works, we put emphasize on not only the total number of PMs but also the duration each PM stay active when calculating the energy consumption by PMs. In this paper, therefore, we aim to find a VM placement that minimize the total cost of the PMs actively used over the whole period.

We first define the notations used in this problem (Table 1).

Table 1. Notations

Notation	Description
V	The set of VMs
n	The number of VMs
v_i	The ith VM in V
R_i	The resource demand of v_i
Su_i	The submission time of v_i
L_i	The running time of v_i
P	The set of PMs
p_i	The ith PM in P
St_i	The start time of p_i
E_i	The execution time of p_i
y_{jk}	p_i is used in time k if $y_{jk} = 1$
x_{ijk}	v_i is allocated on p_i in time k

We study the VM placement problem for minimizing the cost caused by the utilization of PM, we called PM cost. We assume that all PMs are homogeneous with unit capacity and normalize the resource demand of the VM to be a fraction of the capacity. For example, when VM needs 10% of the memory of the host PM, its size can be defined as 0.1.

Obviously, the PM cost is not only proportional to the number of PMs but also the total execution time of all PMs. In this problem, the customers submit their request of a set of VMs, where each VM is define as the resource size and the running time length.

Without loss of generality, we assume that the request from customers and the response measures from cloud system will be completed in every unit time. For simplicity, we set the cost rate of unit as 1, and the unit time is also indicated by 1. The problem seeks to determine a set of PMs that can accommodate the requested VMs with the minimum PM cost. Accordingly, the object of the VM placement problem is to minimize the total execution time of all the PMs. We use E_i to represent the executing time of p_i, and the problem can be formulated as:

$$\text{minimize} \sum\nolimits_{i=1}^{n} \sum\nolimits_{j=1}^{n} \sum\nolimits_{k=\min_{v_i \in V} Su_i}^{\max_{v_i \in V}(Su_i + E_i)} x_{ijk}$$

Subject to:

$$\sum_{i=1}^{n} R_i \cdot x_{ijk} \leq y_j, j = \{1, 2, 3 \ldots n\}, k = \{St_j, St_j + 1, \ldots St_j + E_j\}$$

$$x_{ijk} \in \{0, 1\}, y_i \in \{0, 1\}$$

This problem is NP-hard because, when the submission time and the running time of each VM are identical, the problem then becomes a variant of the classical one dimensional bin packing problem, which is one of the well-studied NP-hard problem. Note that here we are minimizing the total execution time of the PMs, rather than the number of the PMs in bin packing problem. The result is that our problems are more complicated. Therefore, we prepare to present heuristic algorithms to find an acceptable solution.

4   Algorithm

4.1   Offline Problem Analysis

In this section, we analyze the offline VM placement, which aims to minimize the total execution time of all the PMs. We know all the information about the requested VMs set a priori, including the number of VMs, the resource requirement and the submission time of each VM. Because of the order in submission time, the VMs may not be placed arbitrarily. Thus, we need to sort the VMs increasingly by submission time at the beginning of our algorithm. Then we describe the placement strategy to place the VMs one by one in the time order.

Our basic heuristic idea is based upon the resource utilization rate in a period of time. In order to explain the idea more clearly, we use an example here. We assume that in the time T, there is an active PM p. Let the residual capacity of resource in p be 0.4, and the planned execution time is 4. At the same time two requested VMs v_1 and v_2 have same resource demand of 0.3. Then we have two possible options below according to their running time.

(1)   The running time of v_1 is 3, which is smaller than the planned execution time of p, the resource utilization rate is *(0.6 * 4 + 0.3 * 3)/4 = 0.825.*

(2) The running time of v_2 is 5, which is larger than the planned execution time of p, the resource utilization rate is *(0.6 * 4 + 0.3 * 5)/5 = 0.78.*

Thus, under this circumstance we will place v_1 onto p, since that it can achieve a higher resource utilization rate in the coming period of time.

Let $U(p, v)$ represents the resource utilization rate when placing v onto p, C represents the residual capacity of resource in p, E represents the planned execution time of p, R represents the resource demand of v, L represents the running time of v. Then the calculation formula of resource utilization rate is given.

$$U(p, v) = \frac{(1 - C) * E + R * \max(E, L)}{\max(E, L)}$$

The resource utilization rate $U(p, v)$ can describe the resource usage in the next period of time. Higher rate means that more resources are more likely to be fully utilized. Based on this we propose a heuristic algorithm to reduce the total execution time of all the PMs (Table 2).

Table 2. Offline algorithm.

Algorithm 1 Heuristic algorithm for offline VM placement.
Input: V={v_1, v_2, v_3...v_n}
Output: P={p_1, p_2, p_3...p_m}, x
1: P'← ∅
2: Sort V increasingly by Su and then decreasingly by R and L
3: for i =1 to n do
4:   if ∃p_i ∈ P', C_i ≥ R_1 then
5:   p ← $arg_{p_i∈P}$max $U(p_i, v_i)$
6:   allocate v_i to p
7:   update P' and x
7:   else
8:   open new_p and allocate v_i to new_p
9:   add new_p to P'
10:   update P' and x
11:   end if
12: end for
13: return P' and x

4.2   Online Problem Analysis

In this section, we present a greedy strategy for VM placement under the online scenario. In this problem, the requests of VMs are submitting one by one. Hence, we should place VM one after another according to the current usage of the PM. Similar to the dynamic bin packing problem, we first study two commonly used packing algorithms, First Fit and Best Fit. Their main idea will be described below.

Each time when there is a new VM request submitted, First Fit seeks to allocates it to earliest opened PM that can accommodate it, while Best Fit seeks to allocate it to the suitable PM with the minimum residual capacity. And if all of the opened PMs are available for the new requested VM, the cloud will open a new PM to accept it. When there are no VMs running, the PM can be switched to a low power mode or shut down.

It should be noted that under the online scenario, the opened PM may have a planned execution time obtained by the allocated VMs. Our greedy algorithm is learned from the Best Fit algorithm, and the basic idea is discussed then. When a new requested VM submits, suppose that there are a set of opened PMs with the information of remaining resources R and planned execution time E. The greedy algorithm helps to develop the strategy to allocate the VM. We have three cases to analyze: (1) There are opened PMs that can accommodate the VM and the planned execution time of the PMs is longer than its running time, then we allocate the VM to the suitable PM that with the minimum residual capacity. (2) There are opened PMs that can accommodate the VM but none of them has a longer execution time than the VM's running time, then we allocate the VM to the PM with a longest planned execution time, if there are multiple eligible PMs, choose the one with the minimum residual capacity. (3) Otherwise, open a new PM and put the VM in it.

Table 3. Online algorithm.

Algorithm 2 Greedy strategy for online VM placement .

Input: v = (R, L): The new requested VM.

   P={p_1, p_2, p_3...p_m}: The set of opened PMs.

Output: p: A suitable PM to accommodate v.

1: Find the VM v_i with the maximum residual capacity.

2: if R > C_i then

3:   open p_{m+1} and allocate v to p_{m+1}

4:   return p_{m+1}

5: else

6: $P' = arg_{p_i \in P} min\ Inc(v, p_i)$

7:   p = $arg_{p_i \in P'} min\ C_i$

8: end if

9: return p

According to above analysis, we let *Inc (p, v)* represents the cost increment when VM v is allocated to PM *p*, where the resource demand is *R*, running time is *L* and planned execution time of *p* is *E*. In case (1) and (2), *Inc (p, v)* is 0 and *L-E*, while in case (3) it is undefined. Based on this strategy we design a greedy algorithm and the details can be described as following (Table 3):

5   Experiments

In this section, simulation experiments are conducted to evaluate the performance of the proposed algorithms under offline and online scenario. And we evaluate the performance of the algorithms by comparing the results to other common used bin packing algorithms.

5.1   Simulation Settings

The experiments are conducted on Intel(R) Core(TM) i7 processor @ 12.0 GB using C++. Originally, we generate several problem instances with different sizes of requested VMs. We also conduct groups of experiments for VM placement under offline and online scenario respectively. In our simulations, the size of the VMs range from 0 to 1, the submission time range from 1 to 10, and the running time range from 1 to 20. Our goal is to find a VM placement in this period to reduce the total execution time of all the PMs.

For the purpose of comparison, we also implement the common bin packing algorithms, such as FF (First Fit), BF (best fit), FFD (First Fit Decreasing), BFD (Best Fit Decreasing). And there are two evaluation standards for the algorithms in this problem. One is the number of used PMs, and the other is the total execution time.

5.2   Simulation Results

The simulations are divided into three case studies. In Case 1 and Case 2, we evaluate the performance of the VM placement algorithms under offline and online scenario respectively. We compare the offline placement algorithm with FFD and BFD, while the online placement algorithm with FF and BF. And the comparison sets of simulations are conduct in Case 3.

Case 1: Offline Placement Algorithm
In the first case, we conduct four groups of simulations, in which the default number of VMs are 200, 400, 600, and 800 respectively. We compare our proposed heuristic algorithm (PHA) with FFD and BFD by the two standards.

Figure 2 shows the number of PMs that the three different algorithms produce respectively under the offline scenario. The x-coordinate is the number of the requested VMs, while the y-coordinate indicates the number of PMs used to host the VMs. We can see that in the same case, BFD has fewer PMs than FFD, while our algorithm has the minimum PMs.

Figure 3 shows the total execution time of all the used PMs that the three different algorithms produce respectively under the offline scenario. The x-coordinate is the total execution time, besides, Fig. 3 has the same format as Fig. 2 As the result, BFD and FFD produce more execution time than our proposed algorithm.

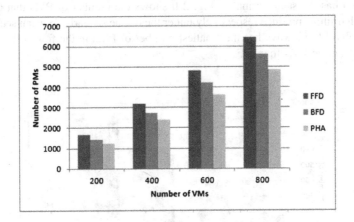

Fig. 2. Number of PMs under offline scenario

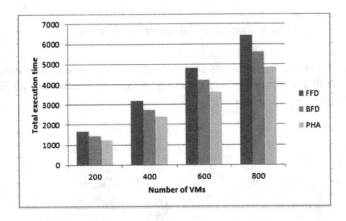

Fig. 3. Total execution time of PMs under offline scenario

According to the results, we have the observation that our proposed heuristic algorithm has a better performance than both BFD and FFD. This is because a request VM is more likely to be placed on the PM with less resource left while taking the running time into account, which enhances the resource utilization rate during the placement period.

Case 2: Online Placement Algorithm

In the second case, we evaluate our proposed algorithm under the online scenario, which is also a modified best fit algorithm (MBF). We compare MBF with FF and BF by the two standards. We use the same problem instances in case 1 to evaluate the performance of the online placement algorithm.

Figure 4 has the same format as Fig. 2 It shows the number of PMs that the three different algorithms produce respectively under the online scenario. We can see that our proposed algorithm always has the smallest number of PMs in the four instances, and FF has a bit more PMs than BF.

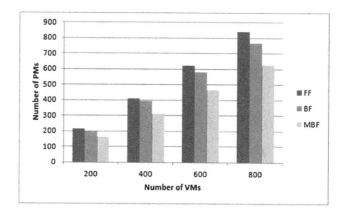

Fig. 4. Number of PMs under online scenario.

Figure 5 has the same format as Fig. 3. It shows that BF and FF produce more execution time than our proposed algorithm under the online scenario, and FF always produce the most execution time.

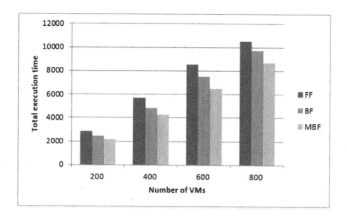

Fig. 5. Total execution time of the PMs under online scenario.

According the results, we can make a conclusion that our proposed heuristic algorithm has a better performance than both BF and FF under the online scenario. This is because when a new requested VM is coming, our placement strategy tries to choose a PM to place on with the minimum execution time increasing.

Case 3: Analysis of Offline and Online Algorithm
Finally, in case 3, we conduct the experiments to compare both of our proposed algorithms. Figures 6 and 7 shows the number and the total execution time of the PMs of our proposed offline and online algorithms. The online algorithm has more (more than about 1.8 times) PMs and execution time than the offline algorithm. Furthermore, the online algorithm brings more PMs and execution time than the offline algorithm as the number of requested VMs increasing. This is due to the characteristic of the problem that we know all the information about the requested VMs under the offline scenario, while the online algorithm only gives the only coming VM.

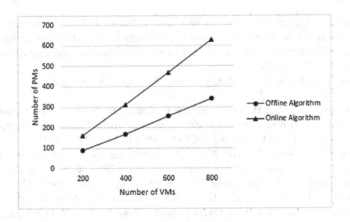

Fig. 6. Number of PMs.

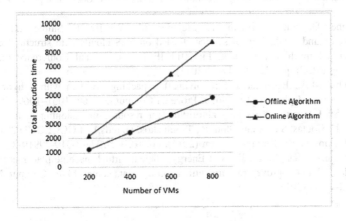

Fig. 7. Total execution time of the PMs

6 Conclusion

In this paper, we study the VM placement problem for minimizing the cost caused by the PMs in cloud data center. We aim to reduce the total execution time of the used PMs. We formulate a new optimization problem for deciding the placement of VMs in cloud data center. Moreover, we propose heuristic algorithms for both off-line and on-line scenarios to solve the problem. To evaluate the performance, simulation experiments are conducted to observe the placement of VMs. In the future, we plan to further optimize the performance of our algorithm, and give some theoretical proof.

Acknowledgement. This work is supported by Research Initiative Grant of Sun Yat-sen University under Project 985 and Australian Research Council Discovery Project DP150104871.

References

1. Amarante, S.R.M., Roberto, F.M., Cardoso, A.R., Celestino, J.: Using the multiple knapsack problem to model the problem of virtual machine allocation in cloud computing. In: 2013 IEEE 16th International Conference on Computational Science and Engineering (CSE), pp. 476–483. IEEE (2013)
2. Anderson, T., Peterson, L., Shenker, S., Turner, J.: Overcoming the internet impasse through virtualization. Computer **38**(4), 34–41 (2005)
3. Dong, J., Jin, X., Wang, H., Li, Y., Zhang, P., Cheng, S.: Energy-saving virtual machine placement in cloud data centers. In: 2013 13th IEEE/ACM International Symposium on Cluster, Cloud and Grid Computing (CCGrid), pp. 618–624. IEEE (2013)
4. Fukunaga, T., Hirahara, S., Yoshikawa, H.: Virtual machine placement for minimizing connection cost in data center networks. In: 2015 IEEE Conference on Computer Communications Workshops (INFOCOM WKSHPS), pp. 486–491. IEEE (2015)
5. Greenberg, A., Hamilton, J., Maltz, D.A., Patel, P.: The cost of a cloud: research problems in data center networks. ACM SIGCOMM Comput. Commun. Rev. **39**(1), 68–73 (2008)
6. Hage, T., Begnum, K., Yazidi, A.: Saving the planet with bin packing-experiences using 2D and 3D bin packing of virtual machines for greener clouds. In: 2014 IEEE 6th International Conference on Cloud Computing Technology and Science (CloudCom), pp. 240–245. IEEE (2014)
7. Jayasinghe, D., Pu, C., Eilam, T., Steinder, M., Whally, I., Snible, E.: Improving performance and availability of services hosted on IaaS clouds with structural constraint-aware virtual machine placement. In: 2011 IEEE International Conference on Services Computing (SCC), pp. 72–79. IEEE (2011)
8. Kaaouache, M.A., Bouamama, S.: Solving bin packing problem with a hybrid genetic algorithm for VM placement in cloud. Procedia Comput. Sci. **60**, 1061–1069 (2015)
9. Kamali, S.: Efficient bin packing algorithms for resource provisioning in the cloud. In: Karydis, I., Sioutas, S., Triantafillou, P., Tsoumakos, D. (eds.) ALGOCLOUD 2015. LNCS, vol. 9511, pp. 84–98. Springer, Cham (2016). doi:10.1007/978-3-319-29919-8_7
10. Li, X., Qian, Z., Sanglu, L., Jie, W.: Energy efficient virtual machine placement algorithm with balanced and improved resource utilization in a data center. Math. Comput. Model. **58** (5), 1222–1235 (2013)

11. Li, X., Wu, J., Tang, S., Lu, S.: Let's stay together: towards traffic aware virtual machine placement in data centers. In 2014 Proceedings IEEE INFOCOM, pp. 1842–1850. IEEE (2014)
12. Mann, Z.Á.: Approximability of virtual machine allocation: much harder than bin packing (2015)
13. Masdari, M., Nabavi, S.S., Ahmadi, V.: An overview of virtual machine placement schemes in cloud computing. J. Netw. Comput. Appl. **66**, 106–127 (2016)
14. Song, W., Xiao, Z., Chen, Q., Luo, H.: Adaptive resource provisioning for the cloud using online bin packing. IEEE Trans. Comput. **63**(11), 2647–2660 (2014)
15. Verma, A., Ahuja, P., Neogi, A.: pMapper: power and migration cost aware application placement in virtualized systems. In: Issarny, V., Schantz, R. (eds.) Middleware 2008. LNCS, vol. 5346, pp. 243–264. Springer, Heidelberg (2008). doi:10.1007/978-3-540-89856-6_13
16. Wang, G., Ng, T.E.: The impact of virtualization on network performance of amazon EC2 data center. In: 2010 Proceedings IEEE INFOCOM, pp. 1–9. IEEE (2010)
17. Wang, W., Li, B., Liang, B.: Dominant resource fairness in cloud computing systems with heterogeneous servers. In: 2014 Proceedings IEEE INFOCOM, pp. 583–591. IEEE (2014)
18. Wang, X., Liu, Z.: An energy-aware VMs placement algorithm in cloud computing environment. In: 2012 Second International Conference on Intelligent System Design and Engineering Application (ISDEA), pp. 627–630. IEEE (2012)

Weighted One-Dependence Forests Classifier

Guojing Zhong and Limin Wang$^{(\boxtimes)}$

Key Laboratory of Symbolic Computation and Knowledge Engineering
of Ministry of Education, Jilin University, Changchun City 130012, China
wanglim@jlu.edu.cn

Abstract. Averaged One-Dependence Estimators (AODE) combines all Super Parent-One-Dependence Estimators (SPODEs) with ensemble learning strategy. AODE demonstrates good classification accuracy with very little extra computational cost. However, it ignores the dependences between attributes. In this paper, we propose aggregating extended one-dependence estimators named Weighted One-Dependence Forests (WODF) which splits each SPODE into multiple subtrees by attribute selection. WODF assigns the weight to every subtree with conditional mutual information. Extensive experiments and comparisons on 40 UCI data sets demonstrate that WODF outperforms AODE and state-of-the-art weighted AODE algorithms. Results also confirm that WODF provides an appropriate tradeoff between runtime efficiency and classification accuracy.

Keywords: Super parent-one-dependence estimators · Aggregating · Conditional mutual information

1 Introduction

Naive Bayes (NB) [1] is the simplest Bayesian classifier and one of the most efficient learning algorithm which uses a simplified Bayesian Network (BN). In the NB, attributes are conditionally independent given the class. However, attributes have complex relationships on many real applications. Therefore, many experts and scholars are devoting to relaxing the attribute independence assumption. Then the "semi-naive Bayesian classifier" is proposed. The basic idea of the semi-naive Bayesian classifier is to properly consider the interdependence information between attributes, so that it does not need to calculate the joint probability completely and does not ignore the strong attribute dependency completely. "One-Dependent Estimator (ODE)" is the commonly used strategy for semi-naive Bayesian classifiers. It is assumed that each attribute is up to one parent except the class node. The direct approach is to assume that all attributes are dependent on the same attribute. This method is called Super-Parent ODE (SPODE) [2]. Averaged One Dependent Estimator (AODE) attempts to make each attribute as a super parent node for building multiple different structures [3]. Similar to the naive Bayesian classifier, AODE does not require model selection, so it can achieve fast and accurate classification. In recent years, many scholars focused on assigning different weights on each SPODE in AODE. In this paper, each SPODE is extended to more

© Springer Nature Singapore Pte Ltd. 2017
G. Chen et al. (Eds.): PAAP 2017, CCIS 729, pp. 366–375, 2017.
DOI: 10.1007/978-981-10-6442-5_33

ODE structures by adjusting the arc between attribute and its super parent node and the final Bayesian Network is constructed by ensemble learning.

The paper is organized as follows: we briefly make a description of BNs in the next section. In Sect. 3, we introduce a new algorithm for structure learning of BNs. The results of experiments on UCI data sets are given in Sect. 4. Finally, we draw conclusions for our study and describe the future research in Sect. 5.

2   Representation of Bayesian Networks

A BN consists of a directed acyclic graph. Each attribute in the graph is denoted by a collection of conditional probability tables [4]. The nodes in the graph correspond to the attributes in the domain, and the arcs between nodes represent casual relationships among the corresponding attributes. When two nodes are joined by an arc, the causal node is called the parent of the other node, and the other one is called the child. How one node influences another is defined by conditional probabilities for each node given its parents [5]. Suppose a training set $D = \{X_1, X_2......X_n\}$ where X_i denotes both the attribute and its corresponding node; C denotes the class variable. BN classifier outputs a class label c as follows:

$$c = \underset{c}{\arg\max}\, P(c|x_1, x_2,......x_n) = \underset{c}{\arg\max} \frac{P(c, x_1, x_2,......x_n)}{P(x_1, x_2,......x_n)} \propto P(c, x_1, x_2,......x_n) \quad (1)$$

2.1   SPODE

SPODE belongs to typical ODE which supposes all attributes to depend on the same attribute. SPODE has the same efficiency as NB but with a higher classification accuracy [6]. The predicted class label c in SPODE is defined as follows:

$$c = \underset{c}{\arg\max} \sum_{i=1}^{n} P(x_i|c, x_j)P(c) \quad (2)$$

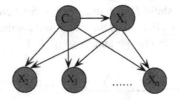

Fig. 1. The SPODE model

The structure of SPODE is as Fig. 1.

2.2   AODE

The AODE classifier uses an average strategy to aggregate all SPODEs as Eq. 3. It has got lots of attention because of improving NB's accuracy with only a slight increase in time complexity [7]. Zheng and Webb proves that AODE has better performance on error reduction compared to the rest of semi-naive techniques [8].

$$c = \sum_{j=1}^{n} P(c, x_j) arg \ \max_{c} \ \prod_{i=1}^{n} P(x_i|c, x_j) P(c) \tag{3}$$

2.3   WAODE

Many researches extended AODE by assigning attributes different weight values, which is referred to as weighted SPODEs (WSPODE/WAODE) [9]. Many weighting methods have been used, such as gain ratio [10], correlation-based algorithm [11], mutual information [12] and Relief attribute ranking algorithm [13]. Here is the

$$c = \sum_{j=1}^{n} w_j P(c, x_j) arg \ \max_{c} \ \prod_{i=1}^{n} P(x_i|c, x_j) P(c) \tag{4}$$

where w_j is the weight of the SPODE for attribute x_j.

3   The Proposed Algorithm

AODE demonstrated better performance because of the ensemble learning. However, every SPODE in AODE requires that all attributes has the same super parent node ignoring the degree of dependency between parent node and attribute node. To solve this problem, we proposed a new algorithm called WODF. The main idea of WODF is deleting the arc connecting super parent node and attribute node when their dependence is less than the threshold K and attempting to choose another node as new parent node for the attribute node with condition mutual information as Eq. 5. Once there is an arc has changed in SPODE, a new network structure is built. We call each SPODE based-structure. Then, assigning different weight values to these new structures for ensemble learning. The threshold K is defined as Eq. 6 where α denotes the minimum condition mutual information, β denotes the maximum condition mutual information, η controls the number of arcs to change and $0 \leq \eta \leq 1$.

$$I(X1; X2|C) = \sum_{x1 \in X} \sum_{x2 \in X} \sum_{c \in C} P(x1, x2, c) \log_2 \frac{P(x1, x2|c)}{P(x1|c)P(x2|c)} \tag{5}$$

$$K = \alpha + \eta(\beta - \alpha) \tag{6}$$

where α denotes the minimum condition mutual information, β denotes the maximum condition mutual information, η controls the number of arcs to change and $0 \leq \eta$

1. When $\eta = 0$, the arc with the minimum condition mutual information will be changed; when $\eta = 1$, all arcs in BN will be adjusted. For example, data set contact-lenses consists of $\{X_1, X_2, X_3, X_4, C\}$. Figure 2(m) shows the conditional mutual information between attributes and the mutual information between class label an attribute. From Fig. 2(m), when X_1 is considered as super parent node, $I(X_1; X_2| C) = 0.1611$, $I(X_1; X_3|C) = 0.0042$, $I(X_1; X_4|C) = 0.0742$, $\alpha = 0.0042$; $\beta = 0.1611$. Suppose $\eta = 0.5$, then $K = 0.0042 + 0.5 * (0.1611 - 0.0042) = 0.0827$. Because $0.0042 < K$ and $0.0742 < K$, the arc connecting X_1 and X_3 and the arc connecting X_1 and X_4 will be removed. Firstly, removing the arc connecting X_1 and X_3 and choosing another node as parent for X_3 based on the conditional mutual information. As $I(X_2; X_3| C) = 0.0182 > I(X_4; X_3|C) = 0.0085 > 0.0042$, X_2 is considered as the parent of X_3 in Fig. 2(a−1). Similarly, attempting to remove the arc connecting X_1 and X_4. However, I $(X_1; X_4|C)$ is the largest among $I(X_1; X_4|C)$ and $I(X_2; X_4|C)$ and $I(X_3; X_4|C)$, so X_1 is still the parent of X_4. Thus, Fig. 2(a) is ultimately derived from one structure as Fig. 2(a−1). Similarly, when X_2 is considered as super parent node, X_1 replaces X_2 as the parent of X_4; when X_3 is considered as super parent node, Fig. 2(c) is ultimately derived from two structures as Fig. 2(c−1) in which X_2 replaces X_3 as the parent of X_1 and Fig. 2(c −2) in which X_1 replaces X_3 as the parent of X_4; when X_4 is considered as super parent node, Fig. 2(d) is ultimately derived from two structures as Fig. 2(d−1) in which X_1 replaces X_4 as the parent of X_2 and Fig. 2(d−2) in which X_2 replaces X_4 as the parent of X_3. Finally, six different structures are extended by four SPODEs.

The second step of WODF is integrating all new structures by assigning different weight values to every structure. The external difference in all new structures is the super parent node of based-structure. The inner difference in all new structures is the newly added arc. Thus, considering the mutual information between super parent and class variable as the first layer weight and the conditional mutual information of the newly arc as the second layer weight. The weights for each layer are individually normalized. The weight for every structure is obtained by multiplying the first layer weight and the second layer weight. For example, the weight of Fig. 2(a−1) is as Eq. 7

$$Wla-1 = \frac{I(X1; C)}{I(X1; C) + I(X2; C) + I(X3; C)}$$

$$W2a-1 = \frac{I(X2; X3|C)}{I(X2; X3|C)} = 1 \tag{7}$$

$$Wa-1 = Wla-1 * W2a-1 = 0.0258$$

Similarly, weights of other structures are as follows:

$$W_{b-1} = W1_{b-1} * W2_{b-1} = 0.0193 * 1 = 0.0193$$
$$W_{c-1} = W1_{c-1} * W2_{c-1} = 0.3799 * 0.6847 = 0.2601$$
$$W_{c-2} = W1_{c-2} * W2_{c-2} = 0.3799 * 0.3153 = 0.1198 \tag{8}$$
$$W_{d-1} = W1_{d-1} * W2_{d-1} = 0.5751 * 0.8985 = 0.5167$$
$$W_{d-2} = W1_{d-2} * W2_{d-2} = 0.5751 * 0.1015 = 0.0584$$

$I(X_i;X_j|C)$	X_1	X_2	X_3	X_4	$I(X_i;C)$
X_1	*	0.1611	0.0042	0.0742	0.0236
X_2	0.1611	*	0.0182	0.0086	0.0177
X_3	0.0042	0.0182	*	0.0085	0.3486
X_4	0.0742	0.0086	0.0085	*	0.5278

(m)   Conditional mutual information and mutual information

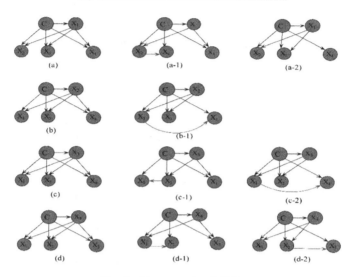

Fig. 2. The example of WODF

The learning procedure of the WODF learning algorithm can be summarized as follows.

- Compute the mutual information $I(X_i;C)$ for every attribute and the conditional mutual information $I(X_i; X_j|C)$ for all pairs of attributes.
- Use each single attribute as a super parent to build an SPODE.
- When the conditional mutual information $I(X_i;X_j|C)$ between attribute X_i and super parent node X_j is less than K, attempting to find a new parent node X_k whose conditional mutual information with X_i is greater than $I(X_i;X_j|C)$ for an extended structure.
- Calculate the first layer weight with mutual information between super parent node and class label and the second weight with the conditional mutual information between new parent node and attribute.
- Assign every new structure a weight which is the product of the first layer weight and the second weight.
- Compute the posterior probability by weighted new extended structures.

4   Experiments

We implement the proposed algorithm using 40 well-known data sets, downloaded in the UCI (University of California, Irvine) repository of machine learning databases [14]. A brief description of the data sets is given in Table 1.

Filling the missing values with modes and means from the training data, respectively. Qualitative attributes are discretized by MDL discretization [15].

We conduct an empirical comparison for the AODE, the WAODE and the proposed algorithm(WODF) in terms of 0–1 loss defined by Kohavi and Wolpert [16]. In all cases we have used 10-fold cross validation. We report the averaged 0–1 loss over the ten test folds. As it seems to be no formal method to select an appropriate value for η, we conduct an experiment to find it. We present the zero-one loss results from $\eta = 0.1$ to $\eta = 0.8$ with an increment of 0.1. Table 2 shows the detailed 0–1 loss for WODF, WAODE (weighted with mutual information) and AODE.

Firstly, we use a non-parametric measure, Friedman test, to evaluate the null hypotheses. Friedman test first ranks the algorithms with each sample (data set). Then it sums the ranks for all algorithms. The null-hypothesis is that all the algorithms are equivalent. We can compute the Friedman statistic

$$Fr = \frac{12}{Nt(t+1)} \sum_{j=1}^{t} R_j^2 - 3N(t+1)$$

by using the chi-square distribution with $t - 1$ degrees of freedom, where $Rj = \sum_i r_i^j$ and r_i^j is the rank of the j-th of t algorithms on the i-th of N data sets. Thus, for any selected level of significance α we reject the null hypothesis if the computed value of Fr is greater than χ_α^2, the upper-tail critical value for the chi-square distribution having $t - 1$ degrees of freedom. The critical value of χ_α^2 for $\alpha = 0.05$ is 2.08. The friedman statistic with 40 data sets, is 23.5309. And $P < 0.0001$ for all cases. Hence, we reject the null-hypotheses.

Mean zero-one loss Averaged results with all data sets shows a simplistic general method of relative performance. We present the averaged zero-one loss AODE and WAODE and WODF across 40 data with η in Fig. 3. From the experimental results, we can see that: For all settings of η, WODF outperforms WAODE and AODE. And WODF has the best performance with $\eta = 0.8$.

Zero-one loss Table 3 shows the win/draw/loss results of zero-one loss for WODF against WAODE and AODE. We evaluate a significant improvement with a 95% confidence level. Each entry w/t/l in the table means that WODF win on w datasets, tie on t datasets, and lose on l datasets, compared to AODE, WAODE respectively. From our experimental results, we can see that: WODF has great superiority to AODE and WAODE in terms of the zero-one loss when $0.1 \leq \eta \leq 0.8$. Furthermore, the proposed algorithm is based on SPODE so that it offers a good training efficiency as SPODE.

These results lead us to the conclusion that the proposed algorithm could significantly improve the performance of AODE and WAODE.

Table 1. Data sets

No.	DataSet	#Instance	Attribute	Class
1	contact-lenses	24	4	3
2	post-operative	90	8	3
3	promoters	106	57	2
4	iris	150	4	3
5	teaching-ae	151	5	3
6	glass-id	214	9	3
7	hungarian	294	13	2
8	primary-tumor	339	17	22
9	syncon	600	60	6
10	soybean	683	35	19
11	anneal	898	38	6
12	tic-tac-toe	958	9	2
13	vowel	990	13	11
14	german	1000	20	2
15	contraceptive-mc	1473	9	3
16	car	1728	6	4
17	mfeat-mor	2000	6	10
18	segment	2310	19	7
19	hypothyroid	3163	25	2
20	splice-c4.5	3177	60	3
21	kr-vs-kp	3196	36	2
22	sick	3772	29	2
23	spambase	4601	57	2
24	phoneme	5438	7	50
25	page-blocks	5473	10	5
26	satellite	6435	36	6
27	mushrooms	8124	22	2
28	thyroid	9169	29	20
29	pendigits	10992	16	10
30	sign	12546	8	3
31	nursery	12960	8	5
32	seer	18962	13	2
33	magic	19020	10	2
34	letter-recog	20000	16	26
35	adult	48842	14	2
36	shuttle	58000	9	2
37	waveform	100000	21	3
38	localization	164860	5	11
39	census-income	299285	41	2
40	poker-hand	1025010	10	10

Table 2. Experiment results of 0–1 loss

DataSet	AODE	WAODE	AODF_0.1	AODF_0.2	AODF_0.5
adult	0.149	0.1445	0.1396	0.1398	0.1397
anneal	0.0089	0.0089	0.0078	0.0078	0.0078
car	0.0816	0.0885	0.0671	0.059	0.0613
census-income	0.1008	0.0884	0.0873	0.0874	0.0877
contact-lenses	0.375	0.375	0.3333	0.3333	0.3333
contraceptive-mc	0.4942	0.4922	0.4895	0.4834	0.4874
german	0.248	0.24	0.242	0.234	0.237
glass-id	0.2523	0.257	0.2477	0.243	0.2523
hungarian	0.1667	0.1565	0.1531	0.1531	0.1497
hypo	0.0095	0.0101	0.0098	0.0093	0.0101
iris	0.0867	0.0867	0.0867	0.0733	0.0867
kr-vs-kp	0.0842	0.0576	0.0566	0.0563	0.0557
letter-recog	0.0883	0.0853	0.0853	0.0844	0.0857
localization	0.3596	0.3566	0.3473	0.351	0.3495
magic	0.1752	0.1762	0.1718	0.1739	0.1729
mfeat-mor	0.3145	0.313	0.3055	0.3055	0.3035
mushroom	0.0001	0.000	0.000	0.000	0.000
nursery	0.073	0.0708	0.0658	0.0662	0.0656
page-blocks	0.0338	0.0347	0.0323	0.0338	0.0333
pendigits	0.02	0.0199	0.0201	0.0197	0.0197
phoneme	0.2392	0.2308	0.2247	0.2236	0.2153
poker-hand	0.4812	0.1758	0.1677	0.1677	0.1677
post-operative	0.3333	0.3222	0.3444	0.3444	0.3556
primary-tumor	0.5752	0.5752	0.5723	0.5693	0.5752
promoters	0.1321	0.1415	0.1132	0.1321	0.1226
satellite	0.1148	0.1148	0.1127	0.1131	0.1124
seer	0.2328	0.2315	0.2313	0.2315	0.2307
segment	0.0342	0.0338	0.0333	0.032	0.0329
shuttle	0.0008	0.0009	0.0008	0.0008	0.0008
sick	0.0273	0.0244	0.0244	0.0239	0.0244
sign	0.2821	0.2768	0.2716	0.268	0.272
soybean	0.0469	0.0483	0.0498	0.0483	0.0498
spambase	0.0672	0.0648	0.0654	0.0648	0.0652
splice-c4.5	0.0365	0.0365	0.0353	0.0362	0.0359
syncon	0.01	0.01	0.01	0.01	0.01
teaching-ae	0.4901	0.4503	0.4437	0.4901	0.4901
thyroid	0.0701	0.0655	0.0639	0.0642	0.0643
tic-tac-toe	0.2651	0.2724	0.2651	0.2495	0.2526
vowel	0.1495	0.1949	0.201	0.1808	0.198
waveform	0.018	0.0181	0.0179	0.0178	0.018

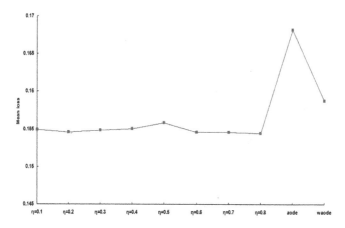

Fig. 3. AODE and WAODE and WODF with η

Table 3. Win/Draw/Loss of 0–1 loss comparison

W/D/L	AODE	WAODE
WODF_0.1	15/23/2	7/23/1
WODF_0.2	14/22/1	8/30/2
WODF_0.3	17/22/1	10/28/2
WODF_0.4	18/20/2	10/28/2
WODF_0.5	14/22/4	8/30/2
WODF_0.6	16/22/2	12/25/3
WODF_0.7	16/22/2	11/26/3
WODF_0.8	17/21/2	12/25/3

5   Conclusions and Future Study

In this paper, a new Bayesian classifier algorithm, called WODF, is proposed. Unlike the current weighted AODE methods, we propose that weighting with new structures which are extended from each SPODE in AODE. The proposed method is implemented and tested with 40 UCI data sets and validations. The experimental results shows that the proposed method has better performance than its counterpart algorithms in terms of classification accuracy.

For the future work, we will attempt to other weighting method for the ensemble learning and apply the idea expanding the structure to other classification algorithms, such as tree augmented naive Bayes (TAN) and k-dependence Bayesian classifier (KDB). Furthermore, with the rapid growth of data size in application, We will apply the proposed algorithm with massive data sets.

Acknowledgements. This work was supported by the National Science Foundation of China (Grant No. 61272209) and the Agreement of Science & Technology Development Project, Jilin Province (No. 20150101014JC).

References

1. Chien, Y.: Pattern Classification and Scene Analysis, pp. 462–463. Wiley, Hoboken (1985)
2. Yang, Y., Webb, G.I., Cerquides, J., et al.: To select or to weigh: a comparative study of linear combination schemes for superparent-one-dependence estimators. IEEE Trans. Knowl. Data Eng. **19**(12), 1652–1665 (2007)
3. Webb, G.I., Boughton, J.R., Wang, Z.: Not so naive Bayes: aggregating one-dependence estimators. Mach. Learn. **58**(1), 5–24 (2005)
4. Taheri, S., Mammadov, M.: Structure learning of Bayesian networks using unrestricted dependency algorithm. In: The Second International Conference on Advances in Information Mining and Management, pp. 54–59 (2012)
5. Sahami, M.: Learning Limited Dependence Bayesian Classifiers, pp. 335–338 (1998)
6. Yang, Y., Korb, K., Ting, K.M., Webb, G.I.: Ensemble selection for superparent-one-dependence estimators. In: Zhang, S., Jarvis, R. (eds.) AI 2005. LNCS, vol. 3809, pp. 102–112. Springer, Heidelberg (2005). doi:10.1007/11589990_13
7. Wu, J., Pan, S., Zhu, X., et al.: SODE: self-adaptive one-dependence estimators for classification. Pattern Recogn. **51**, 358–377 (2015)
8. Zheng, Z., Webb, G.I.: Lazy learning of Bayesian rules. Mach. Learn. **41**(1), 53–84 (2000)
9. Jiang, L., Zhang, H.: Weightily averaged one-dependence estimators. In: Yang, Q., Webb, G. (eds.) PRICAI 2006. LNCS, vol. 4099, pp. 970–974. Springer, Heidelberg (2006). doi:10.1007/978-3-540-36668-3_116
10. Zhang, H., Sheng, S.: Learning weighted naive Bayes with accurate ranking. In: IEEE International Conference on Data Mining, pp. 567–570 (2004)
11. Hall, M.: A decision tree-based attribute weighting filter for naive Bayes. J Knowl.-Based Syst. **20**(2), 120–126 (2007)
12. Jiang, L., Zhang, H., Cai, Z., et al.: Weighted average of one-dependence estimators. J. Exp. Theor. Artif. Intell. **24**(2), 219–230 (2012)
13. Robnik-Šikonja, M., Kononenko, I.: Theoretical and empirical analysis of ReliefF and RReliefF. J. Mach. Learn. **53**(1–2), 23–69 (2003)
14. Frank, A., Asuncion, A.: UCI Machine Learning Repository, vol. 23 (2013)
15. Fayyad, U.M., Irani, K.B.: Multi-interval discretization of continuous-valued attributes for classification learning. In: Proceedings of the International Joint Conference on Artificial Intelligence, pp. 1022–1029 (1993)
16. Kohavi, R., Wolpert, D.: Bias plus variance decomposition for zero-one loss functions. In: Proceedings of European Conference on Machine Learning, Finland, pp. 275–283 (1996)

Research and Realization of Commodity Image Retrieval System Based on Deep Learning

Cen Chen, Rui Yang, and Chongwen Wang[⊠]

Beijing Institute of Technology,
No. 5, South Street, Zhongguancun, Haidian District, Beijing, China
wcwzzw@bit.edu.cn

Abstract. This paper proposed a commodity image retrieval system based on CNN and ListNet sort learning method. CNN contained two convolutional layers, two pooling layers and two innerproduct layers. ReLu function was used as the activation function after the convolutional layer, achieving the sparsity and preventing the disappearance of the gradient. The pooling layer used stochastic pooling method and improved the generalization ability of the model. In addition, softmax regression was used for classification. Innerproduct layer adopted dropconnect method, which is more powerful than the generalization of dropout, and it can effectively prevent the occurrence of the overfitting. What's more, the feature extraction of the network was optimized by stochastic gradient descent (SGD) algorithm. And we combined the learn to rank algorithm of the text retrieval domain. We used ListNet algorithm to combine a variety of feature vectors, solving the problem of the image retrieval.

Keywords: Deep learning · Convolutional network · Learn to rank · ListNet

1 Introduction

How to extract useful information from massive images is one of the hottest issues in nowadays. With the concept of deep learning proposed by Hinton et al. In 2006, deep learning has been widely concerned by academia and industry.

2 Design of Commodity Image Retrieval Scheme

The commodity image retrieval system framework is shown in Fig. 1.

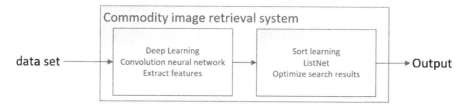

Fig. 1. Commodity image retrieval system framework combined CNN and ListNet

© Springer Nature Singapore Pte Ltd. 2017
G. Chen et al. (Eds.): PAAP 2017, CCIS 729, pp. 376–385, 2017.
DOI: 10.1007/978-981-10-6442-5_34

3 Deep Learning Feature Extraction

3.1 Network Structure of CNN

The flow chart of the CNN extraction feature is shown in Fig. 2.

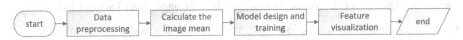

Fig. 2. Flow chart for extracting features with CNN

3.2 Data Preprocessing

Grab picture data with crawler, read jpg format images, and use the output leveldb format data as training set as well as test set.

Create list files for the training set and the test set. The file format is the picture address and the corresponding tag label. Read the pictures into Datum, serialize the datum into string and store it in the WriteBatch class as key-value pairs, Finally, WriteBatch was written to leveldb.

3.3 Calculate the Image Mean

First, get the amount of input data or the minimum number of batches when the stochastic gradient drops, channel number and height and width of the picture.

Then, enumerate the training set and the test set to superimpose the data. Finally, set the mean and write the result to the binary file.

3.4 Model Design and Training

Data Read. The Datalayer was used to read the data in Leveldb format. Batchsize is needed, and extract data of batchsize to deal with each time. If cropsize was set, you need to cut the input picture. If the mirror image was set, the image can be mirrored after the data was read.Import the mean file, and substract the image from the mean file. In forward transmission, datalayer each time pull a batch of data to the cpu or gpu to be processed, and then open a new thread to continue to obtain data.

Convolution Feature Extraction. The parameters to be set in the convolution layer include blobs_lr(learning rate), kernel_size(filter size), num_output(number of output), bias_filler(biased initialization method), stride(filter step) and so on.

Pooling. This paper adopted Stochastic pooling. First get the probability of each pixel, Then get the weighted average of element matrix and probability matrix to obtain the results.

Activation Function. The ReLU function was adopted [1]. And the formula is as follows.

$$f(x) = max(0, x) \tag{1}$$

ReLU can achieve sparse purpose without regular or other ways and will not lead to gradient disappearance.

Innerproduct Layer. In this paper, DropConnect was used to prevent overfitting in the innerproduct layer. Its generalization ability is stronger than Dropout [4].

Cost Function and Optimization. Softmax cost function is defined as follows.

$$\sigma_i(z) = \frac{exp(z_i)}{\sum_{j=1}^{m} exp(z_j)}, i = 1, \ldots, m \tag{2}$$

we used the SGD as optimization algorithm, the formula is as follows.

$$delta_{i+1} = \mu * delta_i - \text{lr} * \nabla L(\omega_i) \tag{3}$$

$$\omega_{i+1} = \omega_i + delta_{i+1} \tag{4}$$

ω_i represents the parameter at time i and $delta_i$ is the increment at time i, lr is the learning rate, μ is the previous weight.

Feature Visualization. This paper used the trained model to test the pictures and visualize the resulting features. The flow chart is shown in Fig. 3.

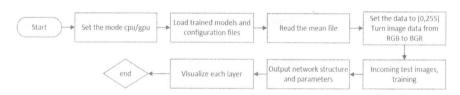

Fig. 3. Flow chart of feature visualization

Define a function, visualize the pixels with the specified width and height in the grid. First display the image data, For the processed results of each layer, Through the net.params to obtain the data spread to the layer from the input, then the data was regularized, The image data was then visualized in the grid. Then call the function at different network layers, visualize the features extracted by each layer, through which to find the optimized method.

4   ListNet Sort Learning

The architecture of sort learning includes learning system and sorting system. Learning system determined the cost function by selecting sorting learning method, The model was optimized by minimizing the cost function to get the model parameters, The model parameters were then sent to sorting system, The sorting system predicted the scores of the query input for test, The system architecture is shown in Fig. 4.

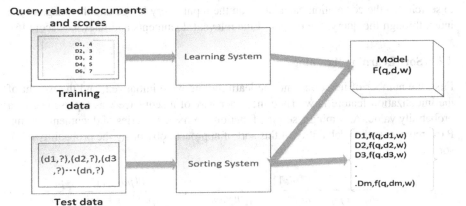

Fig. 4. System architecture diagram

4.1 Build Training Sample Set

The format of the input sample is shown below.

Label queryId: Id featureId: featureValue featureId: featureValue featureId: featureValue ⋯ # pidId.

Each of which is a sample, label represents the correlation between the query image and the target image, The label can be artificially marked, the correlation can be divided into completely related, relevant, part of the relevant, irrelevant and other levels. "#pid" indicates that the row sample is a correlation sample of the query image corresponding to the image Id. So it is necessary to mark the relevant documents related to the query.

In this paper, we used open source framework Lire to do automatic labeling on the relevance of the image, and at the same time obtain some of the basic characteristics of the image, Such as color histogram, texture feature, shape, color and edge directivity descriptor (CEDD), scale invariant feature transform (SIFT), Integrating multiple features for image retrieval [3].

4.2 Indexing

Get all the pictures in the directory;

Create a document generator DocumentBuilder class, index images, the use of CEDD and other features;
Create Lucene writing index tool IndexWriterConfig, loop to index the image.

The process of creating an index was to read the image into memory BufferedImage, build the Document, and add it to IndexWriterConfig.

4.3   Relevance Label

The establishment of the index reader IndexReader, through ImageSearcherFactory create a picture querier, Initialize the feature querier, set the number of relevant images to search, Do the correlation calculation on the input query image and the image of the index through the query, the output of the relevant documents and relevance score [6].

4.4   Sort Learning Training

First, the number of iterations and the learning rate were initialized, and the weight of the initialization feature is w. Then the order way of a sequence was represented as a probability value, Assuming a scoring function f, α means a series of document sorting, P (α) represents the probability of this sort of approach, obviously the probability of all sort is 1.

$$P(\alpha) = \frac{\theta(pic1)}{\theta(pic1) + \theta(pic2) + \ldots + \theta(picn)} * \ldots * \frac{\theta(pic_{n-1})}{\theta(pic_{n-1}) + \theta pic_n} * \frac{pic_n}{pic_n} \quad (5)$$

The time complexity of the formula is N, so the probability of the sequence was represented by computing the probability of Top-k.

The gradient descent algorithm was used to calculate the delta. The formula is as follows.

$$\Delta\omega = -\sum P\frac{\partial f}{\partial\omega} + \frac{1}{\sum exp(f) * \sum exp(f)\frac{\partial f}{\partial\omega}} \quad (6)$$

The parameters of the model were obtained by updating the weight of delta. And then through the parameters of the model to forecast the sort of the query later.

5   Experimental Results Analysis

5.1   Data Set Construction

In this paper, we built a data set to implement the dynamic crawler tool, using PhantomJS. Analysis of Taobao commodity detail page, you need to grab the field as shown in Table 1.

The process of crawling is shown in Fig. 5.

The file name for the Taobao pictures crawled is its commodity id, crawling the product image, as a training set and test set for subsequent neural network extraction. Training set includes four categories, 2500 pictures, test set includes 500 pictures.

5.2   Convolution Neural Network Feature Extraction [5]

Configuration file cloth_solver.prototxt, set the number of iterations and learning rate and other training parameters. As follows.

Table 1. Taobao product details table

Field name	Type	Field meaning
productId	Long	Product id
productUrl	String	Product address
productImgUrl	String	Product image url
productName	String	Product name
productPrice	String	price
productShop	String	shop
productStatus	String	Product status

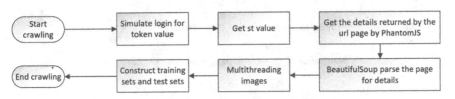

Fig. 5. Process of crawling

```
train_net : "cloth_train.prototxt"

test_net : "cloth_test.prototxt"

# Test batch

test_iter: 100 # When the test data amount is 10000, if the number of batches is
100, test_iter is 100

test_interval: 500  # Each iteration test_interval rounds for a test

# Initial learning rate, momentum, weight decay

base_lr: 0.001

momentum: 0.9

weight_decay: 0.004
lr_policy: "step"
gamma:"0.1"
stepsize:2000  #  Reduce learning rate every stepsize iterations
display: 100  # Display once every 100 times
max_iter: 10000  # Maximum number of iterations
snapshot: 1000  # Display status per iteration snapshot times
snapshot_prefix: "cloth_full"
solver_mode: GPU   # GPU run
```

The training profile cloth_train.prototxt is basically the same as the test profile cloth_test.prototxt, The difference is that the latter also has an accuracy level that was used to test the accuracy of the classification, There are cloth_deploy.prototxt files used to do the prediction, there was no input layer, only specify the size of the input. The structure of the network is shown in Fig. 6.

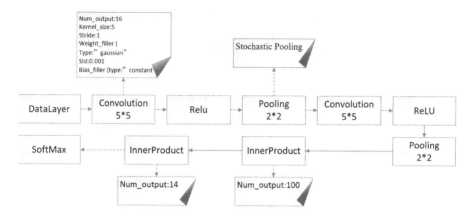

Fig. 6. Convolution neural network structure

During the training, the learning rate lr and the loss function loss were displayed every iteration 100 times, and the test was performed every 500 times, output accuracy score0 and the test loss function score1.

Figure 7 shows the classification of four categories of clothes (A dress, big dress, female suit, shirt) by convolution neural network [7], It can be seen that the stochastic pooling can preserve the image features and enhance the generalization ability of the model, the adjustment of network initialization parameters can also improve the accuracy of network classification [8].

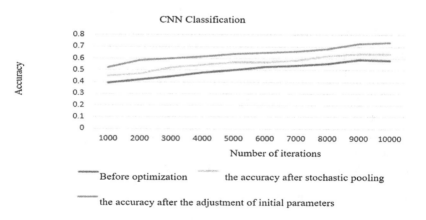

Fig. 7. CNN training accuracy chart

The original image is shown in Fig. 8. The features extracted by SIFT algorithm and SURF algorithm are shown in Figs. 9 and 10.

Fig. 8. Original image

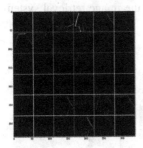

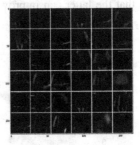

Fig. 9. SIFT feature extraction

Fig. 10. SURF feature extraction

The effectiveness of the convolutional neural network [5] in feature extraction was proved by comparing the results.

Next, the extracted features of the convolutional neural network were visualized, The extracted image features are shown in Figs. 11 and 12.

Fig. 11. Results after the first convolution layer

Fig. 12. Results after the second convolution layer

From the results of the middle layers of the convolution neural network, it can be seen that the CNN network structure can successfully extract the contour and wrinkle characteristics of the clothes [9].

5.3   Image Retrieval Performance Analysis

Taking query image 37566558911.jpg as an example, The number of images associated with the query image is 36, Using the extracted features extracted by the convolution neural network method to retrieve the top 20 related images of the query, 14 of these were associated with an accuracy rate of 70% and a recall of 39%.

As shown in Fig. 13 is the effect of convolution features for image classification, The average accuracy and repetition rate of traditional feature extraction method and the convolution neural network feature extraction method after ten experiments were expressed respectively, It shows that the convolution neural network has positive significance to the feature extraction of image high level semantics [2].

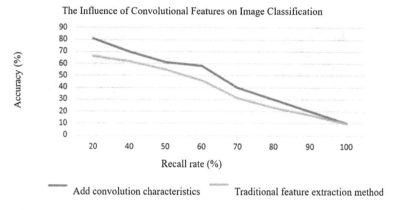

Fig. 13. Effect of convolution characteristics on image classification

The result of comparing individual feature methods with the ListNet method combined with convolution features and other basic features is shown in Fig. 14. It can be seen that the latter can improve the accuracy of the image retrieval system [10].

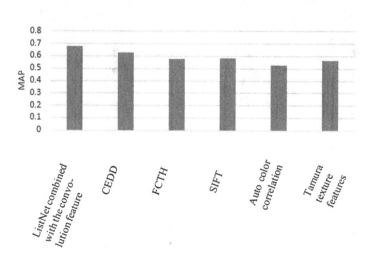

Fig. 14. ListNet algorithm optimization results

6 Summary and Future Work

In this paper, we proposed a commodity image retrieval system framework based on CNN and ListNet. Extract image features by using stochastic pooling, softmax cost function, DropConnect and SGD optimization method, and use ListNet to combine various features to solve the problem of sorting in image retrieval.

For the future work, the following aspects need further study.

- we need to dig out richer high-level semantics [2].
- Introduce a new index mechanism, speed up the retrieval speed of commodity images, and realize real-time retrieval [3].
- Using enhanced learning methods to reduce human intervention in learning and forecasting processes, The model generates the input sequence according to the data characteristics, and the model is trained by the fuzzy supervisory method.
- Introduction user feedback mechanism, and carry out incremental learning, to improve the accuracy of the retrieval.
- Combine the advantages of different models (Auto-encoder, RBM, CNN) to produce better algorithms.

References

1. He, K., Zhang, X., Ren, S., et al.: Delving deep into rectifiers: surpassing human-level performance on imagenet classification. arXiv preprint arXiv:1502.01852 (2015)
2. Zhao, F., Huang, Y., Wang, L., et al.: Deep semantic ranking based hashing for multi-label image retrieval, pp. 1556–1564 (2015)
3. Liu, H., Wang, R., Shan, S., et al.: Deep supervised hashing for fast image retrieval. In: IEEE Conference on Computer Vision and Pattern Recognition, pp. 2064–2072. IEEE (2016)
4. Srivastava, N.: Improving Neural Networks with Dropout (2013)
5. Ng, Y.H., Yang, F., Davis, L.S.: Exploiting local features from deep networks for image retrieval, pp. 53–61 (2015)
6. Bell, S., Bala, K.: Learning visual similarity for product design with convolutional neural networks. ACM (2015)
7. Babenko, A., Lempitsky, V.: Aggregating deep convolutional features for image retrieval. In: Computer Science (2015)
8. Morère, O., Veillard, A., Lin, J., et al.: Group Invariant Deep Representations for Image Instance Retrieval (2016)
9. Gordo, A., Almazán, J., Revaud, J., Larlus, D.: Deep image retrieval: learning global representations for image search. In: Leibe, B., Matas, J., Sebe, N., Welling, M. (eds.) ECCV 2016. LNCS, vol. 9910, pp. 241–257. Springer, Cham (2016). doi:10.1007/978-3-319-46466-4_15
10. Radenović, F., Tolias, G., Chum, O.: CNN image retrieval learns from BoW: unsupervised fine-tuning with hard examples. In: Leibe, B., Matas, J., Sebe, N., Welling, M. (eds.) ECCV 2016. LNCS, vol. 9905, pp. 3–20. Springer, Cham (2016). doi:10.1007/978-3-319-46448-0_1

An Improved Algorithm Based on LSB

Yali Luo, Xiulai Li, Chaofan Chen, and Mingrui Chen[✉]

Hainan University, Haikou 570228, China
mrchen@hainu.edu.cn

Abstract. LSB algorithm is a common image information hiding method. Based on Logistic chaotic sequence, an improved LSB algorithm is proposed. With the XOR operation to complete the encryption of embedded information. The simulation results show that the algorithm has good safety performance and can resist the pollution of salt and pepper noise. It is a kind of practical algorithm.

Keywords: Logistic · LSB · Security · Information hiding

1 Introduction

In recent years, information security has become a hot topic of concern. Information hiding is one of the hottest research points. With the development of the network, information transmission becomes simple and convenient. But in the convenience also brought a new question: how to protect the information in the transmission process does not suffer illegal theft and tampering? How to protect the intellectual property rights of electronic media? In order to solve these problems, information hiding gradually developed. Information hiding is the rise of a discipline in 1996. In the rapid development of network technology today, information hiding technology research is more practical significance [1]. Its basic features are imperceptible, robust and hidden capacity. Since the 90s of last century to the present, information hiding theory has gradually matured, and applied to data confidential communications, identity authentication and many other areas.

Digital watermarking is a method of information hiding. The main step is to embed the watermark information into the carrier (image, audio, text) in a certain way, and does not affect the use value of the carrier, nor is it easy to be detected or modified. Its main features are security, concealment, robustness and sensitivity. The LSB algorithm is a simple algorithm based on airspace that can implement digital watermarking. It is embedded in a simple way, hidden capacity, embedded pictures and the original picture in the visual almost no difference.

This paper presents a new LSB algorithm based on logistic sequences. Experiments show that the algorithm has a high simple and practical, with good security.

2 Chaotic Sequence

A chaotic sequence refers to an unpredictable activity in determining a system. The resulting sequence is sensitive to the initial value, nonperiodic. If you take the same initial value, you can easily copy the same sequence. Therefore, the chaotic sequence is

G. Chen et al. (Eds.): PAAP 2017, CCIS 729, pp. 386–392, 2017.
DOI: 10.1007/978-981-10-6442-5_35

applied to the encryption and decryption, which can improve the system security and improve the search space of the secret key, and the secret key is easy to manage [2].

The chaotic sequence used in this paper is Logistic chaotic sequence. The Logistic sequence was first proposed by the ecological mathematician R. May. The mapping formula is:

$$x(k+1) = u_0 x(k)(1 - x(k)) \tag{1}$$

where the sequence is in a chaotic state during the interval $(0,1)$ and u_0 in $(3.5699456, 4]$. The sequence generated at this time is nonperiodic, does not converge, and is very sensitive to the initial value. At that time, and $0 < u_0 < 3.5699456$, the resulting sequence was periodic [3].

3 LSB Algorithm

For a computer, a picture is a matrix of values that mark the pixel brightness. For gray-scale images, the value range of 0–255, take each bit pixel value of the 8-bit binary, can constitute a three-dimensional histogram. The same bit of the 8-bit binary position of each pixel forms a plane called a "bit plane". The highest bit plane represents the main information of the image, and the lowest bit plane is generally the redundant part of the image [4]. LSB algorithm is to be hidden information into binary, and these binary information hidden in the low plane.

LSB replacement process: If you want to hide the binary value of [0 1 0 0 0 1], the carrier information shown in Fig. 1. Replace the least significant bit of the original 8-bit binary with the original pixel. The specific algorithm flow is as follows:

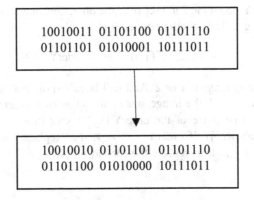

Fig. 1. The algorithm flow

The LSB algorithm is simple and easy to implement. Since it is replaced on the low plane, the algorithm has good imperceptibility. In addition, the hidden capacity of the algorithm is large. Is a very practical algorithm.

4 Specific Process to Improve the Algorithm

4.1 Embedding Algorithm

In order to hide and encrypt the transmission of information, we will use two one-dimensional Logistic chaotic sequence, the specific operation process is as follows:

Step 1:   Take the Logistic chaotic sequence mapping formula to produce two one-dimensional chaotic sequences equal to the size of the image to be embedded. Arranging the chaotic sequences from small to large or from large to small, and storing the position information of the two sequences in the original array to obtain a new position sequence, orderX, orderY

Step 2:   The sequence of the sequence to do on the 4 to take the operation, and then add the results to get 5. The specific formula is as follows:

$$\mathrm{orderX}'(1, j) = \mathrm{mod}(\mathrm{orderX}, 4) + 5 \tag{2}$$

After the operation orderX', the value ranges from 5-8 to an integer. Then we will raise the dimension operation into a two-dimensional matrix that is the same as the size of the embedded image [5].

Step 3   Select the carrier image I, the size of which, using the array, select the corresponding image of the carrier plane value, get tmp_img. The specific operation is as follows:

$$\mathrm{tmp\_img}(i, j) = \mathrm{bitget}(I(i, j), \mathrm{orderX}'(i, j)) \tag{3}$$

where the bitget function gets the Xth bit in a binary.

Step 4   Embed the information embedded in the previous step tmp_img and the information to be embedded in the encode_image

Step 5   The orderY do on the 2 to take over the operation, and in the recovery results on the 1, get. The specific operation is as follows:

$$\mathrm{orderY}'(1, j) = \mathrm{mod}(\mathrm{orderY}, 2) + 1 \tag{4}$$

The resulting range is 1 or 2. And will be raised dimension operation, it will be embedded with the image size of the same two-dimensional matrix.

Step 6   Select the first plane of the orderY'(i, j) vector image and replace it with encry_image (i, j). If orderY'(i, j) = 1, then replace the first bit plane, if orderY'(i, j) = 2, then replace the second bit plane

4.2 Extraction Algorithm

Extract the inverse process of the embedded algorithm. The specific process is as follows:

Step 1: Calculate the specific chaotic sequence by x_0, u_0, y_0, u_1, in order to sort, get the corresponding X, Y sequence. The X sequence to do 4 to take additional

5 operation, get the sequence. The Y sequence is subjected to a 2-to-1 addition to the sequence.

Step 2: Select the image H that embeds the secret information. Select the corresponding bit value in the pixel binary bit and store it as the tmp_low matrix.

Step 3: Extract the bit of the pixel value corresponding to the H image of the embedded secret information and store it as the tmp_high matrix.

Step 4: Traverse tmp_low, the value of tmp_low (i, j) is different from the value of tmp_high (I, j), and the exclusive OR result is the embedded information.

5 Experimental Results and Analysis

5.1 Visual Performance Analysis

Unsuitable means that the camouflage vector after embedding the secret information should not significantly reduce the quality of the source vector, and the visual effect is not significantly changed. In the simulation experiment, 256 × 256 Lena images were used as hidden vectors, and binary images with embedded image sizes of 150 × 148 and 256 × 256 were respectively respectively. The results are shown in Figs. 2(a) (b) and 3(a) (b), it can be seen that there is no difference in the image and the carrier image of the embedded information. Compared with Figs. 2 and 3, it is found that there is no difference in the embedded image and the original hidden image, which indicates that the algorithm has better imperceptibility [6].

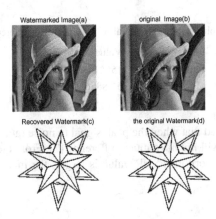

Fig. 2. Embedded image (150 × 148) and original image

In order to evaluate the concealment of the image, we use the peak signal to noise ratio to measure the difference between the original image and the embedded image. The specific formula is as follows:

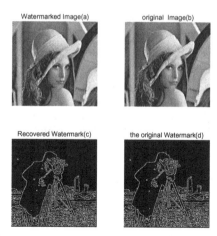

Fig. 3. Embedded image (256 × 256) and original image

$$PSNR = 20log_{10}\frac{MAX}{\sqrt{MSE}} \qquad (5)$$

where MAX represents the maximum value of the image point color and MSE represents the mean square error between images. Respectively, the size of the embedded image is 150 × 148 and 256 × 256 peak signal to noise ratio, the specific results shown in Table 1:

Table 1. Embeds the peak signal to noise ratio of different images

Title	Value	Value
Inserted image	150*148	256*256
PSNR	51.8856	47.1567

It is generally believed that when the peak signal to noise ratio is greater than 35 db, the image in the visual will not have much difference [7]. From the peak signal to noise ratio of the value, the algorithm PSNR value is larger, the image has a better hidden effect as Fig. 3.

5.2 Safety Analysis

Digital watermark security number refers to the word watermarking system should be illegal to extract a strong immunity, against unauthorized removal, embedding and testing, thus protecting the digital products. Generally, it uses the classic algorithm of cryptography to ensure the security of the key. The chaotic system has a particularly sensitive to the initial conditions. On the chaotic system, select the difference and its small initial value, after a long time calculation, can still output completely different results. Therefore, the use of chaotic system can be a good way to improve the security of the system.

The algorithm introduces two one - dimensional Logistic chaotic sequences, which are sensitive to the initial value due to the chaotic sequence. Even if the two difference is very small initial sequence, still can not extract the final embedded information. As shown in Fig. 4, its initial value and the original value of the final difference between the extraction results and can not show any information about the original map.

Recovered Watermark(a)

Fig. 4. The initial value of $10^{\wedge}$ (−9) extracted from the embedded information

As can be seen from Fig. 4, the algorithm is more secure. Can further improve the security of LSB algorithm, more convenient in the field of information security.

5.3 Robustness Analysis Unsuit

Robustness refers to the protection of information in digital watermarks after some changes, such as transmission, filtering operations, re-sampling, coding, lossy compression, etc., embedded information should maintain its integrity, can not be easily removed, and with a certain correct probability to be detected [8].

Recovered Watermark(a)

Fig. 5. The embedded image is extracted from the embedded image

Noise pollution can have an indelible effect on the quality of the image. In this paper, the image of the watermark is embedded into the salt and pepper noise pollution, and then the corresponding watermark information is extracted. The results are shown in Figs. 4 and 5. From the restoration of the watermark can be seen that the algorithm can be a good resistance to salt and pepper noise pollution. Robustness is better.

6   Conclusion

Based on the Logistic chaotic sequence, an improved LSB algorithm is proposed by using the residual sum and the operation. The simulation results show that the algorithm is simple and efficient, has no good perceptibility, and has high safety performance, and the algorithm is effective for resisting salt and pepper noise.

References

1. Cheddad, A., Condell, J., Curran, K., Mc Kevitt, P.: Digital image steganography: survey and analysis of current methods. Signal Process. **90**(3), 727–752 (2010)
2. Katzenbeisser, S., Petitcolas, F.: Information Hiding Techniques for Steganography and Digital Watermarking, vol. 15(4). Artech House, Boston (2000)
3. Hamid, N., Yahya, A., Ahmad, R.B., Al-Qershi, O.M.: Image steganography techniques: an overview. Int. J. Comput. Sci. Secur. **6**(3), 168–187 (2012)
4. Wang, S.J.: Steganography of capacity required using modulo operator for embedding secret image. Appl. Math. Comput. **164**(1), 99–116 (2005)
5. Chandramouli, R., Memon, N.: Analysis of LSB based image steganography techniques. In: Proceedings of the International Conference on Image Processing, vol. 3(2), pp. 1019–1022 (2001)
6. Zhang, X., Wang, S.: Vulnerability of pixel-value differencing steganography to histogram analysis and modification for enhanced security. Pattern Recogn. Lett. **25**(3), 331–339 (2004)
7. Provos, N.: Defending against statistical steganalysis. In: Proceedings of the Usenix Security Symposium, vol. 10(2), pp. 323–336 (2001)
8. Hsiao, J.Y., Chan, K.F., Chang, J.M.: Block-based reversible data embedding. Signal Process. **89**(4), 556–569 (2009)

A Report on the Improvement of Information Technology Capability of Teachers in Primary and Middle Schools in Hainan Province

JingYu Luo, Zhao Qiu$^{(\boxtimes)}$, JianZheng Hu, and XiaWen Zhang

College of Information Science and Technology,
University of Hainan, Haikou, China
499450006@qq.com, qiuzhao73@hotmail.com,
1694789915@qq.com, 1072148351@qq.com

Abstract. With the rapid development of our technology, information technology capacity has become the necessary professional ability for teachers. In June 2014, the Ministry of Education promulgated the "primary and secondary school teacher information technology application ability standard (Trial)", opened a new round of primary and secondary school teachers in information technology application ability improve training. In order to understand the current level of primary and secondary school teachers in application of information technology capabilities in Hainan for the implementation of capacity training to provide the basis. This paper take 856 primary and middle school teachers in Hainan Province as the research object, and carries out a more comprehensive research on the status quo of its application technology from the standard angle.

Keywords: Primary and secondary school teachers · Application of information technology · Competency standard

1 Introduction

This study is based on dozens of typical primary and secondary schools in Hainan Province. Based on the random sampling survey, this paper analyzes and studies the effective strategies to promote the educational informationization ability of primary and secondary school teachers. Through the implementation of this study, this paper puts forward the countermeasures to solve the problem of insufficient informationization ability of teachers' education at present stage, improve the informationization ability of teachers' education, construct a high-quality specialized teachers, promote the education of school education, realize the modernization of education and promote the quality Education and basic education curriculum reform in-depth development, improve the quality and efficiency of education, to enhance the level of professional development of teachers has a very important practical significance.

G. Chen et al. (Eds.): PAAP 2017, CCIS 729, pp. 393–399, 2017.
DOI: 10.1007/978-981-10-6442-5_36

2 Research Content and Method

2.1 Research Content

a. To investigate the level of application ability of information technology in primary and secondary school teachers in our province, and to examine the current situation of application training and capability test of information technology in primary and secondary schools in our province.

b. This paper analyzes the factors that affect the use of information technology in primary and secondary school teachers by investigating the status quo of information technology application ability of primary and secondary school teachers in Hainan Province, and explores how to stimulate the enthusiasm of teachers to apply information technology to teaching practice and how to improve teachers' use of information Technical ability.

c. Depth analysis of the "Ministry of Education primary and secondary school teachers in the application of information technology standards (Trial)" and other national primary and secondary school teachers in the application of information technology to enhance the relevant documents and norms, research through the establishment of constraints and incentives, teaching practice and competition measures to improve Primary and secondary school teachers in the application of information technology capabilities, the formation of teachers to promote the application of information technology to enhance the management system, provincial and municipal counties and schools to enhance the application of teachers information technology implementation program.

2.2 Research Method

We use the questionnaire survey, literature analysis and case analysis, the most important of which is the questionnaire survey. On the basis of several dimensions of teachers' informationization teaching ability, this questionnaire has compiled the questionnaire and distributed the questionnaire to primary and secondary school teachers in various areas of Hainan Province. Through the statistical analysis of the questionnaire, we can find out the existing problems in the teaching ability of middle school teachers in our province, and provide the scientific and objective basis for the related research of the follow-up teachers' informationization teaching ability promotion strategy.

3 Survey Design

From two aspects of "computer knowledge to master and use" "to receive information technology training" and six dimensions "understanding and attitude, technical literacy, planning and preparation, organization and management, assessment and diagnosis, learning and development "And" the basic situation "to form a questionnaire. Questionnaire using the scale method, the options were "very disagree" "do not agree" "basic approval" "agree" "very much agree", followed by 1, 2, 3, 4, 5. The higher the score, the stronger the ability of teachers to apply information technology. The scale is as follow (Table 1):

Table 1. Questionnaire scale

Primary index	Secondary index
Basic information	Education
	Tech age
	Teach title
Computer knowledge to master and use	The basic operation of computer technology to master the situation
	The frequency of use information technology to find teaching resources
	The degree of attention to information technology
	The frequency of the use of information technology in the class
	The use of multimedia tools
	The use of information technology and students to communicate the specific situation
	The master situation of Information technology software to
	The choice of prepare a lesson
Acceptance of information technology training	The frequency of training
	The direction of information technology training

4 Questionnaire Analysis

The survey was conducted on the Internet, and the number of primary and secondary school teachers was relatively small, and the qualifications of the teachers were basically concentrated in undergraduate and college. The age of the teachers who received the survey was from 5 years to 21 years, and the distribution of teachers at each stage of education was relatively average. With the help of the network, the questionnaire was distributed and recovered well, a total of 856 copies were recovered. The following table is a detailed sample of the study sample (Table 2):

4.1 Discriminator

The basic principle of discriminant analysis is to find the critical value of each topic in the scale, and delete or modify the CR value without reaching a significant level. The specific method is to obtain the total score of each test scale, then take the upper and lower 27% for the high and low group, the subject of independent sample T test to detect the high and low subjects in each subject average difference between the significant If the question of the CR value of 0.05 to reach the significance level, that the subject has discrimination, on the contrary, it means that the subject does not have discrimination, consider deleting or modifying the item, so that the quality of the questionnaire can be improved.

Table 2. Sample description

	Number	Percentage
Gender		
Male	361	42.2
Female	495	57.8
Education		
Technical secondary school	11	1.3
Junior college	228	26.6
Bachelor	586	68.5
Master and above	31	3.6
Teach age		
5 years or less	143	16.7
5 to 10 years	177	20.7
11 to 15 years	128	14.9
16 to 20 years	154	18.0
More than 21 years	254	29.7
Title		
Professor	7	0.8
Senior teacher	164	19.1
First-grade teacher	331	38.7
Second-grade teacher	279	32.6
Third-grade teacher	75	8.8
Grade		
Primary school	422	49.3
Secondary school	341	39.8
High school	93	10.9
Subject		
Chinese	230	26.9
Math	164	19.2
English	78	9.1
Physics	36	4.2
Chemistry	17	2.0
History	20	2.3
Politics	17	2.0
Geography	14	1.6
Biology	18	2.1
Art	38	4.4
Other	224	26.2
Location		
Haikou or Sanya	451	52.7
County	127	14.8
Town and countryside	278	32.5

Based on this principle, the statistical analysis of the questionnaire was carried out by using the statistical software SPSS19.0. According to the results of the independent sample T test, seven items were found to have reached the significance level and deleted.

4.2 Reliability

Reliability refers to the same method used to repeat the measurement of the same object when the degree of consistency of the results. The general reliability is divided into three types: the re-test reliability, the complex reliability and the internal consistency of the reliability of the three kinds of re-test reliability is the same questionnaire on the same group of respondents interval time to repeat the test, The correlation coefficient of the two test results. The copy of the reliability is to let the same group of respondents once fill two copies of the questionnaire to calculate the correlation coefficient of the two replicas. The most common is the consistency of the internal consistency, that is, with the Cronbach α coefficient to measure the consistency of the scores of the title, Cronbach α coefficient is best in 0.8 above, 0.7–0.8 acceptable; subscale reliability coefficient Preferably above 0.7, 0.6–0.7 is acceptable. Cronbach's alpha coefficient if the 0.6, the following should consider modifying the scale. In this paper, Cronbach α coefficient method is used to measure the reliability of the questionnaire. The measurement results are as follows (Table 3):

Table 3. Reliability analysis

	Cronbach's alpha	Conclusion
Whole questionnaires	0.856	Good reliability
Computer knowledge to master and use	0.877	Good reliability
Acceptance of information technology training	0.766	Good reliability

From the test results, the reliability coefficient of the questionnaire in the reliability test results is above 0.7, which indicates that the questionnaire has high internal consistency and conforms to the data requirements of this study.

4.3 Validity

Validity means that the measurement tool or means can accurately measure the degree of things required to measure, the measurement results and the content to be observed more consistent, the higher the validity; the other hand, the lower the validity. Validity generally includes content validity and structural validity. The validity of the general questionnaire refers to its structural validity. The following is a test of the structural validity of the questionnaire. According to the standard given by the statistician Kaiser, the value of KMO is above 0.9, indicating that the data collected are very suitable for factor analysis; between 0.8 and 0.9, it is suitable for factor analysis; 0.7–0.8 between, do factor analysis is still; between 0.6–0.7, barely; 0.5–0.6 between, is not suitable for factor analysis; when the KMO value of 0.5 below, it is not suitable for the factor analysis (Table 4).

Table 4. Validity analysis

Kaiser-Meyer-Olkin measure of sampling adequacy		0.887
Bartlett's test of sphericity	Approx. Chi-Square	6497.357
	df	171
	Sig.	0.000

From the above data, we can see that the KMO values of the two scales are all greater than 0.8 and pass the Bartley ball test with significance level of 0.05, indicating that the structural validity of the questionnaire is in accordance with the research requirements. So the reliability and validity is as follow (Table 5):

Table 5. Reliability and validity test result

Primary index	Secondary index	Factor loading	Combined reliability
Basic information	Education	0.420	0.663
	Tech age	0.430	
	Teach title	0.765	
Computer knowledge to master and use	The basic operation of computer technology to master the situation	0.681	0.877
	The frequency of use information technology to find teaching resources	0.628	
	The degree of attention to information technology	0.336	
	The frequency of the use of information technology in the class	0.582	
	The use of multimedia tools	0.504	
	The use of information technology and students to communicate the specific situation	0.418	
	The master situation of Information technology software to	0.804	
	The choice of prepare a lesson	0.352	
Acceptance of information technology training	The frequency of training	0.437	0.766
	The direction of information technology training	0.559	

5 Summary and Prospect

The process of research has some unavoidable problems. The following is summarized:

(1) Many of our school hardware/software facilities are not perfect, after the teacher training, it is difficult to display training to learn the information technology, over

time, easy to forget the training of knowledge learned, so that the training effect greatly reduced.

(2) Many teachers on the concept of education and understanding of the concept of understanding is not accurate enough to change the concept of difficulty, resulting in their teaching in the "wear new shoes to go the old road" phenomenon. So that the use of information technology to become formal.

(3) In the development of the use of questionnaire tools, there are still many deficiencies, how the development of tools and research objectives, research content to form a strict logical relationship, it seems professional lack of, but also need to strengthen this learning and improvement.

Acknowledgment. This work is partially supported by Educational Commission of Hainan Province of China (No. HNKY2014-03). Thanks to Professor Zhao Qiu, the correspondent of this paper.

References

1. Mansheng, R.: Primary and secondary school teachers in information technology application ability training. Hunan Normal University, Changsha (2007)
2. Yi, Z.: Primary and secondary school teachers in the application of information technology capacity assessment. China Audio-Visual Educ. **8**, 2–7 (2014)
3. Shangjun, C.: To find out the teacher's needs to establish an effective teacher incentive mechanism. J. Xiaogan Univ. **26**(2), 105–107 (2006)
4. Li, S.: Improve the training of teachers and students in the application of information technology skills training practice. China Audio-Visual Educ. **337**, 129–133 (2015)
5. Fang, S.: Henan province primary and secondary school teachers in the application of information technology capabilities and enhance the measures. Comput. Age **5**, 95–96 (2016)
6. Jianliang, H.: Primary and secondary school teachers to enhance the application of information technology capabilities. China Audio-Visual Educ. **357**, 23–25 (2016)
7. Cui, K.: Primary and secondary school teachers in information technology ability training strategy. Mod. Prim. Second. Educ. **139**, 52–54 (2005)
8. Zhu, C.: Analysis of rural primary and secondary school teachers in information technology application ability promotion strategy. Inventor Innov. (Educ. Inf.) 29–32 (2014)
9. Zhang, H.: Primary and secondary school teachers in the application of information technology to enhance the effectiveness of countermeasures. Prim. Second. Schools 20–22 (2005)
10. Tian, X.: Primary and secondary school teachers training status, problems and countermeasures. Hebei Normal University (2015)

The Research of the Airport Retail Layout Based on the Location Model

Han-tao Yang[✉]

Sanya Aviation and Tourism College, Sanya 572000, Hainan, China
371750371@qq.com

Abstract. With the airport authorities realizing the importance of the non-aviation revenue, the airport retails are taken as one of the most important non-aviation business, how to give it a reasonable layout is becoming the most important point for the airport commercial plan. Firstly, it set the model hypothesis considering the characteristics and structure of airport terminal, and carried out the location model of airport model. Secondly, the airport retails shops of the landside hall and airside hall were relayouted and improved the efficiency and productivity of retails shops. Finally, it drew a conclusion that airport layout of airport retail is very complex, layouting the airport retail using airport retail location model is efficient but should think about safety policies, passenger procedure and passenger number further.

Keywords: Non-aviation revenue · Airport retail · Location model · Airport retails layout

1 Introduction

In recent years, with the rapid development of China's civil aviation transportation industry and the improvement of people's living standards, the flight journey has gradually become a popular way of travel. The passengers have required the diversification and individuation in retail industry of airport. However, the needs of aviation business are given priority to the space of departure hall. Due to insufficient business development of non-aviation business, the space isn't enough and the layout isn't reasonable. As an important part of the non-aviation business in airport, airport retail is significant to its reasonable planning and layout.

In the actual layout of the airport retail, due to the airport security and passenger convenience, the space for layout retail and the choice of retail shops to the travel routes are subject to great restrictions. How to distribute the airport retail in a limited space result in the profit maximization of airport retailing, which has become a key to commercial layout in airport.

Nevertheless, a comparatively systematic methodology guidance is short of the airport retail currently. We can only base on feeling and fulfill the layout of making a decision, which can't certainly master the final results [1]. As a theoretical model of spatial structure in industrial economy, regional model is widely used in the practical

© Springer Nature Singapore Pte Ltd. 2017
G. Chen et al. (Eds.): PAAP 2017, CCIS 729, pp. 400–409, 2017.
DOI: 10.1007/978-981-10-6442-5_37

production and life. Regional model is composed of four variables, such as population, organization, environment, and technology. Various factors of the space layout model in other fields are concluded to 4 variables, so as to describe the relationship among all the variables by the simple form [2]. Thus, it can be used as an important tool to solve the spatial arrangement of the airport retail. In paper [3], Considering the particularity of the food industry production process, the optimized layout solution of the multi-floor pastry plant is obtained, which solves the existing problems effectively. In paper [4], this paper enumerates four kinds of methods, their advantages and disadvantages are analyzed. In paper [5], determining the distribution of EV charging stations must consider the uncertainties in electric vehicle short driving range and long charging time as well as the long-term sustainable planning for electric vehicle growth. Considering construction continuity in time and space, this paper proposes a continuous step-by-step optimization method that compares alternative layout plans based on the service capability under different scenarios. The characteristics and applicability of proposed model is discussed through case study. The results show that a high continuity for charging station distribution at network level is achieved with the optimized method, which can help to plan the construction of charge stations in stages. In paper [6], based on the systematic analysis of the layout of civil aviation airport, a two-stage method of airport layout planning based on GIS is proposed. Based on the discussion of the theory of facility layout and demand distribution, a multi-objective model of airport layout optimization is established, which provides theoretical decision support for civil aviation airport layout planning. In paper [7], Based on the neoclassical growth theory, agglomeration theory accessibility theory and trade theory, the paper used panel data from 2003 to 2013 analyzed 46 different size airports' regional economic influences. Also, I tested the effects of regional economic growth on airport. In paper [8], the impact mechanism of the airport to the regional economy was analyzed, and the evaluation model of the regional economic contribution degree was established, then the empirical research to the model was carried on.

2   Analysis for Regional Correlation Degree Between the Airport Retail and Departure Hall

The flight action of aviation passengers will cause the desire for buying the related goods in travel. But, whether the purchase desire can be converted into a reality of purchase behavior is to create the buying inclination of travelers. Meanwhile, the goods should also be accessible. The passengers can walk closer, and they can more easily see these goods, which has a direct relationship between the type of retail goods and their spatial position. The location of the retail goods in the airport are needed to be accessible, and the commercial value of the departure hall is related to a variety of factors. In general, the price of goods and the location is the most basic factor, then how is the relationship among the price of goods, location and the purchase behavior of passengers?

3   Basic Assumption of Overall Layout Locations of Retail at the Airport

The location model is the significant method for analyzing product differentiation. In the location model, the product is depicted in two parameters, one is Price (P) and another is position (X). Different positions represent the products of different brands. The longer the distance between two brands, the bigger their differentiation will be. The consumer's position (Y) represents their most favored brand. If $Y = X$, it means product X is most favored by consumer Y [9].

1. The entrance of safety inspection and the board gate are linear, the former is at the right end of the straight line, and the latter is at the left end of the straight line, and the length between is 1 unit. It is specified as the closer the distance to the boarding gate is, the rent of the retail shop will be, i.e., the rent and the distance are in the linear relationship (the entrance of safety inspection is the starting point), thus, in the model, the rent of the retail shop can be represented by solving the distance of the retail shop.

2. There are two retailers A and B of which the cost is zero on the straight of which the unit is 1. The distance between the retail shop A and the left end is a, and the distance between the retail shop B and the right end is b (as shown in Fig. 1)m wherein $a + b \leq 1$, $a \geq 0$, $b \geq 0$. Without loss of generality, the standard of consumer number is 1, and assume the consumers are evenly distributed on the linear straight line, and everyone has the unit demand of the goods.

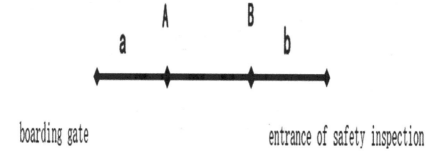

Fig. 1.   Basic assumption of layout of retail shops at the airport

3. The consumer's favor can be expressed by the consumer's buying transportation cost $T(z)$, i.e., assume the consumer would like to pay more walking cost for one goods, and it means the consumer has more favors on the goods. $T(z)$ is the linear function of the distance z between the consumer and the shop, i.e., $T(z)=Tz$, wherein t is the unit transportation cost. Assume the prices of the retailers A and B are respectively P_A and P_B.

4 Construction and Solution of the Location Model of the Overall Layout of Retail at the Airport

The effect function of the consumer x can be expressed as follows according to the above assumption:

$$U_X = -P_A - t|X - a| \tag{1}$$

If buys from the retailer A;

$$U_X = -P_B - t|1 - X - B| \tag{2}$$

If buys from the retailer B.

Wherein:

U : consumer's effect;

P : price of the retail product;

t : unit transportation cost.

Make $\overline{X}$ represent the marginal consumer who has no buying differentiation between A or B, and then:

$$-P_A - t(\overline{X} - a) = -P_B - t(1 - b - \overline{X}) \tag{3}$$

Thus, the demands of the retailer A and the retailer B are respectively:

$$D_A = \overline{X} = \frac{P_A - P_B}{2t} + \frac{1 - a + b}{2} \tag{4}$$

$$D_B = 1 - \overline{X} = \frac{P_A - P_B}{2t} + \frac{1 - a + b}{2} \tag{5}$$

Wherein:

D : consumer's demand on the retail goods;

$\overline{X}$: the consumer who has no differentiation of the retail goods

To meet the transverse differentiation of the two types of products, when the prices of the two retailers are same, different consumers have different favors on different goods, thus, we must ensure $a \leq \overline{X} \leq 1 - b$, or:

$$|P_A - P_B| \leq t(1 - a - b) \tag{6}$$

The intuitive meaning is the prices of the two retailers shall not exceed their transportation cost. According to the analysis above, we can know the profit of the retail shop A is:

$$\pi_A(P_A \bullet P_B) = \begin{cases} \frac{P_A P_B - P_A^2}{2} + \frac{(1-b+a)P_A}{2}, \ when |P_A - P_B| \le t(1-a-b); \\ P_A, \ when \ P_A < P_B - t(1-a-b); \\ 0, \ when \ P_A > P_B + t(1-a-b). \end{cases} \quad (7)$$

wherein

π: sales profit of the retailer

In the first condition, the two retailers segment the market, in the second condition, the whole market is supplied by the retailer A, and in the third condition, the whole market is supplied by the retailer B (shown in Fig. 2).

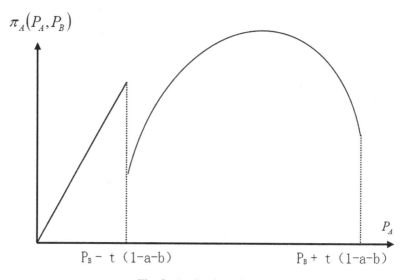

Fig. 2. Profit of retailer A

According to symmetry, the profit of the retailer B is:

$$\pi_B(P_B \bullet P_A) = \begin{cases} \frac{P_B P_A - P_B^2}{2} + \frac{(1-b+a)P_B}{2}, when |P_A - P_B| \le t(1-a-b); \\ P_B, \ when \ P_B < P_A - t(1-a-b); \\ 0, \ when \ P_B > P_A + t(1-a-b). \end{cases} \quad (8)$$

At first, according to the first condition: $|P_A - P_B| \le t(1-a-b)$, when it requires the two retailers' profits to be maximized, the follows can be obtained according to the phase 1 condition of the two retailers' maximized profits:

$$\begin{cases} P_A^* = \frac{t(3-b+a)}{3} \\ x^* = \frac{3-b+a}{6} \\ P_B^* = \frac{t(3+b-a)}{3} \end{cases} \quad (9)$$

$$\begin{cases} \pi_A^* = \overline{x^*} \bullet P_A^* = \frac{t(3-b+a)^2}{18} \\ \pi_B^* = (1 - \overline{x^*}) \bullet P_B^* = \frac{t(3+b-a)^2}{18} \end{cases} \quad (10)$$

During the above assumption when $|P_A - P_B| \leq t(1 - a - b)$, the equilibrium exists, and then the corresponding result is further obtained, however, the existence of equilibrium cannot be sufficiently ensured only by this condition. The two retailers' position parameters a and b are given, and then the necessary and sufficient conditions for price equilibrium (P_A^*, P_B^*) are as follows:

$$\begin{cases} \left(1 + \dfrac{a-b}{3}\right)^2 \geq \dfrac{4}{3}(a+2b) \\ \left(1 + \dfrac{b-a}{3}\right)^2 \geq \dfrac{4}{3}(b+2a) \end{cases} \quad (11)$$

Because at first assume the equilibrium exists, then indicated by (11):

$$P_A^* = \frac{t(3 - b + a)}{3}, P_B^* = \frac{t(3 + b - a)}{3} \quad (12)$$

$$\pi_A^* = \overline{x^*} \cdot P_A^* = \frac{t(3 - b + a)^2}{18} \quad (13)$$

However, if retailer A deviates from equilibrium at this time and makes the price which is a little lower than $P_A = P_B - t(1 - a - b)$, he will seize the whole market and get the profit $\pi_A = P_A = \frac{2t}{3(2b+a)}$. Thus, the necessary and sufficient condition for the retailer A not deviate from the equilibrium strategy is $\pi_A^* \geq \pi_A$. After putting in order, the first condition (11) can be obtained. According to symmetry, mutually exchange a and b, and then the second condition (11) can be obtained, thus, known from the process, the price equilibrium is unique.

The necessary and sufficient condition indicates the differentiation equilibrium can only be generated when the distance of the two retail shops are far. When the positions of the two retailers are symmetric, the condition for the differentiated product equilibrium is:

$a = b \leq \frac{1}{4}$, it requires the two retailers to be respectively in $[0, \frac{1}{4}]$ and $[\frac{3}{4}, 1]$, and their distance is at least $\frac{1}{2}$.

When it is put practically in the airport, to be convenient for planning, we can specify the distance of every three boarding gates is 1. Indicated by the final results of the internal location model, we can know two shops can be arranged in the distance of which the unit is 1, and their distance shall at least be $\frac{1}{2}$, i.e., one shop is arranged between every two boarding gates. The distance of two contiguous shops at least shall be the distance of two boarding gates, only in this way it can ensure the two profit revenues are maximized.

5 Overall Layout of Retail of an Airport

The retail layout areas of the air terminal of an airport are mainly centralized in the departure hall. Because the streams of people are dense, the arranged retail shops are relatively more, and the types of the retail products are also more (as shown in Fig. 3).

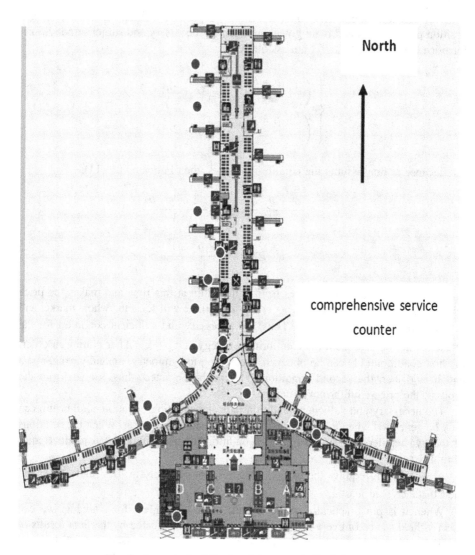

Fig. 3. Layout of retail of real departure hall of airport

The layout of the retail shops of the departure hall is shown as the circle in the figure. The retail types include gift shops, restaurants, convenience stores, etc.

Most of the retail shops in the isolating area are centralized near the comprehensive service counter, wherein the gift shops, the clothing shops, etc., are located at the right side of the comprehensive service counter, the distances of the shops are close. Compared with the principle of $\frac{1}{2}$ of the internal location model, the shop layout is crowded, which will confuse the consumers. There are many restaurants arranged at the north side of the comprehensive service counter, and the distance of every two shops conforms to the internal location model, but the retail shops arranged at the register offices at the east and west sides of the comprehensive service counter are fewer, and also there is no restaurant arranged.

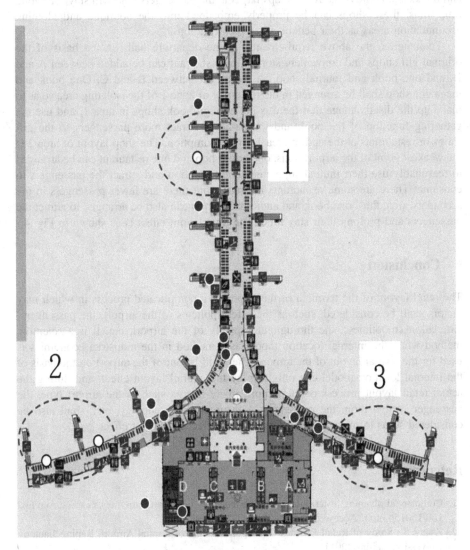

Fig. 4. Layout of retail of real departure hall of airport (Color figure online)

The number of the retail shops arranged in the departure hall is a little few, and all of the shops are in C and D areas, which doesn't conform to the principle of shopping convenience. The passengers entering from gate 1 may lose the interest of shopping because of the long walking distance.

According to the current condition that the retail shops arranged at the east and west sides of the comprehensive service counter in the isolating area of the terminal of the airport, and also the streams of people are not as many as the triangle collection part of the comprehensive service counter, the retail shops of restaurants, books and audio-visual products, convenient stores, etc., can be considered so as to meet the passengers' immediate demands at different areas at one hand, and at the other hand, the restaurant and books and audiovisual retail shops target at the passengers who will stay for a long time, thus, these shops can be properly embedded into the boutique and clothing accumulation areas as their beneficial supplementation.

To integrate the above requirements: in the departure hall, on the basis of the original gift shops and convenient stores, one restaurant can be added between A and B, and one book and journal shop can be added between B and C. One book and magazine shop shall be arranged in the red circle of zone 1 of the isolating area so as to make up the disadvantage that there is not enough book shops in area 1, and use the gathering function of the book and magazine to attract more passengers to the gift shops or restaurants at the end of area 1 for consumption. The shop layout of area 2 is the weakest point of the airport, thus, one book shop and one restaurant can be arranged to reasonably use their mutual supplementation function and attract the passengers to consume. There are some restaurants in area 3, but there are fewer passengers in the secondary area, thus, one book and audiovisual shop can also be arranged to attract the passengers and prolong their stay time. The final layout effect is as shown in Fig. 4.

6 Conclusion

The retail layout of the terminal of the airport is a complicated process in which may factors shall be considered, such as the safety policies of the airport, the passengers' circulation smoothness, the throughput capacity of the airport, etc. It is a scientific method when the internal location model as a method in the industrial economics is used for the spatial layout of the airport. The retail layout of the airport on the basis of the internal location model can only display the overall layout effect, and during the actual retail layout process of the airport, many factors such as the airport type, the passenger composition, the passengers' consumption psychology, etc., shall also be considered so as to make detailed layout and perfect the layout effect.

References

1. Commercial planning of terminal under the tendency of airport. http://news.carnoc.com/list/160/160154.html. Accessed 21 May 2017
2. Zhang, J.: Study of Retail Business Management Plan of Capital Airport. Beijing Jiaotong University, Beijing (2011)

3. Lv, P., Li, G.: The application of SLP in food industry's multi-floor plant layout optimization. J. Guangxi Univ. Sci. Technol. **28**(1), 108–115 (2017)
4. Cui, R.: Review on the application of operations research in logistics distribution center layout optimization. Logist. Eng. Manage. **11**, 58–59 (2015)
5. Liu, K., Li, A., Sun, X.: Optimizing spatial distribution of EV charging stations. Urban Transp. Chin. **3**, 523–528 (2015)
6. Wang, J., Mo, H.: Discussion on layout method of civil aviation airport. J. Civ. Aviat. Flight Univ. Chin. **20**(6), 7–10 (2009)
7. Liu, Y.: Economic Relevance Between Airport and Regional Economic Development and a Coordinated Analysis of MAC. Jinan University, Guangdong (2016)
8. Zhang, T.: Research on Calculation Model of Airport Operation Contribution to Regional Economy. Aviation Flight University of China, Guanghan (2016)
9. Xia, C.: A discussion of urban commercial planning and commercial operation. Chin. Bus. Mark. **16**(4), 23–27 (2002)
10. Park, R.E.: Human communities: The City and Human Ecology. Free Press, New York (2002)
11. Shi, L., Kou, Z.: Industrial Economics (Volume I). Shanghai Sumerian Press, Shanghai (2003)

Research on Model and Method of Relevance Feedback Mechanism in Image Retrieval

Xinying Li[1], Taijun Li[2(✉)], Feng Li[3], and Hongli Wu[3]

[1] Haikou College of Economics, Haikou 571127, Hainan, China
wmdnlxy@163.com
[2] Hainan University, Haikou 570228, Hainan, China
hdltj08@126.com
[3] Hainan Normal University, Haikou 570100, Hainan, China

Abstract. Considering the limitations of image retrieval method based on computer center, the introduction of relevance feedback mechanism increasingly shows its importance in image retrieval. This paper makes a deep research on some commonly used relevance feedback models and feedback methods in image retrieval. The purpose is to improve the query efficiency and retrieval precision of image retrieval.

Keywords: Feedback model · CBIR · Bayesian method

1 Introduction

With the growth of image data in the Internet, it is becoming more and more difficult for users to obtain useful information from large amounts of data. Thus, Image retrieval becomes increasingly important as a tool to help users to obtain information effectively. But a major problem of retrieval is how to understand the user's intention from the user's query and return satisfactory results to the user because of the ambiguity of user queries and the similarity of search results. Moreover, the semantic gap between low-level image representation and high-level semantics, as well as the excessive approximation of images caused by the increase of data increase the difficulty of this problem.

The basic idea of human-computer interaction in image retrieval is consistent with that of text retrieval. Sometimes some ideas and methods in text retrieval can be used in image retrieval after expansion and reconstruction. However, due to the big difference in content and characteristic description of image and the mechanism of relevance feedback and implementing method, we need to study and develop new theories and new methods to meet the needs of image retrieval.

2 The Basic Idea of Relevance Feedback

For different retrieval strategies, the specific process of feedback is also different. Suppose a retrieval strategy is implemented according to the matching function, and the documents and queries that are processed by the system have been represented as

© Springer Nature Singapore Pte Ltd. 2017
G. Chen et al. (Eds.): PAAP 2017, CCIS 729, pp. 410–417, 2017.
DOI: 10.1007/978-981-10-6442-5_38

n-dimensional vector, then only sequential retrieval is considered. The purpose of retrieval is to suppress irrelevant documents by retrieving relevant documents which refers to documents that are relevant to user requirements from the retrieval policy. However, in the process of retrieval, the system answers the query correlation of document instead of answering the question directly. It uses query correlation to approach user relevance which is often unsatisfactory, and hence there must be a feedback process. The feedback process can be divided into 3 categories according to the basic idea: (1) Modifying query vectors or distance discriminant criteria; (2) Adjusting the classification or inter class relationships of an image database;(3) Methods based on Bayesian theory.

3   Relevance Feedback Model of Image Retrieval

3.1   Feedback Model Based on Characteristic Component Factor

In text retrieval, product $TF \times IDF$ of the Keyword frequency TF and the inverse document frequency IDF is an exact estimate of the weight of a keyword in a document. According to the concept of important factors ci (component importance) and important factor of inverse set ici (inverse component importance) of image retrieval, the factor ci reflects the relative importance of a component in a characteristic vector while the factor ici shows the ability to distinguish a feature vector from a feature vector of another set of images.

Suppose the feature vector of image i can be expressed as:

$$F_i = [f_{i1}, \ldots, f_{ik}, \ldots, f_{iN}] \tag{1}$$

In this formula, N refers to the number of features (i.e., the feature dimension) of the image. The different components f_{ik} of the same characteristic vector may have various physical meaning, for example, in a common feature vector of image texture, some components may represent the image contrast, while others represent roughness and so on. The range of values of different components may therefore vary widely.

ci and Component value f_{ik} have very similar meanings. The former represents the relative importance of a component in a vector, while the latter shows the degree of significance of a feature in the image. Another factor is that different f_{ik} defined in different physical categories. In order to eliminate the difference between the components due to different range, the following normalization method is considered to calculate ci:

$$ci_i = [\frac{f_{i1}}{mean_1}, \cdots, \frac{f_{ik}}{mean_k}, \cdots, \frac{f_{jN}}{mean_N}] \tag{2}$$

In the formula above, $mean_k$ refers to the average value of the eigenvectors of component f_{ik} in all images.

Except for the factor ci, the factor ici should also be estimated in image retrieval, which can be done by the following formula:

$$ici_i = [\log_2(\sigma_{i1} + 2), \ldots, \log_2(\sigma_{ik} + 2), \ldots, \log_2(\sigma_{iN_v} + 2)] \tag{3}$$

σ_{ik} is the standard variance of the k th component value of all image's vector ci. It can be seen that if a component value is very close to all images, its standard variance is smaller. On the contrary, if a component value is very different between different images, its standard variance is also great. This Law of standard variance allows factor ici tend to those that have the ability to distinguish between different images and weaken the effect of those that have no resolving power. Thus, standard variance is a good yardstick to measure factor ici.

Also, the weight vector of image i can be calculated by multiplying (the vector of) factor ci and factor ici. Relevance feedback model described in terms of text keyword mode can be directly used to feedback and correct the results of image retrieval after translating feature vector F_i to weight vector W_i.

This shows that the feedback model based on the feature component factor essentially modifies the query vector according to the user feedback, and realizes the relevance feedback by optimizing the query method.

3.2 Feedback Model Based on Related Network Memory Information

In the process of relevance feedback, the acquired knowledge needs to be memorized to continue to improve retrieval efficiency, that is to say, knowledge accumulation should be carried out from different retrieval processes. The challenge of this method is how to memorize the knowledge learned and how to deal with the contradiction between the subjective content of different users and (or) the same user in different retrieval.

The basic idea of explicating memory semantic information to improve the performance of CBIR (content based image retrieval) is to accumulate semantic related information from the image clusters that the user feedback in the related network, that is, to use relevant network to memorize information. Mathematically, the correlation network is described by a correlation matrix called M, defined as follows:

$$M = \begin{bmatrix} w_{11} & w_{12} & \cdots & w_{1N} \\ w_{21} & w_{22} & \cdots & w_{2N} \\ \vdots & & & \vdots \\ w_{N1} & w_{N2} & \cdots & w_{NN} \end{bmatrix} \tag{4}$$

In this matrix, the matrix element (called a weight or coefficient) w_{ij} refers to Semantic correlation between Image cluster i and j.

First of all, all images in database based on the similarity of visual features (such as the use of K- means algorithm) are divided into N group, and at first, images in each group are only similar in selected visual features, which is the same as the classic CBIR system. Moreover, at first, the correlation coefficients of any two different groups are set at 0, i.e., the two groups are totally uncorrelated, whereas any group is completely related to itself. That is, the initial matrix is a unit matrix: $M_0 = I_{N \times N}$.

Next, for a given query requirement, the retrieval starts is based on visual features. Suppose that after a repeated process, the n + m image is displayed as a result, in which the N image is labeled as relevant, and the other m images are marked as uncorrelated. Whether it's related images or irrelevant images, it may come from the same group or from different groups. The system remembers the feedback by updating the correlation matrix in the following manner:

$$M_t = M_{t-1} + \sum_{i=1}^{m} F(q)F(p_i)^T - \sum_{i=1}^{n} F(q)F(n_i)^T \tag{5}$$

Among this correlation matrix, q is the characteristic vector required by query. p_i and n_i are the eigenvectors of positive and negative examples of feedback, and $F(x)$ is a transformation function for determining the update range based on feedback samples. Thus, the relevance between the original group and the positive case group is enhanced. Relevance is used in concurrent retrieval, i.e., visual features and semantic correlations are used to determine the similarity of the retrieved images. Some experiments show that this advanced learning method effectively utilizes the previously retrieved knowledge, thus reducing the number of repetitions and achieving higher retrieval accuracy.

On the one hand, if there are two types of samples of semantically distinct types in an original group, this means that the two classes are mutually negative, and that the underlying network divides the original set into two parts. On the other hand, according to the feedback information, if the two original groups are very close to the feature space and have a high correlation from the correlation matrix, the relevant network will combine the two original combinations into one. That is to say, the related network can not only update the correlation matrix but also dynamically update its structure by learning user feedback.

4 Feedback Method Based on Bayesian Theory

4.1 Image Retrieval Based on Bayesian Theory

CBIR is widely used in the form of example based retrieval (query-by-example), in which the user submits the query example system to obtain query example features, and matches the features of other images in the image library according to a certain algorithm. Once the calculated similarity is greater than a given threshold, it is considered similar and sorted according to the similarity and returned to the user. The user selects some pictures of his satisfaction or dissatisfaction from the display results, and the system shifts to another query according to the user's interaction information until the user finds the target. The nature of the problem can be described as a classification problem. Because the image in the CBIR can be expressed as the form of feature vectors, it can be assumed that the image retrieval is a mapping from the feature vector space F to image collection M:

$$g : F \rightarrow M = \{1, \ldots, K\} \tag{6}$$
$$x \rightarrow y$$

where M is a collection containing K image categories. In this way, image retrieval can be described as follows: Given an image (characterized by x), find the image category Y of which the image belongs, then the image in y can be used as the result of retrieval.

If the feature x is extracted from an image category y, and the $g(x)$ is the category of the X that is calculated according to the above mapping, then the purpose of the search is to find the optimal mapping g to minimize the probability of misclassification $P(g(x) \neq y)$. Based on this retrieval model, the optimal mapping g can be defined by the Bias classifier:

$$g^*(x) = \arg\max_i P(y = i \mid x) = \arg\max_i \{P(x \mid y = i)P(y = i)\} \qquad (7)$$

Among them, $P(x \mid y = i)$ represents the possibility function of the image features of class i images, while $P(y = i)$ represents the prior probability of the class.

In addition to the completeness of the theory, the Bayesian search method has other two additional advantages: First, for different image categories, you can define completely different image features and similarity calculation methods, and all relevant details will be included in the $P(x \mid y = i)$ computing method; Second, the model is based on probability, which provides the basis for designing relevance feedback algorithms based on reliability transfer.

4.2 The Learning and Classification Process Based on Positive and Negative Examples

In fact, relevance feedback retrieval based on content is a gradually refinement procedure, the pictures that the user satisfied and not satisfied with can be respectively as positive training examples and counterexamples of the training data, and further inquiry is a classification process.

Ω^+ is used to represent a partition of the sample space according to the training data provided by the user and the training data of counterexample. $p(\Omega^+)$ and $p(\Omega^-)$ Respectively indicate the prior probabilities of positive class and counterexample class. x_j represents the visual features of the image j. $p(x_j \mid \Omega^+)$ is the class conditional probability density obtained from training data of positive examples.

$p(x_j \mid \Omega^-)$ is the class conditional probability density obtained from training data of counterexamples. $p(x_j \mid \Omega^+)$ indicates the probability that the image j belongs to the positive class, and if it is greater than a threshold, j is considered to belong to Ω^+. According to Bias's theory, the following formula can be used to indicate $p(\Omega^+ \mid x_j)$:

$$p(\Omega^+ \mid x_j) = \frac{p(x_j \mid \Omega^+)p(\Omega^+)}{p(x_j \mid \Omega^+)p(\Omega^+) + p(x_j \mid \Omega^-)p(\Omega^-)} \qquad (8)$$

The above formula transforms the image retrieval problem into Bias classification problem, which means the problem of finding $p(\Omega^+ \mid x_j)$ for each image in the image database. The image is sorted according to the size of $p(\Omega^+ \mid x_j)$, and the retrieval results are obtained. Because the low-level features of images can not reflect the semantic features of images well, the classification results produced by the classifier can not fully meet the needs of users. The image retrieval system here provides an adaptive learning function that allows users to evaluate the current classification results and to improve retrieval accuracy through repeated learning. The positive and negative examples that user specified serve as new training data, and as training data that user provides changes, $p(\Omega^+)$, $p(\Omega^-)$, and $\Omega^{\pm}$ correspondingly change.

In view of the dynamic characteristics of the relevance feedback image retrieval system, the concept of query time is introduced, and the formula (8) is also modified: $\Omega^{\pm}(t)$ is used to represent a partition of the sample space according to the positive and negative examples provided by the t moment; $p(\Omega^{\pm}(t))$ is used to represent the $p(\Omega^{\pm})$ of the t moment; the posterior probability $p(\Omega^{\pm}|x_j)$ of the t moment is represented by $p(\Omega^{\pm}(t)|x_j); p(x_j|\Omega^{\pm}(t))$ represents the class conditional probability density at t moments. After the above amendment, it becomes the following formula:

$$p(\Omega^+(t)|x_j) = \frac{p(x_j|\Omega^+(t))p(\Omega^+(t))}{p(x_j|\Omega^+(t))p(\Omega^+(t)) + p(x_j|\Omega^-(t))p(\Omega^-(t))} \qquad (9)$$

The basic idea of Bayesian method applied in relevance feedback image retrieval is described above. The specific retrieval process can be summarized as: obtaining the corresponding class pattern through the training of the positive and negative examples provided by the user, and calculating the probability $p(\Omega^+(t)|x_j)$ of the positive class of each image and sorting them according to the $p(\Omega^+(t)|x_j)$. In the concrete implementation process, the function similarity() is used to approximate the class conditional probability density.

4.3   Rich Get Richer Method

To further improve retrieval efficiency, we can adopt Rich Get Richer strategy. Specific ideas are as follows:

According to the formula (9), we can compute the probability $p(\Omega^+(t-1)|x_j)$ that picture j belonging to the positive class at the t-1 moment. Obviously, Compared with the positive example training data, $p(\Omega^+(t-1)|x_j)$ is larger, whereas $p(\Omega^+(t-1)|x_j)$ is smaller. In order to make the retrieval results gradually approach the desired state of the user, constant emphasis is placed on promising images which are closer to the counterexample than the counterexample, using $p(\Omega^+(t-1)|x_j)$ and $p(\Omega^+(t-1)|x_j)$ instead of $p(\Omega^+(t))$ and $p(\Omega^-(t))$ in the original form respectively.

$$p(\Omega^+(t)|x_j) = \frac{p(x_j|\Omega^+(t))p(\Omega^+(t-1)|x_j)}{p(x_j|\Omega^+(t))p(\Omega^+(t-1)|x_j) + p(x_j|\Omega^-(t))p(\Omega^-(t-1)|x_j)} \qquad (10)$$

By way of formula (1–11), the system establishes an adaptive mechanism that allows a relatively large $p(\Omega^+(t)|x_j)$ to be quickly obtained with similar pictures of training data given by the user. The user's further interaction process makes

$p(\Omega^+(t) \mid x_j)$ relatively large, while the picture whose content is not consistent with the query target is designated as a counterexample.

Experimental results show that having the advantages of fast convergence and high accuracy, the proposed method can obviously improve the query efficiency.

5   Conclusion

Relevance feedback is a task that improves retrieval performance in CBIR. Almost all the feedback method has been proposed for CBIR are the typical machine learning methods, however, because users usually do not want to or can not provide a lot of feedback sample, so the number of training samples is very small, generally just a few in each round of feedback.In contrast, the characteristic scales in CBIR are usually very high. Therefore, a key issue in implementing relevance feedback in CBIR systems is how to learn from very few training examples in a high-level feature space., Bayesian learning is more prominent in CBIR tasks compared with other learning methods.

In conclusion, it has important theoretical significance and practical value to figure out how to use both theories and methods of digital image processing, pattern recognition, statistical learning, machine vision and so on to build efficient image retrieval model and method combining with traditional database technology, as well as to establish an effective association with high-level semantic information according to the underlying visual attributes of the image to finally obtain image retrieval model and method with good performance and retrieve the desired image of the user.

Acknowledgment. We thanks for the Support of Hainan province natural science fund project (Nos. 614250, 20156219, 20156231, 617175) and Haikou college of economic field research project (Nos. Hjyj2015009, Hjky16-23, Hjkz13-07).

References

1. Liu, B.: Based on feature and relevance feedback image retrieval technology research. University of Electronic Science and Technology of China (2013)
2. Wang, T.: Based on bayesian decision image retrieval relevance feedback method research. J. Eng. Coll. Armed Police Force (2013)
3. Lu, C.: The bayesian classifier related applications in image retrieval. Software (2011)
4. Liu, L.: Relevance feedback correction semantic image retrieval methods of the manifold. J. Chin. Mini-Micro Comput. Syst. (2014)
5. Hu, W.: Based on an active relevance feedback mechanism of image retrieval methods. The Northeast Normal University (2010)
6. Xu, Q.: Content-based image retrieval technology research and related feedback. Nanjing Information Engineering University (2012)
7. Cox, I.J., Minka, T.P., Papathomas, T.V., Yianilos, P.N.: The Bayesian image retrieval system PicHunter: theory, implementation, and psychophysical experiments. IEEE Trans. Image Process. – Special Issue Digital Libr. (2000)
8. Dan, L., Gao, W., Ma, J.: Rich Get Richer in image retrieval, a adaptive method of relevance feedback. Res. Dev. Comput. **38**(8), 960–965 (2001)

9. Vasconcelos, N., Lippman, A.: Bayesian representations and learning mechanisms for content based image retrieval. In: SPIE Storage and Retrieval for Media Databases, San Jose, CA (2000)
10. Li, Q.: Based on color and relevance feedback image retrieval technology research. University of Electronic Science and Technology of China (2010)
11. Zhang, G.: Based on the remarkable regional and SVM relevance feedback image retrieval technology research. Xidian University (2012)

Processing Redundancy in UML Diagrams Based on Knowledge Graph

Yirui Jiang, Yucong Duan$^{(\boxtimes)}$, Mengxing Huang, Mingrui Chen,
Jingbin Li, and Hui Zhou

Hainan University, Haikou, Hainan, China
duanyucong@hotmail.com

Abstract. UML (Unified Modeling Language) is designed for all stages of software development, but it lacks precise semantic information. MDE (Model-driven engineering) takes the model as the primary software product and its main research direction is modeling and model transformation. We propose to explore the entity abstraction scenario where no existing classes fit as the representative entity for other classes through the data, information and knowledge recreation of existing classes and relationships with the introduction of knowledge graphs. In this paper, we proposed a knowledge graph to enhance model design between models.

Keywords: UML · MDE · Redundancy · Knowledge graph · Diagram

1 Introduction

UML (Unified Modeling Language) is a general graphical modeling language for object-oriented development. UML provides a variety of graphical visualization to model stakeholders' intention into a series of models. These formed models comprise model elements. The same model elements can be displayed in a number of graphical models. People can view the modeled stakeholders' intention in the form of expressed models from multiple views. Knowledge map is closer to the natural language and is more expressive than UML models. We use Using knowledge graphs to deal with easier and can collect more information. To manually identify and resolve design inconsistencies are tedious and error prone [1]. UML is formal and expression is limited to the structure. There are some defects in the expression mechanism, We detect redundancy hidden in a diagram to achieve the purpose of accurate modeling and improve modeling quality and efficiency [2]. It is necessary for us to study the transformation between class diagrams and knowledge graph because knowledge graph can enhance the semantic expression of information.

In the paper, we firstly analyze the redundancy existing in modeling in Sect. 2. We elaborate the approach of construction of UML diagram in Sect. 3. In Sect. 4, we elaborate the approach about the transformation between diagram and knowledge graph. Then, we elaborate the method of integrating knowledge graphs in Sect. 5. After that, we analyze the experimental results in Sect. 6. Conclusion is in Sect. 7.

© Springer Nature Singapore Pte Ltd. 2017
G. Chen et al. (Eds.): PAAP 2017, CCIS 729, pp. 418–426, 2017.
DOI: 10.1007/978-981-10-6442-5_39

2 Redundancy in UML Diagrams

UML includes some graphic elements that can be combined with each other. It laws that combine these elements. UML diagrams can be divided into five categories [3]. It includes UseCase diagrams, static diagrams, behavior diagrams, interactived diagrams and implementation diagrams. UseCase diagrams describe the system function from the user's point of view and point out the operator of each function. Static graphs include class diagrams, object graphs and package diagrams. In the UML model diagrams, class diagrams describe the static structure of the class in the system [4]. They not only define classes in the system, but also represent the relationship between classes, such as association, dependency, aggregation. Class diagrams describe a static relationship. Object diagrams are instances of class diagrams that are almost identical to class diagrams. The difference between them is that object diagrams show multiple instances of the classes. Package diagrams are used to describe hierarchical structure of the system. Behavior diagrams describe dynamic models of the system and interactions between components. State diagrams describe all possible states of classes and conditions for transition of the event. State diagrams are complements to class diagrams. Interaction diagrams describe interactions between objects. Sequence diagrams show dynamic cooperative relationship among objects in order to show interaction between objects.

2.1 Redundancy in Data Representation

When requirements are expressed from a different perspective, redundancy is likely to occur. A redundancy occurs when a design artifact (perhaps partial) is represented multiple times, possibly in varying views. Redundancies can occur either in design or data representation [5]. When two design units contain the same elements or information in an expression, it can produce design related redundancy. For instance, structural redundancy refers to the redundancy in a variety of structural and behavioral diagrams in modeling. This type of inconsistency can be viewed as a result of name space mismatch or structural conflict in the requirements specification. When two characteristics share an overlapping description, they will produce state conflict if they have the common usage.

A characteristic of a data representation is opposite in data objects. When R is regarded as the relationship between objects, the relationship has reversed two object positions of the reverse relationship. It can be seen as an inverse relation R^{-1} [6]. The definition of reverse correlation is very complex. So there is data redundancy. The redundancy in UML diagrams is the redundancy of data expression.

3 Transformation to Knowledge Graph

UML is a kind of general visual modeling language and used to describe sofewares, visualize processes and build documents of software system [7]. It supports a whole process of software development from requirements. UML can construct system models according to kinds of diagrams. There are static models and dynamic models in diagrams. Static model diagrams are UseCase diagrams, class diagrams, object diagrams, component diagrams and deployment diagrams. Dynamic model diagrams are sequence diagrams, collaboration diagrams, state chart diagrams and activity diagrams. These diagrams can be used to visualize the system from different angles. UML is a set of graphical expressions used to describe OOAD process. It provides a comprehensive representation of requirements, behaviors and architectures of an object-oriented design [8]. The main function of UML is to help users describe and model the software system. It can describe a whole project that software has been implemented and tested from the requirements. A class diagram shows static views of the system, consisting of classes, relationships between them. Classes are represented by rectangles showing the name of the classes and the name of the operations and attributes. A UseCase diagram can describe user requirements and analyze the function of systems a user's point of view. In a UseCase diagram,an ellipse represents a use case and a human symbol represents a role.

3.1   Transformation from a Class Diagram to a Knowledge Graph

The quality of information transmission is low in a class diagram and the diagram may lose information. It may add invalid or wrong classes. Knowledge graph can make up the limitation of class diagram and enhance semantic expression. A class diagram can be mapped to a knowledge graph. A class diagram can be mapped to a knowledge graph. A class can be mapped to a concept in a knowledge graph. An attribute can be mapped to an entity, and the type of an attribute can be mapped to a concept. An operation can be mapped to a concept. A concept or an entity in a knowledge graph is presented as a node. We give rules for mapping classes and relationship as follows. There are rules for mapping classes.

Rule 1: the node of mapping by a ClassName is named C_ClassName.
Rule 2: the node of mapping by a AttributeName is named A_AttributeName.
Rule 3: the node of mapping by an AttributeTypeName is named A_AttributeTypeName.
Rule 4: the node of mapping by a OperationName is named O_OperationName.

There are rules for mapping relationship.

Rule 1: the representative relationship of "Has a" means that the class ClassName has an attribute named AttributeName.
Rule 2: the representative relationship of "Has o" means that the class ClassName has an operation named OperationName.
Rule 3: the representative relationship of "Is a" means that attribute AttributeName has an attribute named AttributeTypeName.
Rule 4: Relationship types like generalization, aggregation, composition, dependency and association need to be described respectively.

3.2   Transformation from a UseCase Diagram to a Knowledge Graph

A UseCase diagram describes the function of the system from a user's view and abstracts the implementation of each function. A UseCase diagram can not express the requirements clearly. Many details can not describe clearly. We can use a knowledge graph to express completely. We elaborate rules for transforming a use case diagram to a knowledge graph as follows. They are rules for mapping actors.

Rule 1: the node of mapping by an ActorName is named A_ActorName.
Rule 2: the node of mapping by a UsecaseName is named U_UsecaseName.
Rule 3: the node of mapping by a UsecaseTypeName is named U_UsecaseTypeName.

There are rules for mapping relationships.

Rule 1: the representative relationship of "Operate" means that the actor operates a UseCase named UseCaseName.
Rule 2: relationship types like generalization, aggregation, composition, dependency and association need to be described respectively.

3.3   Transformation from a Knowledge Graph to a UML Diagram

A class diagram is a static view that describes classes in a system and relationships between classes. Class Diagrams help us to have a comprehensive understanding about the system before coding. In order to generate codes conveniently, it is necessary for us to transform a knowledge graph to a class diagram. If a knowledge graph satisfies the inverse of rules given previously, it can be transformed to a class diagram as well.

4   Deduplication in UML Diagram

4.1   Deduplication in Class Diagram

If there is no inheritance relationship between the two classes in a class diagram. But they have the same attributes or operations, there are redundant. A solution is to duplicate the attributes or operations. We can construct a new class. It makes a parent class have same attributes or operations, and inherit a parent class.

As shown in Fig. 1, there are two classes with redundant attributes in a class diagram. Librarian and reader have the same attributes like name, ID and Tel. The same attributes are redundant.

We map a class diagram in Fig. 1 to a knowledge diagram as shown in Fig. 2.

Each node can be traversed in a knowledge graph. We define the array Count[] represents traversal times of different nodes.

The initial value of Count[] equals to 0. If a node appears once, then the current value of Count[] = Count[] + 1.

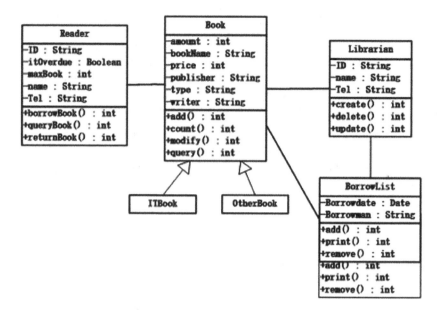

Fig. 1. A class diagram of a book management system

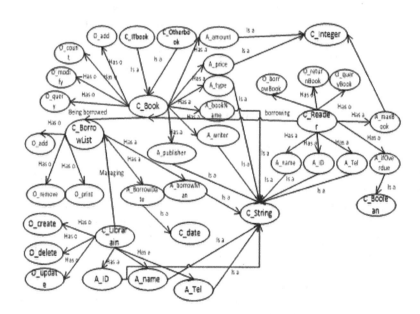

Fig. 2. A knowledge graph of a book management system

Algorithm Dealing with redundancy in knowledge graphs

Input: A knowledge graph array KG[ ]

Output: A refined knowledge graph

For : (each node KG{<KGn>,<R>} in array KG[ ])

      Compute Count[];

        If (Count[]≥*2*)

1. create a new node C_ClassName with the overlapping information

2. delete the same node A_AttributeName/O_OperationName

3.create<A_AttributeName,generalization,C_ClassName>and<O_OperationName,generalizatio n, C_ClassName>

        End If

    End For

4.2   Deduplication in UseCase Diagram

In a UseCase diagram, if different actors have the same use cases, there will be redundant.

    We give a UseCase diagram about a tutor website management system designed by a developer in Fig. 3. The actors are teachers, students and administrators.

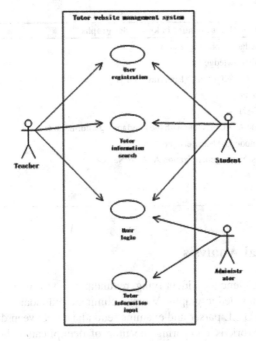

Fig. 3. A use case diagram of tutor website system

We map the use case diagram in Fig. 3 to a knowledge graph in Fig. 4.

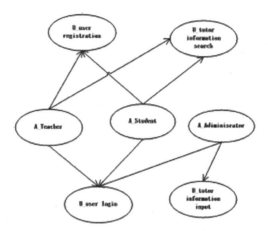

Fig. 4. A knowledge graph of tutor website system

Each node can be traversed in a knowledge graph. We define the array Count[] represents traversal times of different nodes.

The initial value of Count[] equals to 0. If a node appears once, then the current value of Count[] = Count[] + 1.

Algorithm Dealing with redundancy in knowledge graphs

Input: A knowledge graph array KG[ ]

Output: A refined knowledge graph

For : (each node KG{<KGn>,<R>} in array KG[ ])

　　　Compute Count[];

　　　If (Count[]≥2)

1. create a new node A_ActorName with the overlapping information

2. delete the same node U_UsecaseName

3. create<U_UsecaseName,generalization,A_ActorName>

　　　End If

　　End For

---

5   Experimental Analysis

We measure an efficiency of identifying redundancy existing in diagrams through mapping them to knowledge graphs. We can eliminate redundancy between diagrams designed through a UML parser and examine redundancy between diagrams through a UML parser. Our work is comparing the time of deduplication between the use of diagrams and the transformation of knowledge graphs. Experimental tools can eliminate redundancy according to rules we set before in seconds. We measured the median time of performing 1000 times. Our experiment uses for between 1 and 10 diagrams. The time spent in the two groups was measured. In our assessment, the time of eliminating redundancy in the transformation of knowledge graphs is shorter than the time of the use of diagrams (Fig. 5).

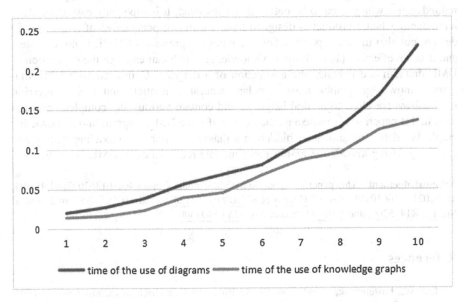

Fig. 5. Median time for eliminating redundancy

6 Related Work

Nowadays, there are two main methods to study the redundancy of UML: the first method is to use formal model. A specific approach is to make change for UML model and allow for easy detection of redundancy. OMG defines the syntax of the UML language through the use of model and the established OCL rules. The precise UML team has a prominent contribution for this. Its goal is to ensure the integrity of each element in the UML. The second method is to eliminate redundancy by establishing good limits. Some researchers have proposed an extension to OCL and claimed that OCL is not sufficient to express all kinds of limitations today. There are also some people who suggest using the rest methods like GCC (graphical consistency conditions) limited description language. The goal of the second approach is to improve the ability to limit the definition of language expression in order to formalize more semantic constraints. In the paper, we transform UML diagram to knowledge graph. Knowledge graph can enhance the expression of UML diagram and enhance the abstraction of UML diagram. This approach has a very good balance between the three languages used in the formal language, the formal language and the redundancy analysis.

7   Conclusion

Model-Driven Engineering (MDE) alleviates the cognitive complexity and effort through the refinement and abstraction of consecutive models [9]. MDE is close to human understanding and cognitive models, especially visual models. Designers can focus on business logics and related information. Changes to a model may introduce redundancies, which need to be detected and resolved. It is especially easy to produce redundancy when a product is designed from a different perspective. It is not only in design, but also in data representation. Language expression of UML is not complete and it is not effective [10]. Through a knowledge graph can enhance the expression of UML diagram and enhance the abstraction of a diagram. Compared with UML diagrams, knowledge graphs have abundant natural semantics, and their expression mechanisms are close to natural language and contain various and complete information. In this paper, we propose a method to transform a UML diagram into a knowledge graph. We detect redundancies hidden in a diagram to improve modeling quality and efficiency. Using knowledge graphs can eliminate redundancy in UML diagram.

Acknowledgement. This paper is supported by NSFC under Grant (Nos. 61363007, 61502294, 61662021, 61661019), NSF of Hainan Nos. ZDYF2017128 and 20156243, NSF of Shanghai No. 15ZR1415200, and STCSM project No. 14YF1404300.

References

1. Liu, W., Easterbrook, S., Mylopoulos, J.: Rule-based detection of inconsistency in UML models. In: Workshop on Consistency Problems in UML-based Software Development, pp. 106–123 (2002)
2. Boehm, B.W.: A spiral model of software development and enhancement. IEEE Comput. **21** (5), 61–72 (1998)
3. Duan, Y., Cheung, S.-C., Fu, X., Gu, Y.: A metamodel based model transformation approach. In: Third ACIS International Conference on Software Engineering Research, Management and Applications, pp. 184–191. IEEE (2005)
4. Sugiyama, K., Tagawa, S., Toda, M.: Methods for visual understanding of hierarchical system structures. IEEE Trans. Syst. Man Cybern. **11**(2), 109–125 (1981)
5. Berardi, D., Calvanese, D., De Giacomo, G.: Reasoning on UML class diagrams. Artif. Intell. **168**, 70–118 (2005)
6. Van Der Straeten, R.: Inconsistency Management in Model Driven Engineering An Approach using Description Logics
7. Chein, M., Mugnier, M.L.: Graph Based Knowledge Representation. Springer, London (2009). doi:10.1007/978-1-84800-286-9
8. Spanoudakis, G., Finkelsteain, A., Till, D.: Overlaps in requirements engineering. Autom. Softw. Eng. **6**(2), 171–198 (1999)
9. Liu, J., Tong, G., Li, M., Zang, F.: Ontology-based semantics reasoning of UML class diagram. Comput. Appl. Softw. **28**(4) (2011)
10. Gogolla, M., Richters, M.: Expressing UML class diagrams properties with OCL. In: Clark, T., Warmer, J. (eds.) Object Modeling with the OCL. LNCS, vol. 2263, pp. 85–114. Springer, Heidelberg (2002). doi:10.1007/3-540-45669-4_6

An Automatic Fall Detection System Based on Derivative Dynamic Time Warping

Hong Yang[✉], Yanqin Yang, Wenchao Xu, and Yuxin Pang

East China Normal University, Shanghai, China
yqyang@cs.ecnu.edu.cn, 51151214036@ecnu.cn

Abstract. Maturation of Internet and rapid development in mobile communication make smart device could bring enhanced services to person especially in health care center. Therefore, we focus on developing a fall detection application running on Android mobile phones. The detector is fit to indoor and outdoor, without confine of its surroundings. We propose a novel method which fuses Derivative Dynamic Time Warping (DDTW) to detect a fall event and algorithm sensitivity is 84.7%, as well as 94% of specificity. Our algorithm is considerable concise and efficient, what's more, it do not intrude on privacy of its users or degrade the quality of life. And above all, the method not only overcomes the shortage of thresholding-based fall detection method, but also applicable to all kinds of people with different weight and height.

Keywords: Fall detection · Derivative dynamic time warping · Mobile phone · Android

1 Introduction

A fall is an event which happens inadvertently and results in a person coming to rest on lower lever, like ground or floor. It is a threat to the health of people, and causes both physical and emotion harm to the individual and their family, as well as a significant financial burden on the healthcare system.

Younger people are at an increased risk of falls, in particular those with a history of falling, neurological conditions, cognitive problems or visual impairments, let alone for older adults, falls remain one of the leading causes of harm in community and care settings. Each year, millions of older people fall, and one out of five falls causes a serious injury such as broken bones or a head injury. In addition, over 800,000 patients a year are hospitalized because of a fall, and the direct medical costs for fall injuries are $31 billion annually [1–3]. Falls and fall-induced injuries in older people are a major public health problem in modern societies with aging populations. As the frequency and severity of falls-related injuries growths dramatically with age, developing a surveillance system which detects falling state in real time and responses immediately is an efficient way to mitigate the effects of falls.

Existing fall detection approaches can be explained and categorized into three different classes: wearable device based, ambience device based and vision based [4]. Most of the existing approaches share the same general framework. Data acquisition merely varies from one sensor to multiple sensors or from one camera to multiple

© Springer Nature Singapore Pte Ltd. 2017
G. Chen et al. (Eds.): PAAP 2017, CCIS 729, pp. 427–438, 2017.
DOI: 10.1007/978-981-10-6442-5_40

cameras. Wearable device based approaches rely on embedded sensors which are fastened on the body to detect the motion and location of human. Ambient based devices attempt to fuse audio and visual data and event sensing through vibrational data. And in-home assistive/care systems, cameras convey multiple advantages over other sensor based systems because cameras are less intrusion and non-wearable.

In this paper, an automatic fall detection system running on Android smart phones are designed and a novel algorithm based on Derivative Dynamic Time Warping (DDTW) is proposed. The rest of the paper is organized as follows. Section 2 introduces related work in thresholding-based fall detection and dynamic time warping. In Sect. 3, we present our fall detection algorithm based on derivative dynamic time warping and framework of our fall detection system. Section 4 elaborately describes detection algorithm which is then evaluated experimentally in Sect. 5. Finally we summarize this thesis and propose some improvement direction for future work.

2   Related Work

2.1   Thresholding-Based Fall Detection

A majority of early fall detection system are based on threshold judgement, up to now, there are still a lot of research in this field. Either the single threshold algorithm in the initial or the improved multi-stage thresholding system, the threshold selection is the crux of algorithm. This kind of system usually divides falling into three stages which are acceleration, impact and inactivity in order. A typical fall is performed as following graph:

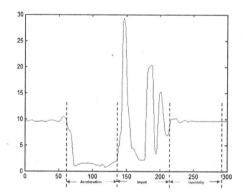

Fig. 1. Acceleration reading of a fall.

A fall causes acutely varieties in acceleration and generally occurs in a short time period with a typical range of 0.4–0.8 s [5].

Acceleration: At the start of a fall, the human body lose balance and consequently a short period of free fall. This causes the acceleration's amplitude to be dropped significantly, even tends toward 0 g. A threshold for this "free fall" is named as $T_{acceleration}$.

$$\sqrt{a_x^2 + a_y^2 + a_z^2} < T_{accleration} \tag{1}$$

If (1) is met, a fall is possible happening.

Impact: Then the body hit onto the ground or other obstacle, the impact results in a significant rise of the acceleration. As shown in Fig. 1, there is a sharp inversion of the polarity of the amplitude in an instant. The extra volatility is caused by elastic sponge cushion where we perform our experiments. In this stage, T_{impact} is a previous determined fixed threshold.

$$\sqrt{a_x^2 + a_y^2 + a_z^2} > T_{impact} \tag{2}$$

Inactivity: After falling, human can return to normal activity or cannot rise immediately and remains in a motionless position for several seconds. On this period, the curve in Fig. 1 flatten out, and the acceleration maintains a steady value of 1 g.

The thresholding-based method is commonly used in fall detection algorithm. Ge and Xu [6] presents an algorithm on adaptive thresholds to recognize falls from acceleration signals. In [7], Nyan et al. use a threshold of absolute peak values of acceleration to determine fall. Lee and Carlisle [8] judge fall event by testing the accelerometer signals exceed upper and lower thresholds. And in [9], the authors benchmark the performance of the thirteen published fall detection algorithms when they are applied to the database of 29 real-world falls.

2.2 Dynamic Time Warping (DTW)

The fall detection algorithm of our work is sequence matching which is a technique commonly employed in data mining and discrete time series analysis to obtain the similarity of two sequences. Similarity search in time series database is essential, because it helps prediction, hypothesis testing in data mining and knowledge discovery [10].

Dynamic Time Warping (DTW) is a common method to measure similarity between two time series which may vary in time or speed, and plays an important part in the speech recognition, machine learning and other aspects. The idea of the algorithm is that through stretching or shrinking along the time axis to find a minimal path between two time series: the test signal and the reference signal. This warping between two time series can be used to determine their similarity [11]. Lots publications describe the DTW algorithms detailed, in this section we briefly introduce it.

Suppose we have two time series, Q and C, of length n and m respectively, where

$$Q = q_1, q_2, \ldots, q_i, \ldots, q_n \tag{3}$$

$$C = c_1, c_2, \ldots, c_j, \ldots, c_m \tag{4}$$

To align two sequence using DTW, we construct an n-by-m matrix where the (i^{th}, j^{th}) element of the matrix contains the distance $d(q_i, c_j) = (q_i - c_j)2$ between the two points q_i and c_j. Then, accumulate distance is measured by:

$$D(i,j) = \min[D(i-1,j-1), D(i-1,j), D(i,j-1)] + d(i,j) \qquad (5)$$

A warping path W defines an alignment between Q and C. Formally, $W = w_1, ..., w_T$ where max $(m, n) \leq T \leq m + n - 1$. W maps the time axis of i of the test signal to the time axis j of reference template nonlinearly.

If DTW attempts to align two sequences that are similar except for local accelerations and decelerations in the time axis, the algorithm is likely to be successful. However, the shortage of DTW is that it may try to explain variability in the Y-axis by warping the X-axis [13]. This can lead to unintuitive alignments where a single point on one time series maps onto a large subsection of another time series as in Fig. 2C.

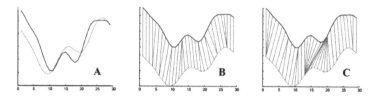

Fig. 2. (A) Two signals. (B) The ideal alignment. (C) The alignment produced by DTW.

In Fig. 2, it's obvious that DTW failed to align the two central peaks because they are slightly separated in the Y-axis.

To prevent this problem, authors of [13] proposed a modification of DTW that does not consider the Y-values of the data points, but rather considers "trend or shape" of data. They called the new algorithm Derivative Dynamic Time Warping (DDTW) as they obtain information about shape by using the first derivative of the sequence.

3   Proposed Algorithm and System

In simulated cases, the correct warping can be known by warping a time series and attempting to recover the original.

With Derivative Dynamic Time Warping (DDTW), the distance of two series d (q_i, c_j) is not based on Euclidean but rather the square of the difference of the estimated derivatives of q_iand c_j. There exist sophisticated methods for estimating derivatives, however, we use the following estimate for simplicity and generality.

$$x[q_i] = \frac{(q_i - q_{i-1}) + ((q_{i+1} - q_{i-1})/2)}{2} \qquad (6)$$

This method is more robust to outliers than any estimate considering only two data points as it is the average of the slope of the point in question and its left neighbor, and the slope of the left neighbor and the right neighbor.

Then replacing d (q_i, c_j) in Eq. (5) with d $(x[q_i], x[c_j])$. Note that in Eq. 6, this estimate is not defined for the first and last points of the sequence, and use the estimates of the second and penultimate elements for equivalent substitution. The subsequent calculations are same as DTW.

The difference between Fig. 3.3 and 3.4 clearly illustrates the superiority of the DDTW.

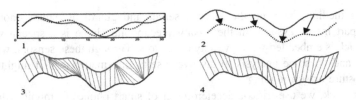

Fig. 3. (1) Two signals. (2) The feature to feature warping alignment. (3) The alignment produced by DTW. (4) The alignment produced by DDTW.

Fusing the DDTW into our fall detection algorithm, we successfully develop a system which consists of three major elements, server, target audience and smart phone. The relationship of three is depicted in Fig. 4.

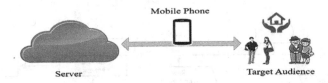

Fig. 4. Framework of fall detection system

3.1 Server

As the control center, server enables to relate organically between each module, and they together add up to the integrity of the system. It collects and analyses all the information coming from the individual users and healthcare institutions. For user's convenience query of their historic record, we uploads their daily activity information at a special time which is predefined every day.

3.2 Target Audience

Target audience of our system is three groups, namely the elderly, guardians and healthcare institutions. Their connection, information sharing and changing are realized by server.

After install, our system runs on phones, continuously monitors health condition and physical safety of the elderly. In addition, we develop a web client to help guardians who are away from their parents check their situation all the time. Message or call will be automatically trigged in the elderly client if a fall is detected.

The larger idea is to develop a more complete system which can be applied to healthcare institutions to help health workers monitor patients' physical condition and this will lessen their work burden.

3.3 Smart Phone

To be the medium between the elderly, server and guardians, smart phone is an essential part in our system. In the meanwhile, mobile phone is a special wearable device which is embedded with a variety of sensors. Through these sensors, we collect their data and do some analysis to discover useful information which might include human posture or surrounding.

In our work, we use tri-axis accelerometer of smart phones to monitor the acceleration change of human body.

4 Execution Procedure of Fall Detection Algorithm

The actual flow of the fall detection algorithm is given out in Fig. 5, and whole algorithm can be divided into two branches. The left half of the flow chart is sample sequence extraction, as well as the right side is query sequence extraction.

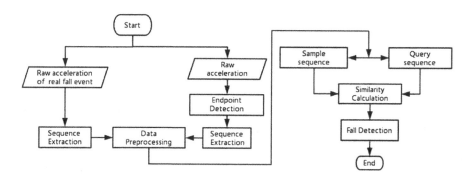

Fig. 5. Flow chart of fall detection algorithm.

To utilize DDTW in fall detection algorithm, a sample sequence of real fall event is required. We collect the raw x, y, and z data of a tri-axis accelerometer which is integrated in a smart phone when volunteers perform fall events. Before using these data as a sample, they need to be preprocessed to make the data more reasonable and with low noise.

The method to get query sequence is similar with the way mentioned above, except that there is a step called endpoint detection before storing raw acceleration.

4.1 Endpoint Detection

The aim of endpoint detection is reducing unnecessary computational work and improve algorithm speed. According to the difference between falling and daily activities, as well as acceleration change of falling, a threshold value T can be determined. If the acceleration is bigger than T, it belongs to daily activities. Otherwise, it is predicted as a falling state which needs a further decision.

4.2 Sequence Extraction

Through testing the different combination of sampling frequency of accelerometer and sample size, we set the length of query and template sequence are both 300 sampling points and frequency is set as "SENSOR_DEALY_NORMAL". The same length of sample and query sequences makes the DDTW algorithm in this system is simple and efficient.

4.3 Data Preprocessing

Not just for obtaining useful and reasonable information, but for eliminating the influence of orientation, we first transform the raw data into traditional magnitude of signal vector (Msv). The value of Msv is calculated by the following formula:

$$M_{sv} = \sqrt{x^2 + y^2 + z^2} \tag{7}$$

where x, y, z represent the data of three axis of accelerometer respectively. The significance behind Msv is orientation insensitive, in other words the position of mobile phone (e.g., vertical, horizontal, slant) have no influence on this value.

The next step of preprocessing is median filter which is an effective method that can, to some extent, remove the isolated noise points, and thus can improve the results of later processing.

4.4 Similarity Calculation

Figure 6 is a real practice of DDTW in our fall detection system. Curve A is the acceleration of a real fall, namely sample sequence. And curve B is a query sequence which we predict it might be a fall. It is obvious that the two sequence have the approximately the same component shapes, but the shapes do not line up in x-axis or

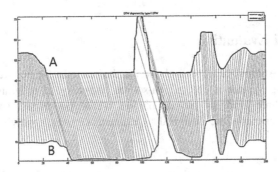

Fig. 6. The alignment produced by DDTW between two real acceleration data of fall event.

timeline. However, through DDTW, we find a warping path and align the points "feature to feature".

The similarity value is measured as the ratio of the minimum distance of two sequence and maximum size value between them, therefore smaller value means higher similarity.

4.5   Fall Detection

With the help of DDTW, we successfully get the match degree of two sequences, and the subsequent process of fall judgement is described in Fig. 7.

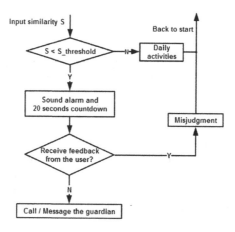

Fig. 7. The flow chart of fall detection.

First, we predefine a threshold S_threshold to divide activities into daily activities and potential falls. Second, if there is a potential fall, mobile phone sounds alarm and waits for 20 s for users to respond it. Third, guardians will receive a call or message from users' phone if no response is taken within the countdown. Otherwise, user cancel alarm in time, the system regards this potential fall is a misjudgment, and algorithm is back to its start.

5   Result and Evaluation

We have chosen Android-based cell phones as the platform for our study because the Android operating system is free, open-source and easy to program. We perform experiments in three kinds of mobile phones whose parameters are given out in Table 1. The experiments are designed to evaluate system sensitivity as well as its specificity.

Table 1. Parameters of three phones

Brand	Model	OS	ROM	RAM
Samsung	Galaxy Note Three	4.3	16 GB	3 GB
HUAWEI	P8 Lite	5.0	16 GB	2 GB
Moto	M XT1662	6.0	32 GB	4 GB

5.1 Experiment Setup

Considering safety issue and risk of our experiments, we made the choice not to ask elderly people to perform fall event. We recruited 10 healthy participants aged from 20 to 25 years old, four of whom are female and others are male. They are all asked to perform fall and normal ADLs in an outdoor environment. The participants stand upright and perform the fall down activity while the phone is brought in specific position. As shown in Fig. 8, we fix the mobile phone at their waist in this way.

Fig. 8. The position of mobile phone.

Before experiment, we verify that there is no difference in algorithm performance between the three kinds of phones. Therefore we use Samsung Galaxy Note Three whose configuration is the minimum to complete our experiment.

The whole experiment can be divided into two parts. First, a sticky sponge cushion which is shown in Fig. 9 is used for Forward-Fall, Backward-Fall and Side-Fall and

Fig. 9. Fall experiments in a sponge cushion

these experiments are repeated 5 times for each category, thus 150 fall samples are obtained. At the meanwhile, we record the right alarm times and fail times to compute true positive and false negative. Next, we ask the participants to perform ADL behaviors which are walking, running, climbing up/down, sit down, squat down and stand up for 10 times and record the alarm times to compute the false positive.

5.2   Fall Detection Result

The test result are shown in Table 2.

Table 2. Experiment result

Event	Fall	ADLs
Number of samples	150	100
Alarm times	127	4
TP = 91.3%, FN = 8.7%, TN = 95%, FP = 5%		

Sensitivity is the capacity to detect a fall

$$Sensitivity = \frac{TP}{TP + FN} = 0.847 \qquad (8)$$

Specificity is the capacity to detect only a fall

$$Specificity = \frac{TN}{TN + FP} = 0.96 \qquad (9)$$

where TP = true positives (detected fall), FN = false negatives (undetected fall), TN = true negatives (normal movement, not alert), FP = false positives (normal activity, alert).

According to Table 2, our system recognition rate is 84.7%, and the specificity is 96% which means that our system distinguish fall from ADLs very well. Furthermore, during experiment, we notice that the performance of our algorithm doesn't be influenced by changes of fall direction as well as the weight and height of volunteers.

The recognition rate is far below theoretical value, through analyzing experiment data, this is due to the fact that our algorithm doesn't perform well when following occurs, mobile phone thrown from body and caused extra significant change of acceleration. The solution is improving algorithm's universality or add sample sequence of complex situation.

6   Conclusion and Future Work

In this paper, an automatic fall detection based on derivative dynamic time warping method is proposed. Our recognition algorithm fundamentally differs from most existing fall detection methods which use tri-accelerometer with threshold algorithms.

Our method extracts a query sequence of human motion by using accelerometer embedded in mobile phone. Then DDTW is used to match with sample sequence of real fall events acceleration. The experiments show that our system can detect fall statue efficiently with low failing rate. And also shows a dramatic decrease in false-positive rates to a near zero level for all non-failing activities while exerting no negative effect in the fall detection rate.

We will enhance our system in the following respects. One is enhancing the universality of sample sequence to make the fall detection algorithm performs well on complex situation. The other is using DDTW to recognize other human activities and establishing a better health system.

The method in the paper not only overcomes the shortage of the thresholding-based fall detection method, but also applicable to all kinds of people with different weight and height. DDTW is insensitive to surrounding environment and changes in body size of user. It is almost impossible for tradition fall detection method to achieve the sensitivity and specificity as we do in a sample and efficient algorithm.

Acknowledgment. The research work was supported by the National Natural Science Foundation of China under Grant Nos. 61300043, 61373156 and 91438121, and the Science & Technology Commission of Shanghai Municipality under grant no. 14DZ2260800.

References

1. Kannus, P., Niemi, S., Parkkari, J.: Continuously increasing number and incidence of fall-induced, fracture-associated, spinal cord injuries in elderly persons. Arch. Intern. Med. **160**, 2145–2149 (2000)
2. Song, F.X., Zhang, Z.J., Gao, F., Zhang, W.Y.: An evolutionary approach to detecting elderly fall in telemedicine. In: 2015 First International Conference on Computational Intelligence Theory, Systems and Applications (CCITSA), Yilan, pp. 110–114 (2015)
3. Centers for Disease Control and Prevention, National Center for Injury Prevention and Control, Division of Unintentional Injury Prevention. Important Facts about Falls. https://www.cdc.gov/HomeandRecreationalSafety/Falls/adultfalls.html. Accessed 24 July 2016
4. Medrano, C., Igual, R., Plaza, I., Castro, M., Fardoun, H.M.: Personalizable smartphone application for detecting falls. In: IEEE-EMBS International Conference on Biomedical and Health Informatics (BHI), Valencia, pp. 169–172 (2014)
5. Lin, C.W., Ling, Z.H., Chang, Y.C., Kuo, C.J.: Compressed-domain fall incident detection for intelligent homecare. J. Sig. Process. Syst. **49**(3), 393–408 (2007)
6. Ge, Y., Xu, B.: Detecting falls using accelerometers by adaptive thresholds in mobile devices. J. Comput. **9**(7), 1553–1559 (2014)
7. Nyan, M.N., Tay, F.E., Manimaran, M., Seah, K.H.: Garment-based detection of falls and activities of daily living using 3-axis MEMS accelerometer. J. Phys: Conf. Ser. **34**, 1059–1067 (2006)
8. Lee, R.Y., Carlisle, A.J.: Detection of falls using accelerometers and mobile phone technology. Age Ageing **40**(6), 690–696 (2011)
9. Kostopoulos, P., Nunes, T., Salvi, K., Deriaz, M., Torrent, J.: F2D: a fall detection system tested with real data from daily life of elderly people. In: 2015 17th International Conference on E-health Networking, Application & Services (HealthCom), Boston, MA, pp. 397–403 (2015)

10. Lee, S., Kwon, D., Lee, S.: Efficient similarity search for time series data based on the minimum distance. In: Pidduck, A.B., Ozsu, M.T., Mylopoulos, J., Woo, C.C. (eds.) CAiSE 2002. LNCS, vol. 2348, pp. 377–391. Springer, Heidelberg (2002). doi:10.1007/3-540-47961-9_27

11. Muda, L., Begam, M., Elamvazuthi, I.: Voice recognition algorithms using mel frequency cepstral coefficient (MFCC) and dynamic time warping (DTW) techniques. TTPS 2 (2010)

12. Keogh, E., Ratanamahatana, C.A.: Exact indexing of dynamic time warping. Knowl. Inf. Syst. 7(3), 358–386 (2005)

13. Keogh, E.J., Pazzani, M.J.: Derivative Dynamic Time Warping (2001)

14. Vo, V., Hoang, T.M., Lee, C.M., et al.: Fall detection for mobile phone based on movement pattern. 인터넷정보학회논문지 13(13), 23–31 (2012)

15. Jia, H., Li, M., Ning, Y., Liang, S., Li, H., Zhao, G.: Implementation of Android-based fall-detecting system. In: 2016 IEEE 13th International Conference on Signal Processing (ICSP), Chengdu, China, pp. 1323–1328 (2016)

On Signal Timing Optimization in Isolated Intersection Based on the Improved Ant Colony Algorithm

Huang Min[(⊠)]

Hainan Tropical Ocean University, Sanya 572022, Hainan, China
Huangmin198100@126.com

Abstract. The unreasonable allocation of traffic lights is liable to trigger traffic jam. The ant colony algorithm is a universal random optimization algorithm, which can solve the problem of traffic signal timing. Aiming at the problem of signal timing optimization in isolated intersection, the paper comes up with the improved ant colony algorithm, which is showed by the experimental simulation results that it is better than the traditional way in signal timing.

Keywords: Ant colony algorithm · Isolated intersection · Signal timing · Traffic lights

1 Introduction

Nowadays, traffic jam has become one of the serious problems faced by many cities. However, the main places that cause the traffic jam are the intersections of the city, among which the unreasonable time allocation of the traffic lights is the main reason [1]. The number of the intersection delay and the average stop of vehicle would increase as the growth of the traffic light circle, therefore, the key to solve the traffic jam problem lies in how to allocate the traffic light time in order to reduce the intersection delay and the average stop of vehicle so that the vehicle traffic capacity can be maximized.

In recent years, experts of traffic control areas have been doing a number of researches on the problem of traffic light allocation optimization. For instance, Wan and Wang et al. found the model for the traffic light allocation optimization, aiming to minimize the vehicle delay or stop. Wang et al. also solved the model by simulated annealing algorithm [2, 3]; Yuan and Shi confirmed the best signal cycle by analyzing the conflict of the traffic conditions between the intersections [4]; Combined with the genetic algorithm, Zhang et al. put forward the mixing ant colony isolated intersection signal timing algorithm [5]; Aiming at the minimized waiting time, Wang et al. raised the multi-stage decision model and its prior to the dynamic programming algorithm, and so and so force [6]. Intelligential, multimodal and optimization are what people in the traffic control theory areas pursue and they are bound to become the development direction of the traffic signal control system in the future of our country. Ant colony algorithm, with strong robustness and excellent ability of global optimization, is a swarm intelligent algorithm that simulates ant foraging behavior, which can better solve the traffic dynamic problem and being applied in traffic recently [7]. The article

G. Chen et al. (Eds.): PAAP 2017, CCIS 729, pp. 439–443, 2017.
DOI: 10.1007/978-981-10-6442-5_41

makes some exploratory research on the applying of ant colony algorithm in solving the traffic signal timing problem and comes up with an improved ant colony algorithm focusing on the problem of isolated intersection signal timing optimization.

2 The Foundation of Model

Among the needs of city traffics, there are three main goals for the isolated signal timing optimization: the minimization of vehicle delay; the minimization of average stop; and the maximization of traffic capacity in the intersection. Thus, the problem of isolated signal timing optimization can be transformed into the minimized form of fractional through unified objective function method. Then, the corresponding objective function can be transformed into single objective function containing variables through weight coefficient whose value is closely related to the traffic needs. The model of isolated intersection signal timing optimization is shown as formula (1):

$$\min f(x) = \frac{\sum_{i=1}^{n} (K_i^1 D_i + K_i^2 H_i)}{\sum_{i=1}^{n} K_i^3 Q_i} \tag{1}$$

Among which, $x \in [0.6, 0.8]$; K_i^1, K_i^2, K_i^3 are the weight coefficient of vehicle delay, average stops and intersection traffic capacity respectively, the model is defined as follow:

$$K_i^1 = 2(1 - Y)s_i^{\frac{1}{7}} \tag{2}$$

$$K_i^2 = \frac{10}{9}(1 - Y)s_i^{\frac{1}{4}} \tag{3}$$

$$K_i^3 = \frac{c}{1800}Y \tag{4}$$

Among which, $Y = \sum_{i=1}^{n} \sum_{j=1}^{n} \max y_i^j$, yij refers to the flow ratio of i phase and the direction of j.

The basic parameter settings of the isolated signal timing optimization model [8] are as follow:

(1) The vehicle delay formula:

$$D_i = \frac{cq_i(x - y_i)^2}{2x^2(1 - y_i)} + \frac{x^2}{2(1 - x)}, i = 1, 2, \cdots, n \tag{5}$$

Among which, c means that the signal cycle is long; qi refers to the traffic flow of i phase (pcu/h); x is the saturation of the intersection; yi is the saturation of i phase.

(2) The formula of the average stop of vehicle in i phase is:

$$H_i = 0.9 \frac{c - x_i}{1 - y_i} \qquad (6)$$

(3) The formula of i phase intersection traffic capacity is:

$$Q_i = \frac{s_i x_i}{c} \qquad (7)$$

Among them, sirefers to the saturation of i phase (pcu/h).

3 The Isolated Intersection Signal Timing Optimizations of the Improved Ant Colony Algorithm

The problem of optimization is usually defined as (S, f, Ω), among which S is the searching space, f is the objective function, Ω is the constraint condition, their purpose is to find out the global optimal solution s^*, $s^* \in S$ so as to maximize or minimize f. Ant colony optimization algorithm is the heuristic method adopted in solving more difficult calculation for the combinatorial optimization problems [9]. In the solution space, each node represents a possible solution for the optimization problem. By using the improved ant colony algorithm, the steps for solving the isolated signal timing optimization in traffic are as follow:

Step 1. Initialization. Initializing the relevant parameters in the algorithm, the initializing time t = 0; the algorithm number of iterations N = 0, the largest number of iteration is Nmax; the pheromone of each node is set as a constant, that is $\tau_{ij}(0) = c_0$ (c0 is a constant); $\Delta \tau_{ij}(0) = 0$, ρ is the residual coefficient of the pheromone concentration; Q is the intensity of pheromone; m is the optimal number of ants.

Step 2. Control the number of iteration for the algorithm $N \leftarrow N + 1$;

Step 3. Initialize the number of ants K, and make K = 1, then control the number of ants, so that $K \leftarrow K + 1$;

Step 4. According to the transition probability, the ants determine the choice from node i to node j, and continue to go forward and record the process, among them $j \in allowed_k = \{c - tabu_k\}$;

$$P_{ij}^k(t) = \begin{cases} \dfrac{\tau_{ij}(t)}{\sum\limits_{i=1}^{m} \tau_{ij}(t)} & j \in allowed_k \\ 0 & otherwise \end{cases} \qquad (8)$$

Step 5. If $k < m$, then jump to step 3; Otherwise, when all the ants traverse every node, we form a new solution by the rout of the ants at the end, and calculate the corresponding objective function value f, meanwhile, we renew the maximized value f_{max} and the minimized value f_{min}, and record them.

Step 6. After all the ants traversing, we update the global pheromone through the following formula (2):

$$\tau_{ij}(t+1) = \rho\tau_{ij}(t) + \sum_{k=1}^{m} \Delta\tau_{ij}^{k}(t) \tag{9}$$

$\Delta\tau_{ij}^{k}$ is the pheromone variable quantity that ant k move from node i to node j,

$$\Delta\tau_{ij}^{k}(t) = \frac{af(t)}{f_{max} - f_{min}} \quad \text{(a is a constant)} \tag{10}$$

Step 7. If $N \geq N_{max}$, output the optimal solution, and exit the loop; Otherwise, jump to Step 2.

4 Numerical Experiments

Setting the intersection, which can go straight through east-west and north-south direction only, supposing that, in all imports, the traffic, flow, and saturated flow are shown as Table 1.

Table 1. The intersection signal flow

	East import	West import	South import	North import
Traffic flow q	380	350	760	840
Saturated flow s	1200	1200	2200	2200 ·
Phase saturation yi	0.32	0.29	0.35	0.38
$\max(y_i^1, y_i^2)$	0.32		0.38	
$Y = \sum_{i=1}^{2} \max(y_i^1, y_i^2)$	0.70			

Meanwhile, in the numerical experiments, the algorithm parameter is set to: ant number is 50, the largest number of iterations is 100, $Q = 0.08$, $\rho = 0.79$, $c \in [40, 120]$.

The experiment result between the improved algorithm and traditional method are compared after the simulation calculation. According to the random feature of ant colony algorithm, the comparison is made under the statistical significance, the comparison results are showed in Table 2.

It can be concluded from Table 2: compared with the traditional method, the effect of the improved ant colony algorithm is relatively obvious in solving the isolated intersection signal timing problem in the area of vehicle delay, vehicle stop and traffic capacity.

Table 2. The result comparison between the improved ant colony algorithm and the traditional one

	Cycle	The total delay time	The total number of parking	The total capacity
Improved ant colony algorithm	75	20711.5	1001.8	1317.6
The traditional method	91	25702	1326	1021.3

5 Conclusion

Ant colony algorithm is a new optimization technique based on swarm intelligence. The improved ant colony algorithm can better solve the intersection signal timing optimization problems, and the simulation results show that the improved ant colony algorithm is superior to the traditional one. Because of the strong instantaneity and uncontrollability of transportation, the direction of future research is to consider how to solve these traffic factors in the improved ant colony algorithm.

Acknowledgment. This work was supported by Project of the National Natural Science Foundation of China (No. 61073189) and Project of Natural Science Foundation of Hainan province of China (No. 20166223) and Project of technological collaboration of the college and local government of Sanya city of China (No. 2013YD14) and Scientific Research Foundation of Qiongzhou University (No. QYQN201327).

References

1. Bontoux, B., Feillet, D.: Ant colony optimization for the traveling purchaser problem. Comput. Oper. Res. **35**, 628–637 (2008)
2. Wan, X., Lu, H.: A real time self-adaptive traffic signal setting optimization theory. J. Traffic Transp. Eng. **1**(4), 60–66 (2010)
3. Wang, Q., Tan, X., Zhang, S.: Signal timing optimization of urban single-point intersections. J. Traffic Transp. Eng. **6**(2), 60–64 (2006)
4. Yuan, C., Shi, F.: Using traffic stream characteristics of conflict point to determin signal time of signalized intersection. J. East China Jiaotong Univ. **19**(4), 14–16 (2002)
5. Zhang, L., He, J., Li, Q.: Hybrid ant colony genetic algorithm combining with the single point of intersection signal timing algorithm. Comput. CD Softw. Appl. 91–94 (2014)
6. Wang, L., Zhao, Y., Li, J., Zhang, H., Wang, X.: Multi-stage decision model for signal control problems and its forward dynamic prodramming algorithm. Control Decis. **27**(2), 167–174 (2012)
7. Huang, M., Jin, T., Zhong, S., Ma, Y.: Ant colony algorithm for solving continuous function optimization problem based on pheromone distributive function. J. Guangxi Norm. Univ. (Nat. Sci. Ed.) **31**(2), 34–38 (2013)
8. Yang, J., Yang, D.: Optimized signal time model in signaled intersection. J. TongJi Univ. **29**(7), 789–794 (2001)
9. Ju, J.: Traffic signal control based on ant colony algorithm. Traffic Eng. (2), 198–201 (2014)

Experiments on Neighborhood Combination Strategies for Bi-objective Unconstrained Binary Quadratic Programming Problem

Li-Yuan Xue[1], Rong-Qiang Zeng[2,3(✉)], Wei An[4], Qing-Xian Wang[4], and Ming-Sheng Shang[5]

[1] EHF Key Laboratory of Science, School of Electronic Engineering, University of Electronic Science and Technology of China, Chengdu 611731, Sichuan, People's Republic of China
xuely2013@gmail.com

[2] School of Mathematics, Southwest Jiaotong University, Chengdu 610031, Sichuan, People's Republic of China
zrq@swjtu.edu.cn

[3] Chengdu Documentation and Information Center, Chinese Academy of Sciences, Chengdu 610041, Sichuan, People's Republic of China

[4] School of Information and Software Engineering, University of Electronic Science and Technology of China, Chengdu 610054, Sichuan, People's Republic of China
an-qile@163.com, qxwang@uestc.edu.cn

[5] Chongqing Key Laboratory of Big Data and Intelligent Computing, Chongqing Institute of Green and Intelligent Technology, Chinese Academy of Sciences, Chongqing 400714, People's Republic of China
msshang@cigit.ac.cn

Abstract. Local search is known to be a highly effective metaheuristic framework for solving a number of classical combinatorial optimization problems, which strongly depends on the characteristics of neighborhood structure. In this paper, we integrate the neighborhood combination strategies into the hypervolume-based multi-objective local search algorithm, in order to deal with the bi-objective unconstrained binary quadratic programming problem. The experimental results show that certain combinations are superior to others. The performance analysis sheds lights on the ways to further improvements.

Keywords: Multi-objective optimization · Hypervolume contribution · Neighborhood combination · Local search · Unconstrained binary quadratic programming

1 Introduction

Local Search (LS) is a simple and effective metaheuristic framework for solving a number of classical combinatorial optimization problems. Generally, LS proceeds from an initial solution with a sequence of local changes, which is realized by defining the

© Springer Nature Singapore Pte Ltd. 2017
G. Chen et al. (Eds.): PAAP 2017, CCIS 729, pp. 444–453, 2017.
DOI: 10.1007/978-981-10-6442-5_42

proper neighborhood structure for the considered problem. Then, the neighborhood structure plays an important role in the LS procedure. In order to study the neighborhood combination strategies during the LS process, we present an experimental analysis of neighborhoods using the bi-objective unconstrained binary quadratic programming problem as a case study.

As a classical NP-hard combinatorial optimization problem, the Unconstrained Binary Quadratic Programming (UBQP) problem is capable of representing a wide variety of important problems, such as max-cut problem, maximum clique problem, set partitioning problem, graph coloring problem, among others [4]. The multi-objective extension of UBQP has been proposed in [5], where the multiple objectives are to be maximized simultaneously. In this work, we focus on tackling the bi-objective UBQP problem, which is mathematically formulated as follows [6]:

$$f_k(x) = x'Q^k x = \sum_{i=1}^{n} \sum_{j=1}^{n} q_{ij}^k x_i x_j \tag{1}$$

where $Q^k = (q_{ij}^k)$ is an $n \times n$ matrix of constants and x is an n-vector of binary (zero-one) variables, i.e., $x_i \in \{0, 1\}$ $(i = 1, \ldots, n)$, $k \in \{1, 2\}$.

In [7], Lü et al. studied the different neighborhood combinations for singe-objective UBQP problem, based on the basic one-flip move and two-flip move. The computational results showed that the simple one-flip move based neighborhood performs quite well. Then, they [8] proposed an hybrid metaheuristic approach by incorporating a tabu search procedure into the framework of evolutionary algorithms, where the one-flip move based local search is implemented as a subroutine. With the same neighborhood, the probabilistic GRASP-tabu search algorithm [9] and the backbone guided tabu search algorithm [10] were respectively proposed for singe-objective UBQP problem.

On the other hand, Liefooghe et al. [5] first extended the single-objective UBQP problem into the multi-objective case. They combined an elitist evolutionary multi-objective optimization algorithm with an effective single-objective tabu search procedure using the scalarizing function, in order to solve the multi-objective UBQP problem. In [6], they presented an experimental analysis on stochastic local search based on two scalarizing strategies, Pareto dominance relation and their combinations for bi-objective UBQP problem. Actually, they kept using one-flip move during the neighborhood search process. Therefore, it is very interesting to study different neighborhood combinations with f-flip move in multi-objective case.

In this paper, we integrate the neighborhood combination strategies into the hypervolume-based multi-objective local search algorithm, in order to study the search capability of different neighborhood combinations. Based on the well-known one-flip and two-flip moves, we propose three-flip move and investigate the performance of three metaheuristic algorithms on different neighborhood combinations using these three moves. The experimental results indicate that certain combinations are superior to others. The performance analysis explains the behavior of the algorithms and sheds lights on the ways to further enhance the search.

The remaining part of this paper is organized as follows. In the next section, we briefly introduce the basic notations and definitions of bi-objective optimization. In

Sect. 3, we present the ingredients of hypervolume-based multi-objective local search algorithm with the neighborhood combination strategies for solving bi-objective UBQP problem. Section 4 shows the computational results and comparisons on the benchmark instances of UBQP. The conclusions are provided in the last section.

2   Bi-objective Optimization

In this section, we briefly introduce the basic notations and definitions of bi-objective optimization. Without loss of generality, we assume that X denotes the search space of the optimization problem under consideration and $Z = \Re^2$ denotes the corresponding objective space with a maximizing vector function $Z = f(X)$, which defines the evaluation of a solution $x \in X$ [3]. Specifically, the dominance relations between two solutions x_1 and x_2 are presented below [11]:

Definition 1 *(Pareto Dominance). A decision vector x_1 is said to dominate another decision vector x_2 (written as $x_1 \succ x_2$), if $f_i(x_1) \geq f_i(x_2)$ for all $i \in \{1,2\}$ and $f_i(x_1) > f_i(x_2)$ for at least one $j \in \{1,2\}$.*

Definition 2 *(Pareto Optimal Solution). $x \in X$ is said to be Pareto optimal if and only if there does not exist another solution $x' \in X$ such that $x' \succ x$.*

Definition 3 *(Non-dominated Solution). $x \in S$ $(S \subset X)$ is said to be non-dominated if and only if there does not exist another solution $x' \in S$ such that $x' \succ x$.*

Definition 4 *(Pareto Optimal Set). S is said to be a Pareto optimal set if and only if S is composed of all the Pareto optimal solutions.*

Definition 5 *(Non-dominated Set). S is said to be a non-dominated set if and only if any two solutions $x_1 \in S$ and $x_2 \in S$ such that $x_1 \nsucc x_2$ and $x_2 \nsucc x_1$.*

Actually, we are interested in finding the Pareto optimal set, which keeps the best compromise among all the objectives. However, it is very difficult or even impossible to generate the Pareto optimal set in a reasonable time for the NP-hard problems. Therefore, we aim to obtain a non-dominated set which is as close to the Pareto optimal set as possible. That's to say, the whole goal is to identify a Pareto approximation set with high quality.

3   Neighborhood Combination Strategies

In our work, we integrate the neighborhood combination strategies into the hypervolume-based multi-objective local search algorithm, in order to solve bi-objective UBQP problem. The general scheme of Hypervolume-Based Multi-Objective Local Search (HBMOLS) algorithm [2] is presented in Algorithm 1, and the main components of this algorithm are described in detail in the following subsections.

Algorithm 1 Hypervolume-Based Multi-Objective Local Search Algorithm

---

Input: N (Population size)
Output: A: (Pareto approximation set)
Step 1 - Initialization: $P \leftarrow N$ randomly generated solutions
Step 2: $A \leftarrow \Phi$
Step 3 - Fitness Assignment: Assign a fitness value to each solution $x \in P$
Step 4:
while Running time is not reached **do**
  repeat
    Hypervolume-Based Local Search: $x \in P$
  until all neighbors of $x \in P$ are explored
    $A \leftarrow$ Non-dominated solutions of $A \bigcup P$
end while
Step 5: Return A

In HBMOLS, all the solutions in an initial population are randomly generated, i.e., each variable of one solution is randomly assigned a value 0 or 1. Then, we employ a Hypervolume Contribution (*HC*) indicator proposed in [2] to achieve the fitness assignment for each solution. Based on the dominance relation and two objective function values, the *HC* indicator calculate the hypervolume contribution of each solution in the objective space.

More specifically, a positive value is assigned to each non-dominated solution in the population, while a negative value is assigned to each dominated solution in the population. With the fitness values, a total order in the population is induced by the *HC* indicator, which is used to rank the solutions during the local search process. Afterwards, each solution is optimized by the hypervolume-based local search procedure, in order to obtain a high-quality Pareto approximation set.

3.1 Fast Evaluation and One-Flip Move

In order to deal with the binary problems such as UBQP and knapsack problem, one-flip move is widely used in the local search procedure, which flips a chosen binary variable by subtracting its current value from the opposite one. Considering the UBQP problem, we achieve the one-flip move by flipping a variable from 0 to 1 or from 1 to 0.

$$\Delta_i = (1 - 2x_i) \left(q_{ii} + \sum_{j \in N, j \neq i, x_j = 1} q_{ij} \right) \qquad (2)$$

Let Δ_i be the move value of flipping the variable x_i, and Δ_i can be calculated in linear time by the formula [8] above, more details about this formula can be found in [7]. With such fast incremental evaluation technique, we can calculate the objective function values high efficiently.

3.2   Two-Flip Move and Three-Flip Move

In the case of two-flip move, we can obtain a new neighbor solution by randomly flipping two different variables x_i and x_j. In fact, two-flip move can be seen as a combination of two single one-flip moves. We denote the move value by δ_{ij}, which is derived from two one-flip moves Δ_i and Δ_j $(i \neq j)$ as follows:

$$\delta_{ij} = \Delta_i + \Delta_j \tag{3}$$

Furthermore, three-flip move can be realized by simply flipping three different variables x_i, x_j and x_k $(i \neq j \neq k)$, and the corresponding move value η_{ijk} can be calculated in the following formula:

$$\eta_{ijk} = \Delta_i + \Delta_j + \Delta_k \tag{4}$$

Especially, the search space generated by two-flip move and three-flip move is much bigger than the one generated by one-flip move. In the following, we denote the neighborhoods with one-flip move, two-flip move and three-flip move as N_1, N_2 and N_3 respectively.

3.3   Hypervolume-Based Local Search

In the HBMOLS algorithm, we implement the f-flip $(f \in \{1, 2, 3\})$ move based neighborhood strategy in the hypervolume-based local search procedure. With the fast incremental evaluation techniques mentioned above, we can effectively calculate two objective function values for bi-objective UBQP problem. The exact steps of hypervolume-based local search procedure [2] are presented in Algorithm 2 below.

Algorithm 2 Hypervolume-Based Local Search

Steps:

1) $x^* \leftarrow$ an unexplored neighbor of x by randomly changing the values of the variables of x

2) $P \leftarrow P \bigcup x^*$

3) calculate two objective function values of x^* with fast incremental evaluation techniques

4) calculate the fitness value of x^* in P with the HC indicator

5) update all the fitness values of $z \in P$ $(z \neq x^*)$

6) $\omega \leftarrow$ the worst solution in P

7) $P \leftarrow P \backslash \{\omega\}$

8) update all the fitness values of $z \in P$

9) if $\omega \neq x^*$, Progress $\leftarrow$ True

In this procedure, we generate an unexplored neighbor solution x^* of x by randomly changing the values of the variables of x from 0 to 1 (or from 1 to 0). Then, the fitness value of x^* is calculated by the HC indicator. If x^* is a dominated solution, it is not necessary to update the fitness values of the other solutions. If x^* is a non-dominated solution, the fitness values of the non-dominated neighbors of x^* in the objective space are updated by the HC indicator.

Afterwards, we choose the worst solution ω with respect to the fitness values and delete it from the population P. According to the dominance relation between the solution ω and the population P, we update the fitness values of the non-dominated neighbors of ω. The hypervolume-based local search procedure will repeat until the termination criterion is satisfied, so as to obtain a set of efficient solutions.

4   Experiments

In this section, we present the computational results of 5 different neighborhood combination strategies on 9 groups of benchmark instances of UBQP. All the algorithms are programmed in C++ and compiled using Dev-C++ 5.0 compiler on a PC running Windows 7 with Core 2.50 GHz CPU and 4 GB RAM.

4.1   Parameters Settings

In order to conduct the experiments on the bi-objective UBQP problem, we use two single-objective benchmark instances of UBQP with the same dimension provided in [8][1] to generate one bi-objective UBQP instance. All the instances used for experiments are given in Table 1 below.

Table 1. Single-objective benchmark instances of UBQP used for generating bi-objective UBQP instances.

	Dimension	Instance 1	Instance 2
bo_ubqp_3000_01	3000	p3000.1	p3000.2
bo_ubqp_3000_02	3000	p3000.1	p3000.3
bo_ubqp_3000_03	3000	p3000.2	p3000.3
bo_ubqp_4000_01	4000	p4000.1	p3000.2
bo_ubqp_4000_02	4000	p4000.1	p4000.3
bo_ubqp_4000_03	4000	p4000.2	p4000.3
bo_ubqp_5000_01	5000	p5000.1	p5000.2
bo_ubqp_5000_02	5000	p5000.1	p5000.3
bo_ubqp_5000_03	5000	p5000.2	p5000.3

In addition, the algorithms need to set a few parameters, we only discuss two important ones: the running time and the population size, more details about the parameter settings for multi-objective optimization algorithms can be found in [1]. The exact information about the parameter settings in our work is presented in the following Table 2.

---

[1] More information about the benchmark instances of UBQP can be found on this website: http://www.soften.ktu.lt/-gintaras/ubqopits.html.

Table 2. Parameter settings used for bi-objective UBQP instances: instance dimension (D), population size (P) and running time (T).

	Dimension (D)	Population (P)	Time (T)
bo_ubqp_3000_01	3000	30	300″
bo_ubqp_3000_02	3000	30	300″
bo_ubqp_3000_03	3000	30	300″
bo_ubqp_4000_01	4000	40	400″
bo_ubqp_4000_02	4000	40	400″
bo_ubqp_4000_03	4000	40	400″
bo_ubqp_5000_01	5000	50	500″
bo_ubqp_5000_02	5000	50	500″
bo_ubqp_5000_03	5000	50	500″

4.2   Performance Assessment Protocol

In this paper, we evaluate the efficacy of 5 different neighborhood combination strategies with the performance assessment package provided by Zitzler et al.[2]. The quality assessment protocol works as follows: First, we create a set of 20 runs with different initial populations for each strategy and each benchmark instance of UBQP. Then, we generate the reference set RS^* based on the 100 different sets $A_0, \ldots, A_{99}$ of non-dominated solutions.

According to two objective function values, we define a reference point $z = [r_1, r_2]$, where r_1 and r_2 represent the worst values for each objective function in the reference set RS^*. Afterwards, we assign a fitness value to each non-dominated set A_i by calculating the hypervolume difference between A_i and RS^*. Actually, this hypervolume difference between these two sets should be as close as possible to zero [12].

4.3   Computational Results

In this subsection, we present the computational results on 9 groups of bi-objective UBQP instances, which are obtained by 5 different neighborhood combination strategies. The information about the algorithm with respect to each strategy is described in Table 3.

Table 3. The algorithms for the neighborhood combination strategies.

	Neighborhood description
N_1	One-flip move based local search
N_2	Two-flip move based local search
N_3	Three-flip move based local search
$N_1 \cup N_2$	f-flip move based local search ($f \in \{1, 2\}$)
$N_1 \cap N_3$	f-flip move based local search ($f \in \{1, 3\}$)

---

[2] More information about the performance assessment package can be found on this website: http://www.tik.ee.ethz.ch/pisa/assessment.html.

In this table, N_1, N_2 and N_3 refer to the HBMOLS algorithms using one-, two- and three-flip moves respectively during the neighborhood search process. Besides, $N_1 \cup N_2$ and $N_1 \cup N_3$ refer to the HBMOLS algorithms which select one of the two neighborhoods to be implemented at each iteration during the local search process, choosing the neighborhood N_1 with a predefined probability p and choosing N_2 (or N_3) with the probability $1-p$. In our experiments, we set the probability $p = 0.5$.

The computational results are summarized in Table 4. In this table, there is a value both **in bold** and **in grey** at each line, which is the best result obtained on the considered instance. The values both **in italic** and **bold** at each line refer to the corresponding algorithms which are **not** statistically outperformed by the algorithm obtaining the best result (with a confidence level greater than 95%).

Table 4. The computational results on bi-objective UBQP problem obtained by the algorithms with 5 different neighborhood combination strategies.

Instance	Algorithm				
	N_1	N_2	N_3	$N_1 \cup N_2$	$N_1 \cup N_3$
bo_ubqp_3000_01	**0.322441**	0.455449	0.643901	0.511596	0.567996
bo_ubqp_3000_02	**0.116847**	0.453555	0.482572	0.202315	0.567996
bo_ubqp_3000_03	**0.178425**	0.440120	0.476169	*0.192991*	0.423496
bo_ubqp_4000_01	**0.234291**	0.419082	0.823516	0.264199	0.282936
bo_ubqp_4000_02	**0.274993**	0.483540	0.505930	0.307865	0.314126
bo_ubqp_4000_03	**0.162367**	0.460610	0.748378	*0.171323*	0.220149
bo_ubqp_5000_01	**0.135116**	0.229339	0.450891	*0.137462*	0.160106
bo_ubqp_5000_02	**0.151279**	0.443796	0.630931	*0.159748*	*0.171841*
bo_ubqp_5000_03	**0.189243**	0.323817	0.443202	0.259512	0.278549

From Table 4, we can observe that all the best results are obtained by N_1, which statistically outperforms the other four algorithms on all the instances. Moreover, the results obtain by $N_1 \cup N_2$ and $N_1 \cup N_3$ are close to the results obtained by N_1. Especially, the most significant result is achieved on the instance bo_ubqp_3000_02, where the average hypervolume difference value obtained by N_1 is much smaller than the values obtained by the other four algorithms.

Nevertheless, N_2 and N_3 don't perform as well as N_1, although the search space of them is much bigger than N_1. We suppose that there exists some key variables in the representation of the solutions, which means the values 0 (or 1) should be put in these key variables in order to search the local optima effectively. Two- and three-flip moves change the values of the key variables much more frequently than the one-flip move, then the efficiency of local search is obviously affected by the neighborhood strategy.

On the other hand, $N_1 \cup N_2$ and $N_1 \cup N_3$ have a good performance, especially on the large instances (bo_ubqp_5000_01 and bo_ubqp_5000_02). Actually, randomly selecting the neighborhood strategy between one-flip move and two-flip (or three-flip) move with a predefined probability provides a possibility to keep the values of the key variables unchanged and broaden the search space. Thus, the combination of one-flip move and two-flip (or three-flip) move is very potential to obtain better results.

5   Conclusion

In this paper, we have presented the experimental analysis of neighborhood combination strategies on bi-objective UBQP problem, which are based on one-flip, two-flip and three-flip moves. For this purpose, we have carried out the experiments on 9 groups of benchmark instances of UBQP. The computational results show that the better outcomes are achieved with the simple one-flip move based neighborhood and the neighborhood combinations with two-flip and three-flip moves are very potential to escape the local optima for further improvements.

Acknowledgments. The work in this paper was supported by the Fundamental Research Funds for the Central Universities (Grant No. A0920502051722-53), supported by the West Light Foundation of Chinese Academy of Science (Grant No. Y4C0011001), supported by the Scientific Research Foundation for the Returned Overseas Chinese Scholars (Grant No. 2015S03007) and supported by the National Natural Science Foundation of China (Grant No. 61370150).

References

1. Basseur, M., Liefooghe, A., Le, K., Burke, E.: The efficiency of indicator-based local search for multi-objective combinatorial optimisation problems. J. Heuristics **18**(2), 263–296 (2012)
2. Basseur, M., Zeng, R.-Q., Hao, J.-K.: Hypervolume-based multi-objective local search. Neural Comput. Appl. **21**(8), 1917–1929 (2012)
3. Coello, C.A., Lamont, G.B., Van Veldhuizen, D.A.: Evolutionary Algorithms for Solving Multi-objective Problems. Genetic and Evolutionary Computation. Springer, New York (2007). doi:10.1007/978-0-387-36797-2
4. Kochenberger, G., Hao, J.-K., Glover, F., Lewis, M., Lü, Z., Wang, H., Wang, Y.: The unconstrained binary quadratic programming problem: a survey. J. Comb. Optim. **28**, 58–81 (2014)
5. Liefooghe, A., Verel, S., Hao, J.-K.: A hybrid metaheuristic for multiobjective unconstrained binary quadratic programming. Appl. Soft Comput. **16**, 10–19 (2014)
6. Liefooghe, A., Verel, S., Paquete, L., Hao, J.-K.: Experiments on local search for bi-objective unconstrained binary quadratic programming. In: Proceedings of the 8th International Conference on Evolutionary Multi-criterion Optimization (EMO 2015), Guimãres, Portugal, pp. 171–186 (2015)
7. Lü, Z., Glover, F., Hao, J.-K.: Neighborhood combination for unconstrained binary quadratic problems. In: Proceedings of the 8th Metaheuristics International Conference (MIC 2009), pp. 281–287 (2009)
8. Lü, Z., Glover, F., Hao, J.-K.: A hybrid metaheuristic approach to solving the UBQP problem. Eur. J. Oper. Res. **207**, 1254–1262 (2010)
9. Wang, Y., Lü, Z., Glover, F., Hao, J.-K.: Probabilistic grasp-tabu search algorithms for the UBQP problem. Comput. Oper. Res. **40**, 3100–3107 (2013)
10. Wang, Y., Lü, Z.P., Glover, F., Hao, J.K.: Backbone guided tabu search for solving the ubqp problem. J. Heuristics **19**, 679–695 (2013)

11. Zitzler, E., Künzli, S.: Indicator-based selection in multiobjective search. In: Yao, X., et al. (eds.) PPSN 2004. LNCS, vol. 3242, pp. 832–842. Springer, Heidelberg (2004). doi:10. 1007/978-3-540-30217-9_84

12. Zitzler, E., Thiele, L.: Multiobjective evolutionary algorithms: a comparative case study and the strength Pareto approach. Evol. Comput. 3, 257–271 (1999)

Porting Referential Genome Compression Tool on Loongson Platform

Zheng Du[1], Chao Guo[2], Yijun Zhang[2], and Qiuming Luo[2(✉)]

[1] Shenzhen Cloud Computing Center, Shenzhen 518055, China
duzheng@nsccsz.gov.cn
[2] NHPCC/Guangdong Key Laboratory of Popular HPC and College of
Computer Science and Software Engineering, Shenzhen University,
Shenzhen 518060, China
{guochao,zhangyijun,lqm}@szu.edu.cn

Abstract. With the fast development of genome sequencing technology, genome sequencing become faster and affordable. Consequently, genomic scientists are now facing an explosive increase of genomic data. Managing, storing and analyzing this quickly growing amount of data is challenging. It is desirable to apply some compression techniques to reduce storage and transferring cost. Referential genome compression is one of these techniques, which exploited the highly similarity of the same or an evolutionary close species (e.g., two randomly selected humans have at least 99% of genetic similarity) and store only the differences between the compressed file and well-known reference genome sequence. In this paper, we port two referential compression algorithm to Loongson platform and profiling their performance. And we use multi-process technology to improve the speed of compression.

Keywords: Genomics · Referential compression · Loongson · Optimization

1 Introduction

In 2000 the first human genome was released [1], it took almost 13 years and 3 billion dollars ($1 per base pair) by many scientists form six countries. In the past decade, the rate of sequencing price drop and sequencing throughput growth is surpassing Moore's law. The latest next-generation sequencing (NGS) machine, the Hiseq X Ten by illumina, the cost of a human whole-genome sequencing has dropped to a mere $1000 and delivers > 18,000 human genomes per year [2]. This trend is expected to continue in the following years [3], which contributes to bio-medical and precision medical.

Now there are large scale genome sequencing and analyzing projects to find the common and differences between individual genomes. For example, 1000 Genomes project, International Cancer Genome Project (ICGC) [4], The Cancer Genome Atlas (TCGA) [5] and Encyclopedia of DNA Elements (ENCODE) [6], those research projects yield a strong increase in the amount of genome sequencing data. Managing, storing, and analyzing this quickly growing amount of data is challenging [7].

Data compression is a key technology to reduce the size of gene data and cope with the increasing flood of genome sequences [8, 9]. However, traditional universal

© Springer Nature Singapore Pte Ltd. 2017
G. Chen et al. (Eds.): PAAP 2017, CCIS 729, pp. 454–463, 2017.
DOI: 10.1007/978-981-10-6442-5_43

compression algorithms, gzip (http://www.gzip.org/) and LZMA (http://www.7-zip. org/), are inefficient because they don't take advantage of the biological characteristics [10]. For example, the storage of a complete genome of chinese by 1 byte per base need roughly 2990 MB and using the gzip compression algorithm can reduce the space requirements to 832 MB. The compress ratio is mere 3.6 and we have to spend about 30 min download it via the network, and the realty may be much worse sometimes. There are a lot of genome compressors have been designed by scientists to tackle this problem.

In general, genome compression algorithms can be separated into two types: reference-free and reference-based approaches. Reference-free compression explores the biological characteristics of DNA sequences to compress data. Dictionary-based methods replace repeated substring by references to a dictionary, which is always built offline. Statistical methods use several probabilistic model, which is drives from the input, to predict the next symbols. The classic Reference-free compression algorithms include BIND [11], XM [12] etc. Referential-based compression take advantage of the highly similarity of the same or an evolutionary close species and encode only the differences between the input sequence respect to the reference genome sequence. The representative reference-based compression algorithms include GDC [13], FRESCO [14] and ODI [15].

Loongson microprocessor series are designed and developed by Institute of Computing Technology, Chinese Academy of Science and compatible with MIPS instruction set. The high performance processor of Loongson is 3A and 3B series. Most of the open source compression tool can't work on Loongson platform directly for the lack of some depended-upon packages. So it is urgent to porting some genome compression tools on Loongson platform.

In this paper, we transplant two referential genome compression tools to Loongson platform. Then profiling the performance of the compression tools on two different Loongson high performance processors. At last, we optimize the compression tools by multi-process technology.

The remaining part of this paper is organized as follows. We discuss related work on compression of genome sequences and the bioinformatics software on Loongson platform in Sect. 2. Section 3 presents details about the compression tools migration, the performance bottleneck analysis on Loongson platform and optimization detail of using multi-process technology. We conclude in Sect. 4.

2 Related Work

2.1 Reference-Free Compression

BIND [11], a native bit encoding algorithm, uses binary encoding {00, 01, 10, 11} to express the DNA base symbol{'A', 'T', 'C', 'G'}. The space size of one base symbol reduces from 1 byte to 2 bits. Additional compression is applied on the binary streams to improve the compress ratio. The performance of this compression algorithm is superior the universal compression algorithm (e.g., gzip and LZMA).

Dictionary-based compression algorithms derive a set of phrases which can be used to replace repeated substring economically. Representative examples of dictionary-based compression algorithm are Lempel-Ziv families [16], such as LZ77 or LZ78. For example, A sequence "TTTCGTTTCCAGCC …", and we employ a dictionary {1: TTT, 2:CC, 3:CG, 4:AG…} to compress the sequence as 131242. The compression result depends on the length and frequency of repeats in the genome sequence.

Statistical algorithms create the statistical model by the input DNA sequences and then use the model to predict the next DNA subsequences. Using the arithmetic coding to encode the subsequences based on the frequency of the statistical model. Huffman encoding [17], the most common used statistical algorithm, uses the variable-length code table to encode each possible character. For a sequence "TTTCGTTTCC AGCC…", the Huffman code table maybe {A:111, T:0, C:10, G:110}, the sequence can be compressed as a binary stream 00010110000101011111101010…. Pinho and Ferreira [18] show A DNAEnc3 algorithm based on Markov Chain model. This approach takes into account both the context and palindrome of the given genome sequence, and choose the most effective Markov Chain model for encoding.

We choose three compression from those three compression scheme, and compare their compression performance in Table 1.

Table 1. A comparison of the reference-free compression.

Tool	Scheme	Ratio(:1)	Speed
[19]	Native bit encoding	5.3	20 KB/s
Comrad [20]	Dictionary-based	5.5+	0.3 MB/s
[21]	Statistical	4.7	1.08 MB/s

2.2   Reference-Based Compression

Most of the current sequencing work is to existing species, such as ICGC collect thousands of human genome data on the world. The similarity between the DNA of the same species showed a high similarity, such as the similarity between gorillas and humans up to 99%. The reference-based genome compression algorithm makes use of the high similarity between homologous species to compress the data, which is also the research hotspot to solve the high-throughput sequencing data. The basic principle as shown in Fig. 1, finding out the different parts between the target genome and the reference genome and encode the different parts to achieve the purpose of compression of the target genome.

Fig. 1. The basic principle of referential compression

GRS [22] is a reference-based compression tool for referentially compressing whole genome respect to the well-known reference. It finds the longest common sub-string between the to-be-compressed sequence and the reference at first, and then employ the UNIX diff program to compute a similarity score between the target chromosome and the reference chromosome. If the score exceeds a predefined threshold, the difference between them is compressed by Huffman encoding. If the score is below the threshold, the target and reference are split into smaller blocks and each pair of the block restart the computation. It is to be noted that GRS needs an appropriate reference and the difference between the target and reference is not excessive.

ERGC [23] employs a divide and conquer strategy. ERGC divides both the target sequence and reference sequence into equal size subsequences, and construct a k-mer hash index in the reference subsequences. Target subsequences, in the same part as reference, search and compress in the reference subsequence by the k-mer index. If there are no match region in this region, the reference subsequence diminishes the k to reconstruct the k-mer index and recompute the compression. At last, the Delta encoding and PPMD lossless compression algorithm are employed to compress the variations between the target and reference sequences.

FRESCO [14] is fast to compress the large collections of genomes. FRESCO firstly create the index of the whole sequence of the chromosome, which is measured offline for many targets respect to one reference. And then target sequences find the longest match using the reference index. The Delta encoding and gzip compression algorithm are used to compress the intermediate result.

ODI [15] is designed by Fernando and Vincius and published in a survey published in 2015. ODI is currently the fastest solution to compress a single target sequence considering all the compression process. This method finds the variations by means of four matching algorithms: Direct match, SNP test, Brute-force search, and index lookup. ODI match segments from the target and reference directly by Direct match, and use the SNP test if the previous match ended in a Single Nuleotide Polymorphism. If the previous search methods failed, ODI execute brute-force search to expand the search. When Brute-force search failed, the index of a greater scop was created to find the match quickly. ODI avoidance of building the entire reference sequence, and enables compression to be performed up to one order of magnitude faster.

2.3   Bioinformatic Tools on Loongson

SOAP2 [24] is one of the mainstream alignment tools, but it cannot run on Loongson platform directly. In our previous work [25], we have transplanted and optimize the software of SOAP2 to Loongson platform. The main technical schemes include: (1) Give a paradigm of using the Loongson multimedia instructions to replace the SSE3 instructions, which not only achieve the transplantation, but also improve its performance. (2) Use thread-pool to optimize the SOAP2's IO performance and increase system efficiency. Thread pool is mainly to solve the problem of the high overhead of threads creating. Moreover, we distribute the IO tasks among these threads and reduce the time that waiting main thread to read file data, and make the file IO distributed to obtain higher performance in multi-core machine.

3 Porting and Performance Optimization

3.1 Referential Compression for Loongson

FRESCO and ODI are both outstanding referential compression tools, the FRESCO specializes in gene bank and ODI designed for single or few DNA sequences. We implement the two compression tools on Loongson platform. We randomly selected some gene sequences from 1000 Genome Project [26], and the experiment have ben run on 3B-1500-based and 3A-2000-based services.

Figure 2 is the average time of compression and decompression of a human genome, 22 chromosomes without X and Y, on Loongson 3B-1500 and 3A-2000 processor, C1 is the FRESCO algorithm, C2 is ODI algorithm and D is decompression. The original size of the 22 chromosomes is 2780 MB and compressed into 6 MB by C1 and 5.9 MB by C2. The average compression time of C1 on 3B-1500 is roughly 13 min and only 3.5 min on 3A-2000. And the efficiency of the C2 and D of 3A-2000 is also higher than 3B-1500.

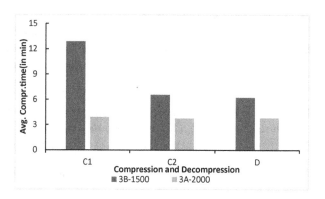

Fig. 2. Comparision of compression

3.2 Performance Profiling

The 3A-2000 CPU has a better performance of single-core than 3B-1500. In Fig. 3, the CPU time of C1 and C2 in 3A-2000 is merely 25% in 3B-1500, but the IO time of them is roughly equal.

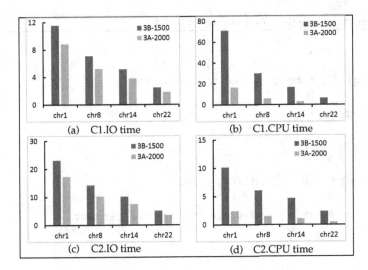

Fig. 3. IO and computing time (in second)

Figure 4 shows the percentage of IO and Computing time; in C1 the ratio of computation is higher than IO but an opposite situation in C2. So, we prepare to use the parallel ideas to speed up the computation and overlap the computation and IO.

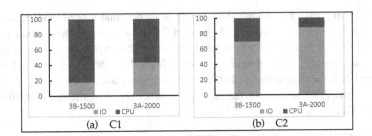

Fig. 4. Percentage of IO and computation time

3.3 Performance Optimization

In our experimental platform, 3B-1500-based service equipped with two 3B-1500 CPUs and each one has 8 cores, 3A-2000-based service equipped with one 4 cores 3A-2000 CPU. We can take advantage of the multi-cores to speed-up the compression. In this paper, we utilize multi-process to make full use of Loongson multi-core to improve the computation and overlap the IO and computing.

For C1 compression, we use a main process to load the target genome sequences and then fork a child process to search matches and compress files. So, the main process undertakes IO tasks and the child processes compute. The algorithm description of multi-process for C1 as Algorithm 1.

Algorithm 1: Description of mutil-process for C1

```
Begin

for seqi in target sequences database

    LoadSqe(seqi);

    pid = fork();

    if(pid  == 0)//child process

        Computing and compressing the target;

        exit;

    else    //main process

        ;       //continue load seq

End
```

The C2 compression is designed for single sequence, and it need load reference and target sequences everytime. As Fig. 4 shows, C2 is IO-wasting compression algorithm. A shared memory was allocated to store the reference sequence. When compress a sequence, the process gets reference sequence from the shared memory and load the target from disk then compress it. If there are multi-processes compressing the targets, we use a semaphore to control the read of targets. The Algorithm 2 describes each process of C2.

Algorithm 2: C2 compression in mutil-process

```
Begin

Get reference sequence from shared memory;

sem_wait(io_sem);

LoadSeq(target);

sem_post(io_sem);

Computing and compressing the target

End
```

The experiment dataset downloads from 100 Genome Project. We choose 10 genomes to test our multi-process compression tool and then we choose different number of genomes to verify the stability of our tool.

The Fig. 5 shows the compression time of 10 genomes, multi-process technology speeds up the compression. The compression speed of C1 on 3b-1500 is improved about 4 times, but on 3A-2000 is merely 25%. The C1 compression with less percentage IO time, so there are more processes running at the same time. The parallelism of 3B-1500-based service is higher than 3A-2000 for the 3B-1500-based service has 16 cores but 3A-2000 has 4 cores. The C2 compression on the two platform are improved about 2 times for their IO time is too long, so the running processes can't up to 4 (Fig. 6).

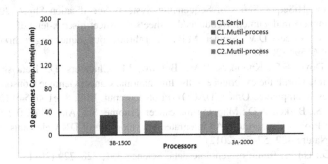

Fig. 5. Genomes compression time in two processor

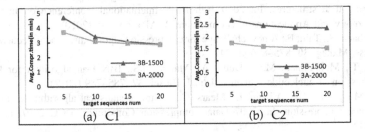

Fig. 6. Average compression time per genome

The number of to-be-compression file was growing from 5 to 10, the compression speed is increased about 20%. With the number continue to grow, the compression speed trends to stabilize.

4   Conclusion

In this paper, we transplant two referential compression tools on Loongson platform. And then we analyze and optimize the performance of compression tool on Loongson platform. The multi-process technology greatly improve the speed of compression for dealing with a lot of targets. It is necessary to analyze the performance of the compression tools before we parallel them.

Acknowledgment. The research was jointly supported by Shenzhen Science & Technology Foundation: JCYJ20150930105133185, National Natural Science Foundation of China: NSF/GDU1301252, and State Key Laboratory of Computer Architecture ICTCA: CARCH 201405. Guangdong Province Key Laboratory Project: 2012A061400024.

References

1. Lander, E.S., Linton, L.M., Birren, B., Nusbaum, C., Zody, M.C., Baldwin, J., et al.: Initial sequencing and analysis of the human genome. Nature **409**, 860–921 (2001)
2. Illumina Int: HiSeq X Series of Sequencing Systems Specification Sheet (2016). https://www.illumina.com/documents/products/datasheets/datasheet-hiseq-x-ten.pdf
3. Reuter, J.A., Spacek, D.V., Snyder, M.P.: High-throughput sequencing technologies. Mol. Cell **58**, 586–597 (2015)
4. Joly, Y., Dove, E.S., Knoppers, B.M., Bobrow, M., Chalmers, D.: Data sharing in the post-genomic world: the experience of the International Cancer Genome Consortium (ICGC) Data Access Compliance Office (DACO). PLoS Comput. Biol. **8**, e1002549 (2012)
5. Collins, F.S., Barker, A.D.: Mapping the cancer genome. Sci. Am. **296**, 50–57 (2007)
6. ENCODE Project Consortium: An integrated encyclopedia of DNA elements in the human genome. Nature **489**, 57–74 (2012)
7. Kahn, S.D.: On the future of genomic data. Science **331**, 728–729 (2011)
8. Nalbantoglu, Ö.U., Russell, D.J., Sayood, K.: Data compression concepts and algorithms and their applications to bioinformatics. Entropy **12**, 34–52 (2009)
9. Antoniou, D., Theodoridis, E., Tsakalidis, A.: Compressing biological sequences using self adjusting data structures. In: 2010 10th IEEE International Conference on Information Technology and Applications in Biomedicine (ITAB), pp. 1–5 (2010)
10. Grumbach, S., Tahi, F.: A new challenge for compression algorithms: genetic sequences. Inf. Process. Manag. **30**, 875–886 (1994)
11. Bose, T., Mohammed, M.H., Dutta, A., Mande, S.S.: BIND–an algorithm for loss-less compression of nucleotide sequence data. J. Biosci. **37**, 785–789 (2012)
12. Cao, M.D., Dix, T.I., Allison, L., Mears, C.: A simple statistical algorithm for biological sequence compression. In: 2007 Data Compression Conference, DCC 2007, pp. 43–52 (2007)
13. Deorowicz, S., Grabowski, S.: Robust relative compression of genomes with random access. Bioinformatics **27**, 2979–2986 (2011)
14. Wandelt, S., Leser, U.: FRESCO: referential compression of highly similar sequences. IEEE/ACM Trans. Comput. Biol. Bioinform. **10**, 1275–1288 (2013)
15. Alves, F., Cogo, V., Wandelt, S., Leser, U., Bessani, A.: On-demand indexing for referential compression of DNA sequences. PLoS ONE **10**, e0132460 (2015)
16. Ziv, J., Lempel, A.: A universal algorithm for sequential data compression. IEEE Trans. Inf. Theory **23**, 337–343 (1977)
17. Huffman, D.A.: A method for the construction of minimum-redundancy codes. Proc. IRE **40**, 1098–1101 (1952)
18. Pinho, A.J., Ferreira, P.J., Neves, A.J., Bastos, C.A.: On the representability of complete genomes by multiple competing finite-context (Markov) models. PLoS ONE **6**, e21588 (2011)

19. Rajarajeswari, P., Apparao, A.: DNABIT compress-genome compression algorithm. Bioinformation **5**, 350–360 (2011)
20. Kuruppu, S., Beresford-Smith, B., Conway, T., Zobel, J.: Iterative dictionary construction for compression of large DNA data sets. IEEE/ACM Trans. Comput. Biol. Bioinform. (TCBB) **9**, 137–149 (2012)
21. Pratas, D., Pinho, A.J.: Compressing the human genome using exclusively Markov models. In: Rocha, M.P., Rodríguez, J.M.C., Fdez-Riverola, F., Valencia, A. (eds.) PACBB 2011, pp. 213–220. Springer, Heidelberg (2011). doi:10.1007/978-3-642-19914-1_29
22. Wang, C., Zhang, D.: A novel compression tool for efficient storage of genome resequencing data. Nucleic Acids Res. **39**, e45 (2011)
23. Saha, S., Rajasekaran, S.: ERGC: an efficient referential genome compression algorithm. Bioinformatics, btv399 (2015)
24. Li, R., Yu, C., Li, Y., Lam, T.-W., Yiu, S.-M., Kristiansen, K., et al.: SOAP2: an improved ultrafast tool for short read alignment. Bioinformatics **25**, 1966–1967 (2009)
25. Luo, Q., Liu, G., Ming, Z., Xiao, F.: Porting and optimizing SOAP2 on Loongson Architecture. In: 2015 IEEE 17th International Conference on High Performance Computing and Communications (HPCC), 2015 IEEE 7th International Symposium on Cyberspace Safety and Security (CSS), 2015 IEEE 12th International Conference on Embedded Software and Systems (ICESS), pp. 566–570 (2015)
26. Genomes Project Consortium: An integrated map of genetic variation from 1,092 human genomes. Nature **491**, 56–65 (2012)

Statistics of the Number of People Based on the Surveillance Video

Zhao Qiu$^{(\boxtimes)}$, ShiYao Lei, JianZheng Hu, and JingYu Luo

College of Information Science and Technology,
University of Hainan, Haikou, China
qiuzhao73@hotmail.com, 565644062@qq.com,
1694789915@qq.com, 499450006@qq.com

Abstract. With the rapid development of economy, large-scale entertainment, sporting events, religious ceremonies and other major activities, there are often a large crowd gathered. In order to avoid accidents occurred, counting the number of development becomes more and more important. And in the daily activities of life relatively slim, for example: the school attendance rate statistics, the convenience given by the statistics of pedestrians should not be underestimated. The number of statistical methods is drawing more and more people's attention.

Keywords: Three frame difference method · Adaboost+haar · Nearest neighbor matching

1 Introduction

With the trend of social intelligence, the number of surveillance video statistics increase in business, civil and industrial applications. Traditional way of artificial monitoring statistics, not only the human resources cost is high, and the man in the condition of long working hours, appears unavoidably concentration causes such as negligence.

The number of computer vision at present, there are a lot of sports statistics technique, but under the condition of the actual number of accurate statistics still faces many difficulties. Transformation mainly includes: the complex background, illumination, the crowd of shade and non-human movement interference with each other.

2 System Structure Design

The main modules of the system: moving target detection module, the head region detection module, target tracking and counting module. What needs to be emphasized is that the result of the moving target detection is a block motion connected domain, this motion connectivity domain number still need to handled by head classifier, in order to identify the same moving target in the each frame, the article has used the target tracking technology, repetitive movements of the target is not counted, new moving object is counted and displayed.

© Springer Nature Singapore Pte Ltd. 2017
G. Chen et al. (Eds.): PAAP 2017, CCIS 729, pp. 464–470, 2017.
DOI: 10.1007/978-981-10-6442-5_44

Module relations are as follows (Fig. 1):

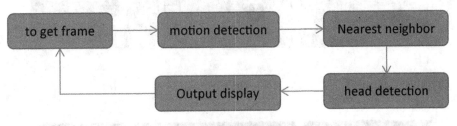

Fig. 1. Module relations

2.1 To Get Frame

a. The core function of frame acquisition is: cvGrabFrame(pCapture)
b. SetTimer(1,33,NULL)and callback function OnTimer() are used to controlled the time of getting frame.

2.2 Motion Detection

Moving object segmentation of video sequences, in general, includes the static background and moving target, in order to separate the moving objects from a static background, two frames of adjacent source image of difference, because of the difference between image frames, the moving target in the middle frame shape contourcan be quickly detected. Using this property, detect intermediate frames of the moving object contour shape, and revise target template on the differential results.

Complete contour is caused by the lack of internal circular edge of the movement of the mobile source video sequence of three consecutive frames image, there is a strong correlation between moving object (rounded rectangle) and the edge of the frame difference image 's motion of the edge. Figure 2(d), $|f_1 - f_2|$ includes the motion edge of first frame and the motion edge of second frame, Fig. 2(E), $|f_3 - f_2|$ includes the moving target of third frame and the moving target of second frame, and Fig. 2(f) is the Complete outline.

2.3 Nearest Neighbor

Algorithm specific implementation are as follows:

① initialization sequence block TrackBlock[50], set the minimum center distance;
② to read the next frame, to obtain all the moving target sequence count (cvFindContours (DST, stor, &cont, sizeof (CvContour))
③ for (; cont; cont->h_next = cont)
   1. Record moving target center avgY, avgX
   2. Calculation, if you update the trajectory guiji[MAX_POINT] axisxy_t, end of the loop. Otherwise, add a new tracking block.
   3. detecting the number of the head of the moving target, counting.

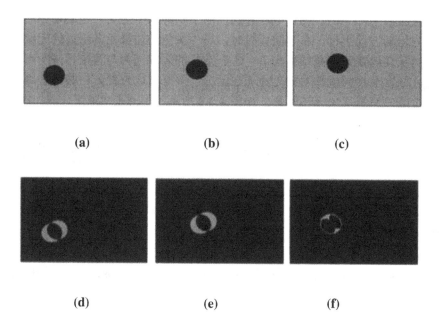

Fig. 2. Schematic diagram of symmetric difference, (a) f_1; (b) f_2; (c) f_3; (d) $|f_1 - f_2|$; (e) $|f_3 - f_2|$; (f) f_2 split profile

④ return to the second step

The following is a structure (Fig. 3):

```
typedef struct AXISXY_S {
    int x;
    int y;
} axisxy_t;

struct AvTrackBlock {              /,
    int Direction;                 //
    int avgX;                      //
    int avgY;                      //
    axisxy_t guiji[MAX_POINT];
    int guiji_index;

}*TrackBlock[50];
```

Fig. 3. Structure

The nearest neighbor matching algorithm flow chart is as follows (Fig. 4):

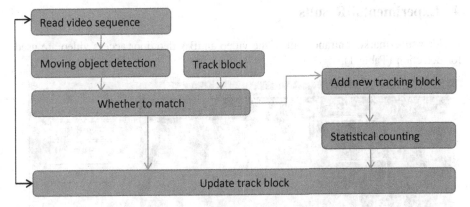

Fig. 4. The nearest neighbor matching algorithm flow chart

2.4 Head Detection

Using Adaboost+haar algorithm to get the head.xml classifier, then the head is detected from the red movement area. The head is out of the white box.

3 The Tools of Writing System

(1) Microsoft Visual Studio
   Visual Studio Microsoft (referred to as VS) is a code integrated development environment.
(2) OpenCV
   OpenCV is an open source computer vision library.

4 Experimental Results

A video supermarket entrance out of the video andB video door access video are used for detection (Table 1).

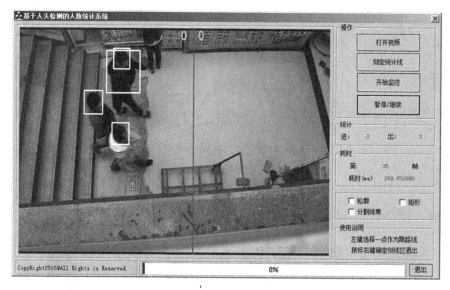

A video

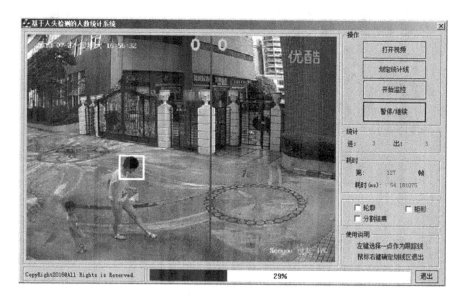

B video

Table 1. Test result

	Actual discrepancy			Test discrepancy		
	Total number	Entry number	Out number	Total number	Entry number	Out number
A video	34	10	24	44	12	33
B video	8	6	2	9	6	2

5 Summary and Prospect

The many problems existing in complex environment surveillance video statistics on the number of: light changes, moving background, target occlusion, non human movement object. Because the author belongs to the early stages of learning, they are the most simple and most basic algorithm, for the existence of many problems, a small part of the solution, most can't solve. The following is summarized:

(1) in this paper through the head classifier the non human motion object is detected, but tracking algorithm based on the nearest neighbor matching algorithm if the tracking target velocity difference, it is difficult to effectively detect the actual number.
(2) the moving object detection using a three frame difference method, only better than two adjacent frame difference algorithm. The algorithm requires the video image background for static background, the background is not the slightest sign of trouble, light is too dark. But the algorithm is fast, and make the procedure easy.
(3) the tracking algorithm is to solve a problem that if the detection target is short time occlusion, it is not to be mistaken, but maybe long time occlusion will be repeated calculation.
(4) head area detection, although the majority of the head can be judged, but the rate of miscarriage of justice is not small.
there is so much things to be improved for the system, for example: head classifier can take better machine learning algorithm to make it more accurate;

Acknowledgment. This work is partially supported by Educational Commission of Hainan Province of China (No. HNKY2014-03).

References

1. Bi, D., Yang, Y., Yu, F., Ju, M.: The video target tracking method. National Defense University Press
2. Zhang, J.: OpenCV 2 Computer Vision Programming Manual
3. Li, J., Wang, J.: Image detection and target tracking technology. Beijing Institute of Technology Press

4. Ding, Y., Yang, H., Fan, J., Jiang, Z., Li, Y.: The detection technology and application of the complex environment moving target. Press
5. Ai, H., Xing, J.L.: Computer vision algorithm and application. Tsinghua University Press
6. Yu, S., Liu, R.: Study OpenCV. Tsinghua University Press
7. Gu, D., Wu, T.: Video image processing based on the number of automatic statistics, pp. 49–51 (2010)
8. Li, P., Lv, X.: Research and implementation of statistical methods in video surveillance, p. 2 (2013)
9. Zhang, Y., Chen, L.: Research on key technologies of real-time statistics based on video, p. 52 (2014)
10. Xue, C., Guo, J.: Statistical system for the number of complex scenes, pp. 6–15 (2014)
11. Huang, J., Jiang, M.: Population density estimation and population statistics in video surveillance, p. 1 (2013)
12. Li, P., Lu, J., Xinqiao: Research and implementation of the number of statistical methods in video surveillance, pp. 3–5 (2014)
13. Zhu, X., Mingwu: Under the condition of complex head detection and counting technology research and implementation of (2009)

Differential Privacy in Power Big Data Publishing

Ping Kong[1], Xiaochun Wang[2], Boyi Zhang[3], and Yidong Li[2(✉)]

[1] State Grid Shandong Power Company, Jinan, China
[2] School of Computer and Information Technology,
Beijing Jiaotong University, Beijing, China
ydli@bjtu.edu.cn
[3] School of Electrical and Electronic Engineering,
The University of Manchester, Manchester M60 1QD, UK
boyi.zhang@postgrad.manchester.ac.uk

Abstract. In recent years, the development of smart grid leads to the explosive growth of power big data. Power companies can analyze these data to provide personalized services to users. However, the analysis of power big data can have the risk of user privacy disclosure. The performance of the traditional algorithms is not satisfied due to the complexity of the power big data on preventing information leakage. Distributed and heterogeneous data generated in the operation, maintenance and other processes of electricity smart grid can cause the complexity of the data. This paper proposes a method of differential privacy to preserve privacy in the power big data publishing. The experimental results shows that the performance of our method is convincing.

Keywords: Differential privacy · Power big data · Privacy protection

1 Introduction

With the rapid development of information technology and the gradual expansion of the use of the Internet, more advanced techniques of information collection, preservation, sharing and comparison have emerged, the collection and processing of personal information brings valuable knowledge and wealth to businesses and countries. At the same time, the privacy and security issues caused by data analysis are becoming increasingly prominent, and data privacy protection is gradually concerned. Traditional privacy methods are k-anonymity, l-diversity, t-closeness and so on. However, due to the uncertainty of background knowledge, privacy leaks cannot be completely avoided during the data release process. The differential privacy protection technology mentioned in this paper is a new method of privacy protection has been proposed in [5]. This is a kind of privacy protection technology based on data distortion, which uses a noise-adding mechanism to distort sensitive data, while maintaining data availability. It can be achieved in the data set to add or delete a record does not affect the output of the query results. So even if an attacker gets a data set with only one record difference. The result of both is the same by analyzing. The attacker could not figure out the hidden

© Springer Nature Singapore Pte Ltd. 2017
G. Chen et al. (Eds.): PAAP 2017, CCIS 729, pp. 471–479, 2017.
DOI: 10.1007/978-981-10-6442-5_45

record. Differential Privacy aims to maximize the accuracy of the query while minimizing the risk of privacy disclosure.

With the application of big data analysis tools in the power enterprise system, the enterprise information, employee information and other relevant information in the network platform may be exposed to the risk of leakage. If the data cannot be effectively controlled, it is easy to cause information disclosure, resulting in a wide range of risks. For example, criminals use electricity business information, and other private information in illegal transactions, profiteering, etc., this will bring the risk for the development and operation of the enterprise. Power big data analytics brings significant risk of user privacy disclosure. The data collected in the electrical power system including the user's home address, telephone, electricity and other sensitive information, the power company provided personalized services to users by publishing these data will also face the risk of disclosure of user privacy. The information of the attacker's background knowledge and the data electricity grid companies published can be used to infer a particular user, or its information used for other commercial purposes. Additionally attacker may be able to infer the user's residence status based on whether a user's idle power consumption, thereby threatening to the user's privacy.

The framework of the paper is as follows. In Sect. 2, we made a survey of related works in privacy preserving. In Sect. 3, we introduced definitions the maximum background knowledge attack model and the differential privacy protection model. Section 4 describes the application of our method in the release of power big data. In Sect. 5, we did a summary of the work and future prospects.

2   Related Works

In recent years, work around homomorphic encryption algorithm is endless in the power big data privacy preserving. [19, 20] study the power data privacy preserving with homomorphic encryption algorithm. In Zhan et al. [19] study of the key technologies of electric power big data and its application prospects in smart grid. In Li et al. [11] put forward a security information aggregation scheme for large data analysis platform based on homomorphic encryption. The aggregation of the encrypted data in this scheme is carried out according to an aggregated tree, but the tree structure is not suitable for large power data representation. In Erkin and Tsudik [7] modified Paillier algorithm, a big power data aggregation scheme, but this algorithm requires that each smart meter be assigned a component of the public key in advance which increases the communications volume. In Wang [16] put forward application of big data visualization technique in power grid enterprise. Another method is security multi-party computation (SMC), in Garcia and Jacobs [8] combine secure multi-party computation and homomorphic encryption, a new privacy protection energy metering program, the method can achieve privacy preserving of user data, It proposed a smart grid with privacy protection of the polymerization program based on pseudonym, homomorphic encryption, identity aggregation signature technology. Method of the power big data polymerization techniques based on cryptography tend to be perfect, but the research on power big data publishing privacy preserving is rare. [6, 10] discuss the differential privacy preserving in big data. In Acs and Castelluccia [1] proposed smart metering system for privacy protection based on the differential privacy. In this scenario, the electricity supplier may be periodically collected from the smart meters and data aggregation, and do not understand the behavior of electricity users. In addition to this method need to add noise to the original data, we also need a valid stream cipher to encrypt the query results, the need for each meter of predefined functions and parameters to calculate the amount too.

A new idea about deep learning differential privacy preserving promoted by Wu [17, 18] is effective and efficient, the algorithm they adapt in their work is functional mechanism (FM). The FM enforces ε-differential privacy by perturbing the objective function of the optimization problem but cannot be applied in electronic data privacy preserving because of deep learning model has not been widely used in the power grid. In Xiong et al. [14] summed up the concept and application of differential privacy. In Clifton and Tassa [3] expound on syntactic anonymity and differential privacy. In Jain et al. [9] discuss to drop or not to drop: robustness, consistency and differential privacy properties of dropout. In terms of data publishing, Li et al. [2, 12, 13] used an anonymous method to preserve privacy of big data. In this paper, we propose a simple power big data releasing scheme based on differential privacy, We add Laplace noise to the query function to achieve the effect of differential privacy protection, making the data query results do not change.

3 Problem Statement

The differential privacy preserving model was originally used in the statistical database security field, designed to preserve the privacy of individuals in the database when publishing statistics, after being widely used in Privacy Preserving Data Release (PPDR) and Privacy Preserving Data Mining (PPDM), and other fields.

As an alternative to the traditional power grid, smart power grid meet the real-time management through frequent data collection, but at the same time it also brings user sensitive information leakage problem, so privacy become a prerequisite for smart grid communication of massive data security. Existing programs for the power big data protection almost based on cryptography tool commonly used method of homomorphic encryption and secure multiparty computation. The face of big data security risks brought about by the information, the power sector must increase the risk management and control efforts, use advanced technology to actively avoid information security issues. Privacy preserving technologies have become a key technology, the current preventive technologies for power system applications include: encryption and key management, digital signature technology, authentication technology. In this paper, we adopt a statistical method to preserve privacy in power big data publishing, it's differential privacy preserving.

Differential privacy is based on the maximum background knowledge attack model, which assumes that the attacker knows the data set other records in addition to a record. Compared to the anonymous method k-anonymity and l-diversity in the differential definition of strict privacy attack model, we can withstand a variety of background knowledge attack, and defines statistical models, easy to use and statistical and mathematical tools prove. The following gives a differential concept of privacy and related definitions.

A. *Maximum background knowledge attack model*

Typically, each of privacy protection method corresponds to a specific attack models, such as the anonymous method, the privacy of each model are based on a specific background knowledge attack. The background knowledge that an attacker has is quasi-identifiers that can be used to infer a set of attributes. The traditional method can only anonymous quasi-identifiers against specific attacks, but cannot deal with other types of attacks. In this paper, we propose a maximum background knowledge attack model MBK Attack, this attack model assumes that if an attacker master of all records except a record, any course cannot be inferred that a hidden record, add or delete a record that is no effect on the data query results.

Definition 1 (*MBK Attack*): Given two data sets D1 and D2 which only differ in one record. In MBK Attack, we assume that the attacker knows all the records in the data set D1 except R1, he can infer the unknown record R according to his knowledge. For example, if an attacker mastered each residential electricity total number of users and users of electricity in addition to the information in a user's home address information, then he can be speculated that the target user is in a residential area

B. *Differential privacy preserving model*

Definition 2 (*Differential Privacy*): Given a random function of A, and two adjacent data sets D1 and D2, R is a subset of the output field random function A, if function A randomized in the data sets D1, D2 on the output satisfies the following inequality

$$\frac{Pr(A(D2)) \in R}{Pr(A(D2)) \in R} \le e^{\epsilon} \tag{1}$$

Random function A is said to provide ϵ-difference privacy. In function (1), Pr representative privacy disclosure risk. A general method for computing an approximation to any function $f(D)$ while preserving ϵ-differential privacy is the Laplace mechanism [14], where the output of f is a vector of real numbers. In particular, the mechanism exploits the global sensitivity of f over any two neighboring data sets (differing at most one record), which is denoted as $S_{f(D)}$. Given $S_{f(D)}$, the Laplace mechanism ensures ϵ-differential privacy by injecting noise n into each value in the output of $f(D)$ where n is drawn from Laplace distribution with zero mean and scale $S_{f(D)}$.

C. *Privacy disclosure risk metrics*

Definition 3 (*Privacy Budget*): ϵ is the differential privacy protection budget used to control the algorithm that provides differential privacy to obtain the same output probability ratio on two adjacent data sets, that is the differential privacy level an algorithm can provide. The smaller ϵ, the higher the intensity of privacy preserving. The value of ϵ depends on the specific circumstances.

Differential privacy preserving can be achieved by adding the appropriate amount of noise to the return value of the query function. Sensitivity is the key parameter that determines the size of the added noise, which refers to the deletion of the maximum change in the query result by any record in the data set. Specific can be divided into global sensitivity and local sensitivity.

Definition 4 (*Global Sensitivity*): Given query function : $D \rightarrow R^d$, The Global Sensitivity of function f is:

$$GSf = \Delta f = max\, D1, D2 \|f(D1) - f(D2)\|_1 \qquad (2)$$

In the function (2), D1 and D2 differs by at most one record, R mapped to the real space, d as the query dimensions of function f, digital L_2 is used to measure the Lp distance of Δf. Noise differential mechanism is an important way to achieve privacy, but the sensitivity and size of the desired query functions closely related to noise, more noise will cause decline in the availability of data, the noise is too small lead to data security decline. When the global sensitivity is large, often require a larger noise in order to ensure privacy and security, which led to a decline in the accuracy of data, often in this case the use of local sensitivity processing.

Definition 5 (*Local Sensitivity*): Given a query function $f : D \rightarrow R^d$. The Local Sensitivity of function f is:

$$LSf = \Delta f = max\, D2 \|f(D1) - f(D2)\|_1 \qquad (3)$$

Function f with the original data set D1 jointly determine the local sensitivity. Local sensitivity is affected by data distribution feature K, is much smaller compared to the global sensitivity. Data distribution characteristics can lead to sensitive data leakage, usually using local sensitivity combined with smooth upper bound to deal with the noise level.

D. *Implementation Mechanics*

In order to make a random selection algorithm can meet the differential privacy protection requirements, for different problems have different implementation mechanisms, the most basic of the two mechanisms is Laplace mechanism and exponent mechanism, Laplace mechanism applicable to numeric protection index mechanism results apply to non-numeric results of protection. The power big data of this study is numerical data, so we use the Laplace mechanism.

Definition 6 (Laplace Mechanism): Given any function $f : D \rightarrow R^d$. Expression A(D) of output satisfies the following equation is satisfied ε-differential Privacy

$$A(D) = f(D) + \left(L\left(\frac{\Delta F}{\varepsilon}\right) \right)^d \qquad (4)$$

In the function (4), $L\left(\frac{\Delta f}{\varepsilon}\right)$ is Laplace noise function, the size of the noise added on the function A is related to Δf and ε.

As is showed in Fig. 1, the algorithm A to provide privacy preserving through output perturbing, While the parameter ε is used to ensure that any records are deleted in the data set, the probability that the algorithm outputs the same result does not change significantly.

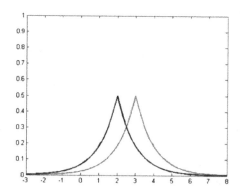

Fig. 1. Random probability algorithm output on the adjacent dataset [4]

4   Differential Privacy in Power Big Data Publication

Differential Privacy is suitable for the power of big data privacy preserving because of its unique advantages talked above. There are 5 approaches to achieve Differential Privacy, the most common approaches are input perturbation, output perturbation and objective perturbation. In addition, exponential mechanism and sample-and-aggregate can also achieve differential privacy. Considering the type of users' electricity data, we use output perturbation and Laplace mechanism to design a differential privacy algorithm.

Existing research shows that differential privacy protection can be applied to interactive statistical query, can also be used in a variety of non-interactive information release occasions. This paper studies the differential privacy mechanism of data release in interactive environment for the characteristics of power big data.

In our scheme, we assume that power companies regularly collect data from smart meters, and then aggregated, statistical and analysis, without understanding what the various activities of the household. For example, a supplier can not tell from the user to track whether or not he when watching TV or cooking. On the other hand, an attacker can not infer the user's private information from published data, such as the behavior of the user of electricity, the user's home address, or infer which housing is idle from the energy used, nor infer the user type from the power billing. Our solution is simple, but efficient and practical.

Algorithm 1 DP-Algorithm

Input: the original data set D, privacy budget ε, Global sensitivity $S(x)$
Output: The result set D'
1: For each $x \in D$
2:       $x' \leftarrow x + Lap\left(\frac{S(x)}{\varepsilon}\right)$; A.Maximum background knowledge attack model
3:       $D' \leftarrow x'$
4: End for
5: Return D'

For example, Table 1 shows a power user dataset D Wherein each record contains the user's type and payment information, as well as users' housing situation(1 means yes, 0 means no). To provide users with statistical data set queries (such as counting queries), but cannot disclose the value of a specific record. Let the user input parameter Si and column number, call the query function $f(i) = count(i)$ in the corresponding column i to meet {"business user"} = 1. Then the number of records, and the function value feedback to the user. Suppose the attacker tend to speculate whether Alice is a business user, and knows Alice in the fifth row of the data set, he can use $count(5) - count(4)$ on the column of business user to launch the correct results, and if an attacker tend to infer whether Jack's house is idle, just using $count(2) - count(1)$ to launch the correct results.

Table 1. Power user data set example

Name	Paid bill	Business user	Idle housing
Jack	1	0	1
Marry	1	1	0
Sara	0	0	1
Bob	1	0	1
Alice	1	1	0

However, if f is a provider of ε-differential privacy query function, such as $f(i) = count(i) + noise$ Wherein noise satisfy the Laplacian distribution. Suppose $f(5)$ possible output from the set 2, 2, 5, 3, then $f(4)$ Almost exactly the same probability will also be output to any 2, 2, 5, 3 is a possible value, so an attacker cannot by $f(5) - f(4)$ to get the desired results. This statistical output for randomized function such that an attacker cannot get the difference between the results of the query, which can ensure the safety of each individual dataset. Specific algorithm described as Algorithm 1.

The time complexity of the Algorithm 1 is $O(n)$, n is the amount of data. In ε under the same circumstances, the higher the sensitivity value, the greater the amount of noise added, the corresponding error is greater. The impact of this algorithm to the data set is ε, with the privacy budget ε, the impact of query result of published data is very small.

5 Conclusion

In this paper, for the problem of user privacy disclosure in power big data publishing, we present a method based on differential privacy preserving, mainly add Laplace noise to the output data set to achieve the effect of ε-differential privacy. Under differential privacy preserving requirements, this method can increase data availability and accuracy. The proposed differential privacy model [4, 5] achieves more comprehensive privacy preserving, which ignoring the background knowledge an attacker obtained, at the same time, providing a measure of privacy mathematical quantitative criteria. In the future, our work will focus on applying differential privacy in distributed power big data analytics platforms, as well as the differential privacy in machine learning, combine the work in [15].

Acknowledgement. This work is supported by National Science Foundation of China Grant #61672088, Fundamental Research Funds for the Central Universities #2016JBM022 and #2015ZBJ007, Open Research Funds of Guangdong Key Laboratory of Popular High Performance Computers.

References

1. Acs, G., Castelluccia, C.: I have a dream! (differentially private smart metering). In: Information Hiding - International Conference, 18–20 May 2011, Prague, Czech Republic, Revised Selected Papers, pp. 118–132 (2011)
2. Chu, D., Li, Y., Wang, T., Zhang, L.: Practical anonymization for protecting privacy in combinatorial maps, pp. 119–123 (2014)
3. Clifton, C., Tassa, T.: On syntactic anonymity and differential privacy. In: IEEE International Conference on Data Engineering Workshops, pp. 161–183 (2013)
4. Dwork, C.: Differential privacy: a survey of results. In: Agrawal, M., Du, D., Duan, Z., Li, A. (eds.) TAMC 2008. LNCS, vol. 4978, pp. 1–19. Springer, Heidelberg (2008). doi:10.1007/978-3-540-79228-4_1
5. Dwork, C., McSherry, F., Nissim, K., Smith, A.: Calibrating noise to sensitivity in private data analysis. In: Halevi, S., Rabin, T. (eds.) TCC 2006. LNCS, vol. 3876, pp. 265–284. Springer, Heidelberg (2006). doi:10.1007/11681878_14
6. Dwork, C., Roth, A.: The algorithmic foundations of differential privacy. Found. Trends? Theor. Comput. Sci. 9(3–4), 211–407 (2014)
7. Erkin, Z., Tsudik, G.: Private computation of spatial and temporal power consumption with smart meters. In: Bao, F., Samarati, P., Zhou, J. (eds.) ACNS 2012. LNCS, vol. 7341, pp. 561–577. Springer, Heidelberg (2012). doi:10.1007/978-3-642-31284-7_33
8. Garcia, F.D., Jacobs, B.: Privacy-friendly energy-metering via homomorphic encryption. In: Cuellar, J., Lopez, J., Barthe, G., Pretschner, A. (eds.) STM 2010. LNCS, vol. 6710, pp. 226–238. Springer, Heidelberg (2011). doi:10.1007/978-3-642-22444-7_15
9. Jain, P., Kulkarni, V., Thakurta, A., Williams, O.: To drop or not to drop: robustness, consistency and differential privacy properties of dropout. Comput. Sci. (2015). arXiv:1503.02031
10. Lei, J.: Differentially private m-estimators. In: Advances in Neural Information Processing Systems (2011)

11. Li, F., Luo, B., Liu, P.: Secure information aggregation for smart grids using homomorphic encryption. In: IEEE International Conference on Smart Grid Communications, pp. 327–332 (2010)
12. Li, Y., Shen, H.: On identity disclosure control for hypergraph-based data publishing. IEEE Trans. Inf. Forensics Secur. **8**(8), 1384–1396 (2013)
13. Li, Y., Li, Y., Xu, G.: Protecting private geosocial networks against practical hybrid attacks with heterogeneous information. Neurocomputing **210**, 81–90 (2016)
14. Xiong, P., Zhu, T.Q., Wang, X.F.: A survey on differential privacy and applications. Chin. J. Comput. **37**(1), 101–122 (2014)
15. Shokri, R., Shmatikov, V.: Privacy-preserving deep learning. In: ACM Conference on Computer and Communications Security, pp. 1310–1321 (2015)
16. Wang, D.: Application of big date visualization technique in power grid enterprise. Jiangsu Electrical Engineering (2014)
17. Wang, Y., Si, C., Wu, X.: Regression model fitting under differential privacy and model inversion attack. In: International Conference on Artificial Intelligence (2015)
18. Wang, Y., Wu, X., Wu, L.: Differential privacy preserving spectral graph analysis. In: Pei, J., Tseng, V.S., Cao, L., Motoda, H., Xu, G. (eds.) PAKDD 2013. LNCS, vol. 7819, pp. 329–340. Springer, Heidelberg (2013). doi:10.1007/978-3-642-37456-2_28
19. Zhan, J., Huang, J., Niu, L., Peng, X., Deng, D., Cheng, S.: Study of the key technologies of electric power big data and its application prospects in smart grid. In: 2014 IEEE PES Asia-Pacific Power and Energy Engineering Conference (APPEEC), pp. 1–4 (2014)
20. Zheng, H., Cong, J.I., Zheng, X., Kunming, L.I.: Analysis technology and typical scenario application of electricity big data of power consumers. Power Syst. Technol. **39**, 3147–3152 (2015)

Parallel Architectures

Parallel Aligning Multiple Metabolic Pathways on Hybrid CPU and GPU Architectures

Yiran Huang[1,2,3], Cheng Zhong[1,2,3(✉)], Jinxiong Zhang[1,2], Ye Li[4], and Jun Liu[1,2,3]

[1] School of Computer and Electronics and Information, Guangxi University, Nanning, China
chzhong@gxu.edu.cn
[2] Guangxi Universities Key Laboratory of Parallel and Distributed Computing, Nanning, China
[3] Guangdong Key Laboratory of Popular High Performance Computers, Shenzhen Key Laboratory of Service Computing and Applications, Shenzhen, China
[4] Data Science of Guangxi Higher Education Key Laboratory, Guangxi Teachers Education University, Nanning, China

Abstract. Metabolic pathway alignment remains an important tool in systems biology, and has become even more important with the growing mass of metabolic pathway data. However, the process of aligning multiple metabolic pathways scale poorly with either the number of pathways or with the size of pathways, and no attempts have been made to exploit the parallelism of the pathway alignments to improve the efficiency. This paper proposes a parallel metabolic pathway alignment method called PMMPA. In PMMPA, we design a commonly used parallel algorithm for the computation of reaction (node) similarity in GPU, and implement a parallel strategy for aligning multiple metabolic pathways in multi-core CPU. The experimental results show that this parallel alignment implementation achieves at most 300 times faster than the single-threaded version, the parallel implementation of aligning metabolic pathways on the hybrid CPU and GPU architecture is promising in improving the efficiency.

Keywords: Metabolic pathway alignment · Reaction similarity · CPU and GPU computing · Phylogenetic tree

1 Introduction

The amount and quality of the metabolic data are increased rapidly in the biological databases such as KEGG [1] and Biocyc [2], which provide strong supports for a deep understanding of metabolism. Analyzing metabolic pathways can help people to reveal the similar connection patterns of metabolic pathways among species [3].

Metabolic pathways are usually denoted as directed graphs, where a node is represented as a molecule which is usually specified as a reaction, compound or enzyme, and the edges are represented as the interactions between molecules [4]. The alignment of metabolic pathways is to find the molecule mappings among the pathways and to

© Springer Nature Singapore Pte Ltd. 2017
G. Chen et al. (Eds.): PAAP 2017, CCIS 729, pp. 483–492, 2017.
DOI: 10.1007/978-981-10-6442-5_46

identify similar connection pattern of metabolic pathways between different species. In recent years, some metabolic pathway alignment methods, such as MPBR [4], METAPAT [5], MetaPathwayHunter [6], GraphMatch [7], SubMAP [8], CAMPways [9], MP-Align [10], MetNetAligner [11], and SAGA [12], have been proposed to find molecule mappings between two pathways.

Owing to the increasing amount of available metabolic pathway data and computational hardness of aligning multiple metabolic pathways, how to improve efficiency of the alignments still remains to be a challenge. A possible solution for this challenge is to implement the metabolic pathway alignment on high performance computing platforms. However, no efforts have been made in parallel aligning metabolic pathways on the hybrid CPU and GPU architectures.

CUDA utilizes multiple GPU cores to perform SIMD instructions. GPU is an important complement to CPU in implementing high-performance parallel algorithms. Due to the fact that the similarity between any two nodes in pathways could be computed in parallel, it will be opportune to improve the efficiency by implementing parallel metabolic pathways alignment algorithm on hybrid CPU and GPU architectures.

In this work, we study the parallelism of metabolic pathway alignments, analyze the improvement and cost of the parallel alignment strategy, and implement a parallel metabolic pathway alignment method called PMMPA. Our main contributions are described as follows. First, we design a commonly used parallel algorithm for the computation of reaction (node) similarity on GPU. Second, we implement a parallel strategy for aligning multiple metabolic pathways on multi-core CPU. To the best of our knowledge, this is the first effort on exploiting the parallelism of metabolic pathway alignment problem on hybrid CPU and GPU architectures. The experimental results show that this parallel alignment implementation achieves at most 300 times faster than the single-threaded version and it will be useful to implement the parallel alignment of metabolic pathway to improve the efficiency.

2   Method

2.1   Preliminaries

We first introduce some definitions and notations. Directed graph $G_p = (V_p, E_p)$ denotes metabolic pathway p, where $V(G_p)$ is the node set of G_p and $E(G_p)$ is the directed edge set of G_p, and each node in $V(G_p)$ denotes a reaction v_i in p, $i = 1, 2, \ldots, |V(G_p)|$. If an output compound of reaction v_i is an input compound of reaction v_j, there is a directed edge from v_i to v_j, $j = 1, 2, \ldots, |V(G_p)|$. If both v_i and v_j are reversible, there is also a directed edge from v_j to v_i. In our previous work [4], we discuss the problem of aligning two metabolic pathways. In this work, we try to solve the problem of parallel aligning a set of metabolic pathways, which is formally stated as follows.

Problem Statement: Given a set of metabolic pathways $G = \{G_1, G_2, \ldots, G_i, \ldots, G_k\}$, the goal is to find the reaction mappings between any two pathways G_i and G_j in G and to compute the similarity between any two pathways G_i and G_j in G based on the found reaction mappings, $i, j = 1, 2, \ldots, k$, and $i \neq j$.

2.2   PMMPA Method

Aligning Multiple Metabolic Pathways
Generally, we can perform three steps to solve the problem stated in Sect. 2.1. Step I: choose two pathways G_i and G_j from G, and compute the similarity between any two nodes (reactions) in G_i and G_j to build a reaction similarity matrix B_{ij}, $i,j = 1,2,...,k$, and $i \neq j$. Step II: employ the Maximum-Weight Bipartite Matching (MWBM) algorithm [13] to extract reaction mappings between G_i and G_j based on B_{ij}, and compute the similarity between G_i and G_j using the reaction mappings. Step III: repeat steps I and II until any two pathways in G have been aligned.

In step I, we adapt the similarity between any two nodes in pathways in our previous work [4] is to compute similarity $S(u,v)$ between node u in G_i and node v in G_j:

$$S(u, v) = \alpha \times Esim(u_e, v_e) + \beta \times Csim(u_{ic}, v_{ic}) + \gamma \times Csim(u_{oc}, v_{oc}) \quad (1)$$

where u_e is the enzyme catalyzing reaction u, v_e is the enzyme catalyzing reaction v, $Esim(u_e,v_e)$ is the similarity between u_e and v_e, u_{ic} and v_{ic} are the input compounds of u and v, and u_{oc} and v_{oc} are the output compounds of u and v. The EC identifier of an enzyme consists of four digits. If all the four digits of the EC identifier of two enzymes are identical, $Esim(u_e,v_e)$ is 1. If the first three digits are identical, $Esim(u_e,v_e)$ is 0.75. If the first two digits are identical, $Esim(u_e,v_e)$ is 0.5. If only the first digit is identical, $Esim(u_e,v_e)$ is 0.25. If the first digit is different, $Esim(u_e,v_e)$ is 0. $Csim(u_{ic},v_{ic})$ is the average similarity score of compounds u_{ic} and v_{ic}, and $Csim(u_{oc},v_{oc})$ is the average similarity score of compounds u_{oc} and v_{oc}. Similar to our previous work [4], the compound similarity scores are obtained from SIMCOMP [14]. We use α, β and γ to control the weights of $Esim(u_e,v_e)$, $Csim(u_{ic},v_{ic})$ and $Csim(u_{oc},v_{oc})$ with the constraint $\alpha + \beta + \gamma = 1$. Here we use $\alpha = 0.4$, $\beta = 0.3$ and $\gamma = 0.3$.

In step (II), each time the MWBM algorithm [13] selects two reactions with the highest reaction similarity as a reaction mapping and extends this reaction mapping to the resulting reaction mapping set. The MWBM algorithm stops when there are no more reaction pairs to be considered. Based on the resulting reaction mapping set, the similarity between pathways G_i and G_j is computed as follows:

$$Sim(G_i, G_j) = NC(G_i, G_j) / \max\{|V(G_i)|, |V(G_j)|\} \quad (2)$$

where $NC(G_i,G_j)$ is the number of correct reaction mappings in the resulting reaction mapping set, and the correct reaction mapping is the mapping that includes two reactions with similarity value larger than 0.9.

Parallel Implementation
The primary issue of implementing the alignment of multiple metabolic pathways on the hybrid CPU and GPU architecture is to evaluate the parallelism in the alignment programs and the merits of the parallelism.

When aligning multiple pathways, the reaction similarity computation is suitable for parallelizing and can be implemented in GPU as it contains a large number of numerical and repeated calculations. On the other hand, the alignment of any two pathways can be executed in parallel and be implemented in multi-core CPU as it requires complex logical control. Next, we will discuss the parallel strategy of aligning multiple pathways on hybrid CPU and GPU architecture, which includes task assignment, data transfer and data storage.

A. Task assignment

When aligning two pathways G_i and G_j, we need to build similarity matrix B_{ij} where $B_{ij}[u,v]$ is the similarity between nodes (reactions) u and v, $u \in V(G_i)$, $v \in V(G_j)$, i, $j = 1,2,...,k$, and $i \neq j$. The computation of the similarity between any two reactions in G_i and G_j can be performed independently. Therefore we can use a GPU thread to compute the similarity between two reactions. When aligning k pathways in G, the number of pairs of pathways is $k(k-1)/2$ and the alignment of two pathways can be assigned to a CPU thread. Therefore we totally use $k(k-1)/2$ CPU threads to perform parallel alignments of k pathways in G.

B. Data transfer and data storage

Because all reaction information will be accessed by each thread and the amount of the reaction information is huge, the reaction information must be stored in the global memory on GPU. In the initialization of the parallel alignment, $|V(G_1)| + |V(G_2)| + ... + |V(G_k)|$ structure information of reactions is transferred from the main memory to the global memory in GPU. For pathways G_i and G_j, since the similarity between any two reactions in G_i and G_j are calculated in GPU, a reaction similarity matrix B_{ij} of size $|V(G_i)| \times |V(G_j)|$ will be transferred from the global memory in GPU to the main memory when B_{ij} is built. For simplicity, assume that the size of reaction similarity matrix for each pair of pathways is q. When aligning k pathways in G, $q \times k \times (k-1)/2$ reaction similarity matrices are transferred from the global memory in GPU to the main memory. Combined with the above mentioned aspects, the parallel strategy of PMMPA and construction of the reaction similarity matrix of two metabolic pathways are shown in Fig. 1.

According to CUDA SDK, the optimal GPU thread number is 512 for single GPU, and the number of threads in a Block is set to 512. When the number of reaction pairs

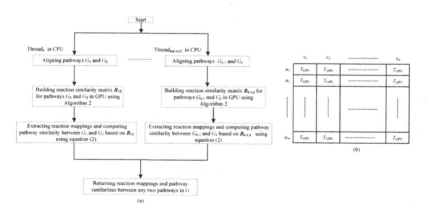

Fig. 1. (a) Procedure of PMMPA method. (b) Constructing the reaction similarity matrix of two pathways by Algorithm 2. For given pathways P and P', T_{GPU} denotes a thread in GPU, $u_1, u_2,...,$ and u_m denote the reactions in pathway P, and $v_1, v_2,...,$ and v_m denote the reactions in pathway P'. The similarity computation of each reaction pair is assigned to a GPU thread T_{GPU}.

exceeds 512, we divided 512 reaction pairs into a group and compute the reaction similarities by group. The last group that may contain less than 512 reaction pairs will be handled as an individual group. This parallel strategy fits well to the CUDA threads and can help to improve the utilization of hundreds of GPU cores. Since the threads in a block must be synchronized, the computation time of each block is the running time of the most time-consuming thread. Therefore, for each block, the computation time complexity is $O(N)$ in the worst case and it is $O(1)$ in the best case. For given pathways G_i and G_j, the frequency of parallel computing reaction similarities is about $|V(G_i)| \times |V(G_j)|/512$.

In the following, we describe our method in detail. Algorithm 1 shows the implementation of PMMPA on hybrid CPU and GPU architecture. Algorithm 2 presents the construction of the reaction similarity matrix of two metabolic pathways in GPU.

Algorithm 1. Parallel Multiple Metabolic Pathway Alignment (PMMPA)
Input: metabolic pathway set $G=\{G_1, G_2, ..., G_i,..., G_k\}$
Output: reaction mappings and similarity between any two pathways in G
Begin
1. Transfer the structure information of reactions for the pathways in G to the global memory in GPU;
2. do steps 3-7 in parallel
3. **for** $i=1$ to k **do**
4. **for** $j=i+1$ to k **do**
5.   Build reaction similarity matrix B_{ij} for G_i and G_j in GPU by Algorithm 2;
6.   Extract reaction mappings between G_i and G_j based on B_{ij} in CPU by MWBM algorithm;
7.   Compute the similarity between G_i and G_j in CPU by equation (2);
   end for
   end for
8. **Return** reaction mappings and pathway similarity between any two pathways in G
End.

In Algorithm 1, the task is assigned to CPU and GPU respectively, and the parallel alignments of k pathways in G are manipulated.

Algorithm 2. Building reaction similarity matrix B_{ij} for pathways G_i and G_j
Input: structure information of the reactions in G_i and G_j
Output: reaction similarity matrix B_{ij} for G_i and G_j
Begin
1. do steps 2-4 in parallel
2. **for** $ti=0$ to $|V(G_i)|-1$ **do**
3. **for** $tj=0$ to $|V(G_j)|-1$ **do**
4.   Compute reaction similarity between the ti-th reaction in G_i and the tj-th reaction in G_j by equation (1);
   Put this reaction similarity value in the ti-th row and tj-th column in B_{ij}
   end for
   end for
5. **Return** B_{ij}
End.

In Algorithm 2, the similarity between any two reactions in two metabolic pathways is computed in parallel in GPU.

3 Results

The experimental evaluations have been conducted on a multi-core computer equipped an AMD Opteron(tm) Processor 6136 containing 8 processing cores and a NVIDIA Tesla C2050/2070 GPU with 32 GB main memory. PMMPA was implemented in C++, OpenMP and CUDA 6.5. The running operating system is a 64-bits Linux with gcc 4.4.5. The metabolic pathway data are retrieved from the KEGG [1].

One of the questions in systems biology is whether it is possible to reconstruct phylogenetic trees by comparing metabolic pathways [10]. In the following we discuss the phylogenetic reconstruction of a set of organisms based on comparing common metabolic pathways of the organisms. Similar to the literature [10], we chose 21 organisms (as detailed in Fig. 2). To build the phylogenetic tree of these 21 organisms, we first performed PMMPA to align the 60 common pathways of these 21 organisms to obtain the similarities of the common pathways. Then we used the pathway similarities to compute the distances of these 21 organisms. The distance $dis(O_i, O_j)$ between organisms O_i and O_j, is computed as follows:

$$dis(O_i, O_j) = 1 - \frac{\sum_{t=1}^{p} Sim(G_{it}, G_{jt})}{p} \tag{3}$$

where p is the number of common pathways between O_i and O_j, G_{i1}, G_{i2}, ..., and G_{ip} represent the p common pathways for O_i, and G_{j1}, G_{j2}, ..., and G_{jp} represent the p common pathways for O_j. Finally, we used PHYLIP [15] to build a phylogenetic tree for these 21 organisms based upon the distances of organisms and used TreeView [16] to show this phylogenetic tree. The NCBI taxonomy is a gold standard for phylogenetic

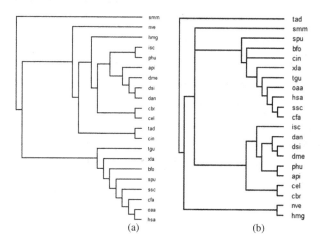

Fig. 2. Phylogenetic trees for 21 organisms: (a) our produced tree T_1; (b) NCBI taxonomy T_2.

trees in this paper. Figure 2 shows our produced tree and NCBI taxonomy [17] for these 21 organisms.

In Fig. 2(a), organisms cel and cbr are grouped together, organisms dsi, dan, and dme are grouped together, and organisms hsa, oaa, cfa, ssc, spu, bfo, xla, and tgu are grouped together. These classifications in Fig. 2(a) are close to the classifications in Fig. 2(b).

In the same way, we chose 8 organisms (as detailed in Fig. 3) and built a phylogenetic tree for these organisms by aligning their 40 common pathways. Figure 3 shows our reconstructed phyletic tree and NCBI taxonomy for these 8 organisms.

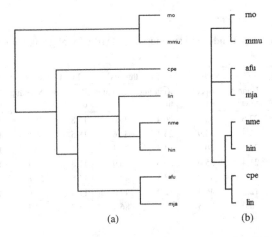

(a)   (b)

Fig. 3. Phylogenetic trees for 8 organisms: (a) our produced tree T_1; (b) NCBI taxonomy T_2.

In Fig. 3(a), organisms rno and mmu are grouped together, organisms nme and hin are grouped together, and organisms afu and mja are grouped together. These classifications in Fig. 3(a) are very similar to the classifications in Fig. 3(b).

The results of Figs. 2 and 3 demonstrate that it is useful to reconstruct phylogenetic trees by comparing metabolic pathways.

Next, we first discuss the performance of PMMPA under different CPU threads. Tables 1 and 2 show the running time and the speedup of PMMPA for aligning the common pathways of the 21 organisms in Fig. 2 and the 8 organisms in Fig. 3, respectively.

As can be seen from Table 1, compared with the serial alignment algorithm, the use of multi-core CPU parallel computing significantly reduces the required time of aligning multiple metabolic pathways. Moreover, from Table 1, we can see that, the parallel execution time decreases with increase of running CPU threads, and the speedup increases with the increase of running CPU threads. When the number of CPU threads becomes 64, the speedup for aligning 20–60 pathways increases to around 70 and 200, and the speedup for aligning 10 pathways becomes 357.5. This demonstrates that, for the multi-core architecture equipped with a CPU of 8 cores, the optimal number of running CPU threads for parallel aligning multiple metabolic pathways is 64. Similar trends can be observed in Table 2. From Table 2, we can see that the use of

Table 1. Running time (seconds) and speedup of PMMPA for aligning common pathways of 21 organisms in Fig. 2

Number of pathways	Serial algorithm running time	8 CPU threads		16 CPU threads		32 CPU threads		64 CPU threads	
		Running time	Speedup	Running time	Speedup	Running time	Speedup	Running time	Speedup
60	13084.0	180.1	72.6	175.5	74.6	164.4	79.6	130.0	100.6
50	9138.0	178.4	51.2	166.4	54.9	157.8	57.9	122.0	74.9
40	6392.0	68.5	93.3	60.5	105.7	54.9	116.4	46.0	139.0
30	5699.0	43.6	130.7	42.1	135.4	41.7	136.7	39.0	146.1
20	4598.0	24.1	190.8	23.6	194.8	21.7	211.9	20.0	229.9
10	3933.0	13.6	289.2	12.4	317.2	11.9	330.5	11.0	357.5

Table 2. Running time (seconds) and speedup of PMMPA for aligning common pathways of 8 organisms in Fig. 3.

Number of pathways	Seiral algorithm running time	8 CPU threads		16 CPU threads		32 CPU threads		64 CPU threads	
		Running time	Speedup	Running time	Speedup	Running time	Speedup	Running time	Speedup
40	925.0	27.6	33.5	26.4	35.0	19.8	46.7	18.8	49.2
35	869.0	26.4	32.9	25.1	34.6	18.5	47.0	16.8	51.7
30	781.0	25.5	30.6	24.3	32.1	18.1	43.1	16.7	46.8
25	750.0	21.4	35.0	20.3	36.9	17.8	42.1	12.6	59.5
20	647.0	13.5	47.9	13.1	49.4	11.8	54.8	11.7	55.3
15	598.0	12.9	46.4	12.2	49.0	11.5	52.0	10.5	57.0
10	501.0	6.0	83.5	5.8	86.4	5.5	91.1	5.0	100.2

multi-core CPU parallel computing greatly reduces the alignment time, and the speedup reaches the optimum when the number of running CPU threads becomes 64.

In PMMPA, we can use GPU to compute in parallel the reaction similarities and align serially the pathways in CPU, and this version of PMMPA is so called GPU parallel version, and the complete version of PMMPA using hybrid CPU and GPU computing is so called CPU/GPU parallel version. To learn the efficiency of GPU parallel computing, Fig. 4 presents the running time for GPU parallel version and CPU/GPU parallel version with 64 CPU threads.

In Fig. 4(a), we can see that CPU/GPU parallel version consumed less time than GPU parallel version in each case. We also see that when the number of aligned pathways is smaller than 30, the advantage of CPU/GPU parallel version over GPU parallel version is not obvious in running time as the data transfer between CPU and GPU take most of the running time. However, when the number of aligned pathways becomes large, the CPU/GPU parallel version has a significant advantage in running time. Similar cases with Fig. 4(b) count. The results from Fig. 4 show that, with the growing number of aligned pathways, the alignment efficiency can be greatly improved by using CPU/GPU parallel version in comparison to using GPU parallel version. This illustrates that hybrid CPU and GPU parallel computing can significantly accelerate the alignment of multiple metabolic pathways.

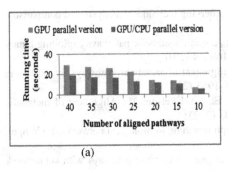 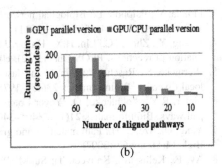

(a)   (b)

Fig. 4. Running time for aligning pathways using GPU parallel version and using CPU/GPU parallel version, respectively. (a) Running time for aligning the common pathways of the 21 organisms in Fig. 2. (b) Running time for aligning the common pathways of the 8 organisms in Fig. 3.

4 Conclusion

In this paper, to improve the efficiency of aligning multiple metabolic pathways, we propose an efficient parallel metabolic pathway alignment method called PMMPA on hybrid CPU and GPU architecture. We design a commonly used parallel algorithm for computation of reaction (node) similarities in GPU, and implement a parallel strategy for aligning multiple metabolic pathways in multi-core CPU. We also analyze and validate the improvement and cost of the proposed parallel strategy for pathway alignments. The experimental results show that this parallel alignment implementation achieves at most 300 times faster than the single-threaded version. This illustrates that the parallel implementation of multiple metabolic pathway alignments on hybrid CPU and GPU architecture can significantly improve the efficiency, and the study of metabolic pathway alignment will benefit from such parallel implementation.

Acknowledgment. This work is supported by the National Natural Science Foundation of China under Grant No. 61462005, Nature Science Foundation of Guangxi under Grant No. 2014GXNSFAA118396, Foundation of Guangdong Key Laboratory of Popular High Performance Computers, Shenzhen Key Laboratory of Service Computing and Applications under Grant No. SZU-GDPHPCL201414, and Data Science of Guangxi Higher Education Key Laboratory.

References

1. Kanehisa, M., Goto, S., Sato, Y., Furumichi, M., Tanabe, M.: KEGG for integration and interpretation of large-scale molecular data sets. Nucleic Acids Res. **40**(D1), D109–D114 (2012)
2. Caspi, R., Foerster, H., Fulcher, C.A., Kaipa, P., Krummenacker, M., Latendresse, M., et al.: The MetaCyc Database of metabolic pathways and enzymes and the BioCyc collection of Pathway/Genome Databases. Nucleic Acids Res. **36**(suppl 1), D623–D631 (2008)

3. Fionda, V., Palopoli, L.: Biological network querying techniques: analysis and comparison. J. Comput. Biol. **18**(4), 595–625 (2011)
4. Huang, Y., Zhong, C., Lin, H.X., Huang, J.: Aligning metabolic pathways exploiting binary relation of reactions. PLOS One **11**(12), e0168044 (2016)
5. Wernicke, S., Rasche, F.: Simple and fast alignment of metabolic pathways by exploiting local diversity. Bioinformatics **23**(15), 1978–1985 (2007)
6. Pinter, R.Y., Rokhlenko, O., Yeger-Lotem, E., Ziv-Ukelson, M.: Alignment of metabolic pathways. Bioinformatics **21**(16), 3401–3408 (2005)
7. Yang, Q., Sze, S.-H.: Path matching and graph matching in biological networks. J. Comput. Biol. **14**(1), 56–67 (2007)
8. Ay, F., Kellis, M., Kahveci, T.: SubMAP: aligning metabolic pathways with subnetwork mappings. J. Comput. Biol. **18**(3), 219–235 (2011)
9. Abaka, G., Bıyıkoğlu, T., Erten, C.: CAMPways: constrained alignment framework for the comparative analysis of a pair of metabolic pathways. Bioinformatics **29**(13), i145–i153 (2013)
10. Alberich, R., Llabrés, M., Sánchez, D., Simeoni, M., Tuduri, M.: MP-Align: alignment of metabolic pathways. BMC Syst. Biol. **8**(1), 1 (2014)
11. Cheng, Q., Harrison, R., Zelikovsky, A.: MetNetAligner: a web service tool for metabolic network alignments. Bioinformatics **25**(15), 1989–1990 (2009)
12. Tian, Y., Mceachin, R.C., Santos, C., Patel, J.M.: SAGA: a subgraph matching tool for biological graphs. Bioinformatics **23**(2), 232–239 (2007)
13. Sankowski, P.: Maximum weight bipartite matching in matrix multiplication time. Theoret. Comput. Sci. **410**(44), 4480–4488 (2009)
14. Hattori, M., Okuno, Y., Goto, S., Kanehisa, M.: Development of a chemical structure comparison method for integrated analysis of chemical and genomic information in the metabolic pathways. J. Am. Chem. Soc. **125**(39), 11853–11865 (2003)
15. Plotree, D., Plotgram, D.: PHYLIP-phylogeny inference package (version 3.2). Cladistics **5**, 163–166 (1989)
16. Page, R.D.: Visualizing phylogenetic trees using TreeView. In: Current Protocols in Bioinformatics, pp. 6.2.1–6.2.15 (2002)
17. Taxonomy - site guide - NCBI. http://www.ncbi.nlm.nih.gov/guide/taxonomy/. Accessed 2 Jun. 2017

Speeding Up Convolution on Multi-cluster DSP in Deep Learning Scenarios

Deng Wenqi, Yang Zhenhao, Lu Maohui, Wang Gai,
Yang JiangPing, and Zheng Qilong[✉]

School of Computer Science and Technology,
University of Science and Technology of China, Hefei, Anhui, China
{dengwq,yangzhh,lumaohui,wanggai,
yjp999}@mail.ustc.edu.cn, qlzheng@ustc.edu.cn

Abstract. Recently, deep learning has achieved great success in artificial intelligent, whose superiority also brought new opportunity for the related research in embedded system. This paper focused on optimizing and speeding the convolution computing, the core operation within convolution neural network based on a multi-cluster digital signal processor, BWDSP. By taking advantage of the BWDSP's architecture and characteristics of convolution computation, a suitable parallel algorithm was designed. Based on features of convolution neural network model structure, an automatic optimization tool for convolution computing with specific arguments was presented as well. The experimental result showed that the parallel algorithm given in this paper is 9.5x faster than GEMM-based algorithm commonly used in GPU and 5.7x faster than the traditional vectorization optimization algorithm. Meanwhile, a comparison was made between the parallel algorithm and tiled-base algorithm widely adopted in system with cache hierarchies, showing that the parallel one could achieve a better performance density of 1.55 times than that of later one, meaning that the work in this paper can make full use of computing resources to make them more efficient.

Keywords: Multi-cluster processor · Embedded system · Deep learning · Convolution neural network · Parallel algorithm

1 Introduction

Recently, deep learning has achieved great success in artificial intelligent [7]. One of the state-of-the-art and most popular deep learning algorithms [3] is CNN [8] (Convolution Neural Network). Its superiority bring new opportunity for the embedded system's artificial intelligence tasks. More and more works focus on how to deploy CNN models on embedded system, such as [12, 13].

One of the challenges to deploy the deep learning model in embedded system is that the huge amount of computation of the model goes beyond the capabilities of many embedded processors. The study [5] proved that convolution operation will occupy over 90% the computation time. In this paper, we studied how to speed convolution computing based on BWDSP [4], a multi-cluster 32-bit digital signal processor.

© Springer Nature Singapore Pte Ltd. 2017
G. Chen et al. (Eds.): PAAP 2017, CCIS 729, pp. 493–503, 2017.
DOI: 10.1007/978-981-10-6442-5_47

Considering the special application scenarios, many special optimization and methods were adopted. By analyzing the characteristics of convolution, we designed a convolution parallel algorithm that is suitable for multi-cluster architecture. We also studied the feature of common convolution neural network structure and developed an automatic optimization tool to optimize convolution computing with specific arguments. Benchmark shows that our algorithm is 9.5x faster than GEMM-based algorithm commonly used in GPU and 5.7x faster than the traditional vectorization optimization algorithm. As BWDSP lack of a cache hierarchy, we thereby compared our algorithm with tiled-based algorithm that widely used in systems having cache hierarchy. In order to compare the two algorithms that tested in different hardware platform. We analyzed the performance density, the average GMACs per multipliers of them. Our algorithm achieved a better performance density of 1.55 times than that of tiled-based algorithm. This indicator means that our algorithm can make full use of computing resources to make them more efficient.

The rest of this paper is organized as follows. Section 2 provides the characteristic and theoretical performance of BWDSP. In Sect. 3, we explain the formula of convolution layer, our parallel convolution algorithm and common convolution computing algorithm used in other hardware platform. In Sect. 4, we introduce the automatic optimization tool that optimizes convolution with specific arguments. The Benchmark and evaluation are placed in Sect. 5. Section 6 is conclusion.

2   BWDSP

2.1   Architecture

BWDSPs are multi-cluster 32-bit float DSP processors designed by The 38th Research Institute of China Electronics Technology Group Corporation. Architecture of BWDSP100 is shown in Fig. 1, they have 4 clusters marked as X, Y, Z and T. In BWDSP100, each cluster has 8 ALUs, 4 multipliers and 2 shifters and 1 special unit and all compute units can share registers in cluster and work in parallelly while there are no resource confliction. Its theoretical throughput is 8 GMACs when working at 500 MHz.

BWDSP100 has three 256 bits width data buses, one write bus and other two read buses. It also contains three memory blocks, so different memory block can be accessed via different bus at same time. With 1.75 GB address space provided, BWDSP100 could satisfy the requirement of large-scale data processing by using external memory.

2.2   Instruction Characteristic

This DSP family adopts VLIW architecture that can launch up to 16 instruction words per cycle. Many instructions are SIMD instructions. The strong readability instruction set provides a flexible control of the calculation components. For example, combination of instructions, XYZTMACC0 += R13*R9 || XYZTMACC1 += R13*R10 || XYZT-MACC2 += R13*R11 || XYZTMACC3 += R13*R12, forces all multipliers in 4 clusters to do MAC (Multiply ACcumulate) operations at same time. The prefixes X,

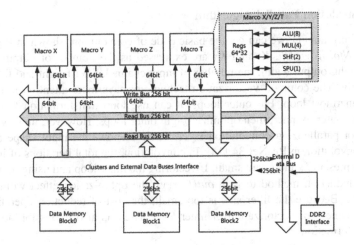

Fig. 1. Architecture of BWDSP100

Y, Z and T indicate which clusters run the instruction. MACC0~4 are the MAC registers in one cluster.

The clusters can't access memory independently. The instruction system provides memory distributed instructions. Instruction XYZTR14:13 = [U1 += U2, U3] reads 4 dual-words from memory and delivers them to registers R13 and R14 of different clusters. The address register U1, U3, U2 stand for base address, stride and offset modification after reading respectively.

3   Convolution Parallel Algorithm

3.1   Convolution

Convolution layer is core of CNN. It uses sliding window filters to extract local, more abstract feature for next layer or final classification. Both input and output of the convolution layer are three-dimensional tensors, which are called input (resp. output) channels, height and width. The parameters of the convolution layer consist of a set of $K_x * K_y$ filters. These filters are sliding in the height and width directions of each input channel, respectively. The sum of the dot products of data of the window at the same position of all input channels and the corresponding filter is an output element. The formula for one output's computing is

$$out(f_o, x, y) = \sum_{f_i=0}^{N_{f_i}-1} \sum_{k_x=0}^{K_x-1} \sum_{k_y=0}^{K_y-1} f(f_i, f_o, k_x, k_y) * in(f_i, x+k_x, y+k_y) \quad (1)$$

where N_{f_i} represents the number of input channel, $in(f_i, x, y)$ (resp. $out(f_o, x, y)$ is input (resp. output) value of channel f_i (resp. f_o) at position (x, y). f is a set of fileters, called kernel. When computing outputs of one output channel, input of different channels uses different filter matrices. So the kernel is a four-dimensional tensor. $f(f_i, f_o, k_x, k_y)$ is a value in the kernel that input channel f_i used to compute output channel f_o.

3.2   Convolution Parallel Algorithm

From the formula (1), we can get the basic code of convolution, shown in Fig. 2. The variables Nfo, h, w, Nfi, Kx and Ky are expressed as the number of output channel, output height, output width, the number of input channel, filter height and filter width respectively. The common vectorization technology pays more attention to optimization of innermost loop. For outer loops, it can use loop transformation to exchange loops from outer to inner, then optimize origin outer loops. However, those work may not suit for parallelizing convolution layer. Table 1 shows the input shape and filter size of convolution in ResNet-34 [14]. In convolution, the total iterations of innermost loops that present k_x and k_y are small. If vectoring the inner loop and using unparalleled part sum reduction method to gain $out(f_o, x, y)$, the optimization effect would not be remarkable. Because the filter size is too small, the cost of reduction operations will offset the benefit of vectorization of inner loop. There is also impossible to perform loop transformation.

```
1 for(int fo=0; fo<Nfo;++fo)
2  for(int x=0; x<h;++x)
3   for(int y=0;y<w;++y)
4    for (int fi=0;fi<Nfi;++fi)
5     for (int kx=0;kx<Kx;++kx)
6      for (int ky=0;ky<Ky;++ky)
7       out(fo,x,y )+=f(fi,fo,kx,ky)*in(fi,x+kx,y+ky);
```

Fig. 2. Pseudo code for convolution computing

Table 1. The input and filter sizes of convolution in ResNet-34

#output channel	#input channel	Input size	Filter size
64	3	224 × 224	7 × 7
{64, 128}	64	56 × 56	3 × 3
{128, 256}	128	28 × 28	3 × 3
{256, 512}	256	14 × 14	3 × 3
512	512	7 × 7	3 × 3

The {x1, x2} means the value may be x1 or x2.

Duo to there are no loop dependences among outer three loops, instead of using conventional methods, we use coarse-grained parallelism to optimize convolution calculation in outer loops and the algorithm is shown in Fig. 3. The computing unites of multi-cluster processor are quite different from vectorization ones. They can not only be combined to carry out vectorization calculations, but also compute individually. In this paper, we use full 16 multipliers throughout 4 clusters to generate 16 outputs at same time, each multiplier is assigned all computation for single output. Benefited from

```
1  for(int fo=0;fo<Nfo;++fo)
2   for(int x=0;x<h;++x)
3    for(int y=0;y<w;y+=16){//parallel computing 16 outs
4     clr_XYZTMACC();//clear multiplier accumulators
5     for(int c=0;c<Nfi;++c)
7      for(int i=0; i < kx; ++i)
8       for(int j=0; i < ky; ++j){
9        //read kernels per 4 clusters
10       if(j%4==0) XYZTfetch-4-kernel-to-tegs();
11       //distribute input for every out's computing
12       XYZTfetch-from-16-conv-window-to-regs();
13       //every multiplier computes one output
14       XYZTMACC0+=R0*R4||XYZTMACC1+=R1*R4
15       ||XYZTMACC2+=R2*R4||XYZTMACC3+=R3*R4;}
16     XTZTCopy-MACCtoR0-3();
17     for(int c=0,k=y;c<4;++c,++k){//write to mem
18      out(f0,x,k)=XRc;out(f0,x,k+4)=YRc;
19      out(f0,x,k+8)=ZRc;out(f0,x,k+12)=TRc;}}
```

Fig. 3. Parallel convolution algorithm used in BWDSP

BWDSP flexible memory access mode, our convolution parallel algorithm can easily distribute the data needed among all multipliers, keeping all multipliers fully working during convolution computing without extra operations.

3.3 Other Parallel Algorithms

There are many other convolution parallel algorithms used in different hardware platform.

The tiled convolution algorithm [3] is widely used in custom deep learning accelerator. It changes convolution's loops stride, computes multiple adjacent outputs and cache input and accumulate values in L1 or L2 cache. The adjacent outputs share input data in cache and significantly reduce the memory accesses. This algorithm needs processor having cache system, not suit-able for BWDSP.

Another algorithm commonly used in GPU is GEMM (GEneral Matrix-matrix Multiplication)–based method [6], shown in Fig. 4. It expands the input and kernel into two matrices, D_m 和 F_m. Each column of D_m consists of all the inputs in a sliding window. Rows of F_m place a set of filters in which the convolution kernel is grouped by the output channel. The dot product of a column in D_m and a row in F_m corresponds to the convolution output calculation. All output compute parallelly by calculating the matrix multiplication of F_m and D_m. We compared its performance with our algorithm in Sect. 5.

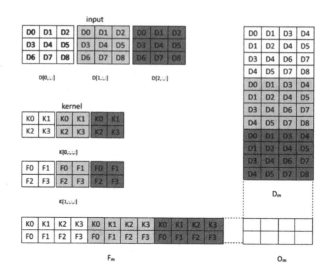

Fig. 4. GEMM-based convolution algorithm

4 Optimized by Specific Layer Arguments

A CNN model normally uses very limited kinds of convolution arguments, even it has more than 100 layers to simplify model design, training and debugging. Table 2 shows some famous CNN models' structure used in ILSVRC [9] (ImageNet Large Scale Visual Recognition Competition). As a typical example, ResNet has 152 layers, but only use three kinds of convolution filter size. This case give us another opportunity to optimize the convolution computing in CNN model deploy. Un-like the general function from algorithm library needs to handle all kinds of argument values, restricting advanced optimization, we can adopt model oriented optimization by tuning convolution computing with specific arguments values. For every kind of filter size, we can provide fully optimized computing function. Since a model normally uses very limited kinds of filter size, it wouldn't product too many functions.

Table 2. Convolution arguments used in IN ILSVRC

Model name	#layers	Filter size
VGG [11]	19	3 × 3
GoogLeNet [10]	27	1 × 1
		3 × 3
		5 × 5
		7 × 7
ResNet-152 [1]	152	1 × 1
		3 × 3
		7 × 7

Considering there is common code framework among convolution with different specific arguments, instead of developing various specify convolution, we use code auto-generated tool to produce optimized code by given convolution arguments for every given model, as Fig. 5 shown. The running time of the inner loops determines the performance of the algorithm, so we pay attention to optimize inner loops. Some optimization methods are listed below.

1. Unrolling inner loops.
   In general convolution function, filter size is un-certain, so we can't unroll inner loops, but in specify version, we can unroll the two innermost loops as shown in Fig. 3 in line 7–8. Since the filter size is small, number of generated code is acceptable. This optimization remove branch instructions and give opportunities to optimize un-rolled loops.
2. Software pipeline.
   In original code in Fig. 3, the innermost loops contain less instructions, so software pipeline did perform well. According to unroll innermost loops, most read input operations, as shown in Fig. 3 in line 12, can combine to last computing and executing in parallely.
3. Taking full advantage of memory bandwidth.
   Because multipliers in same cluster share registers, reading operation can read 16 elements for next 4 iterations. But if filter width is not a multiple of 4, there would be some redundant reading operation. Thus, after unrolling loops, we can insert kernel reading operations in computing sequence and read kernel on demand. As shown in Fig. 5, when filter size is 3 × 3, there are 9 times computation in unroll computing sequence. In the last time, we only need read one kernel element to each cluster.

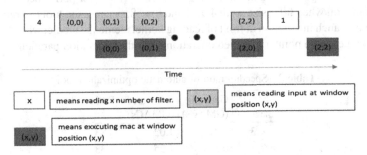

Fig. 5. The automatic optimization of convolution with filter size 3 × 3. Operations in same column can execute parallelly

5 Benchmark and Evaluation

The hardware platform used for our benchmark is BWDSP100 worked at 500 MHz.
We tested the three convolution parallel algorithm, GEMM-based algorithm, vectorized algorithm and our algorithm in BWDSP100 and compared their performances.

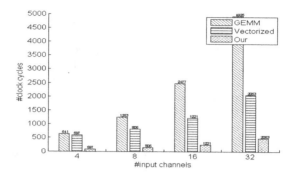

Fig. 6. Convolution computing algorithms' benchmark on BWDSP

The input size is 17*2 and the kernel shape is $[1, N_{if}, 2, 2]$. The result showed in Fig. 6. Our algorithm is better than two other algorithms, 9.5x faster than GEMM-based algorithm and 5.7x faster than vectorized algorithm. Besides reduction operations influenced the performance of vectorized algorithm, the read kernel elements can't share with multipliers and input reading patterns is not suitable for BWDSP's interval memory access instructions. The GEMM-based algorithm performed worst in BWDSP, because in preparation phase, it cost too much to construct two large matrices.

In order to understand the effect of automatic optimization tool. We measured its speedup ratio, shown in Table 3. The input size, number of input channels and number of output channels fixed to 1 × 100, 128 and 1. The automatic optimization achieved speedup ratio is 2.33 to 4.12. Due to memory read instruction cost much more time than other instructions, we read 4 kernel elements for every cluster and computed up to 4 times in innermost loop in general function. When filter size were 4 × 4 and 8 × 8, the general function can take ad-vantage of read kernel elements, so the speedup ratio is lower than other filter size arguments. But the code generated by tool still performed better than general function when filter size were 4 × 4 and 8 × 8, because the tool can remove the conditional branch instruction in front of reading kernel elements, predict when to change pointers and execute pointer assigned operation and other operation parallelly.

Table 3. Speedup ratio of automatic optimization tool

Filter size	General (GMACs)	Generated (GMACs)	Ratio
3 × 3	0.27	1.02	3.83
4 × 4	0.54	1.25	2.33
5 × 5	0.42	1.73	4.12
6 × 6	0.56	1.88	3.33
7 × 7	0.51	1.87	3.66
8 × 8	0.75	1.94	2.58

Figure 7 shows the throughput of our algorithm with different number of input channels and filter sizes. The input size and number of output channels also fixed to

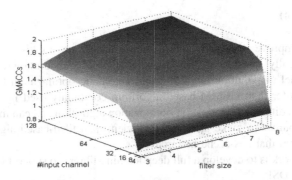

Fig. 7. Our convolution implementation performance with different filter size and number of input channels

1×100 and 1. As the size of the filter increases, the convolution performance continues to increase. The reason is that the larger filter size can make more code for parallel optimization after the innermost loops unrolling. Another factor that determines performance is the number of input channels. Since the number of channels increases, the ratio of the execution time of the non-computed instruction is decreasing. But as the filter size and the number of input channels increase by a certain range, the convolution performance remains stable. The best performance is 1.94 GMACs.

Because BWDSP doesn't have cache, it can't execute the tiled-based algorithm. We compared our algorithm with tiled-based algorithm evaluated on 66AK2H12. Cross-platform benchmarks are often unfair due to the performance of algorithms will be affected by the different hardwares. The 66AK2H12 ships with dual-core ARM (Cortex-A15) and eight-core C66 DSPs with 2 MB shared L2 cache running at 1.4 GHz clock. Every C66 DSP can process 32 16 × 16 integer MAC operations [14] and its processing capacity is 32 × 1.4 GHz × 8 c66 cores = 358.4 GMACs, better than DWDSP's. In order to analyze the two algorithms, we use similar method adopted by [15] to compare the performance density the average GMACs per multipliers. Table 4 presents their comparation. The performance of the tiled-based algorithm is benefited to a certain extent by better hardware and low precision. But our algorithm achieved a better performance density of 1.44 times that of tiled-based algorithms. This indicator means that our algorithm can make full use of computing resources to make them more efficient.

Table 4. Comparing with tiled-based algorithm in different hardwares

	Tiled-based [14]	Ours'
Working processer	C66 DSP 8-core	BWDSP100 4-cluster
Processing capacity	358.4 GMACs	8 GMACs
Precision	16bits fixed	fixed point
#multipliers	256 (16-bit)	16
Performance	20 GMACCs	1.94G MACCs
Performance density	7.81E-2 GMACCs	12.13E-2 GMACCs

6 Conclusion

In order to improve the convolution performance on multi-cluster processor, BWDSP. We fully considered the characteristics of convolution and CNN and de-signed parallel convolution algorithm and automatic optimization tool. Our algorithm is 9.5x faster than GEMM-based algorithm commonly used in GPU and 5.7x faster than the traditional vectorization optimization algorithm and can make full use of computing re-sources to make them more efficient than tiled-based algorithm widely adopted in system that has cache hierarchy.

Our future work is to develop a full deep learning library to support CNN model on deployment BWDSP.

Acknowledgment. This work was supported in part by a grant from China Core Electronic Devices, High-end Generic Chips and Basic Software Major Projects, No. 2012ZX01034-001-001.

References

1. He, K., Zhang, X., Ren, S., et al.: Deep residual learning for image recognition. In: Proceedings of the IEEE Conference on Computer Vision and Pattern Recognition, pp. 770–778 (2016)
2. Gu, J., Wang, Z., Kuen, J., et al.: Recent advances in convolutional neural networks. arXiv preprint arXiv:1512.07108 (2015)
3. Chen, Y., Luo, T., Liu, S., et al.: Dadiannao: a machine-learning supercomputer. In: Proceedings of the 47th Annual IEEE/ACM International Symposium on Microarchitecture, pp. 609–622. IEEE Computer Society (2014)
4. Cetc38.com.cn: BWDSP Product Presentation. http://www.cetc38.com.cn/38/335804/335809/377610/index.html (2017). Accessed 22 Mar 2017
5. Cong, J., Xiao, B.: Minimizing computation in convolutional neural networks. In: Wermter, S., Weber, C., Duch, W., Honkela, T., Koprinkova-Hristova, P., Magg, S., Palm, G., Villa, A.E.P. (eds.) ICANN 2014. LNCS, vol. 8681, pp. 281–290. Springer, Cham (2014). doi:10.1007/978-3-319-11179-7_36
6. Chetlur, S., Woolley, C., Vandermersch, P., et al.: cuDNN: efficient primitives for deep learning. arXiv preprint arXiv:1410.0759 (2014)
7. Bengio, Y., Courville, A., Vincent, P.: Representation learning: a review and new perspectives. IEEE Trans. Pattern Anal. Mach. Intell. **35**(8), 1798–1828 (2013)
8. LeCun, Y., Bottou, L., Bengio, Y., et al.: Gradient-based learning applied to document recognition. Proc. IEEE **86**(11), 2278–2324 (1998)
9. Image-net.org: ImageNet Large Scale Visual Recognition Competition (ILSVRC). http://www.image-net.org/challenges/LSVRC/ (2017). Accessed 22 Mar 2017
10. Szegedy, C., Liu, W., Jia, Y., et al.: Going deeper with convolutions. In: Proceedings of the IEEE Conference on Computer Vision and Pattern Recognition, pp. 1–9 (2015)
11. Simonyan, K., Zisserman, A.: Very deep convolutional networks for large-scale image recognition. arXiv preprint arXiv:1409.1556 (2014)
12. Lane, N.D., Bhattacharya, S., Georgiev, P., et al.: Deepx: a software accelerator for low-power deep learning inference on mobile device. In: 2016 15th ACM/IEEE International Conference on Information Processing in Sensor Networks (IPSN), pp. 1–12. IEEE (2016)

13. Cavigelli, L., Magno, M., Benini, L.: Accelerating real-time embedded scene labeling with convolutional networks. In: 2015 52nd ACM/EDAC/IEEE Design Automation Conference (DAC), pp. 1–6. IEEE (2015)
14. Hegde, G., Ramasamy, N., Kapre, N.: CaffePresso: an optimized library for deep learning on embedded accelerator-based platforms. In: Proceedings of the International Conference on Compilers, Architectures and Synthesis for Embedded Systems, p. 14. ACM (2016)
15. Zhang, C., Li, P., Sun, G., et al.: Optimizing FPGA-based accelerator design for deep convolutional neural networks. In: Proceedings of the 2015 ACM/SIGDA International Symposium on Field-Programmable Gate Arrays, pp. 161–170. ACM (2015)

Optimization Scheme Based on Parallel Computing Technology

Xiulai Li, Chaofan Chen, Yali Luo, and Mingrui Chen[✉]

Hainan University, Haikou, Hainan, China
mrchen@hainu.edu.cn

Abstract. Parallel computing is a high performance technology to solve problems, in order to improve computing efficiency, we use the processor to concurrent execute several parts divided from one problem. Based on the current issues in parallel computing area, both the data processing repetition rate and the parallel computing time depend on the time of the last thread in the task completing. This paper was written to take an overview of the existing parallel computing techniques and structures, and propose a solution of adding an advanced thread or advanced processor to make up the deficiency in parallel computing area.

Keywords: Parallel structure · Parallel computing · Parallel model · Parallel technology

1 Introduction

From the day of the birth of the computer, people continue to redouble efforts to improve the speed of the computer, and has achieved very significant results. However, this effort will not be long before the termination of the limit of the physical device. One of the common characteristics of people in the effort to develop a new generation of computers is the use of parallel technology. Increase in the same time interval the number of operations technology called parallel processing technology; design for parallel processing computer called parallel computers; to solve the problem in parallel computer called parallel computing; in parallel computer implementation of problem solving algorithm called parallel algorithm [1].

Traditionally, the general software design is a serial calculation:

(1) The software runs on a computer with only one CPU;
(2) The problem is decomposed into discrete sequence of instructions;
(3) The instruction is executed by one by one;
(4) At anytime CPU up to only one instruction at run time. The operational principle of CPU is described as Fig. 1.

In the simplest case, parallel computing is to use a number of computing resources to solve the problem.

(1) The purpose of using multi-core CPU to run;
(2) The purpose of the problem is decomposed into discrete parts can be solved at the same time [2];

G. Chen et al. (Eds.): PAAP 2017, CCIS 729, pp. 504–513, 2017.
DOI: 10.1007/978-981-10-6442-5_48

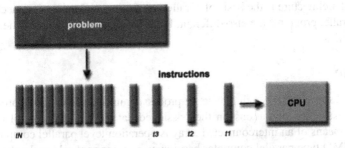

Fig. 1. The operational principle of CPU

(3) The purpose of each part is subdivided into a series of instructions;
(4) In each part of the instruction can be executed simultaneously in different CPU [3]. The operational principle of multi-CPU is described as Fig. 2.

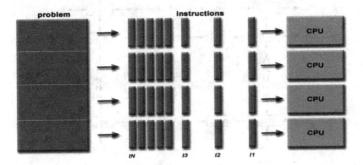

Fig. 2. The operational principle of multi-CPU

A wide range of parallel computing needs, but to sum up there are three types of applications: Compute-Intensive applications, such as large-scale scientific and (Data-Intensive); data intensive applications, such as numerical library, data warehouse, data mining and visualization; network intensive applications, such as collaborative work, remote control and remote medical diagnosis etc. [4].

Parallel computing, said simply that the computation is made in parallel computer, it is often said that the calculation and high performance, super computing is a synonym for any high performance computing and super computing cannot do without parallel technology [5].

2 Parallel Computing Architecture

Since the parallel computing technology since the middle of 60 s, the parallel processing has experienced from the array machine (SIMD), the vector processor, the shared memory vector machine (SMP), massively parallel processing, distributed storage system (MPP) to the workstation (COW) process [6].

Parallel architecture is the basis of parallel computing, and the design mechanism of various parallel programs are also different. It can be roughly divided into the following five categories.

2.1 SIMD

Array processor (SIMD) is a duplicate set processing unit to carry out the provisions of the same instruction operations on their assigned data in a single control unit under control by means of an interconnected array is operation level parallel computer SIMD [7]. The SIMD type parallel computer has played an important role in the development of parallel computer, but due to the development of processor technology since 90s, for science and engineering calculation of the SIMD type parallel machine has basically quit the stage of history. The system of SIMD is described as Fig. 3.

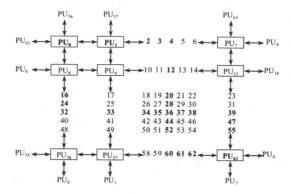

Fig. 3. The system of SIMD

2.2 Vector Machine

Vector Machine can perform high-speed processing of vector operation with a special vector registers and vector flow components, except scalar registers and scalar functions [8]. The system of vector machine is described as Fig. 4.

2.3 SMP

Shared memory processor systems share a central memory, in general there are specialized multi machine synchronous communication components, can support the development of data parallel or control [9]. But the processor number is too much, the processor to the central memory channel will become a bottleneck, limiting the development of the parallel machine, which is one of the main reasons for large-scale distributed memory parallel machine developed. The system of SMP is described as Fig. 5.

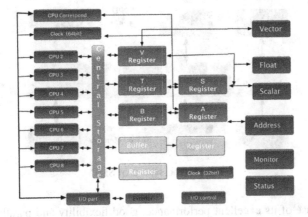

Fig. 4. The system of vector machine

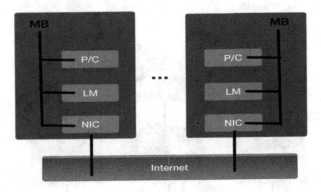

Fig. 5. The system of SMP

2.4   MPP

Distributed memory multiprocessor system which is composed of many parallel nodes, each node has its own processor and memory nodes connected to the interconnection network, parallel development support data also support the control of parallel development [10]. The system of MPP is described as Fig. 6.

2.5   COW

The workstation cluster of workstations (COW) is a collection of all computer nodes interconnected by high performance networks or local area networks [11]. Typically, each node is a SMP server, a workstation or a PC machine, which can be isomorphic or heterogeneous. The number of computers in general is a few to dozens, support for control of parallel and data parallel. Each node has a complete operating system, network software and user interface, can be used as a control node and computing nodes, that is equal between nodes. The cluster system's performance in recent years is

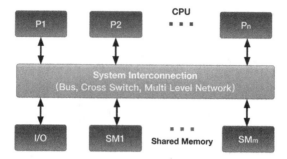

Fig. 6. The system of SMP

striking, because of its excellent performance, good flexibility and parallel processing ability, in addition to widely as a research topic, application development in various industries is also very fast. The system of COW is described as Fig. 7.

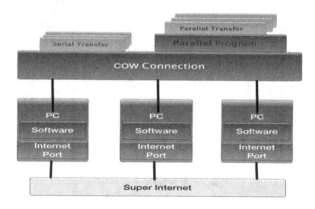

Fig. 7. The system of COW

3 Theoretical Model of Parallel Computing Technology

Please Parallel computing is the process of solving the problem of computing resources at the same time, it is an effective method to improve the computing speed and processing power of computer system [12]. Its basic idea is to use multiple processors to solve the same problem, the problem is decomposed into several parts, and each part is calculated by an independent processor. The parallel computing system can be either a specially designed super computer with multiple processors or a cluster of independent computers which are interconnected in a certain way. Through the parallel computing cluster to complete the data processing, and then return the results to the user.

The theoretical model of structure, the problem will be resolved is divided into N, N computing resources for the N runway, the problem is solved, and a huge problem can also be multiple computing resources to solve the basic model, as follows. In an ideal situation, the time consumed by parallel computing is the formula, that is, each independent computing resource completes the task at the same time, the consumption time is the time to solve the problem. The ideal model of parallel computing is described as Fig. 8.

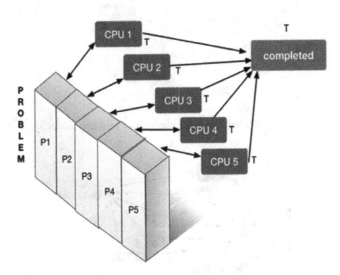

Fig. 8. The ideal model of parallel computing

According to the above parallel computing technology model, it can be known that the time consumed by the parallel computation is the slowest problem modules. The actual calculation may appear in many situations. First of all, the module partition problem, we can not guarantee that every module of the size of the problem is the same, assuming that dealing with computing resources ability is equal, this will lead to the time of computing resources to receive the largest part module significantly longer than other computing resources. It affects the efficiency of parallel computing computing. Assuming that the problem can be evenly divided into N module, if there is a single computing resource because of memory overflow or computational problems, this part module is stopped or delayed, resulting in increase of parallel computing time or can not complete the task. The unreasonable partition of problem model is described as Fig. 9. The abnormal CPU model is described as Fig. 10.

Based on the parallel computing cluster model, the problem to be solved by the main control machine is divided into N problem module, and then assigned to the N computer. In an ideal case, the size of each part module is the same as the computing power of each computer, and the ideal processing time for parallel computing is the time for a single computer to deal with the part module [13]. Assuming that the processing capacity of each computer is the same, but one problem of all is too large, the time consumed by the parallel computing is the time to solve the biggest problem. Assuming the master machine assigned to each computer of the same size, solve the problem in the process, if a computer or abnormal downtime, which leads to the

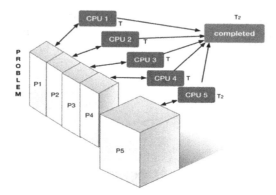

Fig. 9. The unreasonable partition of problem model

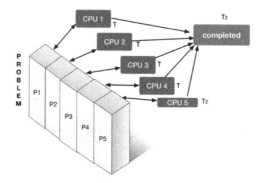

Fig. 10. The abnormal CPU model

problem of processing time is lengthened obviously or in the problem can not be solved in parallel computing.

4   Parallel Computing Technology Optimization

For the problems mentioned in the third chapter, there are many similar problems in the process of the actual parallel computation. Part problem segmentation is not reasonable, resulting in a single independent processor consumes too long, which greatly reduces the efficiency of parallel computing. The process of parallel computing, due to its single processor, the processing speed is relatively slow or midway accident downtime, so the calculation of the time was pro-longed or parallel because some calculation results did not reach a lead to the parallel computing can not be completed. The practical problem, parallel computing an obvious disadvantage is the repeated calculation, by part module problem segmentation in a lot of data and calculation methods are the same, the calculations have been repeated on different computers, it will reduce the efficiency of calculation greatly [14].

In real life, there are many examples about parallel, we can learn from the life of the solution to solve the problems encountered in parallel computing. For example, there is a pile of goods need to be transported from A place to B place, so we prepare a lot of goods vehicles, trucks are loaded cargo weight is not the same, the speed of different trucks carrying is not the same. When the last truck arrives at B place, the task is completed. In the process of transportation, if a truck is loaded with heavy goods, then its speed will be very slow. If one of the vehicles due to their own reasons for slower speed or the middle of the problem then arrives at the B place of time will be late or stop in the road which leads to the task can not be completed. In this situation, we can arrange a fast large trucks, to deal with similar problems occurred in the handling process, to ensure that the task is completed in high efficiency.

Parallel computing technology based on the practical problems, the optimization scheme proposed in this paper, the parallel computation with one or more advanced processor, the processor computing power was significantly faster than that of other processors. Parallel computing process, the main control computer if the layout of a computer to detect the problem is too large, the main control computer will arrange the task of the computer to the advanced computer to continue processing. In the course of parallel computing, if there is a problem with a single computer, the master computer will give the task of the problem computer to the advanced computer. Parallel computing in the process of marking method to calculate more than a certain period of time, the method to replant advanced computer, this time after the node calculation by computer, reduce repeated calculation times so as to improve the efficiency of parallel computing. The super CPU model is described as Figs. 11 and 12.

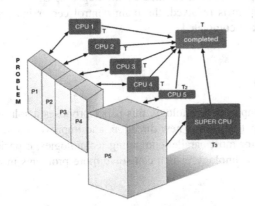

Fig. 11. The super CPU model

Algorithm steps:

(1) Equal cutting problem P, P (0), P (1),..., P (s),..., P (n − 1)
(2) The task is assigned to each CPU, C (0), C (1),..., C (s),..., C (n − 1)
(3) Begin execution, Execute(P)
(4) The control center to monitor the task, If it find a large task P(s) to the Super CPU processing, cycle monitoring processing, until each subtask is almost equal.

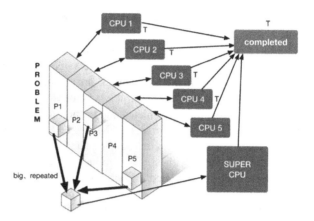

Fig. 12. The super CPU model

(5) The control center real-time monitoring task, Monitor (task), if found in the task P (s) is too large or abnormal CPU during execution, $(P(s))/(C(s)) \gg t^-$, the task is assigned to the Super CPU processing

(6) The main control center searches for duplicate parts in the subproblem, $P'(0) = P'(1) = \ldots = P'(s)$, which is processed by the Super CPU and returns the value to each CPU module

(7) Repeat steps 5 and 6, and the priority $V5 > V6$

(8) Until the last process $P(s)$, handled by the Super CPU

(9) Each sub-problem is resolved, the main control center integrated sub-questions solve, the task is completed

5 Conclusions

For the parallel computing technology, this paper proposes a solution to the problem about parallel computing from a new direction, and improves the original model. The scheme is suitable for most parallel computing technologies, especially for large data parallel computing technology, which can solve more problems in the future.

References

1. Chen, G.L., Sun, G.Z., Zhang, Y.Q., et al.: Study on parallel computing. J. Comput. Sci. Tech. **21**(5), 665–673 (2006)
2. Isard, M., Yu, Y., Birrell, A., et al.: Dryad: distributed data-parallel programs from sequential building blocks. Technical report, Microsoft Research Technical Report, Microsoft Corporation (2006)
3. Asanovic, K., Bodik, R., James, J., et al.: The landscape of parallel computing research: a view from Berkeley. Technical report, Electrical Engineering and Computer Sciences, University of California, Berkeley (2006)

4. Mattson, T.G., Sanders, B.A., Massingill, B.L.: Patterns for Parallel Programming. Prentice Hall, New Jersey (2005)
5. Rajkumar, B., Chee, S.Y., Srikumar, V.: Market-oriented cloud computing: vision, hype, and reality for delivering IT services as computing utilities. In: Proceedings of the 10th IEEE International Conference on High Performance Computing and Communications, 25–27 September 2008, Dalian, pp. 15–22. IEEE CS Press, Los Alamitos (2008)
6. Sun, X.H.: Scalable computing in the multicore era. In: Proceedings of the Inaugural Symposium on Parallel Algorithms, Architectures and Programming, 16–18 September 2008, pp. 1–18. University of Science and Technology of China Press, Hefei (2008)
7. Furtak, T., Amaral, J.N., Niewiadomski, R.: Using SIMD registers and instructions to enable instruction-level parallelism in sorting algorithms. In: Proceeding of the 19th Annual ACM Symposium on Parallelism in Algorithms and Architectures (SPAA) (2007)
8. Pardo, M., Sberveglieri, G.: Classification of electronic nose data with support vector machines. Sens. Actuator B: Chem. 107, 730–737 (2005)
9. Roig, C., Ripoll, A., Senar, M., Guirado, F., et al.: A new model for static mapping of parallel applications with task and data parallelism. In: Proceeding of the International Parallel and Distributed Processing Symposium, pp. 78–85 (2002)
10. Dean, J., Ghemawat, S.: MapReduce: simplified data processing on large clusters. Commun. ACM 51(1), 107–113 (2005)
11. Dean, J., Ghemawat, S.: MapReduce: simplified data processing on large clusters. In: Sixth Symposium on Operating System Design and Implementation, 6–8 December 2004, San Francisco, CA, pp. 10–23. USENIX Association, Berkeley (2004)
12. Komatitsch, D., Goddeke, D., Erlebacher, G.: Modeling the propagation of elastic waves using spectral elements on a cluster of 192 CPUs. Comput. Sci. Res. Dev. 25(1–2), 75–82 (2010)
13. Grama, A.Y., Gupta, A., Kumar, V.: Isoefficiency: measuring the scalability of parallel algorithms and architectures. IEEE Parallel Distrib. Technol. 1(3), 12–21 (1993)
14. Ino, F., Fujimoto, N., Hagihara, K.: LogGPS: a parallel computational model for synchronization analysis. In: Proceedings of the 2001 ACM SIGPLAN Symposium on Principles and Practices of Parallel Programming, PPoPP 2001, Snowbird, Utah, USA, pp. 133–142. ACM (2001)

Research on Client Adaptive Technology Based on Cloud Technology

Xiaojing Zhu$^{(\boxtimes)}$

Hainan Tropical Ocean University, Sanya 572022, Hainan, China
hehe428@163.com

Abstract. With the highly popularity of the Internet, the diversification of the Internet terminal equipment and applications, client development technology and the operation and maintenance of the architecture, development costs and performance are facing more and more challenges. This paper introduces the Web technology in the cloud environment, analyzes the present client adaptive technology and existing problems, put forward a kind of adaptive client cloud environment model, based on Response Web Design, combined with multi template and backend For Frontends (BFF), based on cloud technology, put forward the template fragment combination and on-demand production principle, To solve the adaptive problem of client diversification and intelligent; saving development costs and improve development efficiency.

Keywords: Cloud technology · Client adaptive technology · Template fragment · Server · Backend For Frontends · Response Web Design

1 Introduction

According to the China Internet Network Information Center (CNNIC) thirty-ninth "China Internet development statistics report" [1] pointed out. As of December 2016, China's Internet users reached 731 million, penetration rate reached 53.2%, more than the global average of 3.1 percentage points, more than the Asian average of 7.6 percentage points [2]. A total of 42 million 990 thousand new Internet users, an increase of 6.2%. The size of China's Internet users is equivalent to the total population in Europe. Based on the huge Internet Group, the individual need of the Internet continues to improve, such as diversification of Internet access, access time and the application of diverse needs. All of the current client applications in the development, operation and maintenance are put forward higher requirements. In view of this situation, the development of different applications will be based on different clients, which is likely to cause serious waste of development resources, greatly increasing the cost of development. This paper starts from the cloud technology to explore the characteristics, advantages and disadvantages and the application of two kinds of mainstream client adaptive technology. Finally, based on the two adaptive technologies, a cloud based architecture model is proposed. In order to achieve code reuse, development and maintenance costs have been ideal state.

© Springer Nature Singapore Pte Ltd. 2017
G. Chen et al. (Eds.): PAAP 2017, CCIS 729, pp. 514–521, 2017.
DOI: 10.1007/978-981-10-6442-5_49

2   Cloud Technologies

To understand what cloud technology is, it must first understand what is cloud computing. The idea of "cloud" can be traced back to the origin of utility computing, a concept proposed by computer scientist John McCarthy in 1961. But the definition of cloud computing until 2009 by the National Institute of standards and Technology (NIST) published and widely accepted.

Cloud computing is the development of parallel computing, distributed computing and grid computing, it is a kind of model can be realized easily, according to need, whenever and wherever possible the configurable computing resources shared pool access to resources (such as network, server, storage, applications and services), resources can quickly supply and release, so the workload and service provider management intervention to reduce to a minimum. This cloud model consists of five basic characteristics (network access, flexibility, resource pooling, measurement service and on-demand service), three models (IaaS, PaaS, SaaS) and four deployment models (public cloud, private cloud and hybrid cloud community) [3].

Cloud computing is a business model, the need for information technology support and landing. Cloud technology is referred to support cloud computing and information technology, including network technology, information technology, data center technology, virtualization technology, service technology, management platform and application technology and WEB technology, these technologies can be used independently, but also can be combined with each other, composed of resource pool, on-demand use, flexible convenience.

Due to the fundamental dependence of cloud computing on network connectivity, the universality of Web browsers and the ease of use of Web based service development, Web technology is often used as the implementation medium and management interface for cloud services. For client technology, Web technology is unable to replace.

2.1   Web Technology

WWW is a system of interconnected IT resources accessed via Internet. Its two basic components are the Web client and the Web server. There are other components, such as proxy, cache service, gateway, load balancing, etc., which are used to improve the performance of Web applications such as scalability and security. These additional components are located in the hierarchy between the client and server.

Web technology architecture consists of three basic elements: Uniform Resource Locator (URL), Hypertext Transfer Protocol (HTTP), and Extensible Markup Language (XML). Web browser can request to read, write, update or delete the Web resources on Internet, and identify and locate the URL through the resource. The request sent by HTTP to the resource identified by a URL host, then the Web server location resources and processing the requested operation, the results will be sent back to the browser client, may contain HTML and XML statement processing results.

2.2 Web Applications

The use of distributed applications based on Web technology is often considered as a Web application. Due to the high accessibility, these applications appear in all cloud based environment types. Modern Web applications are usually based on the basic three layer model:

– The presentation layer, used to show the user interface;
– The application layer, used to implement the application logic;
– The data layer, composed of persistent data storage.

The presentation layer is divided into client and server. When the Web server receives the request from the client, according to the application logic, if the request object is a static Web content, the direct access; if it is dynamic content, the indirect access. In response to a request, the Web server interacts with the application server, usually involving one or more underlying databases. The PaaS service model makes it possible for cloud users to develop and deploy Web applications to provide an independent instance of Web servers, application servers, and database storage server environments [4].

3 Client Technologie

The client, also known as the browser side, is a program that provides service to the client. In addition to stand-alone applications, it needs to cooperate with the server running.

According to the classification of the local service architecture, can be divided into Client/Server (C/S) and Browser/Server (B/S) two. The C/S architecture is a typical two layer architecture, its name is Client/Server, namely the client server architecture, the client contains one or more running on the user's computer program, and the server has two kinds, one is the database server, the client through the database connection to access the server data; the other is Socket server, application server and client communication program by Socket.

C/S architecture can also be seen as fat client architecture. Because the client needs to achieve the vast majority of business logic and interface display. This architecture, as part of the client needs to bear great pressure, because the display logic and transaction processing are included in them, through the interaction with the database to achieve persistent data, in order to meet the actual needs of the project.

The full name of the B/S architecture is Browser/Server, the browser/server architecture. Browser refers to the Web browser, a small number of business logic in the client, the main business logic in the server to achieve. This avoids the huge fat client, reducing the pressure on the client. Because the client contains little logic, it has also become a thin client.

With the development of technology and customer machine performance, but also in order to get a better user experience, the application of B/S architecture based on the logic from the server to the client moves more business processing, so that the client becomes "fat", therefore, the fat client and thin client boundaries become increasingly blurred [5, 6].

C/S and B/S is a kind of architecture model, it needs to develop language support, at present the client programming language based on HTML: Hyper Text Markup Language (HTML), Cascading Stylesheets (CSS) cascading style sheets, JavaScript and Extensible Markup Language (XML) is usually based on the application of modern the database development, therefore, in addition to the above client language, there must be a server-side language support. The server-side commonly use C#, JAVA, PHP and Python (see Fig. 1).

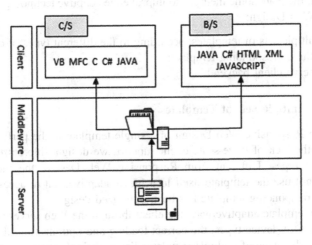

Fig. 1. Client Technology

Because modern Web applications demand and increasing complexity, people from zero Web application development efficiency, in order to improve development efficiency, provide the birth of the Web framework on a variety of functional modules, Ruby on Rails and SaisJS Web framework, it has helped reduce the workload of "common development activities, such as the number of frames provide database access interface, the standard model and session management and improve code reusability, greatly improves the efficiency of development.

4 Client Adaptive Technologie

The most common design of client adaptive technology is response Web design. In 2010, Ethan Marcotte proposed the "Responsive Web Design" [7], the concept of Responsive Web response is the design and development of the page should be based on the equipment environment (platform screen size, screen orientation, etc.) and user behavior (change window size) response and adjustment the corresponding. The specific practice is composed of many aspects, including elastic mesh and layout, pictures, CSS media query, etc. Whether the user is using the PC, tablet computer, or mobile phone, whether or not the full screen full screen display, the screen is either horizontal or vertical, the page should be able to automatically switch the resolution, image size and script function, to adapt to different equipment.

The advantage of Responsive Web Design is designed using the same set of templates for all client design is simple, but it has also very obvious shortcomings, if any modifications will reinvent the wheel, with the expansion of the scale of website, the performance will decline. Therefore, this way is only limited in small website development, website with increased demand and client hardware is more abundant, Responsive Web Design need to be constantly improved, in order to meet the needs of users.

At present, there are some methods to improve the adaptive technology of the client in Response Web Design:

- Design multiple sets of templates: according to the different types of client devices to design the corresponding template;
- Ready to serve client backend.

4.1 Design Multiple Sets of Templates

In response to Responsive Web Design of a single template in the performance, post maintenance, the lack of large-scale client support, we design different templates for different client types. Different from Responsive Web Design, the multi template scheme does not use the template itself to achieve adaptive, but to access the device type to the corresponding template has been optimized design.

This multi template adaptive way, the client through the Web server is responsible for the detection of device types, the correct loading and optimization of the template prepared. The advantage of multiple templates lies in their ability to reuse HTML and JavaScript, reduce the burden on Web servers, and simplify the changes in project management and testing. Since the client template is independent, it can be easily added and modified to support a richer client. The independence of the client module has better adaptability to the deployment environment, such as different templates can be deployed to different cloud environments, in order to obtain better access performance and efficiency. However, much higher than the client template adaptive response design technology, it requires developers to master CSS, HTML and JavaScript technology, the development of adaptive template rich, adapt to different clients (see Fig. 2).

4.2 Backend for Client Services

A common solution to the interface that is more frequently associated with the backend and to provide different content for different devices is to use the server's aggregation interface or API portal. The entry can be arranged for multiple backend calls and provides customized content for different devices. The backend for client service is sometimes called Backend For Frontends (BFF). It allows the team to focus on a given UI at the same time; it will also deal with the server-side component [8].

Multi template method Web server is only responsible for the detection of the client device type, and then loading the ready template, but for the client service is the back-end device detection and template on the Web server in Chengdu. Template is not prepared in advance, according to the results of the test device generated in real time, and then pushed to the client. Obviously this approach can be used for almost all

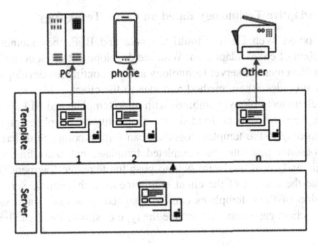

Fig. 2. Multiple sets of templates

devices can generate the corresponding template, allowing more customization. Makes small mobile pages load faster.

However, this method is not suitable for small network platform, in the process of modification and maintenance, the system needs to be changed greatly, which is time-consuming and laborious. When the server overload will bring performance problems, such as when loading a mobile user agent detector, caching mechanism to turn off many deployed on the CDN, which will lead to mobile and desktop visitors user access rate.

As the template and template generation are completed by the server, with the increase of visitors, the server load increased exponentially. At the same time, the requirements for higher back-end programming capabilities (see Fig. 3).

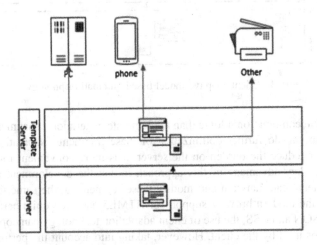

Fig. 3. Backend for Frontends

4.3   Client Adaptive Technology Based on Cloud Technology

From the Response Web Design, multi template and BFF, they cannot completely solve the problem of client adaptation. With the development of cloud technology, the continuous improvement of server technology and the continuous development of Web technology. It provides a new method and idea to the client.

Based on cloud technology, combined with multi template and BFF, in the template design, a complete template is divided into a plurality of template fragments, these fragments according to the template together, can form a complete different template. The server does not generate the completed template, but according to the set of combined logic and client hardware type to push the template fragment. This enables you to generate the content of the client requirements with minimal server resources. The combination of these templates can be designed in advance, and stored on the server side; the client can simply determine the type of simple logic can be determined (see Fig. 4).

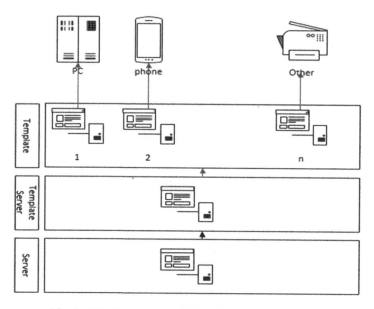

Fig. 4. Client adaptive model based on cloud technology

With simple combinational logic than the template generation consumes less server resources; also can do further optimization of these fragments, separation style and data, to further reduce the burden on the server, it is more convenient for distributed deployment. A key advantage of this approach is that the development and modification of the team can maintain and modify these segments at the same time.

Although the modern browser support for HTML5 has been quite perfect, the use of HTML5, JSON and CSS, the use of client adaptation technology can cope with most of the scenarios used by the client. However, taking into account the performance, big data, the use of the client more intelligent, reduce maintenance funds, etc. The hierarchical design integrated client and server adaptive technology, based on cloud technology, the separation of content and interface, reduce the coupling, increase flexibility, resource reuse, in order to meet the more abundant type of client to client, further more intelligent adaptive.

5   Conclusions

With the continuous development of cloud technology and Web application technology, client adaptive technology has been greatly developed. This paper proposed the client adaptive architecture based on cloud technology, based on Response Web Design, combined with multi template and BFF, based on cloud technology, put forward the template fragment combination and on-demand production principle, To solve the adaptive problem of client diversification and intelligent; saving development costs and improve development efficiency.

Acknowledgment. This paper is supported by the Natural Science Foundation of Hainan Province under Grant No. 20156222.

References

1. China Internet Information Development Report, 31 December 2016. http://www.cnnic.org.cn
2. Global and Asian Internet Penetration Rate, 25 March 2017. http://www.internetworldstats.com/stats.htm
3. Kavis, M.J. (ed.): Architecting the Cloud Design Decisions for Cloud Computing Service Models (SaaS, PaaS, and IaaS). Publishing House of Electronics Industry, Beijing (2016). (Trans. by Chen, Z., Xin, M.)
4. Erl, T., Mahmood, Z., Puttini, R. (eds.): Cloud Computing: Concepts. Technology & Architecture. China Machine Press, Beijing (2014). (Trans. by Yi, G.L., Lin, H., Chong, H.)
5. Xin, L.: Research on Web based fat client cross platform mobile application development technology. Yunnan University (2015)
6. Weihong, R.: Application of Web fat client technology. Appl. Comput. Softw. Cd-Rom **06**, 265–266 (2014)
7. Marcotte, E.: Responsive Web Design, 25 May 2010. https://alistapart.com/article/responsive-web-design
8. Newman, S. (ed.): Building Microservices. Posts & Telecom Press, Beijing (2016). (Trans. by Liqiang, C., Jun, Z.)

Resource Allocation and Energy Management Based on Particle Swarm Optimization

Gang Mei, Mingrui Chen[✉], and Xing Zhen

Hainan University, Haikou 570228, Hainan, China
1341025211@qq.com, mrchen@hainu.edu.cn

Abstract. In recent years, resource allocation and management has become a hot topic among people. Home resource's allocation and management is still a problem when home electricity power as a resource, because the resource control of the load is difficult to achieve. To solve these problems, we must implement a resource allocation and management system for home resource. Particle Swarm Optimization (PSO) algorithm is an effective approach to solve the optimal problem. In this paper, a resource allocation and energy management system (RAAEMS) is proposed to support the resource control based on real-time for home energy resource, and the objective function is solved by using PSO algorithm.

Keywords: Resource allocation and management · Home source management systems · Particle Swarm Optimization (PSO) algorithm

1 Introduction

Internet technology innovation leads to the popularity of smart home, which marks the Internet has entered into the "smart era." The traditional family resource service management model is increasingly unable to meet the needs of users', and causes a lot of waste of resources. How to make full use of idle resources, to achieve a reasonable allocation of resources, is the urgent need to solve the problem. In this context, the allocation of resources was born. With the development of the Internet, smart appliances become more and more popular in the enterprise and the family. Limited resources' reasonable allocation of a region becomes a problem. In order to achieve a reasonable allocation of home energy resources, we must use a good resource allocation algorithm, which can not only improve the calculation and transmission speed, but also reduce resource consumption, making the home energy resources be reasonably used.

Recently, there are rich literatures on resource allocation and management. [1] proposes a resource allocation method based on cloud computing. [2] proposed a dynamic resource allocation approach. [3] study resource allocation and energy management in OFDM-based cellular systems. There are many papers on energy management [4–10]. [4] design and implementation of smart home energy management systems based on ZigBee. [5] analyze a hierarchical architecture for an energy management system. Energy management of multi-carrier smart buildings in [6].

© Springer Nature Singapore Pte Ltd. 2017
G. Chen et al. (Eds.): PAAP 2017, CCIS 729, pp. 522–532, 2017.
DOI: 10.1007/978-981-10-6442-5_50

Simple scheduling algorithm for resource allocation efficiency is very low, and intelligent optimization algorithm can effectively solve this difficulty, so intelligent optimization algorithm is widely used in resource allocation. [11] proposed a genetic algorithm based on the task scheduling model. In [12], a multi-agent genetic algorithm is proposed to improve resource execution efficiency. And there are many intelligent algorithms are applied to solve energy problem. For example genetic algorithm and PSO. In [13, 14], the authors propose a PSO algorithm to reduce the operating cost. Intelligent algorithm other applications in the [15–20]. The advanced intelligent optimization algorithm is simple, efficient and convenient to control, so many scholars use it to optimal problem. All these approaches play an important role in resource allocation and energy management, but the real-time control has not been sufficiently investigated. And the real-time allocation of resources, to meet the dynamic needs of users'. We should take into account of the real-time factors. Only by considering the real-time, resource allocation and energy management can be truly achieved.

In this paper, a resource allocation and energy management system of smart home is proposed based on the particle swarm optimization control strategy and real-time control. The goal is to link the real-time and source allocation, achieve the home energy source management of smart home, reduce energy source consumption, and maximize the users' benefits. Each user has its own demand response and trade - off mechanism of the source consumption, each user maximize their own benefits according to their own trade - off mechanism, and the decision results will be reflected to the system. The resource allocation and energy management system proposed in this paper adopts real-time control model. The system optimizes the arrangement of the household load by real-time control.

This paper is organized as follows: Sect. 2 introduces the system model and the problem formulation. Section 3 presents designed PSO optimal algorithm. Section 4 presents the simulation results of study cases. Conclusions are provided in Sect. 5.

2 System Model and Problem Formulation

The proposed resource allocation and energy management system includes resource layer, control layer and service layer. In this paper, we mainly study the control layer, and the system's structure is shown in the following (Fig. 1):

The management of the resource allocation and energy management system as follows: the resource layer will count the user's power resources, and send the information to the control layer, the control layer through the particle swarm algorithm to control the load of home users to achieve load scheduling, complete resource allocation. When the control layer uses the particle swarm optimization algorithm to complete the allocation of resources, the allocation results sent to the service layer, the service layer to the user resource allocation. And the control flow is as follows (Fig. 2):

In this section, a five-agent system is used in smart home, each module represents an agent, and an edge of graph represents the communication link between two adjacent agents. The topological structure shown in Fig. 3.

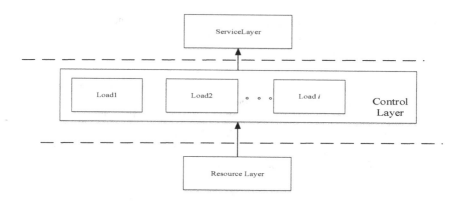

Fig. 1. Structure of the system

As shown in Fig. 1, conclude five loads. Every load will be analyzed by benefit function $C(P)$ and welfare function $U(P)$. And both are subjected to the power constraint that is defined as:

$$P^{\min} \le P \le P^{\max} \tag{1}$$

A. Loads Benefit Function

Each load user has benefit function and also has to pay for consuming energy, so the welfare function of a load user is defined as:

$$U_i(P_i[t]) = C(P_i[t]) - r_i[t] \cdot P_i[t] \tag{2}$$

where $r_j[t]$ is the real-time price of electricity.

Type 1: The fifth type of loads can't be interrupted while it is working. There are two working states-working or not working, such as washing machine, dryer, etc. We define the benefit degree is 100 when load is turned on, and it becomes zero when the load is turned off.

$$C_i = \begin{cases} 100 & working \\ 0 & notworking \end{cases} \tag{3}$$

Type 2: The third type of loads such as TV or computer, the users' benefit function is shown as:

$$C_i(P_i[t]) = b_i \cdot P_i[t] + c_i \tag{4}$$

where b_i, and c_i are the benefit coefficients, and b_i is positive. P_i represents the power consumption of TV or computer i. The type load is in a linear relationship with the load's power consumption.

Type 3: The users' benefit function is very sensitive to the power consumption of the fourth type of loads. People can easily find the optimum point of welfare,

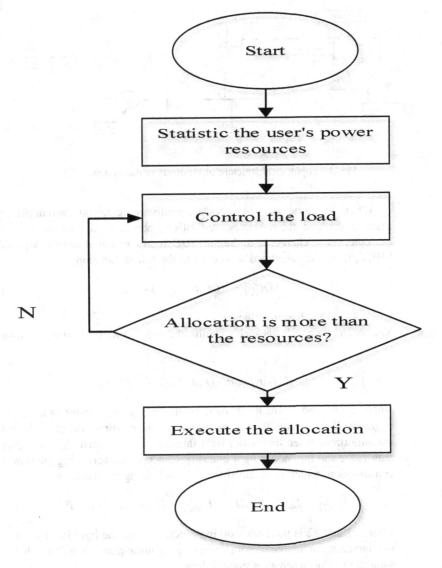

Fig. 2. Control flow of the system

such as plug-in hybrid electric vehicle (PHEV) [12]. The benefit function is defined as:

$$C_i(P_i[t]) = b_i \ln(P_i[t]) + c_i \qquad (5)$$

where b_i, and c_i are the benefit coefficients similarly, and b_i is negative. P_i also represents the power consumption of PHEV i. The plug-in hybrid electric vehicle is a intermittent load, in other words, it is intermittent when they are working.

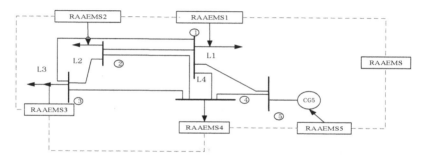

Fig. 3. Topological structure of the loads in this system

When the home user resources are relatively tight, you can choose to interrupt the load, the resources are relatively abundant and then start it, for example, let it charge at midnight. *SOC* represent the state-of-charge of PHEV, it can be calculated according to the follow function:

$$SOC[t] = SOC[t-1] + k_2 \cdot P_i[t] \tag{6}$$

where k_2 is the charging efficiency.

Type 4: The second type of loads such as bulbs, the users' benefit function is shown as:

$$C_i(P_i[t]) = a_i \cdot (P_i[t])^2 + b_i \cdot P_i[t] + c_i \tag{7}$$

where a_i, b_i, and c_i are the benefit coefficients, a_i is positive while b_i is negative. P_i represents the power consumption of bulb i. People will feel uncomfortable when the indoor light intensity is lower than L_{in}^{min} or higher than L_{in}^{max}, and the indoor light intensity I_{in} is updated according the power consumption, which is obtained from the following iteration:

$$L_{in}[t] = L_{in}[t-1] + k1 \cdot (L_{out}[t] - L_{out}[t-1]) + k2 \cdot P_i[t] \tag{8}$$

where $k1$ and $k2$ is the bulb working efficiency, but the light intensity will not increase continuously with the energy consumption, it will reach the saturation value when P_i is enough large.

Type 5: These type of loads consume energy according to indoor temperature such as air conditioners and heaters. We denote the benefit function as the temperature change of the air conditioning or heater consume power. The function is designed as:

$$C_i(P_i[t]) = a_i \cdot (P_i[t])^2 + b_i \cdot P_i[t] + c_i \tag{9}$$

where a_i, b_i, and c_i are the benefit function coefficients, a_i is positive while b_j is negative. P_i represents the power consumption of load i. People will feel uncomfortable when the indoor temperature is lower than T_{in}^{min} or higher than T_{in}^{max}, and the indoor temperature's update according the power consumption, which is obtained from the following iteration:

$$T_{in}[t] = T_{in}[t-1] + k1 \cdot (T_{out}[t] - T_{out}[t-1]) + k2 \cdot P_i[t] \qquad (10)$$

where $k1$ and $k2$ is air conditioning or heater working efficiency.

B. Social Welfare
Social welfare is the summing total benefit of all agents, and it can be defined as follow:

$$Max \sum_{i \in L} U_i \qquad (11)$$

where L are defined as the set of total load users.

3 Proposed Particle Swarm Optimization Algorithm

In this section, we will introduce the design of the PSO optimal algorithm. Home resource allocation and energy management system can control load scheduling by using a PSO algorithm.

Every user is a particle in the energy system. Each particle can find its optimal position in PSO algorithm, and the optimal position of all the particles is global optimum. First, we determine the number of particles, and set the range of control variables. We define the load 1–4 as the control variables, the load five as the state variable and their relationship is as follows:

$$P_{L,5}[t] = P[t] - (P_{L,1}[t] + P_{L,2}[t] + P_{L,3}[t] + P_{L,4}[t]). \qquad (12)$$

In the optimization process, each particle according to their own optimal location and global optimal location update their speed and location, the update process can be shown as:

$$V[t] = W \cdot V[t-1] + 2 \cdot rand \cdot (Xpbest[t] - X[t-1]) \\ + 2 \cdot rand \cdot (Xgbest[t] - X[t-1]) \qquad (13)$$

$$X[t] = X[t-1] + V[t] \qquad (14)$$

where $Xpbest[t]$ and $Xgbest[t]$ is the personal optimal position and global optimal position respectively. $V[t]$ is the current speed when $V[t-1]$ is the speed of last moment. In the same way, $X[t]$ is the current location when $X[t-1]$ is the location of last moment. W is the inertia weight factor and it is very small. The optimization process as shown in the Fig. 4:

According to the flow diagram, we can know the optimization process of the PSO is that:

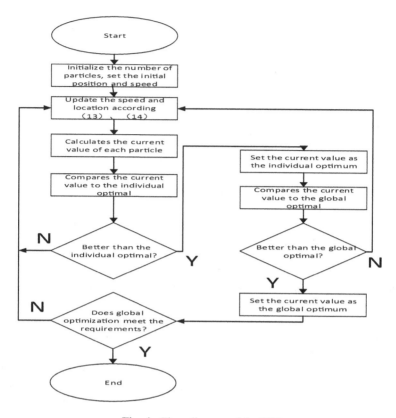

Fig. 4. Flow diagram of the PSO

First, initialize the number of particles, set the initial position and speed. Then, update the speed and location according (13), (14) and calculates the current value of each particle. Compares the current value with the individual optimal and the global optimal value subsequently, Set the current value as the individual optimum and the global optimal value if it better than their value. Otherwise, go back to (13), (14) and update the speed and location. The update process will not be end until the global optimal value meet the setting condition.

4 Simulation Studies

In this section, a 5-agent home resource allocation and energy management system shown in Fig. 1, it is presented to verify the effectiveness of the proposed PSO optimal algorithm. The parameters are summarized as:

The air conditioning's initial temperature is set to 0 °C, and the minmum temperature is 15 °C when the maximun temperature is 30 °C. Lighting has the maximum light intensity, which is 250 lx. Lighting will remain unchanged when the light intensity reaches this value. PHEV's SOC range is [0, 1]. The simulation results are show in Figs. 3, 4 and 5.

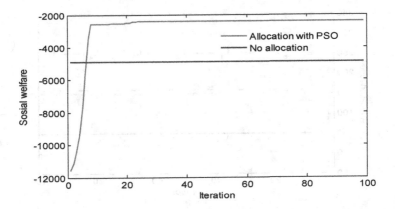

Fig. 5. Social welfare profile based on PSO

According to Fig. 4, it can be seen that the trajectory of each particle is relatively concentrated, which proves that the algorithm has certain convergence, the optimization result can be obtained very quickly through the particle swarm optimization algorithm.

Loads' benefit function has a little fluctuation as shown in Fig. 5, social welfare based on the proposed PSO algorithm is optimized. The effectiveness of proposed approach is validated further when we can see the proposed algorithm converges within a few seconds according to all simulation results.

According to Fig. 6, we can see that trajectory of each particle. We are known that every particle's trajectory is intent, so the proposed approach for this resource allocation and energy management system has a certain convergence, it can be used to solve the system's optimal problem, and the high efficiency is proven by the allocation result according to the Fig. 5.

Figure 7 shows the allocation result of each load to get the energy from the household power resource by the resource allocation and management system, as

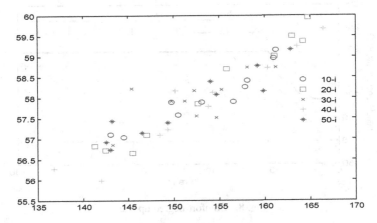

Fig. 6. Trajectory of each particle.

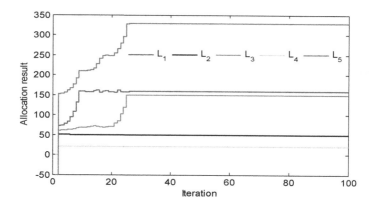

Fig. 7. Allocation result of every load.

shown in the figure, through the allocation of resources with the resource management system, in the case of limited resources, the family load can get the basic power to meet their own needs shown in Table 1. Figure 8 is the family load is the benefit function after the efficiency allocated control.

Table 1. Loads extreme value of power

Bus	Load type	P^{min}	P^{max}
1	Washing machine	65	200
2	TV/computer	35	80
3	PHEV	50	150
4	Bulb	0	50
5	Air conditioning	80	350

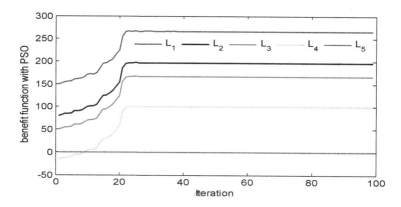

Fig. 8. Comfort degree update

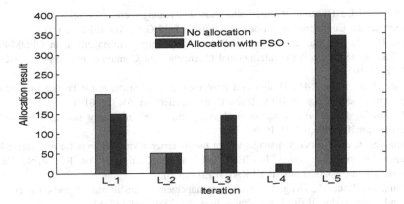

Fig. 9. Allocation result of PSO compare with no allocation

As can be seen from Fig. 9, the smart home's energy resources are extremely uneven when there is no resource allocation, and some appliances may get a high power, resulting in energy waste. And some appliances may be insufficient power supply, so they don't work. The home source distribution is basically even after allocated in smart home's resource allocation and management system, and making the limited resources to meet the needs of various appliances, improve the user's satisfaction.

5 Conclusion

This paper proposes a home's resource allocation and management system that it can reasonably allocate the limited resources of the smart home. The problem of resource allocation is a hot issue, which attracts many domestic and foreign experts and scholars to carry on the research. At present, the intelligent optimization algorithm is widely used in the resource allocation. In this paper, particle swarm optimization (PSO) is applied to resource allocation by deeply studying the characteristics of particle swarm algorithm. And then, make a comparative experiment in the MATLAB simulation platform. The system's optimal result can be got quickly by the proposed PSO algorithm. Simulation results show that the algorithm can effectively solve the energy allocation problem of intelligent households, which can reduce energy waste. The system uses the particle swarm optimization algorithm for optimal control, because the particle swarm algorithm is simple and efficient, convenient to control and so on, is very suitable to solve the convex optimization problem.

References

1. Nahir, A., Orda, A., Raz, D.: Resource allocation and management in cloud computing. In: IFIP/IEEE International Symposium on Integrated Network Management (IM). IEEE (2015)

2. Bradley, T.C.: Dynamic resource allocation and management for level 4 fusion. In: 12th International Conference on Information Fusion FUSION 2009. IEEE (2009)

3. Yu, B., Ma, P., Ma, Y.: Resource allocation and energy management in OFDM-based cellular systems. In: IEEE International Conference on Communication Systems (ICCS). IEEE (2016)

4. Han, D.-M., Lim, J.-H.: Design and implementation of smart home energy management systems based on zigbee. IEEE Trans. Consum. Electron. 56(3) (2010)

5. Piotrowski, K., et al.: A hierarchical architecture for an energy management system. (2016) doi:10.1049/cp.2016.1098

6. Arnone, D., et al.: Energy management of multi-carrier smart buildings for integrating local renewable energy systems. In: IEEE International Conference on Renewable Energy Research and Applications (ICRERA). IEEE (2016)

7. Jiang, Q., Xue, M., Geng, G.: Energy management of microgrid in grid-connected and stand-alone modes. IEEE Trans. Power Syst. 28, 3380–3389 (2013)

8. Ashraf, N., et al.: Active energy management for harvesting enabled wireless sensor networks. In: 13th Annual Conference on Wireless On-demand Network Systems and Services (WONS). IEEE (2017)

9. Singh, A., Kumar, A., Kumar, A.: Network controlled distributed energy management system for smart cities. In: 2nd International Conference on Communication Control and Intelligent Systems (CCIS). IEEE (2016)

10. Gregory, A., Majumdar, S.: Energy aware resource management for MapReduce jobs with service level agreements in cloud data centers. In: IEEE International Conference on Computer and Information Technology (CIT). IEEE (2016)

11. Jain, N., Singh, S.N., Srivastava, S.C.: A generalized approach for DG planning and viability analysis under market scenario. IEEE Trans. Ind Electron. 60(11), 5075–5085 (2013)

12. Peng, D., Qiu, H., Zhang, H., Li, H.: Research of multi-objective optimal dispatching for microgrid based on improved genetic algorithm. In: IEEE 11th International Conference on Networking, Sensing and Control, pp. 69–73, April 2014

13. Peng, X., Peng, L.: Improved particle swarm optimization based economic dispatch of microgrid. Electr. Eng. (10), 7–10 (2011)

14. Selvakumar, A.I., Thanushkodi, K.: A new particle swarm optimization solution to nonconvex economic dispatch problem. IEEE Trans. Power Syst. 22(1), 42–51 (2007)

15. Li, G., Shi, J., Qu, X.: Modeling methods for GenCo bidding strategy optimization in the liberalized electricity spot market–A state-of-the-art review. Energy 36(8), 4686–4700 (2011)

16. Gupta, R., Gajera, V., Jana, P.K.: An effective multi-objective workflow scheduling in cloud computing: a PSO based approach. In: Ninth International Conference on Contemporary Computing (IC3). IEEE (2016)

17. Jain, N.K., Nangia, U., Jain, J.: An improved PSO based on Initial selection of particles. In: IEEE International Conference on Power Electronics, Intelligent Control and Energy Systems (ICPEICES). IEEE (2016)

18. Tan, M.K., et al.: Optimization of urban traffic network signalization using genetic algorithm. In: IEEE Conference on Open Systems (ICOS). IEEE (2016)

19. Ayumi, V., et al.: Optimization of Convolutional Neural Network using Microcanonical Annealing Algorithm. arXiv preprint arXiv:1610.02306 (2016)

20. Zhang, X., et al.: Loading balance of distribution network by applying immune algorithm. In: 2016 IEEE PES Asia-Pacific Power and Energy Engineering Conference (APPEEC). IEEE (2016)

A Parallel Clustering Algorithm for Power Big Data Analysis

Xiangjun Meng[1], Liang Chen[2], and Yidong Li[3(⊠)]

[1] State Grid Shandong Power Company, Jinan, Shandong, China
[2] Shandong Luneng Software Technology, Jinan, Shandong, China
[3] School of Computer and Information Technology,
Beijing Jiaotong University, Beijing, China
ydli@bjtu.edu.cn

Abstract. With the fast development of information technology, the power data is growing at an exponentially speed. In the face of multi-dimensional and complicated power network data, the performance of the traditional clustering algorithms are not satisfied. How to effectively cope with the power network data is becoming a hot topic. This paper proposes a parallel implement of K-means clustering algorithm based on Hadoop distributed file system and Mapreduce distributed computing framework to deal this problem. The experimental results show that the performance of our proposed algorithm significantly outperforms the traditional clustering algorithm and the parallel clustering algorithm can significantly reduce the time complexity and can be applied in analyzing and mining of the power network data.

Keywords: Parallel algorithm · K-means clustering · Power data

1 Introduction

Clustering [5] is one of the most hot issues in data mining research. It is the process of partitioning data objects into subsets. Each subset is a cluster [11], so that the objects in the cluster are similar to each other, but are not similar to the objects in other clusters. A set of clusters generated by the cluster analysis is called a cluster. With the continuous development of the electric power industry and the popularization of database technology, in the electric power industry, a large amount of data [6, 9] is accumulated in different forms. Then, how to store and utilize these data effectively and how to dig out valuable information from the massive data become problems to be solved. In the face with massive data, the existing data mining algorithms have a lot of problems in time complexity and space complexity. For this problem, the solution is to apply the parallel method to the cluster, and to design a clustering algorithm for parallel implementation. Thus improving the performance of clustering algorithm to deal with massive data. To sum up, by using K-means algorithm to cluster analysis of power quality indicators of power enterprises, on the one hand, it can clear the development level of the relevant indicators of the power enterprises, it is conducive for enterprises to continuously improve their own shortcomings, on the other hand, through the cluster analysis, the indicators can not only make a comprehensive analysis and comparison of

© Springer Nature Singapore Pte Ltd. 2017
G. Chen et al. (Eds.): PAAP 2017, CCIS 729, pp. 533–540, 2017.
DOI: 10.1007/978-981-10-6442-5_51

enterprise power management, but also can accurately identify the root causes of the gap between the power supply. Hadoop is an open source cloud computing platform that can be used for parallel processing of large-scale data. It has five characteristics, namely reliability, extensible property, high efficiency, fault-tolerant and low cost. MapReduce [1, 8] is a programming model for the parallel operation of large scale data sets. It specifies a map function, to a set of key value mapping into a new set of keys and specify a reduce function to ensure the mapping of all key value pairs in each share the same set of keys. Hadoop [7] can be widely used in large data processing applications due to its own natural advantages in terms of data extraction, deformation and loading. Hadoop distributed architecture, data processing engine as much as possible near the storage of such as ETL such batch operation relatively appropriate, because such operation of batch processing results can be directly to the storage. MapReduce Hadoop function to achieve a single task break, and will be sent to a number of pieces of mission nodes, and then in the form of a single data set to load into the data warehouse. Hadoop distributed architecture makes data processing engine as much as possible near the storage of such as ETL [10] such batch operation relatively appropriate, because such operation of batch processing results can be directly to the storage. Mapreduce function in the Hadoop achieves to break a single task and send a fragment mission to multiple nodes, then load into a data warehouse in the form of a single data set. So it is the right one to deal with the massive data [5].

2 Related Work

Dundar [3] draw a conclusion that K-means based unsupervised feature learning is a powerful alternative to deep-learning algorithms as well as to conventional techniques that rely on handcrafted features. Wu [11] proposed a K-means algorithm based on Sim Hash, which is used to calculate the feature vectors extracted, and then the fingerprint of each text is obtained, to deal with high dimensional and sparse data effectively and greatly reduce the speed of K-means clustering algorithm. Bai [2] proposes a load model based on the K-means algorithm uses the actual operation data of the power network and voltage static characteristic of load is considered. So it can reflect the actual situation of the power load more clearly. In addition, the clustering algorithm is applied to the processing of load data so that the load characteristic data of each time period can be extracted and thins typical methods. Lee [4] proposes a classification method which combines k-means algorithm and Bayesian inference to build a classifier. The proposed method makes the classifier updated as new data are accumulated and in addition adjusted according the concept drift using a windowing mechanism. To handle big data, the proposed method is realized using the map-reduce paradigm which can be deployed in the big data framework. K-means is a popular clustering algorithm to find the clustering easily by iteration. But the computational complexity of the traditional k-means due to accessing the whole data in each cycle of iterative operations is too great to make it fit for very large data set. This paper presents a new clustering algorithm we have developed, fast k-means clustering algorithm based on grid data reduction (GDR-FKM), by which clustering operations can be quickly performed on very large data set. Application of the algorithm to analysis of the data relativity in TT&C has demonstrated its celerity and accuracy.

From the above researches we can see that only the traditional serial clustering algorithm is studied. But the traditional serial clustering algorithm can not deal with massive data, so we have to study parallel clustering algorithms. In this paper, according to the principle of K-means algorithm and the application of a wide range of MapReduce parallel computing, we propose a parallel implementation of K-means algorithm.

3 Problems and Algorithms

3.1 Power Big Data Mining Problem

Power "big data" concerns power generation, transmission, substation, distribution, electricity and scheduling, it is an analysis and mining of cross-unit, cross-professional and cross-business data. Power big data consists of structured data and unstructured data, with the application of smart grid construction and internet of things, unstructured data is showing a rapid growth momentum, the amount will be larger than that of structured data. The characteristics of power big data meets the five characteristics of large data: Volume, Velocity, Variety, Value and Veracity and there is a high value for the improvement of the profit and control level of electric power enterprises. Big data technology will accelerate the pace of intelligent control of power companies to promote the development of smart grid. For example, we can monitor facilities operation condition dynamically based on the sensor of electric power infrastructure. So that we can effectively change the mode of operation and maintenance. The operation and maintenance fault are eliminated from the bud stage and intelligent operation and maintenance will be achieved.

- **Mapreduce**

The Map/Reduce framework consists of a single job tracker and each cluster node a task tracker. Master is responsible for the scheduling of all tasks that constitute a job, these tasks are distributed on different slaves, master monitors their execution, reimplementation of the failed task. Slave is only responsible for performing tasks assigned by master. The process of Mapreduce is as below. (1) User program in the MapReduce library first divides input file into M blocks (Hadoop default 64M, this parameter can be determined by parameter modification). Then the processing program is executed on the cluster machine. (2) The master control program master assigns the Map task and the Reduce task to the job execution machine worker. A total of Map R tasks and Reduce M tasks need to be assigned. Master will select the free worker and assign these Map tasks or Reduce tasks to the worker node. (3) A Map task assigned by the worker to read and process the relevant input data blocks. From piece of input data parsing out the key, then the key to transfer a user-defined map function, by the map function to generate and output the intermediate key/value key set, these keys to set will be temporarily cached in memory. (4) The key/value in the cache of the partition function is divided into R regions, then periodically written to the local disk. (5) When the worker Reduce program receives the data store location information sent by the master program, it will use the RPC to read the cached data from the host's keyboard from the worker Map host. After reading all of the intermediate data in worker Reduce, the key is sorted by the same key value of the data together. Because many of the different key values are mapped to the same Reduce task, it is necessary that they should be sorted. (6) Worker Reduce traversal sort of intermediate data, for each of the only intermediate key values, worker Reduce program will send the key value and its associated intermediate value of the collection to the user defined Reduce function. The output of the Reduce function is forced to add to the output file of the partition. (7) Finally, after the successful completion of the task, the output of the MapReduce is stored in R output files.

- **Hadoop HDFS Architecture**

Compared with distributed file system architecture based on P2P model, HDFS uses a distributed file system based on Master/Slave. A HDFS cluster contains a single Master node and a number of Slave node servers. The HSDF architecture design is shown in Fig. 1 below. A single Namenode node greatly simplifies the structure of the system. Namenode is responsible for the preservation and management of all HDFS metadata, so the user data does not need to pass namenode, that is to say the file data is read and write directly on the datanode. HDFS stored files are split into fixed size blocks. When you create a block, the namenode server will assign a unique block identifier to each block. Datanode server store the block in the form of linux files on the local hard disk and read data in accordance with the specified block identifier and byte range. For reliability considerations, each block will be copied to multiple datanode servers. By default, HDFS will use three redundant backups. Of course, users can set different file namespace for different copy number. Nematode manages all file system metadata. These metadata include name space, access control information, block mapping information and current block district information. Nematode also manages the activities within the system, such as block rental management, recovery of isolated block. Nematode makes the information cycle and each datanode server communication and send command to each datanode server and receive block information in datanode.

- **K-means Clustering Algorithm**

K-means algorithm accepts the parameter K; and then the prior input of the N data object is divided into K clustering in order to make the obtained clustering to meet: The similarity of objects in the same cluster is higher, and the similarity of objects in different clusters is small. Cluster similarity calculates the mean value of the objects in each cluster to obtain a "center".

3.2   Realization of Parallel K-means Clustering

- **Design of Map Function**

Map function input is the default format of Mapreduce framework, Key is the offset of the current sample relative to the starting point of the input data file. Value is a string

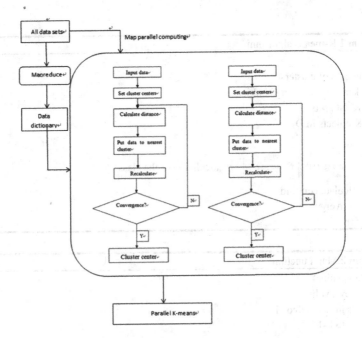

Fig. 1. Parallel realization of K-means algorithm

consisting of the values of each dimension of the current sample. The value of each dimension of the current sample is analyzed from value, then calculate the distance between the center and the K, find the nearest cluster index. The final output is. The parallel realization is shown in Fig. 1 together with pseudo code as follow.

- **Design of Combine Function**

First of all, we parse out the coordinate values of each sample from the list of the string in order. And the corresponding coordinate values of each dimension are added separately. At the same time we record the total number of samples in the list. Pseudo code is shown as follow.

- **Design of Reduce Function**

In the Reduce function, the number of samples processed from each Combine and the coordinates of each dimension of the corresponding nodes are first resolved. Then the corresponding value of each dimension is added separately, and then divided by the total number of samples. That is to get a new center point coordinates. Pseudo code is shown as follow, and the parallel realization of K-means algorithm is shown as follow.

Algorithm 1 K-means algorithm

Input:
  K: Numbers of clusters
  D: A data set containing N objects
Output: K data sets
(1) **For** K objects in D
(2) **Do**

(3) $C^i - \arg\min \left\| x^i \quad u_j \right\|^2$ //Calculation of the class

(4)   Recalculate centroid
(5) Until convergence
(6) **End for**

Algorithm 2 Map Function

Input: <key, value>
Output: <key', value'>
(1) Index is initialized to -1;
(2) **For** i=0 **to** k-1
(3)   **Do**
(4)   dis=instance;
(5)   **if** dis < minDis
(6)   minDis = dis; index = i;
(7)   **End if**
(8) **End for**
(9) Index is regarded as key';
(10) value=value';
(11) **Output** <key', value'>;

Algorithm 3 Combine Function

Input: <key, value>
Output: <key', value'>
(1) Initialize an array;
(2) Initialize variable num;
(3) **While** (V.hasNext())
(4) **Do**
(5)   values in V ;
(6)   add values to array.;
(7)   num++;
(8)   key=key';
(9) We constructs a string that contains information about the num and the various components of the array, which is used as a value';
(10) **Output** <key', value'>;

Algorithm 4 Reduce Function

Input: <key, value>
Output: <key', value'>
(1) Initialize an array;
(2) Initialize V ;
(3) **While** (V.has.Next())
(4) **Do**
(5)   values in V ;
(6)   add values to array.;
(7)   num+=num;
(8)   obtain new center.;
(9)   key=key';
(10) Construct a string that contains information about the value of each dimension of the new center point, which is used as a value.;
(11) **Output** <key', value'>;

---

4   Performance Analysis

4.1   Analysis of Time Complexity

We set the amount of the data set is N, the subset size is SN, the number of the clustering is K, the dimension is D, there are P datanodes. The time consists of communication time and computing time. In an iteration, the computing of namenode consists of: (1) The time of dividing clusters:$P * SN$. (2) The time of receiving clustering result and saving it:SN. (3) The time of dividing remaining data sets:$N - P * SN$. Datanode is mainly responsible for calculating the distance between every data object and clustering center, the time cost is $SN * K * D$. The total time is $O(SN * K * D * Rp + 2 N * Rp)$, Rp is the iteration times of parallel algorithm and $2 N * Rp$ is the communication time.

4.2   Analysis of Space Time Complexity

Given a data set of N, the space that namenode needs is $O(N)$, the space each datanode needs is $O(SN)$, so the total space time complexity of the parallel algorithm is $O(N)$.

4.3   Accelerate Rate Analysis

Let Rs be the iteration times of the serial K-means algorithm, we can calculate the rate is

$$Sp = \frac{O(Rs * K * N * D)}{O(Rp * K * SN * D + 2N * Rp)} = O\left(\frac{1}{\frac{SN}{N} + \frac{2}{K*D}}\right) \quad (1)$$

Because SN is less than N/P, that is to say SN/N is larger than 1/P, so Sp is larger than 1 and is less than P, we can see that the parallel algorithm definitely saves time.

5   Conclusion

This paper mainly introduces the idea of clustering algorithm and parallel computing. The innovation of this paper is proposing the idea of parallel clustering algorithm based on the traditional clustering algorithm, which can effectively deal with the bottleneck of data mining under the current big data environment. However, in this paper, we only propose a framework based on previous studies, we do not realize it, and this is what we will do in the future work.

Acknowledgment. This work is supported by National Science Foundation of China Grant #61672088, Fundamental Research Funds for the Central Universities #2016JBM022 and #2015ZBJ007, Open Research Funds of Guangdong Key Laboratory of Popular High Performance Computers. The corresponding author is Yidong Li.

References

1. Aragues, R., Sander, C., Oliva, B.: Predicting cancer involvement of genes from heterogeneous data. BMC Bioinform. **9**(1), 1–18 (2008)
2. Bai, Z.G., Zhang, H.D.: k-means clustgering algorithm based on mutation. J. Anhui Univ. Technol. **4**, 019 (2008)
3. Dundar, M., Kou, Q., Zhang, B., He, Y.: Simplicity of kmeans versus deepness of deep learning: a case of unsupervised feature learning with limited data. In: IEEE International Conference on Machine Learning Applications (2015)
4. Lee, K.M.: Grid-based single pass classification for mixed big data. Adv. Nat. Appl. Sci. **9** (21), 8737–8746 (2014)
5. Monmarch, N., Slimane, M., Venturini, G.: AntClass: discovery of clusters in numeric data by an hybridization of an ant colony with the Kmeans algorithm (2003)
6. Naimi, A.I., Westreich, D.J.: Big data: a revolution that will transform how we live, work, and think. Information **17**(1), 181–183 (2014)
7. Shvachko, K., Kuang, H., Radia, S., Chansler, R.: The hadoop distributed file system. In: IEEE Symposium on MASS Storage Systems and Technologies, pp. 1–10 (2010)
8. Triguero, I., Peralta, D., Bacardit, J., García, S., Herrera, F.: MRPR: a MapReduce solution for prototype reduction in big data classification. Neurocomputing **150**(150), 331–345 (2015)
9. Varian, H.R.: Big data: new tricks for econometrics. J. Econ. Perspect. **28**(2), 3–28 (2014)
10. Vassiliadis, P., Simitsis, A., Skiadopoulos, S.: Conceptual modeling for ETL processes. In: ACM International Workshop on Data Warehousing and Olap, pp. 14–21 (2002)
11. Wu, G., Lin, H., Fu, E., Wang, L.: An improved k-means algorithm for document clustering. In: International Conference on Computer Science and Mechanical Automation, pp. 65–69 (2015)

Customized Filesystem with Dynamic Stripe Strategies on Lustre-Based Hadoop

Hongbo Li[1,3(✉)], Yuxuan Xing[2], Nong Xiao[1,2], Zhiguang Chen[1,2], and Yutong Lu[1]

[1] School of Data and Computer Science,
Sun Yat-sen University, Guangzhou, China
lihb2113@outlook.com
[2] State Key Laboratory of High Performance Computing,
National University of Defense Technology, Changsha, China
[3] College of Computer Science and Technology,
Jilin University, Changchun, China

Abstract. With large-scale data exploding so quickly that the traditional big data processing framework Hadoop has met its bottleneck on data storing layer. Running Hadoop on modern HPC clusters has attracted much attention due to its unique data processing and analyzing capabilities. Lustre file system is a promising parallel storage file system occupied HPC file system market for many years. Thus, Lustre-based Hadoop platform will pose many new opportunities and challenges on today's data era. In this paper, we customized our *LustreFileSystem* class which inherits from *FileSystem* class (inner Hadoop source code) to build our Lustre-based Hadoop. And to make full use of the high-performance in Lustre file system, we propose a novel dynamic stripe strategy to optimize stripe size during writing data to Lustre file system. Our results indicate that, we can improve the performance obviously in throughput (*mb/sec*) about 3x in writing and 11x in reading, and average IO rate (*mb/sec*) at least 3 times at the same time when compared with initial Hadoop. Besides, our dynamic stripe strategy can smooth the reading operation and give a slight improvement on writing procedure when compared with existing Lustre-based Hadoop.

Keywords: Hadoop · High performance computing · Lustre · Stripe strategy · FileSystem

1 Introduction

Recently, due to the prevalence of Internet, the data grows rapidly. According to IDC reports in [1], the global data generated will exceed 40ZB in 2020, while 33% of the data will contain valuable information. With the increase of the amount of data, the traditional HDFS which used in typical large-scale data processing framework called Hadoop will be a bottleneck in storage. And more and more cases of high performance computing integrated with big data analysis are explored in academic. And the different implementations of MapReduce to solve some data analysis problems become more popular. While at the aspect of back-end storage system, Lustre file system has been

© Springer Nature Singapore Pte Ltd. 2017
G. Chen et al. (Eds.): PAAP 2017, CCIS 729, pp. 541–553, 2017.
DOI: 10.1007/978-981-10-6442-5_52

wildly used in high-performance parallel file systems. According to IDC statistics in May 2014 show that 50% of high-performance computing storage use the Lustre file system. A better performance of data processing can be converted into a growing commercial value. So how to deal with the exponential growth of data in order to make better business decisions is the problem we are considered. Well, through the integration of MapReduce programming model and Lustre file system, we can play both the advantages of better services for large data processing and high-performance computing with no doubt.

The structure of this paper is organized as follows. Section 2 presents background for this paper. And in Sect. 3, we introduce the existing research on Hadoop to improve performance at the aspect of computing and storage. Section 4 presents the implement of our Hadoop on customized Lustre file system and put forward our dynamic stripe strategy. In Sect. 5, we compare and analyze the performance of the proposed Dynamic Stripe Strategy from many aspects. Finally, we conclude the paper with future work in Sect. 6.

2 Background Knowledge

2.1 Hadoop and Lustre File System

Hadoop [2] is an open source large data processing platform originally developed by Googles' GFS [3] and MapReduce [4] which can deal with large-scale data analysis. From the functional point of view, Hadoop is mainly composed of two parts: HDFS serves as a back-end distributed storage system which mainly used for the storage of massive data, and MapReduce acts as a computational framework for processing data. In the initial Hadoop framework, procedures like the output of map and the input of reduce are interact with HDFS. As for the files generated during the other procedures like shuffle, partition which were some temporary files from Linux local file system. In the source code of Hadoop, the developers realized HDFS file system which is composed of a NameNode and a lot of DataNode in the model of Master/Slave. NameNode of HDFS is responsible for storing the file metadata like file path, file permission and so on. And DataNode is the real place for users' contents of files. From the typical composition of Hadoop 1.x (JobTracker/TaskTracker Model) to Hadoop 2.x, YARN [5] framework which separated node and resource in JobTracker, Hadoop has become a highly modular component.

Lustre [6] file system was launched by U.S Department of Energy Office of Science and National Nuclear Security Administration Laboratory nearly ten years ago. Throughout the last decade, it was deployed over numerous medium to large-scale supercomputing platforms and clusters to achieve the expectations of the Lustre user community. According to the latest Top500 list, half of the world's top 30 supercomputers use Lustre as the storage system. Due to the excellent scalability of Lustre file system in architecture, it can be widely deployed in scientific computing, oil and gas, manufacturing, rich media and other fields.

Lustre is composed of three components: Metadata servers (MDSs), object storage servers (OSSs), and clients. All users' data can be divided into some stripes stored in the OSSs. The total data capacity of the Lustre file system is the sum of all individual OST's capacities. Lustre clients can access the data concurrently through the standard POSIX I/O system calls. Files on the Lustre file system can be striped. So this means that the overload can be balanced on all OSTs and we can speed up the procedure of reading concurrently.

2.2 HDFS vs Lustre File System

The traditional MapReduce on HDFS framework faces many challenges. First of all, Hadoop cannot make the task achieve absolute data localization which breaks the Hadoop's principle of "move computing is less expensive than move data". So when some tasks need great input it will cause a huge network overhead of I/O. Second, with less scalability, HDFS is hard to face with upcoming massive data era. Other faults like slower network during shuffle, don't support POSIX will become its bottleneck to develop doubtless. However, Lustre, a kind of parallel file system, has the same metadata server, but its high performance characteristics (scalable, parallel, high speed and so on) can solve the problems caused by HDFS. So our motivation is to design a new Hadoop framework using Lustre as the back-end storage system and to design a dynamic stripe strategy to optimize the performance of Lustre-based Hadoop.

3 Related Work

There are many academic researches to improve the performance of Hadoop as a big-data processing framework which can be generally divided into two categories. The one is to advance the computing abilities to some kind problems from the aspect of computing performance, the other is to enhance the storing performance for serving massive data.

In the period of reduce operation, shuffle and reduce phases are coupling together. And it is not conducive to parallel the map and reduce procedure in multi-tenant scenarios. Guo in [7] decoupled the shuffle and reduce procedure to improve the performance and flexibility of Hadoop. In resource scheduling of Hadoop, traditional FIFO scheduling strategy cannot meet the demands for Hadoop facing variety jobs. Thangaselvi in [8] put forward a scheduling algorithm called SAMR, and Pastorelli in [9] put forward HFSP scheduling algorithm and so on. All of them can perform well in processing Hadoop task scheduling and thus improve the data processing performance. In view of a single kind task, Fujishima in [10] considered the TeraSort job, and tried to optimize the I/O performance during reduce phase through the bitmap to record the TeraSort intermediate data information.

In addition to the above methods, many scholars took the back-end storage system for consideration by contrast the HDFS with other parallel file system wildly used in HPC. For example, Mikami in [11] used Gfarm, a kind of distributed file system to replace the underlying HDFS to combine with MapReduce. And Tantisiriroj in [12] integrated PVFS with HDFS, Maltzahn in [13] used Ceph file system as the back-end storage system as well as Gupta used GPFS in [14] and so on. All this research create numeric application scenes richer and richer, and make the Hadoop ecosystem grad-ually extend to high performance field closer and closer.

For the improvement on data processing, Hadoop has its own limitation for large-scale storage eventually. Except that, many research are just aimed at a certain kind of problems without commonality. A good processing ability cannot do without a strong and effective storing ability. As for PVFS, Gfarm, GPFS and Ceph integrated with MapReduce can also play a very good high performance characteristics. But considered the quota in today high performance file system market, Lustre as a parallel file system becomes more promising in storing data when associated with big data framework like MapReduce. Research on how Hadoop running on Lustre file system has been studied. Intel [15] had explored running Hadoop on Lustre. Their work is to use a kind of transformer to transform data in HDFS to data in Lustre. And also introduce an idea adding block location information to do some optimizations during task allocating. Intel leads our way to use Lustre replace with HDFS in Hadoop, but not to do some further exploration because it actual uses the HDFS on Lustre. Wasi-ur-Rahman in [16–19] used the high-speed network for data communication, and they extended the RDMA ability by realizing the JNI interface on HDFS to reduce network delay. Also they proposed a new architecture of HOMR in [20] which put the temporary data generated during MapReduce on Lustre file system and optimized processing ability of MapReduce on HPC clusters. There is no doubt that RDMA can reach the low latency on data transmission in network greatly. But on the other hand, in fact, the cost on both software and hardware are so expensive that seems do not comply with the original intention of Hadoop running on cheap equipment. In the current study, it seems that there is no research to explore the *FileSystem* class introduced in Hadoop source code. This paper also explores the method to enhance Hadoop from the back-end storage system. In our work, we customize the *LustreFileSystem* class which inherits from *FileSystem* class eventually and optimize the concurrent write and read ability through realizing the dynamic stripe when stored the data on Lustre file system. And study in this article can be also used in all current Hadoop system which is based on Lustre file system.

4   Proposed Method

4.1   Customize the Implementation of *LustreFileSystem* Class

To make full use of Lustre performance during executing the Hadoop jobs, our pro-posed approach makes MapReduce interact only with Lustre file system.

Hadoop defines an abstract file system class called *FileSystem*. Any ordinary file system as long as completes the inheritance of this class, it can replace the HDFS to serve as the back-end storage system of Hadoop. Lustre is transparent to the users, operating on the Lustre file system is as convenient as the user in the local operation of the files. So, we inherit the *RawLocalFileSystem* class which inherits from *FileSystem* class eventually and overwrite the functions involved with file operations (See Table 1). In order to make the Hadoop adapt to the configuration of the customized *LustreFileSystem* class, we list the main attributes and their corresponding values needed to configure Hadoop working environment in Table 2. In the new YARN framework, we only need to execute *start-yarn.sh* script to start the MapReduce service, and the back-end storage system will be the Lustre file system.

Table 1. Mainly overwritten methods in *LustreFileSystem*.

Function	Description
LustreFileSystem	The constructor function
setConf	Initial the configuration according to xml files
moveFromLocalFile	Move local files to Lustre file system
copyFromLocalFile	Copy local files to Lustre file system
rename	Change the file name

Table 2. Mainly configuration items for our *LustreFileSystem*.

File	Property	Value
core-site.xml	fs.lustrefs.impl	org.apache.hadoop.fs.Lustrefs.LustreFileSystem
	fs.defaultFS	lustrefs:///
	fs.lustrefs.mount	/mnt/lustre
	hadoop.tmp.dir	/mnt/lustre/xxx

4.2 Stripe the File/Directory

In Lustre file system, we can configure files/directories stripe size or count on the OSTs when they are creating. And during the procedure of MapReduce, the input of jobs should be uploaded to Lustre file system, and numeric temporary files will be created on the Lustre file system. It's difficult for users to configure the stripe flexibly. So we research the stripe strategy in the rest of the paper.

Some operations like *rename*, *copyFromLocal* and *moveFromLocal* will create a new file in Lustre file system, so our goal is to execute the *setstripe* command during these periods. If we let each file evenly distribute among multiple OSTs as much as possible, we can read the data concurrently in the next time. Fortunately Lustre file system runs on the Linux system, and we can invoke the shell scripts in java JVM easily.

4.3 Dynamic Stripe Strategy/Algorithm

Our dynamic stripe algorithm presented in Fig. 1 includes four steps. And the core idea is how to get the optimized size of files striped in each OST (i.e. *getOptimizedSize* function). As the Lustre Manual in [21] displayed, files stored in Lustre file system must be striped at the size of times of 64 KB. And the size of each temporary files is variable, so we should do some approximate estimations about the optimized stripe size.

Algorithm for adjustStripe

Function: adjust stripe size according pre-size
Input: the path of file, the pre-size
Output: NULL

For Each file wants to create in lustre
do
   | STEP1: coreNum = the number of OST
   | STEP2: stripeSize = getOptimizeSize()
   | STEP3: set the stripe size on lustre fs
   | STEP4: create the file
done

Fig. 1. Dynamic stripe algorithm (funciton *adjustStripe*)

Suppose the variable *size* (KB) represents as the size of the file, and the number of OST is n. In order to achieve load balancing in each OST, we assign a file to these OST fairly. So in theory the amount of information stored on each OST is Eq. (1):

$$avgsize = \frac{size}{n}(modulo) \tag{1}$$

We set the size of stripe is k times of 64 KB, then the big file will be split at the size of *64 * k* (KB) and stored in each OST, and then the remaining part of the file will be stored in next round OST nodes. And we get Eq. (2):

$$stripesize = 64 * k \tag{2}$$

In Eq. (2), variable *stripesize* means the size of stripe, we plot this relation in Fig. 2. And our effective information is the splattering showed in Fig. 2.

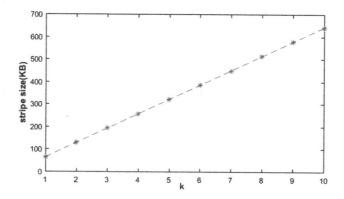

Fig. 2. Relation between OST and stripe size.

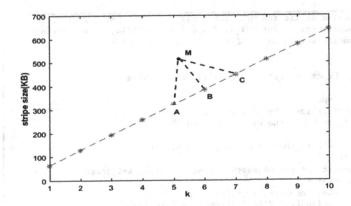

Fig. 3. Multi choices for one kind of file size.

Like the situation shows in Fig. 3 where a file mapped in the Fig. 3 called *M*, and the size of *M* is not exactly the times of 64 KB, so you have different stripe choices to select and you should decide which the suitable stripe size is in *A*, *B* or *C*. Obviously you can choice 64 KB always as the stripe size at any time and to distribute the file evenly. However, it will arise a huge network overhead since you should access the MDS for many times. And this way will increase the pressure of metadata server at the same time.

For Eq. (1), we choice the actual number of OST equal *n* in fact to balance the pressure on OSTs. Suppose we increase the stripe size to *m − 1* times of 64 KB when *m*64* is about going to bigger than *avgsize (the size that should be placed on each OSTs in theory)*. So we get Eq. (3) which represents the number of OST prepared for the remaining data after a round assignment if we choice *(m − 1)*64* as the stripe size and Eq. (4) which represents the number of OST left after store all data if we choice *m*64* as the stripe size correspondingly.

$$y1 = \frac{size}{64 * (m - 1)} - n \tag{3}$$

$$y2 = n - \frac{size}{64 * m} \tag{4}$$

To make full use of our OST and also try to distribute the data evenly on all OSTs. So we regard Eq. (5) as our decision function. And the value of *y* decides our value of *m*. (Something should be informed that the max stripe size currently should not bigger than 4 GB as announced in [21].) In fact, when the *m* is much bigger than 1, and the Eq. (3) is the same with Eq. (4) which means if the file is bigger enough, we can optimize our stripe size by our equation.

$$y = \min(y1, y2) \tag{5}$$

```
public static int MIN_SIZE = 512;   // the best default size
public static int MAX_SIZE = 4 * 1024 * 1024; // max stripe size
public static int TIME = 64; // stripe factor

/*
 * Function: get the optimized stripe size
 *
 * input:
 * fileSize: the size of file
 * coreNum : the number of used OST
 *
 * output: the optimized stripe size(KB)
 */
public int getOptimizeSize(int fileSize, int coreNum){
    int k = 1;
    int result = MIN_SIZE;
    while (k * TIME * coreNum <= fileSize){
        k++;
    }
    k = (1 < k) ? k - 1 : 1;
    k = findK(k);
    if(k * TIME > MIN_SIZE){
        result = k * TIME;
    }
    if(k * TIME > MAX_SIZE){
        result = MAX_SIZE;
    }
    return result;
}
```

Fig. 4. Dynamic stripe strategy to get the optimized stripe size.

So our dynamic stripe strategy relies on the Eq. (5), and Fig. 4 shows our algorithm to get the optimized stripe size.

5 Performance Evaluation

5.1 Experiment Environment

As for our experiments, our clusters consist of 9 nodes of Lustre cluster and the Hadoop cluster consists of 4 nodes. Hardware environment in each node is shown in Table 3. It works well when we pack our code into the Hadoop corresponding directory as the form of plugin.

5.2 Performance Comparison and Analysis

In contrast with the performance of initial Hadoop and other Lustre-based Hadoop, we use the benchmark tools called *TestDFSIO* inner Hadoop to test. We do our benchmark mainly from two aspects: Throughput and Average IO rate when read/write the data from back-end storage system through executing the Hadoop jobs.

Figures in 5, 6 and 8, 9 are our results from comparing Hadoop on HDFS (*HDFS*) with Hadoop on our customized Lustre file system (*Proposed*) and Lustre-based

Table 3. Hardware environment of each node.

Item	Performance
OS	CentOS Linux release 7.2.1511 (Core)
CPU	Intel(R) Xeon(R) CPU E5-2692 v2 @ 2.20 GHz
SSD	Intel DC P3500 (1600.3 GB)
Bandwidth	13.0 Gbits/sec

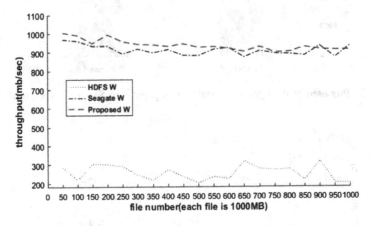

Fig. 5. Throughput comparison in writing operation.

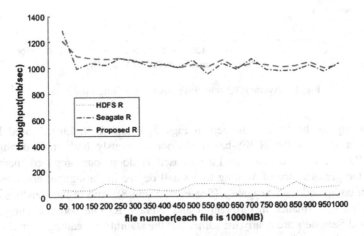

Fig. 6. Throughput comparison in reading operation.

Hadoop introduced by Seagate Inc. (*Seagate*). And in the figures, the abscissa is the number of processed files (the number of files from 50 at the speed of 50 for each step to 1000), where the size of each file is 1000 MB. The ordinate in the figures are throughput and average IO rate of reading and writing correspondingly.

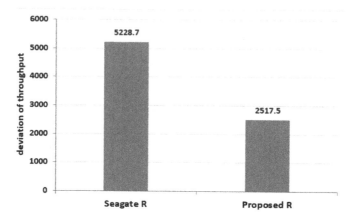

Fig. 7. Proposed method gets more stable throughput than Seagate's when reading.

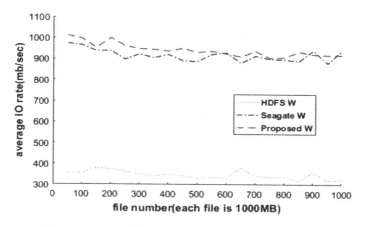

Fig. 8. Average IO rate comparison in writing operation.

According to the figures showed in Figs. 5, 6 and 8, 9, Lustre-based Hadoop performs better than the HDFS-based Hadoop obviously both in throughput and average IO rate of read/write. In Lustre-based Hadoop, our proposed method can improve the performance of Writing to a small degree in throughput and average IO rate when compared with Seagate's (See Figs. 5 and 8). And in the Reading layer of Hadoop, the performance in throughput and average IO rate between our proposed method and Seagate's are nearly the same. But the stability of reading operation in our proposed method is obviously higher than Seagate's (See Figs. 6, 7 and 9, 10).

The main reasons why Lustre-based Hadoop performs superior then the initial one are that: The first factor is when you read data from file system, you can parallelize the read procedure and this key point behave well facing with the large-scale data. And then the protocol of our network is high-speed network protocol which can speed up the transportation of data in network heavily. Compared with our proposed method

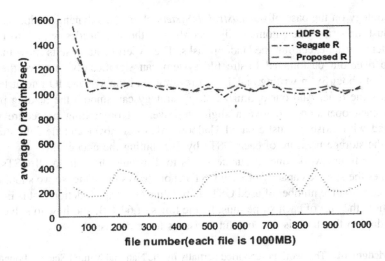

Fig. 9. Average IO rate comparison in reading operation.

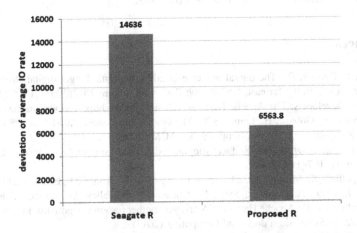

Fig. 10. Proposed method gets more stable average IO rate than Seagate's when reading.

with existing Seagate's Lustre-based Hadoop, our dynamic stripe strategy can write the data concurrently. And with the data distributed to each OST evenly, the read procedure in our proposed method become much smoother than Seagate's. So our Lustre-based Hadoop with dynamic stripe strategy can behave better when facing with the various Hadoop jobs.

6 Conclusion and Future Work

In this paper, we customized our Lustre file system to act as the back-end storage system for Hadoop through inheriting the Abstract *FileSystem* class and overwrite some function associated with file/dir operations. And we put forward our dynamic stripe strategy on the base of our *LustreFileSystem* class. This dynamic stripe strategy can adjust the stripe size automatically according to the size of files written to Lustre file system during executing the Hadoop tasks. The experimental results show that the Hadoop over our customized Lustre file system can significantly improve the data throughput about 3x in writing and 11x in reading, and the average IO rate at least 3 times at same time. And our optimized stripe strategy can smooth the reading proce-dure at each operation and give a slight improvement on writing procedure when compared with existing Lustre-based Hadoop. Also our novel and flexible strategy reduce the storage pressure of each OSTs by distributing the data evenly.

For our future work, due to our design is to distribute the data to all OSTs, and sometimes the storage capacity of each OST is not always the same, so we should also need to consider the number of used OSTs to redefine our stripe function that is Eq. (5). In addition, the size of each stripe must be the times of 64 KB, and how to solve or to better adapt to this limit is also our future issues to be discussed.

Acknowledgment. This work is supported partially by the National Natural Science Foundation of China under Grant nos. U1611261, 61433019, 61402503, and the National Key Research and Development Program under Grant no 2016YFB1000302.

References

1. Gantz, J., Reinsel, D.: The digital universe in 2020: big data, bigger digital shadows, and biggest growth in the far east. IDC iView: IDC Anal. Future **2012**(2007), 1–16 (2012)
2. Hadoop Developers: The Apache Hadoop web site (2005). http://hadoop.apache.org
3. Ghemawat, S., Gobioff, H., Leung, S.T.: The Google file system. ACM SIGOPS Operating Systems Review, vol. 37, no. 5, pp. 29–43. ACM (2003)
4. Dean, J., Ghemawat, S.: MapReduce: simplified data processing on large clusters. Commun. ACM **51**(1), 107–113 (2008)
5. Vavilapalli, V.K., Murthy, A.C., Douglas, C., Agarwal, S., Konar, M., Evans, R., Graves, T., Lowe, J., Shah, H., Seth, S., Saha, B., Curino, C., O'Malley, O., Radia, S., Reed, B., Baldeschwieler, E.: Apache Hadoop YARN: yet another resource negotiator. In: Proceedings of the ACM Symposium on Cloud Computing (2013)
6. Braam, P.J., Zahir, R.: Lustre: a scalable, high performance file system. Cluster File Systems, Inc. (2002)
7. Guo, Y., Rao, J., Cheng, D., et al.: iShuffle: improving Hadoop performance with shuffle-on-write. IEEE Trans. Parallel Distrib. Syst. **28**, 1649–1662 (2016)
8. Thangaselvi, R., Ananthbabu, S., Jagadeesh, S., et al.: Improving the efficiency of MapReduce scheduling algorithm in Hadoop. In: 2015 International Conference on Applied and Theoretical Computing and Communication Technology (iCATccT), pp. 63–68. IEEE (2015)
9. Pastorelli, M., Carra, D., Dell'Amico, M., et al.: HFSP: bringing size-based scheduling to Hadoop. IEEE Transactions on Cloud Computing (2015)
10. Fujishima, E., Yamaguchi, S.: Improving the I/O performance in the reduce phase of Hadoop. In: 2015 Third International Symposium on Computing and Networking (CANDAR), pp. 82–88. IEEE (2015)

11. Mikami, S., Ohta, K., Tatebe, O.: Using the Gfarm file system as a POSIX compatible storage platform for Hadoop MapReduce applications. In: Proceedings of the 2011 IEEE/ACM 12th International Conference on Grid Computing. IEEE Computer Society, pp. 181–189 (2011)
12. Tantisiriroj, W., Patil, S., Gibson, G., Son, S.W., Lang, S., Ross, R.: On the duality of data-intensive file system design: reconciling HDFS and PVFS. In: Proceedings of the International Conference for High Performance Computing, Networking, Storage and Analysis (2011)
13. Maltzahn, C., Molina-Estolano, E., Khurana, A., Nelson, A.J., Brandt, S.A., Weil, S.: Ceph as a scalable alternative to the Hadoop distributed file system. In: USENIX (2010)
14. Gupta, K., Jain, R., Pucha, H., Sarkar, P., Subhraveti, D.: Scaling highly-parallel data-intensive supercomputing applications on a parallel clustered filesystem. In: Proceedings of International Conference for High Performance Computing, Networking, Storage and Analysis (SC) (2010)
15. Castain, R.H., Kulkarni, O.: MapReduce and lustre: running Hadoop in a high performance computing environment. https://intel.activeevents.com/sf13/connect/sessionDetail.ww?SESSIONID=1141
16. Islam, N.S., Rahman, M. W., Jose, J., Rajachandrasekar, R., Wang, H., Subramoni, H., Murthy, C., Panda, D.K.: High performance RDMA-based design of HDFS over InfiniBand. In: Proceedings of the International Conference for High Performance Computing, Networking, Storage and Analysis (SC) 2012
17. Wasi-ur-Rahman, M., Islam, N.S., Lu, X., et al.: A comprehensive study of MapReduce over lustre for intermediate data placement and shuffle strategies on HPC clusters. IEEE Trans. Parallel Distrib. Syst. **28**(3), 633–646 (2017)
18. OSU NBC Lab: High-Performance Big Data (HiBD). http://hibd.cse.ohio-state.edu
19. Wasi-ur-Rahman, M., Lu, X., Islam, N.S., et al.: High-performance design of YARN MapReduce on modern HPC clusters with lustre and RDMA. In: 2015 IEEE International Parallel and Distributed Processing Symposium (IPDPS), pp. 291–300. IEEE (2015)
20. Rahman M.W., Lu, X., Islam, N.S., et al.: HOMR: a hybrid approach to exploit maximum overlapping in MapReduce over high performance interconnects. In: Proceedings of the 28th ACM International Conference on Supercomputing, pp. 33–42. ACM (2014)
21. Oracle, S.: Lustre Software Release 2. x Operation Manual (2011)

Research on Optimized Pre-copy Algorithm of Live Container Migration in Cloud Environment

Huqing Nie[1](✉), Peng Li[1,2], He Xu[1,2], Lu Dong[1], Jinquan Song[1], and Ruchuan Wang[1,2]

[1] School of Computer Science and Technology,
Nanjing University of Posts and Telecommunications, Nanjing 210003, China
`tangziliu@qq.com, lipeng@njupt.edu.cn`
[2] Jiangsu High Technology Research Key Laboratory for Wireless Sensor
Networks, Nanjing 210003, Jiangsu Province, China

Abstract. Some load imbalance problems emerge under the distributed cloud computing platform based on container. And it is necessary to transfer the container in the higher load server to another relatively idle servers. The traditional pre-copy algorithm ignores the characteristics of the memory pages, which causes memory pages with high re-modified rate copying many times in the iterative copy phase. Therefore, this paper proposes an optimized pre-copy algorithm (OPCA) which introduces a Gray-Markov prediction model. Memory pages are added with high re-modified rate into the Hot Workspace by the prediction model, and these memory pages are involved in the stop-copy. The experiments show that OPCA reduces the number of iterations and downtime. Moreover, it also improves the utilization of resources and enhances the user's experience.

Keywords: Cloud computing · Docker container · Live migration · Pre-copy

1 Introduction

With the continuous development of cloud computing, the container is becoming more and more popular. Under the characteristics of micro service, container migration is considered as the key to solve these problems. Generally, the migration of containers can be divided into two types: static and live. In the process of static migration, the container to be moved is immediately shut down, and then a new container is opened in the destination server. The reality requires us to adopt the way of live migration. The basic principle of this approach is that the user can hardly feel any migration process, that is, the migration time can be ignored. The process is transparent to the user. In this paper, pre-copy of live migration is used to optimize the migration.

2 Related Works

The migration of the container is divided into two main directions: one is the determination of the hot container, while the other is the specific migration process. There are some common copy algorithms, such as pre-copy, delayed-copy, Hybrid-copy, pure

© Springer Nature Singapore Pte Ltd. 2017
G. Chen et al. (Eds.): PAAP 2017, CCIS 729, pp. 554–565, 2017.
DOI: 10.1007/978-981-10-6442-5_53

stop-and-copy [1], and pure demand-migration. The migration of containers is mainly including the migration of memory pages. Pre-copy algorithm is to copy the memory page firstly, and then to copy the dirty pages iteratively. On the contrary, delayed-copy firstly turns off the service and copies the state information of the container to the target server. Then, the container is restarted on the target server, and memory pages are transmitted through the network. All of these copies algorithms require some time to prepare necessary resources. And, the performance of delayed copy is mainly dependent on the workload and network link speed. Based on the short life cycle of container, this paper uses the method of pre-copy to optimize the migration.

The paper [1] analyzes the application scenarios of each method by comparing the pre-copy, delayed copy and Hybrid copy. And the pre-copy is the main mode of migration. In the paper [3], the optimization of pre-copy is divided into three stages. To reduce the memory to migrate and the number of memory transfers in the same memory area. At last, it provides a memory compression algorithm. However, the key of migration is to reduce the migrations of same memory page regions. The paper [4] belongs to the confirmation of hot container stage, which predicts the growth trend of container and reduces the number of unnecessary container migrations. The papers [5–8] consider the issue of migration from the point of view of saving energy, which optimize the original dynamic migration algorithm in order to reduce energy consumption as the standard. But the energy consumption standards are not uniform, so it is not easy to make comparison and judgement. The concept of reusing distance is introduced in [9], and the memory pages updated are traced back in every copy. But the complexity of the algorithm is high and it is difficult to implement. In the paper [10], RLE, Huffman Coding, MEMCOM WKdm, and LZ, several common memory compression algorithms are classified and studied from the perspective of memory, and it proposes a new algorithm to reduce the memory transfer time. But there are some limitations to consider from the perspective of memory compression. The papers [11, 12] propose an optimized scheduling model from the point of view of scheduling algorithm. The container will be scheduled on the appropriate server, so as to achieve load balancing. The main application scenarios of dynamic migration are elaborated in paper [13]. In order to achieve disaster recovery of large data structure, it is necessary to design a large scale dynamic migration system based on network conditions, migration policies, controller types and memory distribution. The papers [14, 15] predict the frequently modified memory pages via the time series prediction method, and then transfer them to the destination container in the last copying round.

In summary, a total of four directions can be divided into dynamic migration optimization:

(1) Memory Compression: By compressing the amount of memory to be copied, thereby it reduces unnecessary memory copies and ultimately saves the copy time.
(2) Prediction Mode: The probability that the memory page changed again is predicted. And the page which frequently modified are given transmitting priority.
(3) Scheduling Mode: By optimizing the way of scheduling, the container with high memory and increasing trend will be migrated preferentially, so as to reduce unnecessary container migration.

(4) Energy optimization: Its main purpose is to take the energy consumption into consideration. By establishing the model of energy consumption, high energy consuming stage for migration is optimized.

In this paper, the prediction model is used to predict the probability that the memory page is modified through the grey model and Markov correction.

3   Pre-copy Migration

3.1   Pre-copy Model

Currently, most of the container migration algorithms adopt the method of pre-copy. The process is to copy all of the memory pages from the source container into the target container in the first iteration, and then copy dirty page iteratively. The advantage of pre-copy lies in reliability. When the copy fails, it can restore to the original state.

As shown in Fig. 1, after the preparation of the necessary resources, it will copy N times iteratively. In order to make the user unaware of the pause, the copy is carried on in the state of running. After each copy, the applications which running in the container will modify the memory page in a large probability. So, it is necessary to copy these areas again in the next iteration process. After N round iterations, the source container stops running, and all the dirty pages are copied to the destination container immediately.

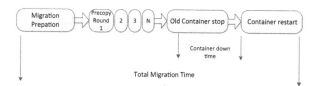

Fig. 1. Process of pre-copy migration

3.2   Optimizing of the Pre-copy Migration Model

On the basis of the traditional migration model, a prediction module is added to modify the memory page. This prediction module is used to collect relevant data as the original data.

As shown in Fig. 2, the source container is migrated from the source server to the destination server. The monitoring module has been collecting the reading and writing status of each memory page in the container. The data obtained by the monitoring module is transmitted to the prediction module, and a prediction is made on the probability that the dirty page is modified again when the memory page is copied every time.

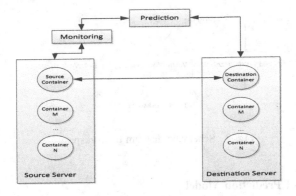

Fig. 2. Optimized pre-copy migration model

4 Optimized Pre-copy Algorithm

From the point of view of prediction, this paper presents an optimized pre-copy algorithm (OPCA), which combines the Gray-Markov Model to predict the probability that dirty pages are modified again. The OPCA selects dirty pages which maintains a high probability of modification within a long time into the hot workspace, leaving pages with lower probability in cold workspace.

4.1 Division of Workspace

In order to improve the efficiency and avoid redundant transmission, the concept of workspace is introduced. Now the workspace is divided into the following:

(1) Free Workspace (FW): a workspace that is read only or does not perform any operations, is not involved in dirty page copying process.
(2) Total Workspace (TW): It is the space that participates in the dirty page iterative copy process. In accordance with the probability of re-modification of the dirty page, the space is divided into three regions: Cold Workspace (CW), Warm Workspace (WW) and Hot Workspace (HW). The CW is the area where the dirty page is modified at a lower probability, and participates in each iteration process. The purpose of the introduction of the WW is to reduce unnecessary errors in the initial state that memory data is small, and also play a role of a hierarchical copy. Memory page with high probability of modification is stored in HW. And these pages are copied in the last round.
(3) Dirty Set: The currently modifies memory page collection.
(4) Send Set: The final collection of dirty pages that are not included in the portion of memory become dirty in the forecast process (Fig. 3).

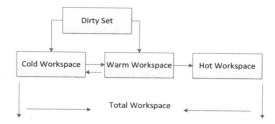

Fig. 3. Schematic diagram of workspace

4.2 Selection of Prediction Model

In this paper, the Gray-Markov model is adopted. With the introduction of Markov theory, grey prediction model is effectively used to solve the problem of low prediction accuracy of random volatility series.

GM (1, 1) gray model is established and the prediction curve is obtained. It sets the original modified dirty page, listed as formula 1.

$$p^{(0)} = \left(p_1^{(0)}, \; p_2^{(0)}, \; p_3^{(0)}, \ldots, \; p_N^{(0)} \right) \tag{1}$$

Among them, $p^{(0)}$ is a collection of data, which has N data. Next, it cumulates processing in the dataset to get a new set.

$$p^{(1)} = \left(p_1^{(1)}, \; p_2^{(1)}, \; p_3^{(1)}, \ldots, \; p_N^{(1)} \right)$$

$$p_j^{(1)} = \sum_{i=1}^{j} p_i^{(0)} \; (1 \leq j \leq N) \tag{2}$$

After the cumulative operation, the volatility of the data will be reduced apparently, then GM (1, 1) model is set up as formula 3.

$$\frac{dp^{(1)}}{dt} + ap^{(1)} = b \tag{3}$$

Next, it is necessary to resolve the value of a and b.

$$A(a, b)^T = P_N \tag{4}$$

Among them, the expression of matrix A is listed as formula 5.

$$A = (\beta_1, \beta_2, \beta_3, \ldots, \beta_{N-1})^T$$

$$\beta_i = \left(-\frac{1}{2} \left(p_i^{(1)} + p_{i+1}^{(1)} \right), 1 \right) \; (1 \leq i < N) \tag{5}$$

The expression of matrix P_N is shown as formular 6.

$$P_N = \left(p_2^{(0)}, p_3^{(0)}, \cdots, p_N^{(0)}\right)^T \tag{6}$$

Finally, the solution of the differential equation is obtained as formular 7.

$$e^{-ak}\left(\frac{b}{a} - p_1^{(0)}\right) = \frac{b}{a} - \hat{p}_{k+1}^{(1)} \tag{7}$$

By the formula 7, the value of $\hat{p}_{k+1}^{(1)}$ is obtained. And the original memory can be calculated according to the formula 8.

$$\hat{y}_k = \hat{p}_{k+1}^{(1)} = \left(p_{k+1}^{(1)} - p_k^{(1)}\right) \tag{8}$$

Based on the prediction curve above, several dynamic state intervals are divided. For a Markov random sequence, y_k, any state Q_k, $Q_k \in [Q_{1k}, Q_{2k}]$. Among them, $Q_{1k} = \hat{y}_k + \varepsilon_k \bar{y}$, $Q_{2k} = \hat{y}_k + \gamma_k \bar{y}$, ε_k and γ_k are the partition coefficients of the corresponding states. After calculating the prediction interval, the intermediate point is achieved. The predicted value is calculated through formula 9.

$$Y_t^* = \frac{Q_{1k} + Q_{2k}}{2} = \hat{y}_t + \frac{\bar{y}}{2}(\varepsilon_i + \gamma_i) \tag{9}$$

To sum up, we have been able to get the probability that the current dirty page is changed again.

4.3   Optimized Pre-copy Migration Mechanism

4.3.1   Traditional Predictive Pre-copy Scheme

In the traditional pre-copy scheme based on prediction, the dirty page set is usually divided into two parts, the cold working set and the hot working set. If the probability of current dirty page to be re-modified is greater than the threshold, the memory page will be directly added into the hot work set. And the pages belong to hot work set involved in stop-copy. Additionally, in the next iteration of the copy process, if the dirty page belongs to hot work set, no judgment is required. Therefore, the number of memory pages in the hot work set is always on the rising.

4.3.2   Optimized Pre-copy Mechanism Based on Prediction

In order to solve the problem of the traditional pre-copy mechanism, we propose the concept of the warm workspace. The warm workspace is to avoid the error operation and play a role of a hierarchical copy.

Phase 1: In earlier period, the data is relatively less. Through the Gray-Markov prediction model, we get the re-modified probability p_i of each memory page. If $p_i < \varepsilon_1$, the memory page is placed in the CW. If $p_i > \varepsilon_1$, we put the memory page into the WW. It puts the pages with high modification rate into the WW temporarily. At this stage, only the dirty pages in the CW are copied.

Phase 2: In the latter period, the data amount is relatively large. In the Nth iteration copy, we still get p_i through the prediction model. And through the ε_1, we determine the current dirty page into CW or WW. For each memory page $M_i \in DS$, if $(M_i \notin WW) \& (p_i > \varepsilon_1)$, we put M_i into WW.

Phase 3: When the size of the workspace is relatively stable, the hot work area is finally determined. And the iteration continues in traditional scheme until the end (Table 1).

We can assume that the iterations are k, the size of DS is m and the size of WW is n. So, the complexity of this algorithm is $O(mn)$. As a result, it takes a little time to finish the computing process.

Now we set the input as DirtySet, which stored the current round of dirty pages. And the total size of the work area is Size(Total Workspace), the size of the application is Size(Application). The iteration threshold is Count, and the output is SendSet.

5 Experiment

5.1 Simulation Parameters

In order to better simulate the real cloud environment, we use two high physical configuration servers as simulation tool in Table 2.

5.2 Parameter Index

In order to measure the effectiveness of our algorithm OPCA, it is measured from three aspects:

(1) Downtime: downtime is the total time of the copy at final round, which needs to stop the Docker container service.
(2) Iteration: The iterations directly affect the efficiency of the copy. If there are many iteration, it means that a memory page is copied many times unnecessarily.
(3) Total copy time: the total time it takes to transfer all the information of source container to destination container.

5.3 Simulation Results

In order to better simulate the actual migration, where we set the memory size of the container was 64, 128, 256, 512, 1024 (MB).

Figure 4 shows the comparison of two algorithms in the case of low dirty page rate. It can be concluded that the number of iterations increases with the increasing of memory. The number of iterations after optimization is slightly less than the traditional way.

Table 1. Pseudo code of OPCA

Input: DirtySet（DS）
Output: SendSet（SS）

1. *While* (Current iterations <Count)
2.   *If* (Size (Total Workspace)/
3.   Size(Application)<α)
4.   *For*(Dirty page $M_i \in$ DS)
5. p_i = Gray_Markov(Q(M_i))
6.   *If*($p_i < \varepsilon_1$)
7.   CW $\leftarrow M_i$
8.   *else*
9.   WW $\leftarrow M_i$
10.   *EndFor*
11.   SS $\leftarrow$ CW
12.   *Return* SS
13.   *else*
14.   *For*(Dirty page $M_i \in$ DS)
15. p_i = Gray_Markov(Q(M_i))
16.   *If*($p_i < \varepsilon_1$)
17.   CW $\leftarrow M_i$
18.   *else*
19.   WW $\leftarrow M_i$
20.   *For*(Dirty page $M_j \in$ WW)
21.   *If*($p_j < \varepsilon_1$)
22.   WW. delete(M_j)
23.   CW $\leftarrow M_j$
24.   *else If*($p_j > \varepsilon_2$)
25.   WW. delete(M_j)
26.   HW $\leftarrow M_j$
27.   *EndFor*
28.   *EndFor*
29. *EndWhile*

Table 2. Parameters of servers

Specification	Server 1	Server 2
Processor	Intel(R) Xeon(R) CPU E5-2620 v3 @ 2.40 GHz	Intel(R) Xeon(R) CPU E5-2620 v3 @ 2.40 GHz
RAM	384G	384G
DISK	975G	975G
OS	Ubuntu 16.04 LTS	Ubuntu 16.04 LTS
Cloud platform	Spark+Docker	Spark+Docker

The comparison in the case of low dirty page rate was shown in the Fig. 5. Compared with Fig. 4, it can be concluded that, in the case of high dirty page rate, the number of iterations is much less than that of the traditional method. Additionally, when the container is large, the advantage is more obvious.

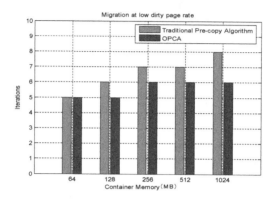

Fig. 4. Iterations at low dirty page rate

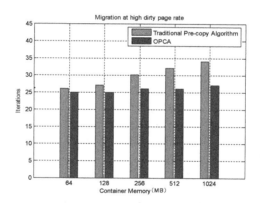

Fig. 5. Iterations at low dirty page rate

Figure 7 shows the comparison of downtime between the traditional pre-copy algorithm and the OPCA in the case of high dirty page rate. We can find that, compared with Fig. 6, the downtime increased slower. And, the downtime is shorter in OPCA than in traditional method.

Figure 8 shows the comparison of the total time. When the container is large, OPCA has little advantage.

According to the experiments, OPCA ensures that total migration time is not greater than traditional method. Moreover, it also reduces the number of iterations and downtime, improves the utilization rate of resources and enhances the user experience.

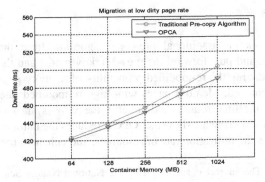

Fig. 6. Downtime at low dirty page rate

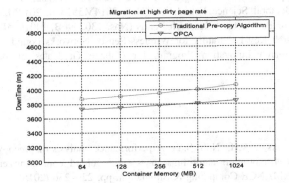

Fig. 7. Downtime at high dirty page rate

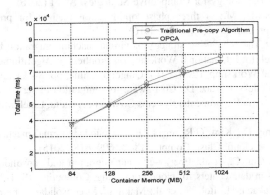

Fig. 8. Total migration time

6 Conclusions

This paper optimizes the traditional pre-copy algorithm from the perspective of reducing downtime, and puts forward the grey-Markov model and the new concept of workspace. By predicting the re-modified probability of memory page, the division of the work area is achieved. Memory pages with high re-modified rate are put into the HW and copied in the final round. Additionally, the WW is introduced to avoid the error operation. The experiments show that OPCA reduces the number of iterations and downtime, improves the utilization rate of resources, and enhances the user experience.

Acknowledgments. The subject is sponsored by the National Natural Science Foundation of P.R. China (Nos. 61373017, 61572260, 61572261, 61672296, 61602261), the Natural Science Foundation of Jiangsu Province (Nos. BK20140886, BK20140888, BK20160089), Scientific & Technological Support Project of Jiangsu Province (Nos. BE2015702, BE2016777, BE2016 185), China Postdoctoral Science Foundation (Nos. 2014M551636, 2014M561696), Jiangsu Planned Projects for Postdoctoral Research Funds (Nos. 1302090B, 1401005B), Jiangsu High Technology Research Key Laboratory for Wireless Sensor Networks Foundation (No. WSNL BZY201508), Research Innovation Program for College Graduates of Jiangsu Province (SJZZ16_0148).

References

1. Shah, S.A.R., Jaikar, A.H., Noh, S.Y.: A performance analysis of precopy, postcopy and hybrid live VM migration algorithms in scientific cloud computing environment. In: High PERFORMANCE Computing & Simulation, pp. 229–236 (2015)
2. Desai, M.R., Patel, H.B.: Performance measurement of virtual machine migration using pre-copy approach in cloud computing. In: International Conference on Information and Communication Technology for Competitive Strategies. ACM (2016)
3. Sharma, S., Chawla, M.: A three phase optimization method for precopy based VM livemigration. SpringerPlus **5**(1), 1–24 (2016)
4. He, H., Cao, B.: The scheduling strategy of virtual machine migration based on the gray forecasting model. In: International Workshop on Frontiers in Algorithmics, Proceedings of 10th International Workshop, FAW 2016, Qingdao, China, pp. 84–91 (2016)
5. Breitgand, D., Kutiel, G., Raz, D.: Cost-aware live migration of services in the cloud. In: SYSTOR 2010: Haifa Experimental Systems Conference, Haifa, Israel, May, pp. 7379–7382 (2010)
6. Jaikar, A., Huang, D., Kim, G.R., et al.: Power efficient virtual machine migration in a scientific federated cloud. Clust. Comput. **18**(2), 609–618 (2015)
7. Li, H., Zhu, G., Cui, C., et al.: Energy-efficient migration and consolidation algorithm of virtual machines in data centers for cloud computing. Computing **98**, 1–15 (2015)
8. Hirofuchi, T., Nakada, H., Itoh, S., et al.: Making VM consolidation more energy-efficient by postcopy live migration. In: Cloud Computing, pp. 195–204 (2011)
9. Alamdari, J.F., Zamanifar, K.: A reuse distance based precopy approach to improve live migration of virtual machines. In: IEEE International Conference on Parallel Distributed and Grid Computing, pp. 551–556. IEEE (2012)

10. Patel, M., Chaudhary, S.: Survey on a combined approach using prediction and compression to improve pre-copy for efficient live memory migration on Xen. In: International Conference on Parallel, Distributed and Grid Computing, pp. 445–450. IEEE (2014)
11. Amani, A., Zamanifar, K.: Improving the time of live migration virtual machine by optimized algorithm scheduler credit. In: International e-Conference on Computer and Knowledge Engineering, pp. 346–351. IEEE (2014)
12. Jin, H., Gao, W., Wu, S., et al.: Optimizing the live migration of virtual machine by CPU scheduling. J. Netw. Comput. Appl. **34**(4), 1088–1096 (2011)
13. Kang, T.S., Tsugawa, M., Matsunaga, A., et al.: Design and implementation of middleware for cloud disaster recovery via virtual machine migration management. In: IEEE/ACM, International Conference on Utility and Cloud Computing, pp. 166–175. IEEE (2014)
14. Hu, B., Lei, Z., Lei, Y., et al.: A time-series based precopy approach for live migration of virtual machines. vol. 42, no. 4, pp. 947–952 (2011)
15. Johnson, J.A.: Optimization of migration downtime of virtual machines in cloud. In: Fourth International Conference on Computing, Communications and NETWORKING Technologies, pp. 1–5. IEEE (2013)

A Cost Model for Heterogeneous Many-Core Processor

Yanbing Li[1(✉)], Qi Wang[1], Yingying Li[1], Lin Han[1], Yuchen Gao[1],
and Qing Mu[2]

[1] State Key Laboratory of Mathematical Engineering and Advanced Computing,
Zhengzhou 450001, China
953390523@qq.com
[2] National Defense University, Beijing 100089, China

Abstract. Heterogeneous many-core processors become an important trend in high-performance computing area, but their sophisticated architecture greatly complicates the programming and compiling issue. The cost model is an important part of optimizing compilers, which is used to analyze the benefits of various program optimizations. This paper constructs a cost model for SW26010 heterogeneous many-core processor, and proposes a dynamic-static hybrid method to analyze benefit based on this cost model. Then these have been implemented in an automatic parallelizing framework for SW26010. The experimental results show that the cost model and the benefit analysis can filter a large number of non-beneficial parallel loops and the performance of the automatically parallelized programs increases significantly.

Keywords: Heterogeneous many-core processor · Cost model · SW26010 · Automatic parallelizing

1 Introduction

Although the CPU develops rapidly following of Moore's Law, the computing power of computer systems still cannot meet Introduction the needs of human society. In the weather forecast, oil exploration, material synthesis, drug research, earth simulation and other areas still need much faster speed to do scientific research. The traditional single-core processors mainly rely on improving the frequency to get faster computing speed. But due to the restrictions of the power and cooling problems, now the frequency has been too difficult to be improved continually. As a result, integrating more processor cores on a single chip becomes the mainstream of processor designs. Currently, the number of cores has increased rapidly. Especially for high-performance computing area, processor has entered the many-core age [1]. Compared with the general multi-core processor, the many-core processor integrates a great number of simplified cores in a chip. In June 2016, the Sunway TaihuLight computer systems were released, and the three key indicators, i.e., the peak performance, continuous performance and performance per watt all rank first in the world. One of the key components of Sunway TaihuLight is the self-designed SW26010 many-core processor [2].

© Springer Nature Singapore Pte Ltd. 2017
G. Chen et al. (Eds.): PAAP 2017, CCIS 729, pp. 566–578, 2017.
DOI: 10.1007/978-981-10-6442-5_54

According to the different types of integrated cores, many-core processors are divided into two categories: homogeneous many-core processor and heterogeneous many-core processor. The homogeneous many-core processors integrate a number of the same structured core in a chip. At present, the mainstream commercial many-core processors are mostly homogeneous, such as Intel's Xeon Phi [3] series processors and NVIDIA GPGPU [4] products. Yet heterogeneous many-core processors integrate some different cores in a ship, usually containing general cores for management and a large number of simplified computing cores for increasing computing power. So heterogeneous many-core processors are able to achieve higher performance, but with lower power consumption.

Hoverer it makes the architecture more complex to integrate a large number of different cores in a ship for heterogeneous many-core processors, which brings great challenge to the development and optimization of parallel programs. There are two basic ways to solve parallel programming issues: one is that programmers write parallel programs manually; the other is to use automatic parallelization tools. The well-optimized hand-written parallel programs often have a high quality, but it has high requirements for programmers and writing parallel programs manually is very inefficient. Automatic parallelization system is a compiling tool that automatically explores the potential parallelism in a program and converts it into a parallel form. It can quickly accomplish the conversion from serial programs to parallel programs, and it has a higher efficiency. But because of the complexity of the automatic parallelization technology, the parallel programs produced by the automatic parallelization tool have a large gap in quality compared with the hand-written parallel programs. The loops often occupy the most time of the program execution in the scientific programs, so the loops are the main goal of automatic parallelization. Because of the overhead of parallel start-up, synchronization, and so on, not all loops can bring the benefits when executing in parallel. Some non-beneficial loops executed in parallel is one of the key factors that affect the performance of automatic parallelizing programs.

From the above analysis, it is necessary to perform the benefit analysis while generating parallel code. There are two main methods to perform benefit analysis at present: one is based on the cost model [5], by assessing the time cost of serial execution and parallel execution; the other is based on machine learning [6], according to the existing experience to determine whether it is beneficial, using the program information of the loop iterations, the number of computing and memory instructions, and so on. The cost model is a commonly used method to determine whether a variety of program optimization is profitable in many excellent compiler, such as GCC [7], LLVM [8] and Open64 [9]. It is a relatively new method to use machine learning to evaluate the benefits. However, since the machine learning method need learning model and a large number of training, and may make mistakes based on the existing experience, the current product-level compilers almost do not use this method.

SW26010 is a high-performance heterogeneous many-core processor [1]. The one chip contains four core groups in SW26010. Each core group integrates a management core and 64 computing cores. On the SW26010, loading the program sections into the computing core will need a certain start up and data transmission costs. So loading some non-beneficial parallel loops into the computing cores may seriously affect the overall performance of the program. This paper constructs a cost model for SW26010 heterogeneous many-core processor, and proposes a combination of static and dynamic method to analyze parallel benefit. And these have been implemented in the SW26010-oriented automatic parallelization framework developed by our research group. The work in this paper can also be used in the compilers of parallel programs to analyze the benefit of the parallel loops in hand-written parallel programs. The main contributions of this paper are as follows: (1) constructing a cost model for SW26010 heterogeneous many-core processor and proposing a parallel benefit analysis method for SW26010-oriented automatic parallelization system; (2) providing a referential resolution for constructing a cost model for heterogeneous many-core processors.

2 Background

2.1 Target Platform

The SW26010 is a heterogeneous many-core processor for high performance computing. The chip is consisted of four core-groups (CG), network on chip (NoC) and system interface (SI), as shown in Fig. 1. Each CG consists of a management processing element (MPE), a computing processing element (CPE) cluster including 64 CPEs and a memory controller (MC). The MPE is a full-featured 64-bit RISC core, which can deal with the serial segments of the program and manage the resources. The CPE is also a 64-bit RISC core, but with limited functions. The design goal of CPE is to achieve the maximum aggregated computing power, while minimizing the complexity of the micro-architecture. The CG of SW26010 is the basic unit that manages the computing resources. The memory system architecture of a single group is shown in Fig. 2. The MPE has L1 and L2 hardware cache. The CPE has a user-controlled scratch pad memory (SPM). The MPE and the CPE share the off-chip main memory. SPM and main memory are addressed unifiedly. The MPE and the CPE can access SPM and main memory, and CPE can realize the bulk data transmission between SPM and main memory through DMA instructions.

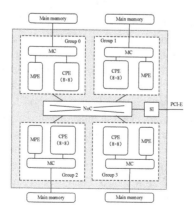

Fig. 1. SW26010 processor structure

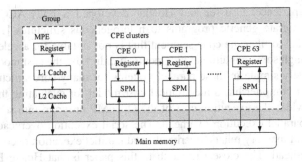

Fig. 2. SW26010 memory hierarchy of each group

It has different parallel programming models and compilation environments to support parallelization at different levels on SW26010. Parallelization among the four CGs can be realized by OpenMP or MPI. Parallelization between the MPE and the CPE cluster in a CG can be realized in OpenACC* [2]. The OpenACC* is an extension of OpenACC 2.0 [10] standard for characteristics of SW26010 processor. OpenACC is a heterogeneous programming model for accelerated device. It loads the specified program segment onto the accelerated device through compiler directives. The other part of the program still is executed on the main processor. The extensions in OpenACC* mainly includes data package copy for fragmented data transmission, data transposition copy for discontinuous memory access and so on. The SWACC is the compilation system for OpenACC*, which supports OpenACC* and MPI mixed programming and OpenACC 2.0 standard completely as well.

The execution model of the OpenACC* program is an accelerated execution model. The MPE cooperates with the CPE collaboratively guided by the MPE. The program starts running on the MPE first, executed as a main thread serially, and then the compute-intensive segments are loaded onto the MPE cluster as an accelerated task under the control of the main thread. The executing steps of accelerated task include that loading the accelerated code to the CPE, transferring the required data from main memory to SPM, the CPE computing the task and sending the results back to main memory, freeing the data space on the SPM.

2.2 Related Work

At present, most of the automatic parallelization research work for the many-core structure aims at the GPU. For example, Lee et al. achieved a source-to-source conversion of automatic transform and optimization framework that transforms OpenMP programs into GPGPU programs [11, 12]. Baskaran et al. implemented an automatic conversion system that converts C programs into CUDA programs directly [13]. The typical research on the Intel Phi processor is the optimization compiler Apricot proposed by Ravi et al. [14], which converts OpenMP programs and serial programs into the programs suitable for MIC structure. It uses a simple cost model to optimize performance, which according to the number of loop iterations, CPU and memory access operations, and the rate of memory access operations to evaluate the cost with the existing experience. Grosser and Hoefler developed a Polly-ACC compilation framework to migrate generic programs to heterogeneous structures [15]. It used a simple cost model to filter the code segment that is not suitable for accelerating on the accelerator through some conditional judgments. It didn't use the cost model or the cost model is relatively simple in these parallel systems for many-core structure.

The Open64 compiler has a hierarchical cost model for evaluating the cost of the loop [16]. It divides the execution cost of the OpenMP program into serial execution overhead and parallel execution overhead. The serial execution overhead depends on the processor and memory related overhead. And parallel execution overhead has extra parallelism overhead. The closest research to this paper is that Huang Pinfeng et al. present a cost model for the Cell processor based on the Open64 cost model [17].

3   Automatic Parallelizing Compilation Framework

The cost model of this paper is based on the SW26010 automatic parallelizing framework developed by our research group. This framework is used to develop the parallelism of applications in the single core-group of the SW26010 processor. It converts the serial programs into a heterogeneous parallel programs by automatically inserting the OpenACC* directive. The automatic parallelizing framework uses C and FORTRAN programs as input and OpenACC* parallel program as output. And it includes task division, data layout, transmission optimization and benefit analysis module, as shown in Fig. 3, which is implemented based on the open-source compiler Open64.

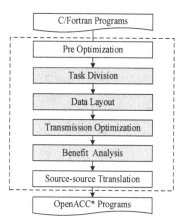

Fig. 3. Parallel compilation framework

It requires pre-optimization before automatic parallelization. Pre-optimization includes constant propagation, inductive variables substitution, dead code elimination, loop normalization and other simple optimization, which is prepared for the next analysis and optimization. The task division module is used to identify the program segments that are suitable for acceleration on the CPE. It is usually a loop with enough computation and can be executed in parallel. The data layout module analyzes the data which needs to be copied to SPM before the calculation and which needs to be copied from the SPM to the main memory after the calculation, based on the result of the data flow analysis. And then it generates the relevant data copy clause. The transmission optimization module optimizes the data transmission between the main memory and the SPM. It minimizes unnecessary data transmission and improves transmission efficiency. It need some cost of startup and data transmission to execute the program in parallel on the CPE cluster. Not all the loops executed in parallel can get benefits, so it needs to analyze the benefit with the benefit analysis module to avoid the loops whose calculation is too small to get benefits being loaded into the CPE. The source-to-source translation process converts the compiler intermediate language representation into a high-level language source program.

While running a program on SW26010 in parallel, loading a small loop with no benefits onto the CPE may result in a serious reduction of program performance, which is an important factor that affects the performance of the program.

4   Cost Model

The cost model is usually composed of a set of sub models which reflect the specific execution details of the programs on the computer system to evaluate the execution overhead. Many compilers and runtime libraries contain a cost model to guide optimizations. The Loop Nest Optimization pass of the Open64 contains a cost model for the singly nested loops, as shown in Fig. 4. It includes two sub-models: the serial model and the parallel model, which is used to evaluate the overhead executing on a single processor and the overhead of parallel execution on multiple processors using OpenMP fork-join mode, respectively. The cost of each sub model can be further divided into fine-grained overhead. The serial overhead depends on the overhead associated with the processor and memory. And the parallel overhead include extra overhead of parallel execution, such as parallel start-up overhead and thread synchronization overhead.

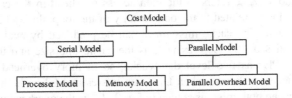

Fig. 4. The structure of cost model in Open64

The cost model for the SW26010 processor also requires two sub models, which are the serial models for execution on a single MPE and the parallel model for execution on the CPE cluster, respectively. The structure of the MPE and the execution model on MPE are similar with the general X86 processor, so the serial cost model can learn from the existing Open64 serial model. Hoverer the parallel execution mode on CPE cluster is quite different from the general X86 multi-core, and the CPE cluster structure is different from X86 multi-core core, such as CPE does not have the traditional cache, so the parallel model needs to be reconstructed.

4.1   Serial Model

The serial execution time is the basis for evaluating the parallel benefit of the loop. On the SW26010 processor, the loop serial execution overhead is clock cycles that the loop runs on MPE, which mainly contains the processor overhead and the memory overhead, as shown in Eq. 1.

$$T_{ser} = T_{processor} + T_{memory} \qquad (1)$$

(1) Processor Overhead

Processor overhead $T_{processor}$ is the overhead that the loop instruction executes on the processor while ignoring the cache and TLB misses cost. The number of every instruction in the loop statement (including the loop header and the loop body) can be obtained by analyzing the intermediate representation in compiling process. According to the number and the cycles of every instruction, the overhead of one iteration can be accumulated, which is named $T_{machine\_per\_iter}$. Then the overhead of whole loop is equal to the loop iterations N_{loop_iter} multiplied by $T_{machine\_per\_iter}$, as shown in Eq. 2.

$$T_{processor} = T_{machine\_per\_iter} \times N_{loop\_iter} \qquad (2)$$

The cycles of every instruction can be obtained by referring to the processor data. The loop iterations N_{loop_iter} is obtained according to the specific situation of the program. When the upper and lower bounds of the loop are constants, it can be parsed at compiling time. When loop iterations are dependent on the input of the program and only can be obtained at runtime, the loop iteration is replaced by a variable and the processor overhead is expressed as a function of the loop iterations.

(2) Memory Overhead

Memory overhead T_{memory} refers to the data access overhead in the case of L1 data cache miss. It is closely related to the data locality of the loop, the cache line size and cache miss overhead. The cache miss times can be predicted by analyzing the data access characteristics of the loop, as well as the cache line size and the cache configuration. The MPE has two level data cache, so memory overhead in SW26010 means the data access overhead when L1 cache misses and L2 cache hits, as well as when two level cache both miss, as shown in Eq. 2. T_{L2} is the overhead when L1 cache misses and L2 cache hits, and T_{main_mem} is the overhead of direct access to the main memory when two level cache both miss. N_{L2} is the L2 cache hit times, N_{main_mem} is the main memory access times, and the sum of N_{L2} and N_{main_mem} is L1 cache miss times. Among them, T_{L2} and Tmain_mem can be obtained by simple test, N_{L2} and N_{main_mem} can be predicted by program analysis.

$$T_{memory} = T_{L2} \times N_{L2} + T_{main\_mem} \times N_{main\_mem} \tag{3}$$

4.2 Target Platform

The parallel execution process of loops on SW26010 is: firstly, loading the program sections and transferring the data to the CPE; then, doing calculation by CPE; finally, returning the result to the main memory. According to the parallel execution model on SW26010, the parallel cost can be decomposed into parallel control overhead, data transmission overhead and parallel computing overhead, as shown in Eq. 4.

$$T_{para} = T_{para\_overhead} + T_{data\_trans} + T_{compute} \tag{4}$$

(1) Parallel Control Overhead

Parallel control overhead is mainly the start-up and management overhead of parallel execution, including the overhead of initializing the parallel computing environment, creating accelerated thread groups, loading acceleration tasks, and starting task execution. The parallel control overhead $T_{para\_overhead}$ can be expressed as Eq. (5), which is a linear function of the number of start-up cores. Among them, T_c is the parallel environment start overhead, T_p is the parallel management overhead of a single CPE core. T_c and T_p can be measured by testing a simple loop.

$$T_{para\_overhead} = T_c + p \times T_p \tag{5}$$

(2) Data Transmission Overhead

The data transmission overhead is mainly dependent on two aspects: one is the amount of data, and the other is the DMA bandwidth. The amount of data can be obtained by analyzing the data transmission clause. The data transfer overhead T_{data_trans} can be expressed by the Eq. (6). T_s is the initialization time of the DMA operation, and T_o is overhead of data packing and transposition in data transmission process. *In* and *out*, respectively, means the data set copying from the memory to SPM before the start of the calculation and copying from the SPM back to memory after the completion of the calculation. X and y is the elements in the in and out, $x.size$ and $y.size$ is the size of the space they occupied respectively. And w is the DMA data transfer bandwidth. T_s can be obtained by a simple test. Data packaging and transpose is operated on MPE core, so T_o can be obtained by calling the serial model.

$$T_{data\_trans} = T_s + T_o + \frac{\sum\limits_{x \in in} x.size + \sum\limits_{y \in out} y.size}{w} \tag{6}$$

(3) Parallel Computing Overhead

Parallel computing overhead is the time it takes for the loop to run on the CPE cluster in parallel. Due to the uncertainty in runtime, this value cannot be accurately predicted at compiling time, so it is approximated using the time running on a single CPE core divided by the number of cores. The computation of the time running on a single CPE is similar to that of the MPE. But because the structure and performance of the CPE are different from the MPE, the calculation method of each part is different. The parallel computing overhead $T_{compute}$ can be expressed by the Eq. (7). $T'_{processor}$ and $T'_{processor}$ are the processor overhead and memory overhead on CPE core.

$$T_{compute} = \frac{T'_{processor} + T'_{memory}}{p} \tag{7}$$

The calculation of $T'_{processor}$ is similar to $T_{processor}$. It only need to replace the execution cycle of each instructions into these on the CPE. Because the CPE has SPM rather than cache, so the calculation of T'_{memory} is different, as shown in Eq. (8). T_{spm} is the SPM access delay, N_{spm} is SPM access times. T_{mem} is overhead for CPE to access memory, N_{mem} is memory access times. Among them, T_{spm} and T_{mem} can be obtained by referring to the processor data, N_{spm} and N_{mem} can be obtained by programs analysis.

$$T'_{memory} = T_{spm} \times N_{spm} + T_{mem} * N_{mem} \tag{8}$$

At present, the parallelizing compilation framework can only deal with the DOALL loops which have no iterative dependencies, so we do not need to consider the inter-core communication overhead.

5 Benefit Analysis

Based on the cost model, it is possible to determine whether the loop executing in parallel is beneficial by comparing the serial cost and parallel cost, as shown in Formula (9) and (10).

$$T_{para} < T_{ser} \tag{9}$$

$$T_{para\_overhead} + T_{data\_trans} + T_{compute} < T_{processor} + T_{memory} \tag{10}$$

For the case that all the loop information can be obtained at compiling time, it is possible to statically determine whether it is beneficial or not at compiling time. For the case that some loop information cannot be obtained at compiling time, such as the loop boundary determined by the program input, we presents a benefit analysis method that combines the static information obtained at compiling time with the dynamic information obtained at run time. Some of the key parameters that are unknown at compiling time are replaced with variables. And we will dynamically analyze the benefit of the loop when the specific values are determined at run time.

As shown in Fig. 5(a), this is a hot loop in program named 462.libquantum from SPEC CPU2006 [18], whose upper bound reg -> size is passed through the function parameter and cannot be determined at compiling time. In this case, the framework generates an if clause that determines whether the loop will be executed in parallel at run time, based on the information collected at compiling time and the information dynamically obtained at run time. Taking the code shown in Fig. 5(a) as an example. The serial model and the parallel model are called respectively, and the serial and parallel cost can be calculated and marked as f (reg -> x) and g (reg -> x). According to Formula (9), the loop running in parallel is beneficial when f (reg -> x) > g (reg -> x), so the critical value X of reg -> x can be obtained. Thus we generate the if clause and decide whether to load the loop on CPE clusters depending on the value of reg -> x at runtime, as shown in Fig. 5(b).

```
for(i = 0;i < reg->size; i++) {
    if(reg->node[i].state & 1 << control1) {
        if(reg->node[i].state & 1 << control2) {
            reg->node[i].state ^= 1 << target;
        }
    }
}
```

(a) original code

```
#pragma acc parallel loop copyin(control1, control2, target)
    copy(reg->node[0:reg->size]) if(reg->size > X)
    for(i = 0; i < reg->size; i++) {
        if(reg->node[i].state & 1 << control1) {
            if(reg->node[i].state & 1 << control2) {
                reg->node[i].state ^= 1 << target;
            }
        }
    }
```

(b) code with if clause

Fig. 5. Example of benefit analysis

Although the models at compiling time used in this paper cannot be precise enough, it can really improve the performance of programs in practical, when taking into account the lack of precision. So GCC, LLVM, Open64 and many other excellent compilers also use the cost model to do benefit analysis.

6   Experimental Results

The cost model and benefit analysis method proposed in this paper have been implemented in automatic parallelizing compilation framework for SW26010 developed by our group. The test platform is Sunway TaihuLight supercomputer, whose computing processing element is SW26010 heterogeneous many-core processor. And it is equipped with complete compilation tools and performance analysis tools. The test programs are selected from the standard benchmark SPEC CPU2006 (V1.2) and NPB3.3.1 [19]. They are 462.libquantum (reference size) and 436.cacatusADM (reference size) from CPU2006, and CG (A size) and BT (A size) from NPB3.3.1. In these four programs, there are a large number of non-beneficial small parallel loops after automatic parallelization. The test method is to compare the number of parallel loops and the speedup of the automatically parallelized programs before and after the benefit analysis.

Table 1 shows the number of parallel loops before and after the benefit analysis. As seen from Table 1, the benefit analysis of this paper can filter the loops that are lack of calculation effectively. The number of parallel loops of the four programs reduced significantly after benefit analysis, such as 462.libquantum and 436.cacatusADM are reduced by more than 80%.

Table 1. Comparison of the number of parallel loops

Programs	Without benefit analysis	With benefit analysis	Reduced proportion
462.libquantum	71	14	80.28%
436.cacatusADM	156	28	82.05%
CG	27	13	51.85%
BT	17	12	29.41

Figure 6 shows the speedup before and after the benefit analysis. As can be seen in Fig. 6, the automatic parallelization performance of the program is improved obviously after the benefit analysis. The program performance of 462.libquantum decreases significantly because a large number of the inner parallel loops whose computation is little are loaded into the CPE cluster. Benefit analysis can filter these loops out, and the speeded up increased by 27.27 times significantly.

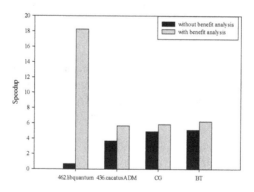

Fig. 6. Comparison of speedup

The test results show that the cost model constructed in this paper and the benefit analysis method based on the cost model can filter a large number of non-beneficial parallel loops and the performance of the automatically parallelized programs increases significantly.

7 Conclusion

The advantages of low power consumption and high performance of the heterogeneous many-core processor make it an important trend in the development of high-performance computing. Compiler optimization for the heterogeneous many-core processor is still in its infancy. The cost model is an important part of the compilation system. This paper focuses on the cost model for heterogeneous many-core processor. And the works are applied in the automatic parallelization system based on SW26010. However, there are still some shortcomings in the cost model of this paper. For instance, the simplification of the data transmission process leads to some deviation in results; it doesn't be considered that the data transmission may overlap with the calculation of the situation, the prediction of cache hits may not be consistent with actual results. These aspects need to be continually improved in future work. Although most of the details in this paper are based on the SW26010 processor, the basic ideas can be referred for the other heterogeneous many-core architecture processor.

References

1. Zheng, F., Yong, X.U., Hongliang, L.I., et al.: A homegrown many-core processor architecture for high-performance computing. Sci. Sin. **45**(4), 523 (2015)
2. Fu, H., Liao, J., Yang, J., et al.: The Sunway TaihuLight supercomputer: system and applications. Sci. Chin. Inf. Sci. **59**, 1–16 (2016)
3. Sodani, A., Gramunt, R., Corbal, J., et al.: Knights landing: second-generation intel xeon phi product. IEEE Micro **36**(2), 34–46 (2016)
4. Wu, G., Greathouse, J.L., Lyashevsky, A., et al.: GPGPU performance and power estimation using machine learning. In: Proceedings of IEEE International Symposium on High PERFORMANCE Computer Architecture, pp. 564–576. IEEE, NJ (2015)
5. Li, Y.B., Zhao, R.C., Liu, X.X., Zhao, J.: Cost model for automatic OpenMP parallelization. Ruan Jian Xue Bao/J. Softw. **25**(2), 101–110 (2014)
6. Wang, Z., Tournavitis, G., Franke, B., et al.: Integrating profile-driven parallelism detection and machine-learning-based mapping. ACM Trans. Archit. Code Optim. (TACO) **11**(1), 2 (2014)
7. Naishlos, D.: Autovectorization in GCC. In: Proceedings of the 2004 GCC Developers Summit, pp. 105–118 (2004)
8. Khaldi, D., Chapman, B.: Towards automatic HBM allocation using LLVM: a case study with knights landing. In: Proceedings of the Third Workshop on LLVM Compiler Infrastructure in HPC, pp. 12–20. IEEE Press (2016)
9. Chakrabarti, G., Chow, F., PathScale, L.: Structure layout optimizations in the open64 compiler: design, implementation and measurements. In: Open64 Workshop at the International Symposium on Code Generation and Optimization (2008)
10. Enterprise C. Cray Inc., NVIDIA and the Portland Group.: The OpenACC application programming interface, v2.0. (2013)
11. Lee, S., Min, S.J., Eigenmann, R.: Open MP to GPGPU: a compiler framework for automatic translation and optimization. ACM SIGPLAN Not. **44**(4), 101–110 (2009)
12. Lee S, Eigenmann, R.: OpenMPC: extended open MP programming and tuning for GPUs. In: Proceedings of the 2010 ACM/IEEE International Conference for High Performance Computing, Networking, Storage and Analysis, pp. 1–11. IEEE (2010)

13. Baskaran, M.M., Ramanujam, J., Sadayappan, P.: Automatic C-to-CUDA code generation for affine programs. Compiler Constr. **6011**, 244–263 (2010)
14. Ravi, N., Yang, Y., Bao, T., Chakradhar, S.: Apricot: an optimizing compiler and productivity tool for x86-compatible many-core coprocessors. In: Proceedings of ICS 2012, 25–29 June 2012, San Servolo Island, Venice, Italy (2012)
15. Grosser, T., Hoefler, T.: Polly-ACC transparent compilation to heterogeneous hardware. In: Proceedings of the 2016 International Conference on Supercomputing. ACM (2016)
16. Liao, C.: A Compile-Time OpenMP Cost Model. University of Houston, Houston (2007)
17. Huang, P., Zhao, R., Yao, Y., Zhao, J.: Parallel cost model for heterogeneous multi-core processors. J. Comput. Appl. **33**(06), 1544–1547 (2013)
18. Henning, J.L.: SPEC CPU2006 benchmark descriptions. ACM SIGARCH Comput. Archit. News **34**(4), 1–17 (2006)
19. Jin, H.Q., Frumkin, M., Yan, J.: The OpenMP implementation of NAS parallel benchmarks and its performance (1999)

Scalable *K*-Order LCP Array Construction for Massive Data

Yi Wu[1], Ling Bo Han[1], Wai Hong Chan[2(✉)], and Ge Nong[1,3(✉)]

[1] Sun Yat-sen University, Guangzhou, China
issng@mail.sysu.edu.cn
[2] The Education University of Hong Kong, Tai Po, Hong Kong
waihchan@ied.edu.hk
[3] SYSU-CMU Shunde International Joint Research Institute, Shunde, China

Abstract. Given a size-n input text T and its suffix array, a new method is proposed to compute the K-order longest common prefix (LCP) array for T, in terms of that the maximum LCP of two suffixes is truncated to be at most K. This method employs a fingerprint function to convert a comparison of two variable-length strings into a comparison of their fingerprints encoded as fixed-size integers. This method takes $O(n \log K)$ time and $O(n)$ space on internal and external memory models. It is also scalable for a typical distributed model consisting of d computing nodes, where the time and space complexities are evenly divided onto each node as $O(n \log K / d)$ and $O(n/d)$, respectively. For performance evaluation, an experimental study has been conducted on both external memory and distributed models. From our perspective, a cluster of computers in a local area network is commonly available in practice, but there is currently a lack of scalable LCP-array construction algorithm for such a distributed model. Our method provides a candidate solution to meet this demand.

Keywords: Fingerprint function · K-order LCP-array construction · External memory and distributed models

1 Introduction

Suffix array (SA) [1] is a fundamental data structure for many string processing applications. The SA construction problem has been intensively studied in the past two decades, resulting in various algorithms proposed for internal memory model [3–9], see [10] for a comprehensive survey. Among them, the SA-IS algorithm [9] is fast in practice and has a linear time complexity in the worst case. Recently, this algorithm has been extended for computing both suffix and LCP arrays simultaneously on internal and external memory models [11]. With the emerging applications of ever-increasing massive data, three novel SA construction algorithms eSAIS [12], EM-SA-DS [16] and EM-SA-IS [17] have been proposed to adapt SA-IS for sorting suffixes in external memory. As reported, these disk-based variants achieve substantial performance gains against the previous state-of-the-art [18]. In particular, eSAIS can produce both suffix and LCP arrays. In combination with the LCP array, an SA can be applied to emulating a bottom-up or top-down traversal of the corresponding suffix tree [2].The research on

© Springer Nature Singapore Pte Ltd. 2017
G. Chen et al. (Eds.): PAAP 2017, CCIS 729, pp. 579–593, 2017.
DOI: 10.1007/978-981-10-6442-5_55

constructing LCP-array has also attracted much attention. Kasai et al. proposed the first linear time algorithm for building the LCP-array using T, the SA and the inverse SA [13], Manzini et al. improved this algorithm for a better space efficiency [14]. Kärkkäinen et al. presented an alternative that first constructs the permuted LCP-array and then transforms it into the LCP-array in linear time [15]. Given SA already known, the LCPscan algorithm can build the LCP-array from the permuted LCP-array and the inverse SA [19]. LCPscan outperforms eSAIS when T is not very long, but the former suffers from a performance degradation when T is considerably larger than the available internal memory capacity. By relaxing T from a single string to a collection of strings, algorithms with further performance improvements can be designed for constructing the suffix and LCP arrays, e.g. eGSA [20] and exLCP [21]. The GSA algorithm can compute both the suffix and LCP arrays for T consisting of many variable-length strings using a multi-way merger. The exLCP algorithm is a lightweight LCP-array construction algorithm for a collection of sequences, where the Burrows-Wheeler transform is calculated in advance to facilitate the computation.

Although the above algorithms achieve remarkable performance, their designs are quite sophisticated due to the poor locality of memory accesses. As a result, these algorithms are not trivial to be extended for parallel and distributed models to scale the performance by a cluster of computers, for example. It has been observed from [20] that the average LCPs are commonly small for realistic data, or the string T consists of many short strings so that the LCP of any two short strings is upper bounded by the longest size of a short string (e.g. T is a genome database). In these cases, the original full-order LCP-array is actually a K-order LCP-array, where K is the maximum LCP value typically far less than the size of T. This motivated us to design a practical algorithm for computing the K-order LCP-array of T, which is defined as following: given any two neighboring suffixes in the SA of T, the K-order LCP of them is the longest common prefix of their first K characters. In this paper, K is assumed to be far less than T (e.g., $K \leq 2^{13}$ and $T \geq 2^{30}$). Our main contributions in this paper are the following two aspects:

1. We design and implement a K-order LCP-array construction method applicable to internal and external memory models, which builds the LCP-array in $O(n \log K)$ time and $O(n)$ space using the LCP batch querying technique (LCP-BQT) [22] previously proposed for constructing sparse suffix arrays. The program we developed for external memory model is composed of less than 600 lines in C++.

2. We parallelize our disk program in a distributed system consisting of d computing nodes. Different from shared memory models investigated in [23, 24], our distributed model is popular in nowadays computation environments, and hence our method is easy to be employed in practice.

The rest of this paper is organized as below. Section 2 describes the main idea of the proposed method and the RAM algorithm designed by this method. Sections 3 and 4 extend the RAM algorithm to external memory and distributed models, respectively. Section 5 gives the performance evaluation and Sect. 6 the concluding remarks.

2 *K*-Order LCP Computation in RAM

2.1 Notation

Consider an input text $T[0, n-1]$ drawn from a full-ordered constant or integer alphabet, we assume the ending character $T[0, n-1]$ to be unique and lexicographically smaller than any other characters in T. The following notations are used in our presentation for clarity.

- $pre(T, i)$ and $suf(T, i)$: the former denotes the prefix running from $T[0]$ to $T[i]$, while the latter denotes the suffix running from $T[i]$ to $T[n-1]$.
- SA_T: a permutation of $[0, n)$ satisfying $suf(T, SA_T[0]) < suf(T, SA_T[1]) < \cdots < suf(T, SA_T[n-1])$.
- $lcp(i, j)$ and $LCPA_T$: the former denotes the LCP length of $suf(T, i)$ and $suf(T, j)$, while the latter is an integer array such that $LCPA_T[i] = lcp(SA_T[i], SA_T[i-1])$.
- Δ_k: the shorthand notation for $2^{\log n - k - 1}$ (if not otherwise specified, the base of log is assumed to be 2).

2.2 LCP Batch Querying Technique

The LCP batch querying technique was previously proposed for constructing sparse suffix arrays. Given T and a set P consisting of b pairs of position indices, LCP-BQT computes $lcp(i, j)$ for all $(i, j) \in P$ in $O(n \log b)$ time using $O(n)$ space. The underlying idea is to find (i_f, j_f) for (i, j) such that $T[i, i_f - 1] = T[j, j_f - 1]$ and $T[i_f] \neq T[j_f]$. For the purpose, it initially assigns P to P_0 and performs a loop of $\log b$ rounds. In round k, it decides whether or not $lcp(i_k, j_k) \leq \Delta_k$ for each pair $(i_k, j_k) \in P_k$ and generates P_{k+1} as the input for the next round as follows: if $T[i_k, i_k + \Delta_k - 1] = T[j_k, j_k + \Delta_k - 1]$, then insert (i_k, j_k) into P_{k+1}; otherwise, insert $(i_k + \Delta_k, j_k + \Delta_k)$ into P_{k+1}. A naïve method for determining the LCP of two strings is to scan and compare their characters until a difference is detected, this takes $O(2^{\log n})$ time at worst. As a solution for reducing the time overhead, the fingerprint (FP) function presented in [25] is employed herein to transform the involved string comparisons into integer comparisons. The fingerprint of $T[i, j]$, denoted by $fp(i, j)$, can be calculated by the formula $fp(i, j) = \sum_{p=i}^{j} \delta^{j-p} \cdot T[p] \bmod L$, where L is a prime and δ is an integer randomly chosen from $[1, L)$. It should be noticed that two identical strings always have a common fingerprint, but the inverse is not true. Fortunately, it has been proved that the probability of a false match can be reduced to a negligible level when L is set to a large value. By exploiting the use of the fingerprint function, each loop round in LCP-BQT can be done by executing the three steps below in sequence.

- Scan T rightward to iteratively compute the fingerprint of $pre(T, \ell)$ according to the formula $fp(0, \ell) = fp(0, \ell - 1) \cdot \delta + T[\ell] \bmod L$, where $fp(0, \ell)$ is forwarded to a hash table if $\ell \in \{\{i_k - 1\} \cup \{j_k - 1\} \cup \{i_k + \Delta_k - 1\} \cup \{j_k + \Delta_k - 1\}, (i_k, j_k) \in P_k\}$.

- For each $\ell \in \{\{i_k\} \cup \{j_k\}, (i_k, j_k) \in P_k\}$, compute the fingerprint of $T[\ell, \ell + \Delta_k - 1]$ according to the formula $\mathrm{fp}(\ell, \ell + \Delta_k - 1) = \mathrm{fp}(0, \ell + \Delta_k - 1) - \mathrm{fp}(0, \ell - 1) \cdot \Delta \delta^{\Delta_k} \bmod L$, where $\mathrm{fp}(0, \ell - 1)$ and $\mathrm{fp}(0, \ell + \Delta_k - 1)$ is retrieved from the hash table in amortized $O(1)$ time.
- For each $(i_k, j_k) \in P_k$, compare $\mathrm{fp}(i_k, i_k + \Delta_k - 1)$ with $\mathrm{fp}(j_k, j_k + \Delta_k - 1)$. If the two fingerprints are equal, then insert $(i_k + \Delta_k, j_k + \Delta_k)$ into P_{k+1}; otherwise, insert (i_k, j_k) into P_{k+1}.

Notice that there must be $\mathrm{lcp}(i_k, j_k) \leq 2 \cdot \Delta_k$ for any $(i_k, j_k) \in P_k$ at the beginning of round k. After the loop, we have $k = \log b$ and $\mathrm{lcp}(i_{log_b}, j_{log_b}) \leq n/b$ for each $(i_{log_b}, j_{log_b}) \in P_{\log b}$, and it takes $O(n/b)$ time to compute $\mathrm{lcp}(i_{log_b}, j_{log_b})$ by scanning $\mathrm{suf}(T, i_{\log b})$ and $\mathrm{suf}(T, j_{\log b})$ rightward. Given $\mathrm{lcp}(i_{log_b}, j_{log_b})$ already known, $i_f = i_{\log b} + \mathrm{lcp}(i_{log_b}, j_{log_b})$ and $j_f = j_{\log b} + \mathrm{lcp}(i_{log_b}, j_{log_b})$. The time complexity of LCP-BQT is dominated by the loop of $\log b$ rounds. Each round takes $O(n)$ time for iteratively computing the fingerprints of $\mathrm{pre}(T, \ell)$ and $O(b)$ time for computing and comparing the fingerprints of strings indicated by the suffix and LCP arrays. The space in need is limited to $O(b)$ words ($\log \lceil n \rceil$ bits per word) for maintaining the hash table. As a result, the LCPs of any b pairs of suffixes in T can be correctly computed in $O(n \log b)$ time using $O(b)$ RAM space with a high probability. In the following paragraphs, if not otherwise specified, we reuse $LCPA_T$ and $\mathrm{lcp}(i, j)$ to represent the K-order LCP-array of T and the K-order LCP of $\mathrm{suf}(T, i)$ and $\mathrm{suf}(T, j)$, respectively.

2.3   Details

The K-order $LCPA_T$ construction problem can be converted into computing $\mathrm{lcp}(i, j)$ for $(i, j) \in \{(SA_T[1], SA_T[0]), (SA_T[2], SA_T[1]), \ldots, (SA_T[n-1], SA_T[n-2])\}$. We design an internal-memory algorithm called lcpa-ram based on this idea. Before the presentation, we introduce some notations below, where $i \in [0, n)$ and $k \in [0, \log K)$.

- CP_k and PP_k: integer arrays for caching the starting and ending position indices of target substrings. The algorithm lcpa-ram computes and compares the fingerprint of $T[CP_k[2i], CP_k[2i+1]]$ and $T[PP_k[2i], PP_k[2i+1]]$ to check equality of these substrings.
- ICP_k and IPP_k: integer arrays generated from CP_k and PP_k, by sorting i with $CP_k[i]$ and $PP_k[i]$, respectively.
- HT: a hash table for storing and retrieving the fingerprints of $\mathrm{pre}(T, p)$, where $p \in \{\{CP_k[i]\} \cup \{PP_k[i]\}, \in [0, 2n)\}$.

In line 2, Algorithm 1 computes CP_0 and PP_0 as follows: (1) $CP_0[2i] = SA_T[i] - 1$ and $CP_0[2i+1] = SA_T[i] + \Delta_0 - 1$; (2) $PP_0[2i] = SA_T[i-1] - 1$ and $PP_0[2i+1] = SA_T[i-1] + \Delta_0 - 1$. Then, it performs a loop of $\log K$ rounds in lines 4–9, where the key operation in round k is to sort the entries in CP_k and PP_k for iteratively computing the fingerprints of target substrings. Afterward, lcpa-ram computes and compares $\mathrm{fp}(CP_k[2i] + 1, CP_k[2i+1])$ and $\mathrm{fp}(PP_k[2i] + 1, PP_k[2i+1])$ to produce

CP_{k+1} and PP_{k+1}. If $\mathsf{fp}(CP_k[2i]+1, CP_k[2i+1]) = \mathsf{fp}(PP_k[2i]+1, PP_k[2i+1])$, then increase $CP_k[2i]$ and $PP_k[2i]$ by Δ_k and $CP_k[2i+1]$ and $PP_k[2i+1]$ by Δ_{k+1}; otherwise, decrease $CP_k[2i+1]$ and $PP_k[2i+1]$ by Δ_{k+1}. Finally, in lines 10–11, it takes $O(n)$ time to compute the *K*-order LCP-array. This leads to the lemma stated below.

Lemma 1. Given T and SA_T, the *K*-order $LCPA_T$ can be computed correctly in $O(n \log K)$ time using $O(n)$ words RAM space with a high probability.

Algorithm 1: Compute *K*-Order $LCPA_T$ in RAM

---

1 lcpa-ram(T, SA_T, n, K, HT)
2 Scan SA_T rightward to produce CP_0 and PP_0.
3 Let $k = 0$.
4 **while** $k < \log K$ **do**
5   Radix-sort CP_k and PP_k to produce ICP_k and IPP_k.
6   For $i \in [0, n)$, scan T rightward to compute the fingerprint of $\mathsf{pre}(T, i)$ and let
        $\mathsf{fp}(0, i) = HT[i]$ if $i \in \{ICP_k[j] \cup IPP_k[j], j \in [0, 2n)\}$.
7   For $i \in [0, n)$, scan CP_k and PP_k rightward to compute and compare
        $\mathsf{fp}(CP_k[2i]+1, CP_k[2i+1])$ and $\mathsf{fp}(PP_k[2i]+1, PP_k[2i+1])$ for generating
        CP_{k+1} and PP_{k+1}.
8   Let $k = k + 1$ and clear HT.
9 **end**
10 For $i \in [0, n)$, scan T, $CP_{\log K}$ and $PP_{\log K}$ rightward to compute
    $\Upsilon_i = \mathsf{lcp}(CP_{\log K}[2i], PP_{\log K}[2i])$.
11 For $i \in [0, n)$, let $\mathsf{lcp}(SA_T[i], SA_T[i-1]) = CP_{\log K}[2i] + \Upsilon_i - SA_T[i]$.

---

3 *K*-Order LCP Computation in External Memory

A hash table is employed in Algorithm 1 to store the fingerprints for quick lookups. This is feasible in RAM model, but will not be practical when n becomes too large and the table cannot be wholly accommodated in internal memory any more. To overcome the problem, we reformulate lcpa-ram to design a disk-friendly external memory algorithm called lcpa-disk.

3.1 Notation

Given the internal memory capacity M in our external memory model, T and SA_T are evenly partitioned into $d = n/m$ blocks, where each block is of size $m = O(M)$ and thus can be processed as a whole in RAM. We extend CP_K and PP_K for lcpa-ram to define their siblings ECP_k and EPP_k for lcpa-disk, both of them consist of $2n$ entries and each entry is a triple consisting of the following components:

- *idx*: $ECP_k[i].idx = EPP_k[i].idx = i$
- *pos*: The starting or ending position index of a substring in T.
- *fp*: The fingerprint of $\mathsf{pre}(0, pos)$.

Notice that $\mathsf{fp}(ECP_k[2i].pos + 1, ECP_k[2i+1].pos)$ is computed by the formula $ECP_k[2i+1].fp - ECP_k[2i].fp \cdot \delta^{\Delta_k} \bmod L$. We also compute $\mathsf{fp}(ECP_k[2i].pos + 1, ECP_k[2i+1].pos)$ in the same way.

Algorithm 2: Compute K-Order $LCPA_T$ in Disk

---

1 lcpa-disk(T, SA_T, n, K)
2 Scan SA_T rightward to produce ECP_0 and EPP_0.
3 Let $k = 0$.
4 **while** $k < \log K$ **do**
5   Radix-sort ECP_k and EPP_k by pos to produce $IECP_k$ and $IEPP_k$.
6   For $i \in [0, n)$ and $j \in [0, 2n)$, scan T rightward to iteratively compute the fingerprint of $\mathsf{pre}(T, i)$ and assign $\mathsf{fp}(0, i)$ to $IECP_k[j].fp$ or $IEPP_k[j].fp$ if $IECP_k[j].pos = i$ or $IEPP_k[j].pos = i$, respectively.
7   Radix-sort $IECP_k$ and $IEPP_k$ by idx to reproduce ECP_k and EPP_k.
8   For $i \in [0, n)$, scan ECP_k and EPP_k rightward to compute and compare each pair of
         $(\mathsf{fp}(ECP_k[2i].pos+1, ECP_k[2i+1].pos), \mathsf{fp}(EPP_k[2i].pos+1, EPP_k[2i+1].pos))$
         for generating ECP_{k+1} and EPP_{k+1}.
9   Let $k = k + 1$.
10 **end**
11 For $i \in [0, n)$, scan T, $ECP_{\log K}$ and $EPP_{\log K}$ rightward to compute
      $\Upsilon_i = \mathsf{lcp}(ECP_{\log K}[2i].pos, EPP_{\log K}[2i].pos)$.
12 For $i \in [0, n)$, let $\mathsf{lcp}(SA_T[i], SA_T[i-1]) = ECP_{\log K}[2i].pos + \Upsilon_i - SA_T[i]$.

3.2 Details

Different from lcpa-ram, lcpa-disk performs external-memory sorts to arrange fixed-size entries by an integer key of $\log n$ bits during each loop round. As illustrated in Algorithm 2, the entries of ECP_k and EPP_k are first sorted by the pos field to produce $IECP_k$ and $IEPP_k$ in line 5, and the entries of $IECP_k$ and $IEPP_k$ are then sorted by the idx field to regenerate ECP_k and EPP_k in line 7. During the first sort, we compute the fingerprints of all the prefixes in T and record the computation results in the fp field of each entry. After the second sort, we sequentially retrieve the fingerprints in ECP_k and EPP_k to produce ECP_{k+1} and EPP_{k+1} following the way of producing CP_{k+1} and PP_{k+1} from CP_k and PP_k in lcpa-ram. According to the discussion, we draw the result stated in Lemma 2.

Lemma 2. Given T and SA_T, the K-order $LCPA_T$ can be computed correctly in $O(n \log K)$ time using $O(n)$ words disk space with a high probability.

3.3 Optimization

The time complexity of lcpa-disk is dominated by the while-loop of $\log K$ rounds. A possible solution for reducing the running time is to decrease the number of loop rounds. We modify the procedure in lines 8-9 of lcpa-disk as below to generate ECP_{k+2}/EPP_{k+2} from ECP_k/EPP_k in a single round.

1. Compute $\mathrm{fp}[ECP_k[2i].pos + 1, ECP_k[2i + 1].pos]$ and $\mathrm{fp}[EPP_k[2i].pos + 1, EPP_k[2i+1].pos]$. If they are different, goto 3.
2. Compute $\mathrm{fp}[ECP_k[2i + 1].pos + 1, ECP_k[2i + 1].pos + \Delta_{k+1}]$ and $\mathrm{fp}[EPP_k[2i + 1].pos + 1, EPP_k[2i + 1].pos + \Delta_{k+1}]$. If they are different, increase $ECP_k[2i].pos$ and $EPP_k[2i].pos$ by Δ_k, increase $ECP_k[2i + 1].pos$ and $EPP_k[2i].pos$ by Δ_{k+2}, goto 4; otherwise, increase $ECP_k[2i].pos$ and $EPP_k[2i].pos$ by $\Delta_k + \Delta_{k+1}$, increase $ECP_k[2i + 1].pos$ and $EPP_k[2i + 1].pos$ by $\Delta_{k+1} + \Delta_{k+2}$.
3. Compute $\mathrm{fp}[ECP_k[2i].pos + 1, ECP_k[2i].pos + \Delta_{k+1}]$ and $\mathrm{fp}[EPP_k[2i].pos + 1, EPP_k[2i].pos + \Delta_{k+1}]$. If equal, increase $ECP_k[2i].pos$ and $EPP_k[2i].pos$ by Δ_{k+1}, decrease $ECP_k[2i + 1].pos$ and $EPP_k[2i + 1].pos$ by $\Delta_{k+1} + \Delta_{k+2}$.
4. Let $ECP_{k+2} = ECP_k$ and $EPP_{k+2} = EPP_k$.
5. Let $k = k + 2$.

This method can be generalized to merge any number of successive rounds into a single one. However, it leads to a side effect on the disk use, because each entry of ECP_k/EPP_k is extended to hold more fingerprints. We will demonstrate in Sect. 5 the performance of the adapted algorithm, namely lcpa-disk-m, in comparison with its prototype.

4 K-Order LCP Computation in a Distributed System

In this section, we describe how to parallelize lcpa-disk and lcpa-disk-m in a distributed system shown in Fig. 1, where the system consists of d computing nodes $\{N_0, N_1, \ldots, N_{d-1}\}$ interconnected by a high-speed switch operating in the full-duplex mode

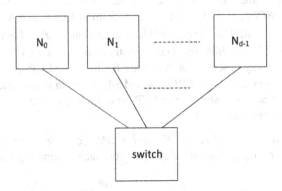

Fig. 1. The distributed system of d computing nodes.

4.1   Notation

Assume each node is equipped with a disk of size E. We evenly partition T and SA_T into $d = n/e$ blocks of size e $= O(E)$ and allocate one block of T and SA_T to each node. The entries of ECP_k/EPP_k and $IECP_k/IEPP_k$ are also evenly divided and scattered to the computing nodes. Some symbols and notations are given below, where $i \in [0,d)$, $j \in [0,e)$ and $k \in [0,\log K)$.

- T_i: $T_i[j] = T[ie+j]$. T_i resides on N_i.
- SA_{T_i}: $SA_{T_i}[j] = SA_T[ie+j]$. SA_{T_i} resides on N_i.
- $ECP_{i,k}/EPP_{i,k}$: $ECP_{i,k}[j] = ECP_k[ie+j]$ and $EPP_{i,k}[j] = EPP_k[ie+j]$.
- $IECP_{i,k}/IEPP_{i,k}$: $IECP_{i,k}[j] = IECP_k[ie+j]$ and $IEPP_{i,k}[j] = IEPP_k[ie+j]$.
- $SB1_{i,j}$ and $SB2_{i,j}$: N_i maintains two buffers $SB1_{i,j}$ and $SB2_{i,j}$ for every other node $N_j (j \neq i)$, the former caches the entries of $ECP_{i,k}/IECP_{i,k}$ destined for N_j and the latter caches the entries of $EPP_{i,k}/IEPP_{i,k}$ destined for N_j.
- $RB1_{i,j}$ and $RB2_{i,j}$: N_i maintains two buffers $RB1_{i,j}$ and $RB2_{i,j}$ for every other node $N_j (j \neq i)$, the former caches the entries of $ECP_{j,k}/IECP_{j,k}$ received from N_j and the latter caches the entries of $EPP_{j,k}/IEPP_{j,k}$ received from N_j.

4.2   Details

Recall that, lcpa-disk and lcpa-disk-m compute $ECP_{\log K}$ and $EPP_{\log K}$ by two runs of external-memory sort. To emulate the sorting processes in our distributed system, we set up a group of buffers on each node for data transmission. In lines 7–15 of Algorithm 3, lcpa-ds computes the fp field of the entries of $ECP_{i,k}$ and $EPP_{i,k}$ using the sending and receiving buffers. To be specific, for each entry x in $ECP_{i,k}$, node N_i dispatches x to $SB1_{i,j}$ if j $= x.pos/e$ and delivers it to $RB1_{i,j}$ on N_j (lines 7–8). Then, N_j caches entries from the other nodes in $\{RB1_{j,0}, RB1_{j,1}, \ldots, RB1_{j,d-1}\}$ and sorts them by their pos field to generate $IECP_{j,k}$ (lines 9–10). Afterward, N_j scans T_j rightward to compute the fingerprints of prefix and records the results in the entries of $IECP_{j,k}$. In lines 12–15, N_j moves entries of $IECP_{j,k}$ to $SB1_{j,i}$ and forwards them to $RB1_{i,j}$, and N_i sorts the entries in $\{RB1_{i,0}, RB1_{i,1}, \ldots, RB1_{i,d-1}\}$ by their idx to regenerate $ECP_{i,k}$. Finally, $ECP_{i,k+1}$ is produced from $ECP_{i,k}$ in line 16. The algorithm computes $ECP_{i,k+1}$ from $EPP_{i,k}$ following the same way described above. After the loop, the LCP-array of $T[ie, ie+e)$ can be computed from $ECP_{i,\log K}$, $EPP_{i,\log K}$ and SA_{T_i} on N_i in linear time. Therefore, we can simply collect and concatenate the LCP-array on each node to produce the final $LCPA_T$. Hence, we get this result:

Lemma 3. Given T and SA_T, the K-order $LCPA_T$ can be computed correctly in $O(n/d \cdot \log K)$ time using $O(n/d)$ disk space on each computing node with a high probability.

Algorithm 3: Compute K-Order $LCPA_T[ie, ie + e)$ on N_i

---

1 lcpa-ds(T_i, SA_{T_i}, e, K)

2 Scan T_i rightward to compute $ECP_{i,0}$ and $EPP_{i,0}$.

3 Let $k = 0$.

4 For $j \in [0, d)$, create sending buffers $SB1_{i,j}$ and $SB2_{i,j}$.

5 For $j \in [0, d)$, create receiving buffers $RB1_{i,j}$ and $RB2_{i,j}$.

6 while $k < \log K$ **do**

7   For $j \in [0, d)$ and $p \in [0, 2e)$, scan $ECP_{i,k}$ and $EPP_{i,k}$ rightward to cache $ECP_{i,k}[p]$ in $SB1_{i,j}$ or $EPP_{i,k}[p]$ in $SB2_{i,j}$ if $ECP_{i,k}[p].pos$ or $EPP_{i,k}[p].pos$ belongs to $[je, je + e)$.

8   For $j \in [0, d)$, send entries in $SB1_{i,j}$ and $SB2_{i,j}$ to $RB1_{j,i}$ and $RB2_{j,i}$ on N_j.

9   For $j \in [0, d)$, cache entries from $SB1_{j,i}$ and $SB2_{j,i}$ on N_j in $RB1_{i,j}$ and $RB2_{i,j}$.

10   Radix-sort entries in $RB1_i$ and $RB2_i$ by pos to produce $IECP_{i,k}$ and $IEPP_{i,k}$.

11   For $p \in [0, e)$ and $q \in [0, 2e)$, scan T_i rightward to iteratively compute the fingerprint of $pre(T, ie + p)$ and assign $fp(0, ie + p)$ to $IECP_{i,k}[q].fp$ or $IEPP_{i,k}[q].fp$ if $IECP_{i,k}[q].pos = ie + p$ or $IEPP_{i,k}[q].pos = ie + p$.

12   For $j \in [0, d)$ and $p \in [0, 2e)$, scan $IECP_i$ and $IEPP_i$ rightward to cache $IECP_{i,k}[p]$ in $SB1_{i,j}$ or $IEPP_{i,k}[p]$ in $SB2_{i,j}$ if $IECP_{i,k}[p].pos$ or $IEPP_{i,k}[p].pos$ belongs to $[je, je + e)$.

13   For $j \in [0, d)$, send entries in $SB1_{i,j}$ and $SB2_{i,j}$ to $RB1_{j,i}$ and $RB2_{j,i}$ on N_j.

14   For $j \in [0, d)$, cache entries from $SB1_{j,i}$ and $SB2_{j,i}$ on N_j in $RB1_{i,j}$ and $RB2_{i,j}$.

15   Radix-sort entries in $RB1_i$ and $RB2_i$ by idx to reproduce $ECP_{i,k}$ and $EPP_{i,k}$.

16   For $p \in [0, e)$, scan $ECP_{i,k}$ and $EPP_{i,k}$ rightward to compute and compare the fingerprints of $T[ECP_{i,k}[2p].pos + 1, ECP_{i,k}[2p + 1].pos]$ and $T[EPP_{i,k}[2p].pos + 1, EPP_{i,k}[2p + 1].pos]$ for generating $ECP_{i,k+1}$ and $EPP_{i,k+1}$.

17   Let $k = k + 1$.

18 end

19 For $p \in [0, e)$, scan T_i, $ECP_{i,\log K}$ and $EPP_{i,\log K}$ rightward to literally compute $\Upsilon_{i,p} = \mathsf{lcp}(ECP_{i,\log K}[2p].pos, EPP_{i,\log K}[2p].pos)$.

20 For $p \in [0, e)$, let $\mathsf{lcp}(SA_{T_i}[p], SA_{T,i}[p - 1]) = ECP_{i,\log K}[2p].pos + \Upsilon_{i,p} - SA_{T_i}[p]$.

It is noteworthy that the available RAM capacity on each node is partitioned into two parts, where the first part is for establishing communication buffers to compensate data transmission delay and the second part is for establishing I/O buffers to amortize the I/O overhead.

5 Experimental Results

A series of simulation experiments are conducted on a real data set collected from the web site http://download.wikimedia.org/enwiki/ for performance evaluation of our C++ programs, in terms of time and space consumption. We employ lcpa-disk-m as a baseline for performance and scalability assessment of lcpa-disk. The hardware platform for lcpa-disk and lcpa-disk-m is a computer equipped with 2 Intel Xeron E3-1220 CPUs, a 4 GB DDR3 main memory and a 2 TB 7200RPM disk, while the one for lcpa-ds is a distributed system consisting of 4 identical computers interconnected with a gigabit switch. We use gcc 4.8.4 and mpicc to compile the programs for lcpa-disk/lcpa-disk-m and lcpa-ds under Ubuntu 14.04 operating system, respectively.

STXXL [26] is a C++ STL library designed for efficient computations in external memory, freely available at http://stxxl.sourceforge.net/. Instead of to develop a radix sorter specific for our purpose, we use STXXL to perform the external memory sorts in our programs. Specifically, a priority queue provided by STXXL is employed for sorting the entries in $ECP_k = EPP_k$ by pos to form $IECP_k = IEPP_k$ and another for sorting them back by idx. Benefiting from the powerful priority queues, the programs for lcpa-disk, lcpa-disk-m and lcpa-ds are less than 400, 600 and 700 lines, respectively. Each result in the following tables and figures is a mean of two runs of the programs, where the running time and peak disk use are collected by shell command "time" and "stxxl::block manager" provided by STXXL, respectively.

5.1   Performance of the External Algorithms

The following parameters and metrics are used for the experiments:

- S: corpora size.
- ST: total running time
- MT: average running time spent in processing a character per loop's round.
- PD: peak disk use.
- LR: number of loop's rounds.
- W: every W successive loop's rounds in lcpa-disk are merged into one in lcpa-disk-m.
- H: internal memory allocated to each priority queue when sorting the entries in external memory.

To show the effect of W, we demonstrate in Table 1 the experimental results of lcpa-disk and lcpa-disk-m. As can be seen, lcpa-disk-m has a smaller ST than lcpa-disk when $W = 2$, while the latter outperforms the former in terms of MT and PD. Specifically, given $K = 8192$, the speed of lcpa-disk-m for $W = 2$ is nearly 1/3 faster than that of lcpa-disk when S varies from 200 MB to 2 GB. However, the total running time of lcpa-disk-m for $W = 3$ grows rapidly as S increases and exceeds that of lcpa-disk when S approaches 2 GB. This can be explained as follows. As described in Sect. 3, lcpa-disk-m can merge every W successive loop's rounds for decreasing LR by computing all the fingerprints possibly involved in these W successive loop rounds. For example, given $W = 2$ and $W = 3$, lcpa-disk-m maintains 3 and 7 fingerprints in each entry of ECP_k and EPP_k, respectively, and updates them during each round of the loop. This leads to a sharp growth in the computation and I/O overhead per round against lcpa-disk and becomes a performance bottleneck. Figure 2 illustrates the variation trend of MT with increasing S. The nonlinear relationship between the two parameters (i.e. MT and S) violates the assumption that MT is linear proportional to S.

Table 1. The performance results of the external memory algorithms.

S(MB)	lcpa-disk			
	LR	*MT*	*ST*	*PD*
200	13	1.03	2803	30.94
400	13	1.11	6034	31.50
600	13	1.18	9650	31.58
800	13	1.34	14557	31.52
2048	13	1.72	48021	46.04
	lcpa-disk-m (*W* = 2)			
	LR	*MT*	*ST*	*PD*
200	7	1.33	1950	47.36
400	7	1.40	4108	47.36
600	7	1.48	6928	47.36
800	7	1.80	10547	47.36
2048	7	2.47	37029	60.48
	lcpa-disk-m (*W* = 3)			
	LR	*MT*	*ST*	*PD*
200	5	2.43	2541	77.52
400	5	2.20	4605	78.58
600	5	3.14	9860	130.86
800	5	3.06	12827	103.50
2048	5	4.59	49280	94.40

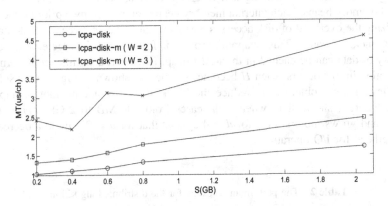

Fig. 2. *MT* for lcpa-disk and lcpa-disk-m with *S* varying from 200 MB to 2 GB, given $K = 8192$ and $W \in \{2, 3\}$.

This behavior is partially due to the use of STXXL, where the priority queue provided by the library takes logarithmic time to sort elements in external memory.

As reported in [19], the disk space requirements for eSAIS and LCPscan are 65*n* and 21*n* bytes, respectively, while the peak disk use of our implementation for lcpa-disk-m (*W* = 2) rises up to 61*n* bytes for processing a 2 GB corpora. This indicates that lcpa-disk-m has a space requirement comparable to that of eSAIS, but around 3 times as that of LCPscan. However, as will be seen in the following part, our method is much easier to be implemented and strongly scalable for the parallel and distributed model. In particular, the communication overhead on each node of the distributed model is balanced as $O(n/d)$. For $d \geq 3$, the space required by each node is less than that of LCPscan.

5.2 Performance of the Distributed Algorithm

The algorithm lcpa-ds is strongly scalable, in terms of that it can be naturally executed in parallel except for the external memory sorts. We have described in Sect. 4 how to emulate the sorter by using a group of sending and receiving buffers. The communication overhead of each computing node in Fig. 1 is upper bounded by $O(e)$. According to Amdahl's Law, the theoretical speed of lcpa-ds is $d - 1$ times faster than that of lcpa-disk, where d is the number of computing nodes in the distributed system. To validate this point, we adopt lcpa-disk-m as a baseline to evaluate the performance of lcpa-ds.

As observed from Table 2, given $K = 8192$, $W = 2$ and $d = 4$, lcpa-ds outperforms lcpa-disk-m with respect to time and space consumption, where MT/ST and PD for the former are 1/2 and 1/4 times of that for the latter, respectively. In particular, MT for lcpa-ds increases from 0.71 to 1.24 when S increases from 200 MB to 2 GB. However, the performance gain does not fully meet our expectation for the ideal case. The main reason lies on the limited internal memory capacity available on each computing node. In our implementation, each computing node spends its internal memory on not only the sending and receiving buffers, but also the I/O buffers and internal memory heaps. Each internal memory heap is employed by a priority queue for amortizing the overhead of disk accesses when swapping data between the internal and external memory banks. This swapping process is fast when H is large enough, as the majority of data can be cached in the RAM heap. However, a significant performance degradation in MT occurs when H becomes smaller, as shown in Fig. 3. One solution for relieving the problem is to reduce the block size e by adding more computing nodes. Figure 4 shows that, when S increases from 200 MB to 2 GB, MT is much smaller and grows more slowly for $d = 4$ against that for $d = 2$, due to a decrease in the overhead for I/O operations.

Table 2. The performance results for the distributed algorithm.

S(MB)	LR	lcpa-ds ($N = 4$)			lcpa-disk-m			ST Ratio (%)
		MT	ST	PD	MT	ST	PD	
200	7	0.71	1038	13.50	1.33	1950	47.36	0.54
400	7	0.72	2113	13.50	1.40	4108	47.36	0.52
600	7	0.74	3235	13.75	1.48	6928	47.36	0.50
800	7	0.78	4547	13.82	1.80	10547	47.38	0.43
2048	7	1.24	18617	16.21	2.47	37029	60.48	0.50

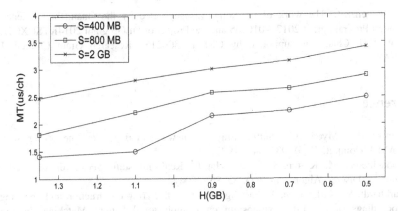

Fig. 3. *MT* for lcpa-disk-m with *H* varying from 1.35 MB to 0.5 GB, given *K* = 8192, *W* = 2, *d* = 4 and *S* ∈ {400, 800, 2048} MB.

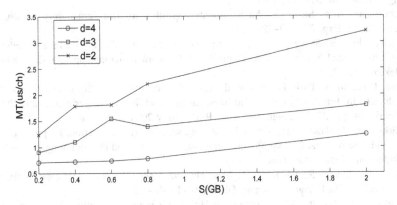

Fig. 4. *MT* for lcpa-ds with *S* varying from 200 MB to 2 GB, given *K* = 8192, *W* = 2 and *d* ∈ {2, 3, 4}.

6 Conclusion

We present in this paper a practical *K*-order LCP-array construction method that can be easily applied on both internal and external memory models. Our program for lcpa-disk-m is less than 600 lines when using STXXL to implement the external sorts. We also show that the proposed method is straightforward to be extended for running on a typical distributed system of a cluster of *d* computing nodes, where the time and space complexities are evenly divided onto each node as $O(n/d \cdot \log K)$ and $O(n/d)$, respectively. A cluster of computers in a local area network are commonly available in practice, but there is currently a lack of scalable LCP-array construction algorithms for such a distributed model. In this sense, our algorithms provide a candidate solution to meet the demand. For performance improvement of the algorithms, we are investigating techniques to exploit the use of GPUs for accelerating the computation of fingerprints.

Acknowledgement. The work of G. Nong was supported by the Guangzhou Science and Technology Program grant 201707010165 and the Project of DEGP grant 2014KTSCX007. The work of W.H. Chan was supported by GRF (18300215), Research Grant Council, Hong Kong SAR.

References

1. Manber, U., Myers, G.: Suffix arrays: a new method for on-line string searches. SIAM J. Comput. **22**(5), 935–948 (1993)
2. Abouelhodaa, M., Kurtzb, S., Ohlebuscha, E.: Replacing suffix trees with enhanced suffix arrays. J. Discret. Algorithms **2**(1), 53–86 (2004)
3. Burkhardt, S., Kärkkäinen, J.: Fast lightweight suffix array construction and checking. In: proceedings of the 14th Symposium on Combinatorial Pattern Matching, pp. 55–69, Morelia, Mexico, May 2003
4. Manzini, G., Ferragina, P.: Engineering a lightweight suffix array construction algorithm. Algorithmica **40**, 33–50 (2004)
5. Schürmann, K.B., Stoye, J.: An incomplex algorithm for fast suffix array construction. Softw. Pract. Exp. **37**, 309–329 (2007)
6. Ko, P., Aluru, S.: Space efficient linear time construction of suffix arrays. In: Proceedings of the 14th Annual Symposium on Combinatorial Pattern Matching, pp. 200–210, Morelia, Mexico, May 2003
7. Kim, D.K., Jo, J., Park, H.: A fast algorithm for constructing suffix arrays for fixed-size alphabets. In: Proceedings of the 3rd International Workshop on Experimental and Efficient Algorithms, pp. 25–28, Angra dos Reis, Brazil, May 2004
8. Kärkkäinen, J., Sanders, P.: Simple linear work suffix array construction. In: proceedings of the 30th International Colloquium on Automata, Languages and Programming, pp. 943–955, Eindhoven, Netherlands, June 2003
9. Nong, G., Zhang, S., Chan, W.H.: Two efficient algorithms for linear time suffix array construction. IEEE Trans. Comput. **60**(10), 1471–1484 (2011)
10. Puglisi, S.J., Smyth, W.F., Turpin, A.H.: A taxonomy of suffix array construction algorithms. ACM Comput. Surv. **39**(2), 1–31 (2007)
11. Fischer, J.: Inducing the LCP-array. Algorithms Data Struct. **6844**, 374–385 (2011)
12. Bingmann, T., Fischer, J., Osipov, V.: Inducing suffix and LCP arrays in external memory. In: Proceedings of the 15th Workshop on Algorithm Engineering and Experiments, pp. 88–102 (2012)
13. Kasai, T., Lee, G., Arimura, H.: Linear-time longest-common-prefix computation in suffix arrays and its applications. In: proceedings of the 12th Annual Symposium on Combinatorial Pattern Matching, pp. 181–192, Jerusalem, Israel, July 2001
14. Manzini, G.: Two space saving tricks for linear time LCP array computation. In: Proceedings of the 9th Workshop on Algorithm Theory, pp. 372–383, Humlebaek, Denmark, July 2004
15. Kärkkäinen, J., Manzini, G., Puglisi, S.J.: Permuted longest-common-prefix array. In: Proceedings of the 20th Annual Symposium on Combinatorial Pattern Matching, pp. 181–192, Lille, France, June 2009
16. Nong, G., Chan, W.H., Zhang, S., Guan, X.F.: Suffix array construction in external memory using D-critical substrings. ACM Trans. Inf. Syst. **32**(1), 1–15 (2014)
17. Nong, G., Chan, W.H., Hu, S.Q., Wu, Y.: Induced sorting suffixes in external memory. ACM Trans. Inf. Syst. **33**(3), 1–15 (2015)

18. Dementiev, R., Kärkkäinen, J., Mehnert, J., Sanders, P.: Better external memory suffix array construction. ACM J. Exp. Algorithmics **12**(3), 1–24 (2008)
19. Kärkkäinen, J., Kempa, D.: LCP array construction in external memory. In: Proceedings of the 13th International Symposium on Experimental Algorithms, pp. 412–423, Copenhagen, Denmark, June 2014
20. Louza, F., Telles, G., Ciferri, C.: External memory generalized suffix and LCP arrays construction. In: Proceedings of the 24th Annual Symposium on Combinatorial Pattern Matching, pp. 201–210, Bad Herrenalb, Germany, June 2013
21. Bauer, M., Rosone, A.C.G., Sciortino, M.: Lightweight LCP construction for next-generation sequencing datasets. In Proceedings of the 12th International Workshop on Algorithms in Bioinformatics, pp. 326–337, Ljubljana, Slovenia (2012)
22. Bille, P., GØrtz, I.L., Kopelowitz, T., Sach, B., VildhØj, H.W.:. Sparse suffix tree construction in small space. In Proceedings of the 40th International Colloquium on Automata, Languages, and Programming, pp. 148–159, Riga, Latvia, July 2013
23. Shun, J.: Fast parallel computation of longest common prefixes. In Proceedings of the 40th International Conference on High Performance Computing, Networking, Storage and Analysis, pp. 387–398, New Orleans, LA (2014)
24. Deo, M., Keely, S.: Parallel suffix array and least common prefix for the GPU. In: Proceedings of the 18th ACM SIGPLAN Symposium on Principles and Practice of Parallel Programming, pp. 197–206, New York, USA, August 2013
25. Karp, R., Rabin, M.: Efficient randomized pattern matching algorithms. IBM J. Res. Dev. **31** (2), 249–260 (1987)

A Load Balancing Strategy for Monte Carlo Method in PageRank Problem

Bo Shao[1,2], Siyan Lai[1,2], Bo Yang[1,2], Ying Xu[1,2],
and Xiaola Lin[1,2(✉)]

[1] School of Data and Computer Science, Sun Yat-Sen University, Guangzhou,
China
{shaobo2,laisy2,yangb65,xuying63}@mail2.sysu.edu.cn,
linxl@mail.sysu.edu.cn
[2] The Key Laboratory of Machine Intelligence and Advanced Computing,
Sun Yat-Sen University, Ministry of Education, Guangzhou, China

Abstract. PageRank algorithm is key component of a wide range of applications. Former study has demonstrated that PageRank problem can be effectively solved through Monte Carlo method. In this paper, we focus on efficiently parallel implementing Monte Carlo method for PageRank algorithm based on GPU. Aiming at GPU, a parallel implementation must consider instruction divergence on the single instruction multiple data (SIMD) compute units. Due to the fact that low-discrepancy sequences are determined sequences, we adopt the low-discrepancy sequences to simulate the random walks in PageRank computations in our load balancing strategy. Furthermore, we allocate each thread of a block to compute a random walk of each vertex with a same low-discrepancy sequence. As a result, no idle thread exists in the PageRank computations and warp execution efficiency is up to 99%. Moreover, our strategy loads the low-discrepancy sequences into shared memory to reduce the data fetch cost. The results of experiments show that our strategy can provide high efficiency for Monte Carlo method in PageRank problem in GPGPU environment.

Keywords: Monte Carlo method · PageRank · GPGPU · Parallel computing · Load balancing

1 Introduction

The key issue of search engines is to list the relevant pages of queries in proper order. PageRank algorithm is developed by Larry Page and Brin et al. to address this issue [16]. PageRank is first introduced as a principle criteria with which Google ranks the importance of webpages on the Internet. It's a stationary vector of a random walk which simulates the process of surfing the Internet [15]. Therefore, it can be interpreted as the frequency that a random surfer browses the web page, to some extent, it determines the popularity of the page [1, 2]. PageRank algorithm has been a key component of a wide range of applications.

To define the PageRank problem, firstly, denote the total number of pages on the Web as n, and define the $n \times n$ hyperlink matrix P in the following way:

© Springer Nature Singapore Pte Ltd. 2017
G. Chen et al. (Eds.): PAAP 2017, CCIS 729, pp. 594–609, 2017.
DOI: 10.1007/978-981-10-6442-5_56

$$P_{ij} = \begin{cases} \frac{1}{k}, & \text{if } i \text{ has } k > 0 \text{ } \textit{outgoing links and } j \text{ is one of the links;}\\ \frac{1}{n}, & \text{if } i \text{ has no outgoing link;}\\ 0, & \text{otherwise.} \end{cases}$$

If a page has no outgoing link, the surfer will start navigating another page randomly to ensure that the process continues, therefore, the probability is spread among all pages of the web. As for those pages which have $k > 0$ outgoing links, it is assumed that the surfer will chose an outgoing link randomly with probability c or jump to a random page with probability $(1 - c)$. Thus, the process can be interpreted as a Markov chain whose state space is the set of all Web pages, and the transition matrix is

$$\widetilde{P} = cP + (1 - c)(1/n)E. \tag{1}$$

Here E is a matrix whose entries all equal one and the damping factor $c \in (0, 1)$ is typically chosen to be 0.85. The PageRank is defined as a stationary distribution of the Markov chain, that is, there exists a unique row vector π such that

$$\pi\widetilde{P} = \pi, \quad \pi\underline{1} = 1, \tag{2}$$

where $\underline{1}$ is a column vector of ones.

Consider that there's no analytical formulas for general directed graphs, it's difficult to compute the PageRank distribution exactly [12]. However, there are many methods can be used to estimate the PageRank distribution. All these methods can be roughly classified into two kinds: the linear algebraic formulations and the Monte Carlo methods. The linear algebraic techniques like Power Iteration method compute the PageRank vector iteratively by $\pi^k = \pi^{(k+1)}\widetilde{P}$, starting from a uniform distribution vector $\pi^0 = (1/n)\underline{1}^T$, it won't stop until the precision $\in$ is achieved. The numbers of flops needed for the method to converge is of the order $\frac{\log \varepsilon}{\log c}nnz(P)$, where $nnz(P)$ is the number of non-zero elements of the matrix P, and it's approximately linear in n [17–21]. With Monte Carlo methods, the PageRank of important pages can be computed accurately after the first iteration.

The prime advantage of the Monte Carlo methods is that they can determine the PageRank of important pages with high accuracy after the first iteration. Also they are highly parallelizable, and allow continuous update of the PageRank. In order to keep up with rapid changes in the hyperlink structure of the web, Google and some social networks like Facebook have to update its PageRank very frequently. The problem is that the matrix P regularly has a huge size of billions by billions, makes the PageRank update an intricate task. Instead of the power iteration method employed initially by Google and the linear system formulation noticed by Moler and Moler [4], Bianchini et al. [5], and Breyer and Roberts [6], it turns out that the Monte Carlo methods are more efficient in term of PageRank computation.

On the other hand, GPU which has a large number of compute units attracts increasing interest in high performance computing applications as its potential high raw compute power. In this Paper, we study on how to efficiently parallel implement Monte Carlo method to solve PageRank problem under GPGPU environment. Due to the specific architecture of GPGPU, some factors affect the performance of implementation, such as instruction divergence and load balancing between the threads of a warp, memory access conflicts between threads, cache performance among threads, and local memory utilization. We will consider the factors to optimize the implementation.

As GPGPU adopts SIMD model, threads sequentially execute when branch divergence occurs in a warp. As Monte Carlo method simulates absorbing Markov chains for PageRank problem, the load balancing problem is caused by instruction divergence when pseudo random numbers are applied. In order to avoid the instruction divergence between threads, we harness the low-discrepancy sequences to improve the performance. The low-discrepancy sequences are applied to replace the pseudo random sequence in many fields to reduce the variance of sampling in Monte Carlo methods [9–11]. One of the most important features is that low-discrepancy sequences are determined sequences, so the random walks of each vertex are simulated by the same low-discrepancy sequences. We propose a divergence avoidable strategy to allocate the threads of a block to compute a random walk of each vertex with identical low-discrepancy sequence. Hence, the threads of a block will simultaneously execute same instruction to compute next state of Markov chain or terminate. We can address the instruction divergence by our strategy. Furthermore, we consider a static load balancing strategy and a dynamic load balancing strategy in the specific implementation based on instruction divergence avoidable strategy. The static load balancing strategy pre-computes the length of each Markov chain to reduce the cost of absorbing state determination. The dynamic load balancing strategy dynamically determines the length of Markov chains to avoid the cost of loop condition determination. We will use experiments to test the efficiency of two strategies.

Another optimization is that we load the low-discrepancy sequences into shared memory to speed up data fetch operations. Since fetching data from on-chip memory can be about 10X faster than off-chip memory in GPU memory architecture [13]. Due to the fact that the repeated fetch operation of same low-discrepancy sequences from the off-chip GPU global memory for different vertices in a block, we take advantage of on-chip shared memory to reduce the time of fetch operation. The results of experiments demonstrate that our optimizations can significantly improve the performance of PageRank computations on GPU based on Monte Carlo method. The experiments indicate the bottleneck of our strategy is the memory access conflicts between threads.

The remainder of the paper is organized as follows. Section 2 provides background of Monte Carlo Method for PageRank Problem on Sect. 2.1 and a basic implementation in GPGPU environment on Sect. 2.2. Section 3 describes the optimizations we propose. Section 4 presents evaluation methodology and results. Section 5 gives conclusion and direction for future work.

2 Implementation of Monte Carlo Method in PageRank

2.1 Monte Carlo Method for PageRank Problem

We briefly introduce the core concept of Monte Carlo method of PageRank computation. The Monte Carlo method to compute PageRank problem was introduced by

Avrachenkov et al., who has also discussed the correctness of method in theory [8]. The Except for the absorbing Markov chains that explore in the former research, there are also several different interpretations for PageRank through expectations. For instance, the PageRank can be interpreted as the average number of surfers navigating a given page at a given time instant. A surfer can continue navigating by chosen a random outgoing link with probability c or cease navigating and jump to each page on the Web with probability $(1 - c)$ [3]. This interpretation is helpful in understanding but it involves the time component which makes it hard to use in practice. While the interpretation via absorbing Markov chains is easier to implement for simulation algorithms of PageRank computation.

Assuming a robot crawls the web from a chosen page, at each step, it either terminates crawling with probability $(1 - c)$ or crawls a randomly chosen outgoing link 7. This process can be interpreted as a discrete-time absorbing Markov chain, in which a transition from a transient state to an absorbing state occurs with probability $(1 - c)$. Each state in the Markov chain corresponds to the web in the Internet. Then there comes the Monte Carlo algorithm. First, for each transient state i in the Markov chain, run m independent Monte Carlo simulations, and cm/k simulations through each of k outgoing links of state i, use the empirical mean number of visits to state j (excluding the visit to absorbing state) as an estimator for state i to j. Then use the sum of all the transient states' empirical mean number of visits to state j as an estimator for state j. So we can roughly define the PageRank of state j as a fraction of estimator for j divided by estimator for all the states. The accurate formulation is

$$\pi_j = \frac{estimator\ for\ j + 1}{\sum_{l=1}^{n}(estimator\ for\ l + 1)}, \tag{3}$$

Therefore we use the vector $\pi = (\pi_1, \ldots, \pi_n)$ as the estimate of PageRank weighted value π.

2.2 Basic Implementation

In this subsection, we describe how to implement the Monte Carlo method on GPGPU, including storage format and the details of basic parallel implementation.

2.2.1 Storage Format
Since real world graph are always very sparse, each row only contains a few non-zero element, in this case we need to choose a format to decrease the storage space needed. In this method, we store the adjacency matrix of a graph as the Compressed Sparse Row (CSR) format because it's space-efficient. Instead of explicitly store the row indices of the non-zero elements, the non-zero elements in rows are stored contiguously in a vector, along with a vector that contains the offset for the first of each row's elements in the vector. The CSR representation of Fig. 1 is shown in Fig. 2, none zero elements of row i and the corresponding column indices are stored respectively in the data and column index vectors at index r: *RowOffset [i] $\leq$ r < RowOffst [i + 1]*. Then as a Graph(V, E), the space complexity is $O(|V| + |E|)$.

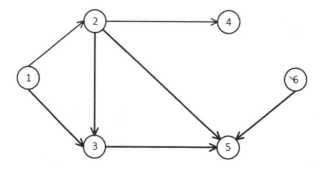

Fig. 1. An example of directed Graph

Row offset

1	2	3	4	5	6	
1	3	6	7	7	7	8

Colum index

| 2 | 3 | 3 | 4 | 5 | 5 | 5 |

Fig. 2. An example for CSR format

2.2.2 Basic Implementation

Since each random walk can be simulated independently, we can easily assign the computation of each walk for each thread. Furthermore, the all random walks begin from same node will be assigned to one block and all threads in a block simulate all random walks starting from one vertex in parallel. Figure 3 illustrates the computing model of kernel function. We need to apply sum to $|V|$ blocks to run the algorithm. Each block computes one vertex k in block k correspondingly. Threads loop several times to simulate the random walks in each block.

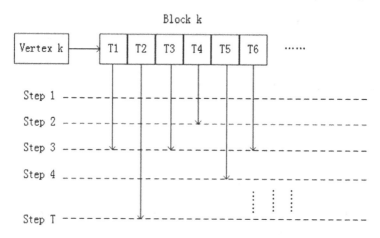

Fig. 3. The computing model of basic implementation

In Algorithm 1, we implement Monte Carlo method for PageRank computation on GPU. N is the number of random walks and T is the max length. The block id is equal to id of the start vertex of random walks, so we can get the start vertex s (line 1). Then we can get the random work number from thread id (line 4). And in line 4, k represents the id of current random walk. Then the inner loop simulates the random walk up to T steps. In each step, we first load the random numbers in line 6, and then choose the next state of Markov chain according to p. Finally, record the result in the pagerank_value array at the end of each loop.

Algorithm 1 basic_kernel

Input:

    1.n: the number of vertexes in the graph;

    2.N: the number of random walks;

    3.T: the max length of each random walk;

    4.R: the array of pre-generating random sequence in global memory

    5.alpha: damping factor

Output:

    1.pagerank_value: PageRank weighted value of vertices.

    1.tid←the current thread id in a block

    2.bid←the current block id

    3.s = bid

    4.for k = tid to N, k+=blockdim:

    5.   for i = 0 to T:

    6.   p = R[N*i+ k]

    7.   If p>aloha return

    8.   load the child nodes of s into Vc

    9.   s = Vc[p/alpha*| Vc |]

    10.   Atomic_add(pagerank_value[s],1)

Host: basic_kernel<N, blockdim>(…)

In this algorithm, we need to store the information of graph as CSR format, which has two arrays. The row offset array contains $|V|$ column offset number of all vertices. Then $|E|$ column indexes store in the other array consecutively. Therefore, the space complexity is O $(|V| + |E|)$.

3   Optimization

In this section, we will analysis the defect of basic implementation and then counter to these disadvantages. Low-discrepancy sequence are adopted to avoid threads divergence, and to achieve load balancing between threads in a block to improve the performance on GPU.

As showed in Fig. 3, we can see some drawbacks in basic implementation. Firstly threads in a block have serious problem of loading imbalance. As threads in a block repeatedly simulate different random walks of same vertex and every step for each random walk has a termination probability of $(1 - c)$, then the length of all random walks will from 0 to $T*r$, where r is the number of random walks simulated by a thread. Then the time performance of a block will be critically influenced by the thread with longest random walks. So we try to use the determinacy of quasi random numbers to solve this problem and propose some computing model and some other tricks to improve the performance.

3.1 Static Load Balancing Strategy

We want to solve problem of instruction divergence which is ubiquitous in basic implementation. The reason of branch divergence is that different random walks stop at arbitrary steps. Then we need to determine the end condition at each step, which will cause serious instruction divergence, and bring seriously load imbalance problem.

We have defined some parameters above, the number of random walks N, the max length T, and damping factor c. In first step, the expectation of random walk of the same start vertex randomly skipping to other node is $N*c$, the other $(1 - c)*N$ walks finish. Similarly after next step, there will remain $N*c^2$. Then we can get the expectation of the random walks' number R_k after k step:

$$R_k = N * \alpha^k \tag{5}$$

Then the expectation number of random walks whose length is k can be computed by R_k:

$$W_k = \begin{cases} R_{k-1} - R_k & k \leq T \\ R_k & k = T \end{cases} \tag{6}$$

According to Eq. (6), we can set the constant number W_k of each step k. Then we can allocate corresponding thread number to compute the random walks. Therefore we can call T kernels which have constant number of threads, W_k, each kernel computes k steps so that random walks in block have same length, in other words under same workload.

Figure 4 shows the kernel functions' execution in device. For different length of random walks, each stream call different kernel functions to compute in parallel. Each kernel will have W_k blocks. Furthermore, in each block i of the same kernel function, it will compute random walks of *blockDim* vertices with the same quasi random numbers. *blockDim* is the number of threads in a block, we always set to 1024, that means we compute 1024 vertices in one time. And we use CudaStream to call different kernels to execute for many times to compute all vertices. Then all threads in a block have same workload, as a result, the warp execution efficiency will be improved.

We utilize the low-discrepancy sequences to reduce the times of fetching the random sequence from global memory in GPU. Since the quasi-random sequence of numbers is deterministic, random walks started from different vertices can share the

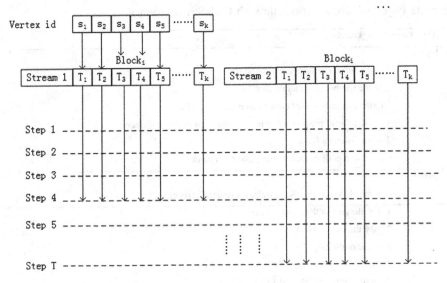

Fig. 4. Static method implementation

same sub-sequences. Therefore, we assigned each random walk using same random sequence in a block. In Fig. 4, we can see the mapping relationship clearly. We compute *blockDim* vertices in a block, then each block only need to load the random numbers from global memory one time.

Algorithm 2 Host driver
Input:
1. n: the number of vertexes in the graph;
2. N: the number of random walks;
3. T: the max length of each random walk;
4. R: the array of pre-generating random sequence in global memory
5. alpha: damping factor
1. R[0] = N
2. for i =1 to T-1:
3.   R[i] = R[i-1]*alpha
4.   W[i] = R[i-1]-R[i]
5. W[T] = R[T-1]*alpha
6. for i =0 to N/blockDim:
7.   for j = 1 to T:
8.   static_kernel<<<W[j],blockDim,newStream>>>()

In Algorithm 2, before we call kernel functions, we first compute array *R* and *W* (line1 to line5). Then we can get the size of each kernel. We solve *blockDim* vertices in a loop. According to different length of random walks, we use variant kernels to execute PageRank computations through CudaStream control.

Algorithm 3 static_kernel
Input:
1.n: the number of vertexes in the graph;
2.N: the number of random walks;
3.L: the current length of each random walk in the block;
4.R: the array of pre-generating random sequence in global memory
5.alpha: damping factor
6.s: the first node of each vertices group in a block
Output:
1.pagerank_value: PageRank weighted value of vertices.
1.tid←the current thread id in a block
2.bid←the current block id
3.shared probility[]
4.if tid<T
5.   probility[tid] = R[N* tid + bid]
6.synchronization
7.s = s + tid
8.for i = 0 to L:
9.   p = probility[i]
10.   load the child nodes of s into Vc
11.   s = Vc[p *| Vc |]
12.   Atomic_add(pagerank_value[s],1)

In algorithm 3, each kernel computes the *blockDim* random walks individually in GPU. And each thread simulates the walk in L step, L is from 1 to T. Different random walks in the same blocks belong to different vertices, then they share the same quasi random sequence loaded in *probility*. From 3 to 6, we load the quasi random numbers in shared memory. Then get the start vertex index by the thread id in a block. Finally simulate the random walks for L times, which is the walks' length in its kernel function.

3.2 Dynamic Load Balancing Strategy

The static strategy has successfully solved load balancing problem, nevertheless, the overhead of CudaStream control should be considered. The cost of calling additional kernels also increases in static strategy as well.

Since random walks in a block share the same random sequence in static strategy, we use the quasi random numbers to judge whether to stop the random walks. In algorithm 2, we load a random number p at each step, then if $p > c$, the random walk will stop. Otherwise also use p to determine skipping next vertex after uniform p in range 0 to 1. And write back the result into the PageRank weighted value.

As shown in Fig. 5, s_i is the vertex's index, T_i is thread id in a block. Then we need N blocks in total, each block compute random walks of k vertices using the same quasi random numbers. We needn't allocate workload for each block, when kernel function execute, the random numbers loaded by each block dynamically decide the length of random walks in the execution. Then all random walks in a block have the same length because they use same quasi random numbers to decide termination conditions.

Algorithm 4 dynamic_kernel

Input:

    1.n: the number of vertexes in the graph;

    2.N: the number of random walks;

    3.T: the max length of each random walk;

    4.R: the array of pre-generating random sequence in global memory

    5.alpha: damping factor

    6.s: the first node of each vertices group

Output:

    1.pagerank_value: PageRank weighted value of vertices.

    1.tid ← the current thread id in a block

    2.bid ← the current block id

    3.shared probility[]

    4.if tid<T

    5. probility[tid] = R[N* tid + bid]

    6.s = s + tid

    7.for i = 0 to T:

    8. p = probility[i]

    9. If p>alpha return

    10.   load the child nodes of sinto V_c

    11.   s = V_c[p/alpha*| V_c |]

    12.   Atomic_add(pagerank_value[s],1)

Host: for i =0 to N/block_dim:

        dynamic_kernel <<N, blockDim >>(…)

In Algorithm 4, we improve the performance from the static method. Instead of using CudaStream to assign the length of random walks to different block in different kernel functions, we only call one kernel function and utilize the quasi random numbers to determine the step length dynamically (line 8). At host, we simulate a group of vertices in a loop, the number of a group is equal to the *blockDim*, and always be 1024 on GPU same as static method. From line 3 to line 5, we fetch the random numbers in shared memory in order to all random walks of each vertex in a group needn't load from global memory. Then each block computes the random works of the vertices group with the same random sequence. Similar to algorithm 1, simulate a Markov chain from line 7 to line 11, save the result. In basic implement, to fetch random number, the

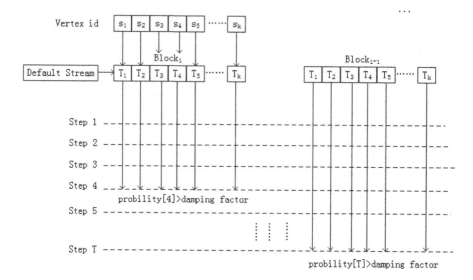

Fig. 5. Dynamic method implement

kernel function need to load from global memory for $T * N * |V|$ times, in static kernel, the number of accessing global memory decreases to $T * N * |V|/blockDim$.

We also utilize some other features of GPU to accelerate the performance of dynamic load balance method. The Graph data stored by CSR format needn't be modified in computation, so we can use "restrict" key word to limit it in order to decrease the number of instructions and then improve the performance. And the cost of loading random numbers in GPU is shared by all threads in one block. We also keep this read operation has the best locality to accelerate the global memory access.

4   Experiment

We report the numerical experiments to examine the efficiency of our strategies for Monte Carlo method for PageRank computations in this section. All the numerical results are obtained with NVidia GTX 1070. NVidia GTX 1070 has 1920 CUDA cores, and including 15 SMs which have 128 CUDA cores individually. We use several graphs of real world to analyze the efficiency of our strategies in GPGPU environment. Table 1 shows the characters of SNAP datasets [22] we use in the experiment (Fig. 6).

4.1   Performance

In this test, we can see the running time of static strategy and dynamic strategy compared to basic implementation. It is obviously that both static strategy and dynamic strategy can achieve better performance than basic implementation. For all the graphs, the two optimizations can get average 2X speed up ratio, on average. As RoadNet-CA dataset has most vertices, the load imbalance problem is most serious in basic

Table 1. Kernel analysis between basic method and dynamic method.

Datasets	Vertexes	Edges	Average degree
cit-hepPh	34546	421578	12.203
Douban	154908	654188	4.2230
Amazon0302	262111	1234877	4.7112
dblp	317080	1049866	3.3110
web-Stanford	281903	2312497	8.2031
Wiki-Vote	7115	103689	14.573
youtube	1134890	2987624	2.6325
RoadNet-CA	1965206	5533214	2.8160

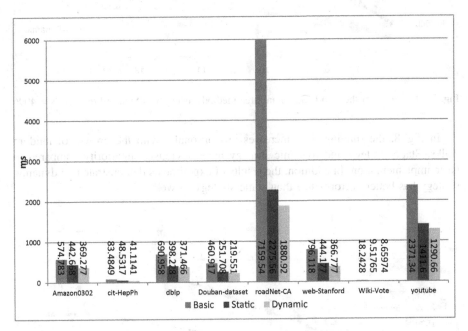

Fig. 6. Performance of Monte Carlo method for PageRank computation, N = 2048, T = 10

implementation, so optimizations have best speed up ration in this dataset. Compared dynamic strategy to static strategy, the performance of former strategy is superior to the latter one. We deduce that overhead of stream control is larger than termination condition decision in PageRank computing.

4.2 The Effect of Random Walk Lengths and Numbers

As illustrated in Fig. 7, with the growth of step length, the increment run time of basic method are much faster than static strategy and dynamic strategy. As running time of each block is decided by thread with longest length, the performance will be critically influenced by random walk length T. Nevertheless, since the number of random walks with k steps R_k is always a small number with growth of k. As a result, the length of random walk won't affect the performance of two optimization strategies too much.

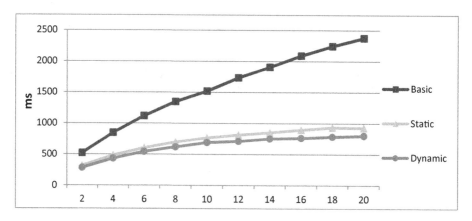

Fig. 7. The effect of the step length T in three methods, using web-Stanford dataset, N = 4096

In Fig. 8, the running time increases proportionally with the growth of random walks. Static strategy and dynamic strategy have obviously superiority compared to basic implementation. In addition, the results of experiments demonstrate that dynamic strategy has better performance than static strategy as well.

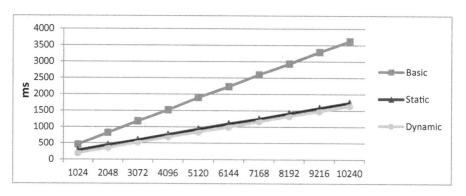

Fig. 8. The effect of the random walk's number N in three methods, using web-Stanford dataset T = 10

We also make this test in all datasets and the result is same as web-Stanford, which verifies our deduction.

4.3 Visual Profiler

In Table 2, we show the specifics of memory utilizations and warp executions. Since static strategy applies CudaStream to execute several kernels one time, so we just compare basic implementation and dynamic strategy. Obviously, dynamic strategy utilizes the L2 cache and shared memory to improve the speed of accessing the quasi random sequences. Furthermore, dynamic strategy has better space locality in global memory reading and writing. As a result, dynamic strategy has better warp execution efficiency 99% than 48.2% of basic implementation. The kernel of dynamic strategy has lower memory dependency and higher execution dependency too. Thus the warp execution efficiency up to 99% means that compute resources are nearly fully utilized when all threads in a warp have the same branching and predication behavior.

Table 2. Kernel analysis between basic method and dynamic method.

	Basic method	Dynamic method
Shared memory utilization	0	42.108 GB/s
L2 cache utilization	814.722 GB/s	923.155 GB/s
L1 cache/Texture utilization	644.496 GB/s	469.417 GB/s
Global memory read/write	0.9642 GB/s	1.843 GB/s
Memory dependency	78.64%	62.36%
Execution dependency	15.90%	23.89%
Warp execution efficiency	48.2%	99%
Active warps	62.3	57

4.4 The Bottleneck of Dynamic Strategy

In this test, we investigate the portion of pure computing time in dynamic strategy. We separate the PageRank computations as two parts, including pure computing time and the recording time of the vertices visits in Fig. 9. It's obvious that the random access global memory to record vertices visits at each step is the prestige bottleneck in this method.

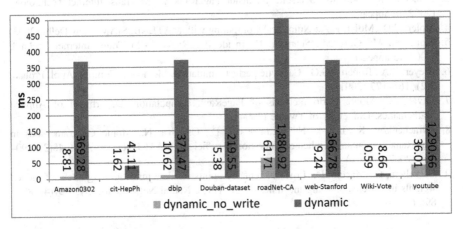

Fig. 9. The effect of write back operation for performance, N = 2048, T = 10

In this section, we verify the efficient of our strategies through a series of experiments. The results of experiments demonstrate that static and dynamic method has a great advantage than basic method for all kinds of graphs. In addition, we analyses the performance with different random walks' number and length. We also use visual profile to learn the run status of kernel functions, and confirm that our dynamic method has solved nearly perfect. Finally we also find the bottleneck of the dynamic method.

5   Conclusion

In this paper, we have studied efficient implementation of Monte Carlo method for PageRank problem in GPGPU environment. According our analysis, a load balancing problem is caused by instruction divergence between warp when pseudo random numbers are applied. We adopt low-discrepancy sequences to overcome this problem in our optimization. Specially, we allocate each thread of a block use identical low-discrepancy sequence to simulate a random walk for a vertex correspondingly. Based on the above optimization, we propose a static load balancing strategy and a dynamic load balancing strategy in the specific implementation. Moreover, the low-discrepancy sequences are organized to load in the on-chip shared memory to accelerate fetching quasi random numbers. The results of experiments show that the dynamic strategy can achieve about 2X speed up ration, on average. The experiments indicate that the bottleneck of our load balancing strategy is memory access conflicts of recording the visits of vertices. We are going to address this issue in the future work.

References

1. Brin, S., Page, L.: The anatomy of a large-scale hypertextual web search engine. In: 7th International World Wide Web Conference, Brisbane, Australia (1998)
2. Avrachenkov, K., Litvak, N., Nemirovsky, D., Osipova, N.: Monte Carlo methods in pagerank computation: when one iteration is sufficient. SIAM J. Numer. Anal. **45**(2) (2007)
3. Bianchini, M., Gori, M., Scarselli, F.: Inside PageRank. ACM Trans. Internet Technology (2002, to appear)
4. Moler, C.D., Moler, K.A.: Numerical Computing with MATLAB. SIAM, New Delhi (2003)
5. Bianchini, M., Gori, M., Scarselli, F.: Inside PageRank. ACM Trans. Internet Technol. (2002, to appear)
6. Breyer, L.A., Roberts, G.O.: Catalytic perfect simulation. Methodol. Comput. Appl. Probab. **3**(2), 161–177 (2001)
7. Litvak, N.: Monte Carlo methods of PageRank computation. Department of Applied Mathematics, University of Twente (2004)
8. Avrachenkov, K., Litvak, N., Nemirovsky, D., Osipova, N.: Monte Carlo methods in PageRank computation: when one iteration is sufficient. SIAM J. Numer. Anal. **45**(2), 890–904 (2007)
9. Cervellera, C., Macciò, D.: Low-discrepancy points for deterministic assignment of hidden weights in extreme learning machines. IEEE Trans. Neural Netw. Learn. Syst. **27**(4), 891–896 (2016)

10. Gan, G., Valdez, E.A.: An empirical comparison of some experimental designs for the valuation of large variable annuity portfolios. Dependence Model. **4**(1) (2016)
11. Zapotecas-Martínez, S., Aguirre, H.E., Tanaka, K., et al.: On the low-discrepancy sequences and their use in MOEA/D for high-dimensional objective spaces. In: 2015 IEEE Congress on Evolutionary Computation (CEC), pp. 2835–2842. IEEE (2015)
12. Sarma, A.D., et al.: Fast distributed PageRank computation. Theoret. Comput. Sci. **561**, 113–121 (2015)
13. Andersch, M., Lucas, J., Lvlvarez-Mesa, M.A., et al.: On latency in GPU throughput microarchitectures. In: 2015 IEEE International Symposium on Performance Analysis of Systems and Software (ISPASS), pp. 169–170. IEEE (2015)
14. Langville, A.N., Meyer, C.D.: Deeper inside PageRank. Preprint, North Carolina State University (2003)
15. Yang, W., Zheng, P.: An improved PageRank algorithm based on time feedback and topic similarity. In: 2016 7th IEEE International Conference on Software Engineering and Service Science (ICSESS), Beijing, pp. 534–537 (2016)
16. Page, L., Brin, S., Motwani, R., Winograd, T.: The PageRank citation ranking: bringing order to the web. Stanford Digit. Libr. Working Paper **9**(1), 1–14 (1999)
17. Boldi, P., Santini, M., Vigna, S.: Page rank: functional dependencies. ACM Trans. Inf. Syst. **27**, 1–23 (2009)
18. Chung, F.: The heat Kernel as the PageRank of a graph. Proc. Natl. Acad. Sci. U.S.A. **104**, 19735–19740 (2007)
19. Langville, A.N., Meyer, C.D.: Google's PageRank and Beyond: The Science of Search Engine Rankings. Princeton University Press, Princeton (2006)
20. Meyer, C.D.: Limits and the index of a square matrix. SIAM J. Appl. Math. **26**(3), 469–478 (1974)
21. Vigna, S.: Spectral Ranking, November 2013
22. Leskovec, J., Krevl, A.: {SNAP Datasets}:{Stanford} Large Network Dataset Collection (2015)

Experiences of Performance Optimization for Large Eddy Simulation on Intel MIC Platforms

Zhengxiong Hou[1]($^{\boxtimes}$), Chengwen Zhong[2], Christian Perez[3],
Qing Zhang[4], and Yunlan Wang[1]

[1] Center for High Performance Computing,
School of Computer Science and Technology,
Northwestern Polytechnical University, Xi'an 710072, China
houzhx@gmail.com
[2] School of Aeronautics, Northwestern Polytechnical University,
Xi'an 710072, China
[3] INRIA/LIP, 46 allee d'Italie, 69364 Lyon Cedex, France
[4] Inspur Corporation, Jinan, China

Abstract. Large Eddy Simulation (LES) is a mathematical model for turbulence used in Computational Fluid Dynamics (CFD). We have implemented LES on multi-core CPUs and General Purpose Graphics Processing Units (GPGPUs). In this work, we port and optimize LES on Intel Many Integrated Core (MIC) platforms. On Intel MIC co-processor (KNC), we implement LES using the main execution modes, including native, offload and symmetric execution modes. The newly emerging second generation of Intel MIC processor (Knights Landing, i.e. KNL) acts as an independent multi-core computing node, it is more convenient to port the application. On both of the MIC platforms, some important performance optimization techniques are implemented and evaluated, such as parallelization with OpenMP threads and MPI processes, single-instruction-multiple-data (SIMD) vectorization, memory access optimization, threads scheduling, etc. The experimental results demonstrate that performance optimization techniques are very important when porting applications on MIC platforms.

Keywords: LES · Xeon Phi · MIC · KNC · KNL · Performance optimization

1 Introduction

Large eddy simulation (LES) is a mathematical model for turbulence used in computational fluid dynamics (CFD). Previous work has been proposed for a compute unified device architecture (CUDA)-based simulation solution of LES using General Purpose GPU (GPGPU) [1]. The solution adopts the "collision after propagation" lattice evolution way and puts the misaligned propagation phase at global memory read process. In order to make use of multiple GPGPUs, the whole working set is evenly partitioned into sub-domains. The experimental results show that it scales well on multiple NVIDIA TeslaC2070 GPUs.

© Springer Nature Singapore Pte Ltd. 2017
G. Chen et al. (Eds.): PAAP 2017, CCIS 729, pp. 610–625, 2017.
DOI: 10.1007/978-981-10-6442-5_57

Intel Many Integrated Core (MIC) is newly emerging as one of the main many-core platforms for high performance computing (HPC). In the recent list of TOP 500 supercomputers [2], there are already dozens of supercomputers using Intel MIC platforms, which is comparable with those using GPGPUs. Especially, there are already some supercomputers equipped with the newly second generation of Intel MIC Knights Landing (KNL) processor [3]. With the increasing number of cores in the many core architectures, whether application performance scales well with the number of cores is a main concern. For some benchmarks, such as Linpack [4], they may use the maximum number of cores per node and the maximum number of nodes to obtain the optimal performance on a many-core platform. While for real HPC applications, it is very complicated, especially on diverse heterogeneous many-core platforms. When running real applications on a many-core machine, application performance does not necessarily improve as more cores are added [5]. There are some more factors impacting performance, such as vectorization, cache and memory access, etc. A basic question is how to optimize performance and obtain the highest speedup for HPC applications. Automatic optimization in compilers is limited. So, we are interested in the porting and performance optimization of LES on Intel MIC platforms. Porting to Intel MIC platforms is comparatively easy than porting to GPGPU platform. There are also some important approaches for performance optimization on Intel MIC platform. In this work, we present the implementation, performance optimization and evaluation of LES on hybrid CPU and Intel MIC (Xeon Phi) platforms.

The remaining of the paper is organized as follows. In Sect. 2 we introduce related work. The target CPU and MIC platforms and its benchmark results are introduced and analyzed in Sect. 3. Section 4 describes the implementation of LES on hybrid CPU and MIC platform. Section 5 presents the performance optimization techniques on the MIC platforms. Section 6 presents the experiments and performance evaluation for the optimization techniques. In Sect. 7, we give a discussion for the multi-core CPU and many-core processors. Finally, Sect. 8 concludes the paper.

2 Related Work

For the scalability of parallel applications on multi-core and many-core architectures, in the paper [5], the authors perform a scalability analysis of parallel applications on a 64-threaded Intel Nehalem-EX based system. They found that applications which scale well on small number of cores, exhibit poor scalability on large number of cores. A lot of applications are limited by memory bandwidth on many-core platforms. In the paper [6], the authors conducted an experiment on real commodity Chip-Multi-Processors (CMP) machines, using a CMP benchmark suite to investigate the influence of cache sharing and memory bandwidth on the scalability of emerging parallel applications. And in the paper [7], on many-core processors, the authors showed that under-subscription or not utilizing all of the CPU cores, often yields significant increases in both performance and energy efficiency.

For large eddy simulation on many-core machines, there are some related works on GPGPUs [1, 8–11]. Although there are a lot of related work about performance optimization of various parallel applications on GPGPU and MIC, there are few related work about performance evaluation and optimization of LES on MIC. There are some related works of CFD on MIC. For example, Che et al. [12] comparatively evaluated the micro-architectural performance of two representative CFD applications on Intel MIC co-processor, and the Intel Sand Bridge (SNB) processor. Both of the CFD applications were solving the Navier-Stokes equations on structured grids. And in the paper [13], the authors simulate stencil-based application on future Xeon Phi processor, i.e. KNL. The results show that key architectural features of KNL processor will positively impact the performance of the application.

In this work, we present an implementation and performance optimization of LES on MIC platforms, including areal KNL processor.

3   Target Experimental Platforms

The target experimental platforms include Intel Xeon multi-core CPU, a hybrid platform with multi-core CPU and Intel MIC co-processor (KNC), as well as a KNL platform. There are two kinds of computing nodes, pure CPU nodes and KNC nodes. For pure CPU nodes, each is equipped with one Intel Xeon CPUE5-2670 (16 cores), which has a peak double precision performance of 332.8 GFLOP/s and a measured memory bandwidth of 58.86 GB/s. Each KNC node is equipped with one Intel Xeon CPU E5-2670 and two Intel Xeon Phi 7120Pcards (61 cores). For the detailed architecture of a KNC card, there are already some literatures describing it, e.g. [15–17]. The peak double precision performance for one KNC card is 1073.6 GFlops. The measured Linpack performance is 839.9 GFlops. The measured memory bandwidth with stream benchmark [14] is 170 GB/s per KNC card. The bandwidth between hosts via Infiniband switch is much better than that between host and KNC card via PCIe bus.

For the KNL node equipped with Intel Xeon Phi 7250, it acts as an independent computing node. Altogether, there are 68 cores with two cores per tile, two VPUs (Vector Processing Units, AVX-512) per core and each tile shares 1 MB of L2 cache. Rather than the ring topology, the cores in KNL are arranged in a mesh topology using an interconnected fabric. The die has a total of 10 memory controllers-two for DDR4 (supporting three channels each), and then eight for MCDRAM. So, comparing with KNC, there are more cores manufactured with 14-nm technology, which adds performance and power advantages. And there is a large, high-bandwidth, on-package memory. What's more, OmniPath network fabric is integrated.

4   Implementation of LES on Hybrid CPU and MIC Platforms

LES is a promising mathematical model for turbulence used in CFD [1]. LES operates on the Navier-Stokes equation or lattice Boltzmann equation to reduce the range of length scales of the solution, reducing the computational cost. LES resolved large scales of the flow field solution allowing better fidelity than alternative approaches such as Reynolds-averaged Navier-Stokes (RANS) methods. It also models the smallest scales of the solution, rather than resolving them as direct numerical simulation (DNS) does.

We use lattice Boltzmann method for solving LES, i.e. LBM-LES. Just like the basic framework of the LBM, the LBM-LES models the fluid consisting of fictive particles, and such particles perform consecutive propagation and collision processes over discrete lattice meshes. As described in Fig. 1, the approach adopts the "collision after propagation" approach and performed the propagation at global memory reading. The propagation and collision processes iterate again and again until the fluid field achieves convergence condition.

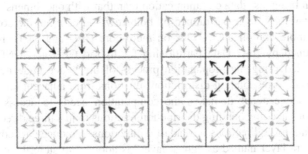

Fig. 1. Propagation is performed at reading

Programming models are important for parallel applications. The common programming models for parallel applications include MPI [18] and OpenMP [19]. As one of the main computing infrastructures is interconnected multi-core machines, applications based on MPI and OpenMP is an active field of research. MPI and OpenMP support two complementary models of parallelism. It must be noted that both models require modifications waved with the domain code. In fact, hybrid MPI and OpenMP is also suitable for hybrid multi-core CPU and Intel MIC platforms.

The most common execution models [18] on KNC include offload, co-processor native, and symmetric execution modes. Using offload execution mode, the host system offloads part or all of the computation from one or multiple processes or threads running on the host to the co-processors, which is similar to GPGPU. Currently, the time consumption for data exchanges between the host CPU and the MIC cards is one of the major performance bottlenecks for the offload execution mode on MIC. An Intel Xeon Phi co-processor has a Linux micro OS running in it and can appear as another machine connected to the host. Using co-processor native execution mode allows the users to view the co-processor as another compute node. In order to run natively, an application has to be cross-compiled for the Xeon Phi operating environment. In the symmetric execution mode, the application processes run on both the host and the Intel Xeon Phi co-processor. They usually communicate through MPI. Both OpenMP and MPI can be used for co-processor native or host native modes. For native modes, OpenMP usually brings a better performance than MPI. MPI or hybrid MPI and OpenMP can be used for the symmetric mode (hybrid CPU and KNC) or multiple nodes. We ported LES on the hybrid CPU and KNC platform using hybrid MPI and OpenMP to reduce data communication. To eliminate the performance bottleneck problem of offload execution mode, KNL acts as an independent compute node, it does not act as a co-processor depending on a host multi-core CPU any longer. So, we mainly evaluate the native and symmetric execution modes.

Domain decomposition and data communication are very important for parallel computation. There is no exception on MIC platforms. For native mode, it is relatively easy to implement the data partitioning. In order to implement and run LES on hybrid CPU and KNC platform with symmetric mode, the global working set should be decomposed among CPU and MICs. LES can be regarded as a two dimensional stencil computation. There are 9 directions on each point. In theory, data partition with two dimensions can bring less data communication than that with one dimension. However, if the input array for LES is partitioned by both row and column, the communication will be much more complicated. Thus, on multi-core CPU and KNC, input data for LES is partitioned by row. Data partitioning for hybrid CPU and KNC platform is shown in Fig. 2. During the propagation process of the lattice node iteration, every node should read corresponding distribution functions from adjacent lattice nodes. At the boundary between sub-domains, the computation needs to access the adjacent sub-domains for the corresponding distribution fractions in just three directions. So, we just send the data for 3 directions instead of 9 directions to reduce data communication volume. An extra layer named as "ghost layer" is adopted to store the communicated data. Thus, every time before the propagation process, the program should manage to transfer that three distribution fractions of the lattice nodes located at the interface to the corresponding ghost layer at adjacent sub-domain. And data communication for boundary can be overlapped with the computation of inner layers.

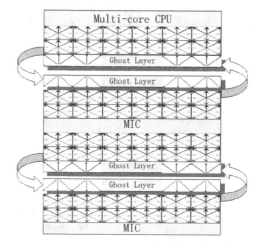

Fig. 2. Data partitioning on hybrid CPU and KNC platform

5   Performance Optimization Techniques

There are some important performance optimization techniques on MIC, such as parallelization, vectorization, and memory access optimization. OpenMP, MPI, as well as hybrid MPI and OpenMP were implemented for parallel computing in different execution modes. IVDEP directive was used for some inner loops to implement vectorization. For memory access optimization, we used streaming data storage and data alignment etc. Further, load balance techniques, overlap computation and communication with asynchronous MPI communication, and threads scheduling were used for performance optimization. Some other common optimization techniques were also utilized, such as reducing redundant computing, compiling optimization, etc.

5.1 Data Partitioning for Parallelization and Load Balance

Let the size of a lattice mesh is (Nx, Ny). If there are M KNC co-processor equipped homogeneous nodes, we adopt a static partitioning method for the homogeneous KNC nodes. For the partitioned data on each KNC node, the length is just Nx, the height is Ny/M. Further, on each KNC node, we adjust the data partitioning according to the real computational capability of CPU and KNC. The ratio for data partitioning on KNC and CPU is the ratio of the optimal computation time on CPU and the optimal computation time on KNC for a specific benchmark. So, a better performance can be obtained due to a better load balancing.

5.2 Optimization of Parallelization

Within a multi-core CPU or a KNL node, we use OpenMP threads for the parallelization. On a hybrid CPU and MIC co-processor (KNC) platform, specific optimization approaches for parallelization include hybrid MPI and OpenMP programming, overlap communication and computation for MPI processes. For parallel computing across CPU and KNC, there are two approaches. One approach is using offload mode. There is a MPI process on the multi-core CPU side. Within the MPI process, some computation work will be offloaded to MIC cards. To fully utilize the computational capability of multi-core CPU, other than the MPI process, we can also use OpenMP threads on the multi-core CPU side. The other approach is using symmetric mode. There will be one MPI process on multi-core CPU and on each MIC co-processor. Within a multi-core CPU and a KNC card, we use OpenMP threads. In this way, hybrid MPI and OpenMP programming is implemented. Because communication only happens for the boundary data exchange, we can divide the computation as two parts for each partitioned data block, one part is boundary rows, the other part is interior rows. Firstly, the boundary rows are calculated quickly, then the communication of boundary data will be executed concurrently with the computation of interior rows using the asynchronous MPI communication. Thus, overlapping communication and computation is implemented.

5.3 Vectorization

Either KNC or KNL has a vector processing unit (VPU), which supports 512-bit instruction. It means that 16 single precision or 8 double-precision operations can be executed at one time. For the vectorization of the inner loop, we used automatic vectorization by adding "#pragma ivdep". Other than the data partitioning for MPI processes, automatic vectorization for the inner loop and OpenMP threads for the outer loop are combined within each partitioned data for one MPI process. In this way, parallel computing with hybrid MPI and OpenMP as well as vectorization are used concurrently for the performance optimization.

5.4   Memory Access Optimization

For memory access optimization on KNC and KNL, we tried a lot of possible approaches. Some of them did not bring a good effect. The main useful approaches include loop unrolling by −O3 compiling parameter for improving cache hit rate, data alignment by mm_malloc, and streaming store etc. Memory alignment is also necessary for the vectorization to use aligned load instructions.

5.5   Threads Mapping Policy

For the runtime performance tuning, we use some typical mapping policies for tasks (threads or processes), including block mapping, cyclic mapping and random mapping [21]. The block mapping policy will distribute tasks to a node such that consecutive tasks use consecutive cores sharing a socket or a node. Cyclic mapping by socket will distribute tasks to sockets such that consecutive tasks are distributed over consecutive sockets (in a round-robin fashion). Random mapping policy will distribute tasks to the cores or the nodes randomly. The mapping policies can be adapted to node-level on a many-core cluster, and socket level or non-uniform memory access (NUMA) node level for a many-core node. For the NUMA nodes, if the application uses shared memory mechanism (like using Pthreads or OpenMP threads), the optimal choice is to regroup the threads and bind them on the same socket. The shared data should be local to the socket and the data will potentially stay on the processor's cache. For MPI processes, the communication is also faster between processes which are on the same socket. So, two processes with large communications shall be bound to the same socket. Using OpenMP threads, there are typically 3 scheduling policies for the iterations, i.e. static, dynamic, and guided scheduling [18]. The simplest policy is static scheduling, which divides the loop into contiguous chunks of roughly equal size. Each thread then executes on exactly one chunk. Dynamic scheduling assigns a chunk of work, whose size is defined by the chunksize, to the next thread that has finished its current chunk. This allows for a very flexible distribution which is usually not reproduced from run to run. Using guided scheduling, threads also request new chunks dynamically, but the chunksize is always proportional to the remaining number of iterations divided by the number of threads. With mapping policies, compact, scatter, balanced policies [15] are provided to bind OpenMP threads to physical processing units by setting "KMP AFFINITY". Scatter policy means scattering across cores before using multiple threads on a given core. Compact policy is usually not wanted since it favors using all threads on a core before using other cores. Especially, balanced policy is only provided on MIC cards. In this case, the threads are affinitized to different cores to distribute computations evenly when the number of threads is less than the maximum number of threads needed to saturate the cores. On KNL, there is another special policy, using Multi-Channel DRAM (MCDRAM). It can be simply implemented by using "numactl –m 1".

5.6   Number of Threads Within a Multi-core Ormany-Core Node

With the increase of the number of cores on a chip, there are dozens or even hundreds of cores within one multi-core or many-core node. While memory bandwidth per socket may remain constant, memory per core may decrease. On our target experimental platforms, there are 16 cores on CPU and 61cores with 244 hardware threads on Intel KNC MIC co-processor. On the newly Intel KNL MIC platform, there are 68 cores with 272 hardware threads. So, how to select the number of threads is a practical issue when running multi-threaded applications on multi-core CPU or MIC, especially for memory bandwidth sensitive applications. On the 16-core CPU, the performance basically can be improved with the increasing number of threads if the number of threads is not greater than the number of cores. On the 61-core KNC co-processor and 68-core KNL, each core can create at most 4 hardware threads. However, one core is usually dedicated for running operating system.

6   Experiments and Results of Performance Optimization Techniques

Computation time and speedup are used as the performance metrics for evaluating the performance of LES on multi-core CPU and MIC. The basis of the speedup is the execution time using one CPU core. Speedup is the result of execution time using one CPU core divided by the execution time using more CPU cores, KNL or hybrid CPU and KNC. The execution time is counted for the main loop kernel, which consumes nearly 100% of the total execution time. We carry out the following sets of experiments for the performance evaluation.

6.1   Stream Memory Benchmark

Gflops of processors, memory bandwidth, I/O, and network are very important for the performance of parallel applications. For the LES application, I/O is not a bottleneck problem. Memory bandwidth is a prominent factor for the performance on multi-core CPU or MIC. We used the stream benchmark [16] to get the basic memory bandwidth data on CPU and MIC. For the stream benchmark, the Triad loop ($a(i) = b(i) + q * c(i)$) has memory access to 3 double precision words(24 bytes) and two floating point operations (one addition and one multiplication) per iteration. We chose Triad as the evaluation metric for memory bandwidth. Figure 3 shows the memory bandwidth in function of threads on the 16-core Intel Xeon CPU. Since each test runs 10 times, only the best time for each is used. There is little difference with repeated experiments. So we do not use error bars in the figures. We used random mapping, cyclic mapping and block mapping policies for the mapping of threads of stream benchmark onto CPU cores. Compared with block mapping, cyclic mapping usually brings a better bandwidth improvement with the increasing number of threads on the multi-core CPU. Groups or sockets may determine the memory bandwidth, e.g. a group of 8 cores sharing L3 cache is determining the memory bandwidth. For the 16-core CPU, memory bandwidth may be saturated before using 16 threads. We also tested all kinds of threads

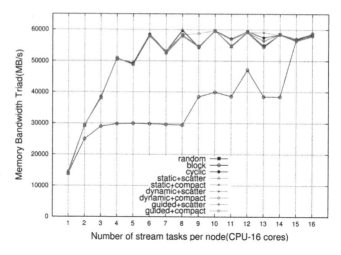

Fig. 3. Memory bandwidth on the 16-core Intel Xeon CPU

scheduling policies and mapping policies of OpenMP threads, i.e. static, dynamic and guided policies for the scheduling of iterations, as well as compact, scatter and balanced policies for threads mapping onto CPU cores. Besides block mapping policy, the memory access is uniformed with other mapping policies spreading over all CPU cores.

Using native mode with OpenMP threads on one KNC card, we also tested the iterations scheduling and threads mapping policies for stream benchmark on MIC co-processor. We found that the compact mapping policy with any scheduling policies is obviously worse than other mapping policies. For all of the other sets of scheduling policies and mapping polices, there were little differences. So, in Fig. 4, we show one typical result of memory bandwidth in function of OpenMP threads on the Intel KNC

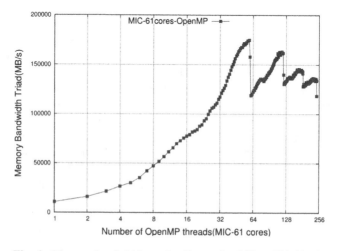

Fig. 4. Memory bandwidth on the 61-core Intel Xeon Phi (KNC)

MIC card. There are 61 cores on one KNC card, each can run 4 hardware threads. We can see that there are just 4 sets of step-wise increase memory bandwidth when using 1–4 threads per core. Altogether there are 8 memory controllers, adding more hardware threads on one core decreases the memory bandwidth mainly due to sharing with the memory controllers. We obtained the optimal memory bandwidth using 60 OpenMP threads (one thread per core) leaving 1 core for running the operating system. For any number of threads (1–4 threads) per core, there was an obvious decrease of memory bandwidth if using all of the cores for OpenMP threads without leaving one core for operating system. On KNL, there is a similar behavior.

6.2   LES Application

Optimization for Serial Computing. Basically, performance optimization on Intel CPU may also take effect on Intel MIC. On the CPU, we firstly optimized the performance of serial computing by reducing redundant computation. For different sizes of arrays from 512 to 8192, the performance gain was upto 3.52% compared with the original serial computing version. Although there was no too much improvement, it is a basic and valuable step for the performance optimization.

Optimization of Parallelization. For the optimization of parallelization, firstly, we use OpenMP threads on the host and on one KNC card or KNL node. On the multi-core CPU, we obtained nearly linear speedup. Using the maximum number of available cores, i.e. 16, the speedup is around 14 for different working sizes. Without vectorization, the performance of parallel computing using OpenMP threads on a 61-core KNC co-processor is worse than that on 16-core multi-core CPU. But the performance on the 68-core KNL node is much better than that on the multi-core CPU even without vectorization.

Optimization of Vectorization. The original code lacks vectorization, the rate of vectorization is 0%. After the optimization of vectorization, the rate of vectorization reached 99.96%. On multi-core CPU, for different sizes, the computation time with vectorization is about 1/3 of that without vectorization. On a MIC card, vectorization brings amazing performance gain. The main performance optimization comes from parallelization using OpenMP threads combined with vectorization. After these steps of performance optimization, the performance on a KNC co-processor is better than that on a multi-core CPU. Combining parallelization with OpenMP threads and vectorization, the speedup increases with the size of array until 2048. Then, the speedup decreases. It was mainly due to the increasing requirement of memory size, while the memory on one KNC card is just 8 GB. It is relatively much smaller than that on CPUs (64 GB). It also demonstrated that it is necessary to make distributed collaborative computation with both of CPUs and KNC co-processors, so the requirements of memory can be decreased on one node or one KNC card. On the KNL node, the performance is also greatly improved by using vectorization. The memory size is not a bottleneck issue any more.

Memory Access Optimization. For the original serial code, the memory read and memory write bandwidth is just 0.62 GB/s. With streaming and data alignment for memory access optimization, there was a better performance improvement on KNC card than that on CPU. For the memory read and memory write bandwidth, it was improved to be between 1.91 GB/s and 31.98 GB/s for different working sizes. For different sizes of arrays from 512 to 8192, the performance gain was up to 14.3% compared with the pure OpenMP version.

Threads Mapping Policy on CPU or MIC. On the 16-core CPU node, there is a good scalability (nearly linear increasing) with the increasing number of OpenMP threads or MPI processes. However, MPI based parallel computing performed worse than OpenMP threads. With 16 OpenMP threads, the best measured speedup was 15.15. On the 61-core KNC card, MPI based parallel computing performed even worse. With 180 OpenMP threads using native mode, the best measured speedup was 61.02. So we just present the experiments using OpenMP threads. The effects of mapping policies are similar for different working sizes. Here, we provide a typical example using 4096 as the size of array. With OpenMP threads, in Figs. 5 and 6 we show the results of different mapping policies on the 16-core CPU and Intel KNC card respectively. The performance gain of mapping policies on KNC is more obvious than that on CPU. On multi-core CPU, there was a little difference among the OpenMP threads scheduling policies. If the number of OpenMP threads is not greater than 8, scatter mapping policy is comparatively better than compact mapping policy. Otherwise, compact mapping policy is comparatively better than scatter mapping policy. It is mainly due to the 2 sockets with 8 cores for each socket. For 16 OpenMP threads, the optimal policy was dynamic scheduling with compact mapping policy. On the MIC architecture, balanced mapping policy is comparatively better than other policies. We should point out that balanced mapping policy is customized for MIC, which is not

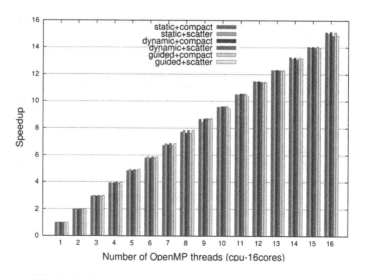

Fig. 5. Performance of mapping policies on the 16-core CPU

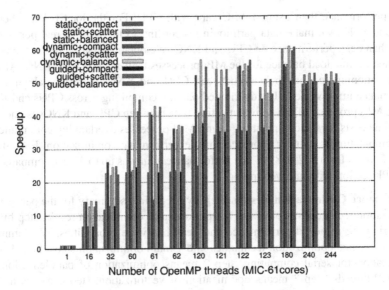

Fig. 6. Performance of mapping policies on the 61-core KNC

provided on multi-core CPUs. The compact mapping policy with any scheduling policies is also obviously worse than other mapping policies, which is in accordance with the stream benchmark. Using 180 threads on the 61-core MIC co-processor, we obtained the best speedup, and the optimal policy was dynamic scheduling with balanced mapping policy. Using 67 threads on the 68-core Intel KNL MIC processor, the optimal policy was usually guided scheduling with balanced mapping policy.

Number of Threads for One Node. On the one 16-core CPU node, the optimal number of OpenMP threads and MPI processes for speedup are both 16. There was only one exception. Regarding the 512 * 512 size, 8 MPI processes brought a little bit better performance than 16 MPI processes. For the small sizes, it was mainly due to the increased communication time when using 16 MPI processes than 8 MPI processes. On the 61-core MIC card, altogether there can be 244 threads. However, for the native mode, the measured best number of OpenMP threads for speedup on the KNC card was 180 (with 3 threads for each core) using dynamic scheduling with balanced mapping policy. 1core is left for running the operating system. Since MPI processes on KNC card behave worse than that on CPU. We did not use pure MPI processes on the KNC card. In fact, with offload mode, 180 was also the measured best number of threads for the speedup of running LES on the MIC node. On the available one 68-core Intel KNL MIC node, 67 threads with 1 thread for each core can usually bring the best performance.

Hybrid CPU and KNC with Symmetric Mode. With hybrid MPI and OpenMP programming for the symmetric mode on hybrid CPU and KNC, we launch one MPI process for each CPU and each KNC card, then, 16 OpenMP threads were used within the 16-coreCPU, and 180 OpenMP threads were used within one KNC card. In this way, we obtained the best speedup. Using 2 MIC cards with CPU can usually bring a better performance than using 1 MIC card with CPU. For the load balance between CPU and KNC, we make data partitioning according to the ratio of real performance rather than peak performance on CPU and KNC. We obtained a good performance gain with the accurate load balance for the MPI processes between CPU and KNC. After the data partitioning for load balance between CPU and KNC, there was about 7–10% performance improvement. For distributed parallel computing across CPUs and KNCs, we use MPI processes. Data communication time between CPU and KNCs is one of the main time costs for the computation. Using MPI processes overlapping communication and computation is another approach for the performance optimization. For different sizes of arrays from 512 to 8192, the performance gain was up to 37.2% compared with the MPI version without overlapping.

Performance Optimization Results Step by Step. In a summary for the performance optimization techniques, we present the performance optimization results step by step, comparing the achieved performance and peak hardware capabilities. We summarize the performance optimization techniques into 7 main optimization steps. Step 1 means optimization for serial computing, step 2 means optimization of parallelization with OpenMP threads, step 3 means optimization of vectorization, step 4 means memory access optimization, step 5 means selecting the optimal number of threads, step 6 means threads optimal mapping policy and affinity, step 7 means hybrid CPU and KNC co-processor. For step 6, a special optimization approach for KNL is using MCDRAM by numactl command.

In Fig. 7, we show execution time and speedup obtained on CPU and KNC (native mode), as well as on KNL step by step. We can see that on the MIC platforms, the most

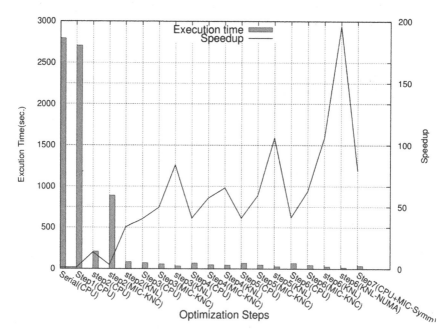

Fig. 7. Time and speedup obtained on CPU and KNC/KNL step by step

obvious performance gain comes from parallelization using OpenMP, vectorization, and load balance. For the original serial code, the original performance on CPU was 1.07 Gflops. After using the optimization techniques, we obtained 25.2 Gflops. It is still less than 10% of the theoretical peak performance or Linpack performance. And the memory access bandwidth reached 27.67 GB/s, which is about 47% of the stream memory bandwidth. It is mainly because that LES is a memory intensive application. How to fully utilize the computing capability of CPU needs a further study. On the 16-core CPU, thanks to the vectorization, the best speedup reached 41.55. On the 61-core KNC MIC platform, the optimal speedup was 62.71 with native execution mode and 79.5 with symmetric execution mode. On the 68-core KNL MIC node, the obtained best speedup was 196.06 using MCDRAM. All of the speedups are calculated on the basis of serial computing using one core on the multi-core CPU.

7   Discussion

In this section, we discuss the comparison between multi-core CPU, MIC and GPGPU, including memory bandwidth, best speedup for the same application. Other than the Xeon Phi7110P co-processor (KNC-MIC) with 61 cores, we also tested the performance of a KNL Xeon Phi processor with 68 cores. For GPGPU, we adopt the same GPGPU platform as described in the previous work [1]. The total memory bandwidth on MIC is the best comparing with that of GPGPU and multi-core CPU. However, the average memory bandwidth per core on KNC-MIC is less than that on CPU. On KNL, memory bandwidth using MCDRAM is greatly improved. For all of the sizes, the performance on 1 KNC card (using native mode) is better than that on 1 CPU. With symmetric mode, it changes for different sizes. For small sizes, the performance on 1 MIC card with native mode is usually better than that of other modes. For large sizes, symmetric mode can usually bring the best speedup. It is because the memory is limited on KNC card. With symmetric mode, large sizes will be partitioned for execution on CPU and KNC cards. And the performance on the GPGPU is better than CPU or hybrid CPU and KNC platform. The performance on KNL is the best. However, with the emergence of new products of Intel MIC processor and GPGPU, it depends on specific scenarios to choose one of them for a better performance. With the increasing number of cores on many-core processors, the maximum number of cores could not obtain the optimal performance. Some improvement may be innovated for the design of processor architecture. So a more compute-memory balanced architecture may be created, or a specific architecture may be designed for one kind of applications, e.g. memory intensive or compute intensive application.

8   Conclusion

In this paper, we ported LES and evaluated some important performance optimization techniques on Intel MIC platforms. Typical execution modes on KNC, such as native, offload, and symmetric mode were implemented and evaluated. The main performance optimization strategies include data partitioning for parallelization and load balance, optimization of parallelization, vectorization and memory access optimization, threads mapping polices, number of threads on multi-core CPU and MIC, as well as overlapping computation with communication for MPI processes, etc. The main programming models for HPC are explored for different execution modes, including MPI, OpenMP threads, hybrid MPI and OpenMP. We also discussed the comparison of performance on multi-core CPU, hybrid CPU and MIC platforms, hybrid CPU and GPGPU platform. Concerning the increasing number of cores in the many-core processors, we argue that some improvement may be innovated for the design of processor architectures. Future work may include the energy efficiency evaluation on KNC and KNL, optimization and evaluation of LES on multiple nodes with MIC and GPGPU. Other programming models, such as OpenCL and PGAS model may also be explored.

Acknowledgments. The authors would like to thank the support from Intel Corporation and Inspur Corporation. The Project Sponsored by the Scientific Research Foundation for the Returned Overseas Chinese Scholars, State Education Ministry. And this work was supported in part by Shaanxi science and technology innovation project plan. No. 2016KTZDGY04-04.

References

1. Li, Q.J., Zhong, C.W., et al.: A parallel lattice Boltzmann method for large eddy simulation on multiple GPUs. Computing **96**(6), 479–501 (2014)
2. Top500 supercomputers. http://www.top500.org. Accessed 10 June 2017
3. Sodani, A., Gramunt, R., et al.: Knights landing: second-generation Intel Xeon Phi product. IEEE Micro **36**(2), 34–46 (2016)
4. Dongarra, J.: The LINPACK benchmark: an explanation. In: Proceedings of 1st International Conference on Supercomputing, pp. 456–474 (1989)
5. Gupta, V., Kim, H., Schwan, K.: Evaluating scalability of multi-threaded applications on a many-core platform. Technical report, Georgia Institute of Technology (2012)
6. Chen, X., et al.: Evaluating scalability of emerging multithreaded applications on commodity multicore server. In: International Conference on Information Technology, Computer Engineering and Management Sciences (ICM), pp. 332–335 (2011)
7. Heirman, W., Carlson, T.E., Craeynest, K.V., Hur, I., Jaleel, A., Eeckhout, L.: Undersubscribed threading on clustered cache architectures. In: 20th IEEE International Symposium on High-Performance Computer Architecture, HPCA 2014, pp. 678–689 (2014)
8. Deleon, R., Jacobsen, D., Senocak, I.: Large-eddy simulations of turbulent incompressible flows on GPU clusters. Comput. Sci. Eng. **15**(1), 26–33 (2013)
9. Wang, W., Shangguan, Y.Q., et al.: Direct numerical simulation and large eddy simulation on a turbulent wall-bounded flow using lattice Boltzmann method and multiple GPUs. Math. Probl. Eng. **1**, 1–10 (2014)
10. Schalkwijk, J., et al.: Weather forecasting using GPU-based large-eddy simulations. Bull. Am. Meteor. Soc. **96**(5), 715–724 (2015)
11. Lopez-Morales, M.R., et al.: Verification and validation of HiFiLES: a high-order LES unstructured solver on multi-GPU platforms. In: 32nd AIAA Applied Aero dynamics Conference (2014)
12. Che, Y.G., et al.: Micro architectural performance comparison of Intel Knights Corner and Intel Sandy Bridge with CFD applications. J. Supercomput. **70**(1), 321–348 (2014)

13. Natarajan, C., Beckmann, C., et al.: Simulating stencil-based application on future Xeon Phi processor. In: Proceedings of 6th International Workshop in Performance Modeling, Benchmarking and Simulation of High Performance Computer Systems, PMBS 2015, SC 2015 (2015)
14. McCalpin, J.D.: Memory bandwidth and machine balance in current high performance computers. In: IEEE Computer Society Technical Committee on Computer Architecture (TCCA) Newsletter (1995)
15. Jeffers, J., Reinders, J.: Intel Xeon Phi Co-processor High-Performance Programming. Morgan Kaufmann, Burlington (2013)
16. Rahman, R.: Intel Xeon Phi Co-processor Architecture and Tools - The Guide for Application Developers. Apress Open, New York (2013)
17. Yang, Y., et al.: Evaluating multi-core and many-core architectures through accelerating the three dimensional Lax-Wendroff correction stencil. Int. J. High Perform. Comput. Appl. **28** (3), 301–318 (2014)
18. Gropp, W., Lusk, E., Skjellum, A.: Using MPI: Portable Parallel Programming with the Message-Passing Interface. MIT Press, Cambridge (1999)
19. Barbara, C., Gabriele, J., van der Ruud, P.: Using OpenMP: Portable Shared Memory Parallel Programming (Scientific and Engineering Computation). MIT Press, Cambridge (2007)
20. Hager, G., Wellein, G.: Introduction to High-Performance Computing for Scientists and Engineers. CRC Press, Boca Raton (2010)
21. Hou, Z.X., Perez, C.: Performance evaluation and tuning of 2D Jacobi iteration on many-core machines. In: 2013 IEEE International Conference on High Performance Computing and Communications, pp. 603–610, IEEE, Zhangjiajie (2013)

Author Index

An, Wei   444

Bai, Yong   201

Cai, Ruiqing   192
Chan, Wai Hong   579
Chen, Cen   376
Chen, Chaofan   386, 504
Chen, Lei   25
Chen, Liang   533
Chen, Mingrui   25, 335, 386, 418, 504, 522
Chen, Xian-yi   268
Chen, Zhiguang   541
Cheng, Haoyu   214
Chu, Tianxing   224
Cui, Baotong   117, 290
Cui, Zelin   343

Deng, Yanfang   201
Dong, Lu   70, 554
Dong, Yawei   82
Du, Shaojie   311
Du, Zheng   454
Duan, Xiaoyu   182
Duan, Yucong   418

Fu, Fang-Fa   51

Gai, Wang   493
Gan, Tian   192
Gao, Liming   41
Gao, Yuchen   152, 566
Guo, Chao   454
Guo, Ruidong   214

Han, Daojun   182, 192
Han, Lin   566
Han, Ling Bo   579
Han, Zhijie   247
Hou, Zhengxiong   610
Hu, JianZheng   393, 464
Hu, Zhuhua   201
Huang, Haiping   108

Huang, Mengxing   201, 418
Huang, Xia   25
Huang, Yiran   483
Hui, Liu   321

Ji, Huihui   117
Ji, Yimu   237
Jia, Wen-Juan   3
Jiang, Yirui   418
JiangPing, Yang   493
Jinyang, Yao   321

Kong, Ping   471
Ku, Junhua   13

Lai, Feng-Chang   51
Lai, Siyan   594
Lao, Bin   279
LAU, Ting Fung   59
Lee, Elizabeth   3
Lei, ShiYao   464
Li, Feng   108, 410
Li, Hang   51
Li, Hongbo   541
Li, Jingbin   418
Li, Ni   95
Li, Peng   59, 70, 554
Li, Taijun   410
Li, Weimin   95
Li, Xinying   410
Li, Xiulai   335, 386, 504
Li, Yanbing   152, 566
Li, Ye   483
Li, Yidong   471, 533
Li, Yingying   152, 566
Li, Yueming   130
Lin, Guo-lan   268
Lin, Han   321
Lin, Xiaola   594
Liu, Gang   258, 301
Liu, Jun   483
Liu, Shangdong   237
Liu, Tao   130

Liu, Wenju   142
Liu, Xia   25
Liu, Xiaodan   82
Liu, Yanbo   224
Liu, Zhenpeng   82
Lu, Huimin   3
Lu, Lili   237
Lu, Yutong   541
Luo, Baozhou   70
Luo, JingYu   393, 464
Luo, Qiuming   301, 454
Luo, Yali   335, 386, 504

Ma, Xiaoxu   142
Mao, Rui   301
Maohui, Lu   493
Mei, Gang   522
Meng, Xiangjun   533
Mi, Yu   70
Min, Huang   439
Mu, Qing   566

Nie, Huqing   554
Ning, Xiaoshuang   224
Niu, Baojun   172
Niu, Na   51
Niu, Yafeng   172
Nong, Ge   279, 579

Pang, Yuxin   427
Perez, Christian   610

Qi, Lingtao   108
Qi, Qi   95
Qi, Wang   321
Qian, Xueming   290
Qilong, Zheng   493
Qiu, Chunmin   311
Qiu, Zhao   25, 393, 464

Shang, Ming-Sheng   444
Shao, Bo   594
Shao, Ying   3
Shen, Hong   343, 353
Shen, Xiajiong   182, 192
Sheng, Mingwei   130
Song, Hongzhi   224
Song, Jinquan   554

Wan, Lei   130
Wang, Chongwen   41, 376
Wang, Dong   152
Wang, Jin-Xiang   51
Wang, Limin   366
Wang, Peng   108
Wang, Qi   566
Wang, Qing-Xian   444
Wang, Ruchuan   59, 237, 554
Wang, Rui   247
Wang, Shuihua   3
Wang, Xiaochun   471
Wang, Yunlan   610
Wang, Ze   142
Wenqi, Deng   493
Wu, Hongli   410
Wu, Jiahuai   353
Wu, Yi   579

Xi, Hongyan   130
Xiao, Nong   541
Xie, Jing Yi   279
Xie, Mingshan   201
Xin, Tong   237
Xing, Yuxuan   541
Xu, He   59, 70, 554
Xu, Jinlong   152
Xu, Wenchao   224, 427
Xu, Ying   594
Xu, Yun   214
Xue, Li-Yuan   444

Yan, YongHang   247
Yang, Bo   594
Yang, Cheng   59
Yang, Han-tao   400
Yang, Hong   427
Yang, Lin   247
Yang, Rui   376
Yang, Yanqin   224, 427
Ye, Cunhuang   301
Yu, Congxiang   108
Yuan, Wanli   182

Zeng, Rong-Qiang   444
Zeng, Xing   335
Zhang, Bin   82
Zhang, Boyi   471
Zhang, Jing   258
Zhang, Jinxiong   483

Zhang, Qing   610
Zhang, XiaWen   393
Zhang, Yijun   454
Zhang, Yongpan   237
Zhang, Yu-Dong   3
Zhao, Bailu   311
Zhao, Xuan   82
Zhen, Xing   522

Zheng, Bing   13
Zhenhao, Yang   493
Zhong, Cheng   483
Zhong, Chengwen   610
Zhong, Guojing   366
Zhou, Hui   418
Zhu, Xiaojing   514
Zou, Dongsheng   172

Printed in the United States
By Bookmasters